PALEONTOLOGY AND GEOLOGY

OF THE MARTINSBURG, SHAWANGUNK, ONONDAGA AND HORNERSTOWN FORMATIONS

(NORTHEASTERN UNITED STATES)

WITH SOME FIELD GUIDES

"Innocent unbiased observation is a myth."

Peter Medawar
Induction and Intuition in Scientific Thought, Chapter 11, Section 2, p. 28, American Philosophical Society, Philadelphia, PA (1969).

"Quite recently the human descent theory has been stigmatized as the 'gorilla theory of human ancestry.' All this despite the fact that Darwin himself, in the days when not a single bit of evidence regarding the fossil ancestors of man was recognized, distinctly stated that none of the known anthropoid apes, much less any of the known monkeys, should be considered in any way as ancestral to the human stock."

Henry Fairfield Osborn
"Osborn States the Case for Evolution," *New York Times* (12 Jul 1925), XX1. Written at the time of the Scopes Monkey Trial, in rebuttal of the anti-evolution position publicized by William Jennings Bryan.

"Why are the bones of great fishes, and oysters and corals and various other shells and sea-snails, found on the high tops of mountains that border the sea, in the same way in which they are found in the depths of the sea?"

Leonardo da Vinci
"Physical Geography," in *The Notebooks of Leonardo da Vinci*, trans. E. MacCurdy (1938), Vol. 1, 361.

PALEONTOLOGY AND GEOLOGY

OF THE MARTINSBURG, SHAWANGUNK, ONONDAGA AND HORNERSTOWN FORMATIONS

(NORTHEASTERN UNITED STATES)

WITH SOME FIELD GUIDES

HOWARD
R. FELDMAN

Library of Congress Cataloging-in-Publication Data:
A catalog record for this book is available
from the Library of Congress.

ISBN 978-1-61811-416-7 (hardback)

ISBN 978-1-61811-417-4 (electronic)

Published by Touro College Press and Academic Studies Press.
Typeset, printed and distributed by Academic Studies Press.
Cover design by Ivan Grave

Touro College Press
Michael A. Shmidman and Simcha Fishbane, Editors
27 West 23rd Street
New York, NY 10010, USA
touropress.admin@touro.edu

Academic Studies Press
28 Montfern Avenue
Brighton, MA 02135, USA
press@academicstudiespress.com
www.academicstudiespress.com

Touro College Press

This book is dedicated to my children, Debra and Brian Belowich, Brian and Baylene Feldman, and my grandchildren, Alexa and Talia Belowich.

CONTENTS

INTRODUCTION

The papers herein not only provide a database for taxonomists, biostratigraphers, sedimentologists, and paleobiogeographers but also enhance our knowledge of marine communities and depositional environments, especially in the Paleozoic. They also increase our ability to correlate Paleozoic formations in the northeastern United States. The Ordovician Martinsburg Formation can be traced into Pennsylvania as can the Silurian Shawangunk Formation, the correlatives of which can be found as far south as Virginia. The Devonian Onondaga Limestone provides insight into formations such as the Anderdon Limestone, Lucas Dolomite, Amhertsburg Dolomite, and Sylvania Sandstone of the Detroit River Group in the Michigan Basin.

The brachiopods from the Paleocene Hornerstown Formation of New Jersey that occur in a biostrome on the New Jersey Coastal Plain help elucidate topoclines, chronoclines, and morphologic variation in the genus *Oleneothyris*. In addition, paleoecological information provides us with information about how these brachiopods lived and under what environmental conditions.

Ordovician-Silurian

The Martinsburg Formation was formed due to Ordovician plate convergence that involved a collision between proto-North America and a Taconic Island Arc as the Iapetus Ocean closed. As the collision progressed, a deep basin formed, into which was deposited a thick sequence of muds and dirty sands that were subsequently consolidated into what we now recognize as the Martinsburg Formation (Feldman et al., 2012). The Martinsburg is generally considered to be representative of a synorogenic basin that deepened as a result of Taconic tectonism (Epstein, 1986). Epstein and Lyttle (1987) consider the Martinsburg Formation in the Wallkill Valley as parautochthonous and Middle to Upper Ordovician in age. However, they believe that the parautochthonous sequence west and southwest of Albany is not contiguous with the rocks of the Wallkill Valley and thus the Shale Bank at Mohonk from which a moderately diverse fauna was collected.

The Martinsburg Formation in the lower mid-Hudson Valley has yielded a *Sowerbyella-Onniella* brachiopod community with the following faunal constituents:

Sowerbyella	*Isotelus*
Onniella	*Conularia* sp.
Dalmanella	*Trocholites*
Ectenocrinus	Corals
Deceptrix	Ostracodes
Cryptolithus	

In addition, numerous sedimentary structures were found: oscillation and interference ripple marks, gutter casts, cross-stratification, and linear to sinusoidal burrows. Based on the fauna and sedimentology, the depositional environment appears to have been relatively shallow water. McBride (1962) suggested that the Martinsburg Formation in the Great Valley was probably deposited at a depth of less than 2,000 m.

The Middle Silurian Shawangunk Formation lies unconformably above the Ordovician shales and graywackes of the Martinsburg. It crops out from near Rosendale, south through Wurtsboro, New York, High Point State Park and the Delaware Water Gap in New Jersey, and at Lehigh Gap, Pennsylvania, after which it continues into Maryland and Virginia (Feldman et al., 2014). The depositional environment of the Shawangunk was one of braided streams with deposition on a coastal plain of alluviation, a linear source toward the southeast, and a marine basin toward the northwest (Epstein et al., 1987; Feldman et al., 2014). *Arthrophycus*, found on the Shawangunk Ridge at Mohonk, near New Paltz, New York, is normally found on the bottom of beds; however, these specimens occur on the top of a bed. The trace maker may have been terrigenous but it is likely that it was formed in a marine environment, possibly estuarine since the basin was just to the northwest. Rises in sea level or tidal ebbs and flows would have enabled marine burrowers to form traces in the conglomerate, a fact supported by the occurrence of (euryhaline) eurypterids in the formation that lived in a wide range of salinities.

Devonian

The first serious attempt to study the fossils of the Onondaga Limestone (Eifelian) in New York was made by Eaton (1832) in which he mentioned twelve Onondagan species. In 1836 Governor Marcy established the Geological Survey of New York shortly after which detailed studies of the Devonian in the state began. In Vanuxem's Fourth Annual Report (1840), he referred to the "Corniferous Rock" of Eaton in which he noted the occurrence of *Cyrtoceras* from the Nedrow Member. In 1842, he published the first significant description of the Onondaga Limestone in which he included the outcrop belt in New York State's Third Geological District. In general the faunal constituents of the Onondaga, especially the brachiopods, were not very well known by workers in the nineteenth century. Hall (1840), in a discussion of the Fourth District, stated that his faunal list represents only a few of the most common and characteristic fossils and that the list will be greatly increased when we have succeeded in ascertaining the names of species.

The papers herein provide a thorough taxonomic treatment of Onondagan brachiopods from central New York through the lower mid-Hudson Valley.

Stratigraphy

Stratigraphically, the Onondaga consists of four members within the study area: Edgecliff, Nedrow, Moorehouse, and Seneca. In the mid-Hudson Valley, the Edgecliff Member can be recognized by abundant light-weathering chert seams with occasional crinoid stems present. It is finer-grained than in central New York and more than twice as thick in the eastern part of the state. Large crinoid stems are typical of the Edgecliff in the Syracuse area. The Nedrow Member in central New York is very shaly resulting in ledges on which brachiopods weather out. In the east, however, the Nedrow becomes thicker-bedded, less argillaceous and coarser-grained (Feldman, 1985). The Moorehouse Member in the Hudson Valley can be recognized by its dark-weathering chert seams and greater thickness. In central New York the Moorehouse is lighter in color with lighter chert. Silicified fossils are found in the east but not the central part of the state. The Seneca Member is an argillaceous limestone with few fossils except for the chonetids found in the pink "Chonetes" (*Hallinetes*) Zone 10 ft above the base in central New York (Feldman, 1985). The Tioga Bentonite separates the Moorehouse from the Seneca in the Syracuse area. The Seneca Member does not crop out east of Cherry Valley, New York.

Community Ecology

The Onondaga Limestone was deposited in a low-energy environment (Feldman, 1980). The hard-shell component of the Onondaga (brachiopod) communities are closely representative of the living communities. There was little post-mortem transport and no significant destruction of the shells due to biopredation of diagenesis. Here I define community as a recurrent association of taxa that were presumably controlled by a set of environmental factors such as: temperature, pressure, current and/or wave action, light penetration, nutrients, water chemistry, dissolved oxygen, and miscellaneous factors (biotic relationships such as symbiosis; mechanical wear, etc.) (Feldman, 1980). It should be noted that there are many varying views on exactly what defines a community. MacArthur (1971), as cited in Boucot (1981), summarized the community definition question well by an appropriate quote from Lewis Carroll: "Humpty Dumpty told Alice, 'when I use a word, it means just what I choose it to mean—neither more nor less' (pp. 189-190)."

The communities described herein are named after the dominant brachiopod taxa. Nine brachiopod communities are recognized in the Onondaga Limestone:

Atrypa-Coelospira-Nucleospira
Atrypa-Megakozlowskiella
Atrypa
Leptaena-Megakozlowskiella
Pacificacoelia
Levenea Community I
Levenea Community II
Amphigenia?
"*Chonetes*" (now *Hallinetes*)

The communities named above belong to Benthic Assemblage 3. A Benthic Assemblage is a group of communities that occur repeatedly in parts of a region in the same position relative to shoreline (Boucot, 1975). Boucot (1975) noted that Benthic Assemblages are probably temperature controlled and highly correlated with depth. Some of the Onondaga communities, such as the *Atrypa-Coelospira-Nucleospira* Community, closely resemble Wang et al.'s (1987) Eifelian *Atrypa-Xystostrophia* Community from the Shan States of Burma in faunal composition as well as depositional environment, that is, a normal, quiet water marine environment (Benthic Assemblage 3). The faunal composition of the *Atrypa-Coelospira-Nucleospira* Community (AMNH locality 3137), in order of abundance, is as follows:

1. *Atrypa*
2. *Coelospira*
3. *Nucleospira*
4. *Megakozlowskiella*
5. *Acrospirifer*
6. *Pentamerella*
7. *Dalejina*
8. *Pentagonia*
9. *Mucrospirifer*
10. *Schizophoria*
11. *Cyrtina*
12. *Elytha* (*Elita*)
13. *Athyris*
14. *Cupularostrum*

Paleocene

In the Paleocene Hornerstown Formation, central New Jersey, two species of large very well-preserved terebratulid brachiopods belonging to the genus *Oleneothyris* (*O. harlani* and *O. subfragilis*) can be found in a biostrome that occurs at the top of the formation. The Hornerstown Formation strikes in a northeasterly direction in a belt along New Jersey's coastal plain (Feldman, 1977a) and is in gradational contact with the overlying Vincentown Formation, whereas the lower contact varies geographically.

Paleoecology

The macrofauna of the Hornerstown Formation is very diverse and includes the following groups:

- Porifera
- Cnidaria
- Brachiopoda
- Bryozoa
- Echinodermata
- Annelida
- Decapoda
- Mollusca
  - Cephalopoda
  - Gastropoda
  - Pelecypoda
- Chordata

The presence of *Cardium* in the upper part of the Hornerstown Formation suggest a sandy bottom, as does the fact that there is an absence of clay-sized particles (Feldman, 1977a). Although deposition in the upper part of the formation was in shallow water with strong wave and current action, post-mortem transport was minimal. The brachiopod community was dominated by one trophic group but the next most dominant species belonged to a different trophic group. The *Oleneothyris* Community conforms to Recent Arctic and Boreal (see Turpaeva, 1957). Morphologic changes in *Oleneothyris* can be correlated with increased turbulence on the sea floor.

Feldman (1977b) described *Oleneothyris fragilis* (subsequently changed to *O. subfragilis* [Feldman, 1985]) also found in the Hornerstown Formation. The species is known only from the biostrome at the top of the formation in New Jersey (Feldman, 1977b) whereas *O. harlani* occurs in Delaware, Maryland, and Alabama.

References

Boucot, A. J. 1975. *Evolution and Extinction Rate Controls.* New York: Elsevier.

Boucot, A. J. 1981. *Principles of Benthic Marine Paleoecology.* New York: Academic Press.

Eaton, A. 1832. Geological text-book, for aiding the study of North American geology: Being a systematic arrangement of facts, collected by the author and his pupils. Albany: Webster and Skinners.

Epstein, J. 1986. The Valley and Ridge Province of eastern Pennsylvania-stratigraphic and sedimentologic contributions and problems. *Geological Journal,* 21: 283-306.

Epstein, J., and Lyttle, P. 1987. Structure and stratigraphy above, below and within the Taconic unconformity, southeastern New York. In R. H. Waines (ed.), *New York State Geological Association, 59th Annual Meeting, Kingston, New York,* November 6-8, 1987, C1-C78. New Paltz, New York, State University of New York, College at New Paltz, Fieldtrip Guidebook.

Feldman, H. R. 1977a. Paleoecology and morphologic variation of a Paleocene terebratulid brachiopod (*Oleneothyris harlani*) from the Hornerstown Formation of New Jersey. *Journal of Paleontology* 51: 86-107.

Feldman, H. R. 1977b. Notes on and description of *Oleneothyris fragilis* (Morton), 1828 (Brachiopoda, Terebratulidae). *American Museum Novitates* 2621: 1-16.

Feldman, H. R. 1980. Level-bottom brachiopod communities in the Middle Devonian of New York. *Lethaia* 13: 27-46.

Feldman, H. R. 1985. *Oleneothyris subfragilis* (d'Orbigny) 1850, a replacement name for the brachiopod *Oleneothyris fragilis* (Morton) 1828. *Journal of Paleontology* 59: 1485.

Feldman, H. R., Smoliga, J., and Feldman, B. A. 2012. Notes on the Geology of the Shawangunk Ridge on the Mohonk Preserve and Environs. Northeast Natural History Conference 2011: Selected Papers 2012. *Northeastern Naturalist* 19, Special Issue, 6: 3–12.

Feldman, H. R., Bartholomew, A., and Kahn, B. 2014. An unusual occurrence of *Arthrophycus alleghaniensis*(?) on the Shawangunk Ridge, lower mid-Hudson Valley. *Geological Society of America Northeastern section meeting, Lancaster, Pennsylvania* 46: 39.

Hall, J. 1840. Fourth annual report of the geological survey of the fourth district. *Geological Survey of New York* 50: 389-455.

MacArthur, R. H. 1971. Patterns of terrestrial bird communities. In D. S. Farner, J. R. King, and K. C. Parks (eds), *Avian Biology*, 189-221. New York: Academic Press.

McBride, E. F. 1962. Flysch and associated beds of the Martinsburg Formation (Ordovician), central Appalachians. *Journal of Sedimentary Petrology* 32: 39-91.

Turpaeva, E. P. 1957. Food interrelationships of dominant species in marine benthic biocoenoses In B. N. Nikitkin (ed.), *Transa*, Institute of Oceanography, Marine Biology USSR Academy Sciences Press 20: 137-148. (Published in the United States by the American Institute of Biological Science, Washington, D.C.).

Vanuxem, L. 1840. Fourth annual report of the geological survey of the third district. *Geological Survey of New York* 4: 335-383.

Wang, Y., Boucot, A. J., Rong, J., and Yang, X. 1987. Community paleoecology as a geologic tool: The Chinese Ashgillian-Eifelian (latest Ordovician through early Middle Devonian) as an example. *Geological Society of America Special Paper* 211: 1-100.

ACKNOWLEDGMENTS

I wish to thank Dr. Arthur J. Boucot of Oregon State University, who mentored me, served as my thesis advisor, and provided many valuable insights into Paleozoic brachiopod taxonomy and paleoecology. My professor and thesis advisor Dr. Richard K. Olsson of Rutgers University constantly challenged me with questions related to stratigraphy and paleontology. I would like to acknowledge the late Dr. Stephen Jay Gould of Harvard University for his thoughts and discussions on paleontological theory when I first began my career. Dr. Mark A. Wilson, of the College of Wooster, deserves special acknowledgment for numerous discussions on the Paleozoic paleontology of the eastern United States, particularly on the Ordovician. Drs. Jack B. Epstein of the United States Geological Survey, John Smoliga of Boehringer-Ingelheim, and the late Russell Waines of the State University of New York, New Paltz, provided invaluable information on the geology of the lower Hudson Valley. I am grateful to John E. Thompson, Director of Conservation Science at the Daniel Smiley Research Center (Mohonk Preserve), for granting access to the Shawangunk and Martinsburg formations on Preserve property. I thank my many colleagues with whom I have shared discussions about paleontology, evolution, paleoecology, cladistics, and biogeography, including Drs. Niles Eldredge and Neil Landman of the American Museum of Natural History. The late Drs. "Gus" Arthur Cooper and Richard Grant of the United States National Museum provided access to the brachiopod collections and contributed many hours of valuable discussion. My colleagues Drs. Carl Brett of the University of Cincinnati, Gordon Baird of the State University of New York, Fredonia, Richard Lindemann of Skidmore College, and the late Dr. Gerald Friedman of the City University of New York, deserve recognition for their thoughtful insights, especially in the field.

Many thanks are due to Mai Reitmeyer, Research Services Librarian at the American Museum of Natural History, who assisted in me in locating hard-to-find publications.

I am very grateful to Dr. Alan Kadish, President of the Touro College and University System (TCUS) for his encouragement and strong support over the past few years. I thank Drs. Simcha Fishbane and Michael Shmidman of TCUS for making this book possible through their guidance and advice. I thank Touro College Press, as well as my editors at Academic Studies Press, Sharona Vedol, Deva Jasheway, Meghan Vicks, and Kira Nemirovsky for critically reading the manuscript and making helpful suggestions for improvement.

Finally, thanks to my family, who still wonders why I have so many "rocks" all over the house, garage, and garden. I recently found an essay written in sixth grade describing how I would become a paleontologist and I remember how my father strenuously argued that I keep paleontology as a hobby and get a "real" career.

It should be noted that a variety of traditional authorities and texts have dealt with the particular issues concerning the age of the universe, and have offered diverse approaches for resolving any apparent conflicts that might arise between Jewish tradition and modern science on this matter.[1] This author follows the lead of these traditional authorities in operating with the accepted principles of the science of paleontology.

---

[1] See, e.g., Babylonian Talmud, standard editions, Hagigah 13b; Midrash Bereshit Rabba, ed. J. Theodor and C. Albeck (Jerusalem, 1996), 3, 5; Rabbi Isaac of Akko, Ozar ha-Hayyim, Ms. Moscow-Russian State Library, Guenzburg 775, 86b-87b; Rabbi Yisrael Lifshitz, Derush Or Ha-Hayyim, in his Tiferet Yisrael on Mishnah, end of Nezikin (Danzig, 1845), 276b-279b.

Chapter One

Notes on the Geology of the Shawangunk Ridge on the Mohonk Preserve and Environs

ABSTRACT

The Shawangunk Formation, a quartz pebble conglomerate of Middle Silurian age, extends from the lower mid-Hudson Valley through New Jersey and into Pennsylvania. It overlies the Ordovician Martinsburg Formation, which is composed of shales and graywackes. The Martinsburg crops out on the Shawangunk Ridge and is quarried by Mohonk Mountain House in New Paltz, New York, in order to prevent erosion and provide good footing on the trails. The quarry, locally known as the "Shale Bank," contains a diverse marine fauna of brachiopods, crinoids, bivalves, ostracodes, corals, trilobites, and conulariids. In this community, the partition of feeding niches results in a reduced competitive trophic structure and therefore increased community stability. Within the Shawangunk Formation, there are rare "pods," domelike structures that are filled with a gray matrix of rounded quartz grains supported by a clay matrix. The pods appeared to have formed along cleavage surfaces. A previously unrecognized metal sulfide deposit has been discovered in the conglomerate along Eagle Cliff. This deposit consists of the Fe-sulfide phases pyrite and marcasite, lesser amounts of the Cu-Fe sulfide chalcopyrite, and trace amounts of anglesite (Pb-sulfate). An outcrop of the Middle Devonian Onondaga Limestone in the Port Jervis Trough contains large crinoid columnals, the coral *Amplexiphyllum,* trilobite fragments, and the brachiopod *Levenea subcarinata.* The Onondaga in this area is part of a carbonate ramp that was a shallow carbonate shelf in the Helderberg-Coxsackie area, a thick accumulation of shelf-margin bryozoan bafflestone between Leeds and Saugerties, and an even thicker accumulation of sparse to packed biocalcisiltites deposited on a carbonate ramp dipping southward into the Port Jervis area.

INTRODUCTION

The Middle Silurian Shawangunk formation lies unconformably above Ordovician graywackes and shales of the Martinsburg Formation. Near Otisville and Port Jervis, the Shawangunk is overlain by the Bloomsburg Red

Beds, which crop out to the west in Pennsylvania and New Jersey (Epstein, 1993). The fluvial deposits of the Tuscarora and Shawangunk formations accumulate to the northwest of uplands lifted during the Taconic Orogeny. Thus, the source area lays to the southeast, and the marine basin lays to the west of the Shawangunk/Bloomsburg fluvial plains. The Shawangunk Formation gradually thins from Port Jervis to its disappearance just north of New Paltz. In the early Paleozoic Era, carbonate banks lay along the east coast of proto-North America. Ordovician plate convergence involving proto-North America formed a deep basin into which a thick sequence of muds and dirty sands accumulated. These deposits were subsequently consolidated to form the Martinsburg Formation. The Martinsburg was deformed during the mountain-building episode known as the Taconic Orogeny. The trend, or strike, of these rocks in southeastern New York is approximately 20 degrees toward the northeast. The intensity of deformation diminishes to the west. The Taconic Mountain system shed coarse sediment that was transported westward as fluvial conglomerates and sandstones of the Shawangunk Formation over beveled folds of the Martinsburg Formation. To the northwest, erosion of the source area was intense and the climate, based on the mineralogy of the rocks, was warm and at least semiarid (Epstein and Lyttle, 1987). The source was composed predominantly of sedimentary and low-grade metamorphic rocks that hosted abundant quartz veins and local occurrences of gneiss and granite. Progressive erosion of the source regions caused the steep, braided streams of the Shawangunk to give way to the lower-gradient meandering streams from which the Bloomsburg Red Beds accumulated. The abundance of vein quartz could explain the abundant conglomerate of the Shawangunk Formation. A thin diamictite (a poorly sorted, noncalcareous, land-derived sedimentary rock that contains a wide range of particle sizes) with exotic pebbles records a brief geologic episode of colluvial deposition that occurred during the Taconic hiatus (Epstein, 1989). As the mountains eroded, finer clastic sediments and even carbonates accumulated more or less continuously through the Middle Devonian Period. Clastic influx during the Middle Devonian records another, later mountain-building episode, the Acadian Orogeny.

REGIONAL OVERVIEW

The highest point in the Shawangunk Mountains (698 m [2289 ft]) lies near Sam's Point (Figure 1). On a clear day, one can see (from southeast to north) the New York highlands underlain by Precambrian rocks thrust on top of Cambrian and Ordovician carbonates and shales of the Wallkill Valley. West of the highlands are Schunnemunk and Bellvale mountains underlain by conglomerates and sandstones of the Middle Devonian Schunnemunk Conglomerate in the Green Pond Outlier. The rocks of the outlier are in fault and sedimentary contact with the Precambrian. The Shawangunk and Kittatinny mountains, held up by the Shawangunk Formation, trend to the southwest, with Tristates Monument marking the highest elevation in New Jersey at High Point (550 m [1803 ft]). The Shawangunk at Sam's Point dips very gently to the northeast, near the broad crest of the Ellenville Arch. Tough cross-bedding is well exposed and indicative of current trends ranging between 80SW and 70NW. Glacial striae with chattermarks on the bedding surfaces indicate that the Wisconsinan glacier flowed over the mountains moving 16SW. The lower 24 m (80 ft) of the Shawangunk here consists of medium- to thick-bedded conglomerate with quartz pebbles as much as 5 cm (2 inches) long. Channel cuts are common. Nowhere do we see any pebbles from the underlying Martinsburg, a peculiarity that exists throughout New York, New Jersey, and eastern Pennsylvania, and one that eludes a good sedimentologic explanation. The Shawangunk is separated into blocks tens of feet wide that have moved apart along the soft shales of the underlying Martinsburg, probably forced apart by gelifraction, wedging of boulders that fall into the cracks, and block sliding. At the Ice Caves, 0.8 km (0.5 mi) to the east, the joints parallel the cliff face of the mountain (Feldman and Thompson, 2008), and cold air trapped in the maze of blocks and snow may persist throughout the summer, hence their name.

In Devonian times, there was a shallow carbonate shelf in the Helderberg-Coxsackie area that graded into a shelf-margin bryozoan bafflestone near Leeds and Saugerties. Further southwest toward Port Jervis, the sediments consisted of sparse to packed biocalcisiltites deposited on a carbonate ramp. This ramp occupied the northern margin of a structural basin depocenter, located to the east of and not directly related to the topographic basin of central New York, which was centered in the Tristates vicinity from the Late Silurian until the early Middle Devonian (Lindemann and Feldman, 1987).

THE "SHALE BANK"

The Martinsburg Formation (Figures 2-3) crops out about 1.6 km (1 mi) from the entrance to Mohonk Mountain House (on Mountain Rest Road).

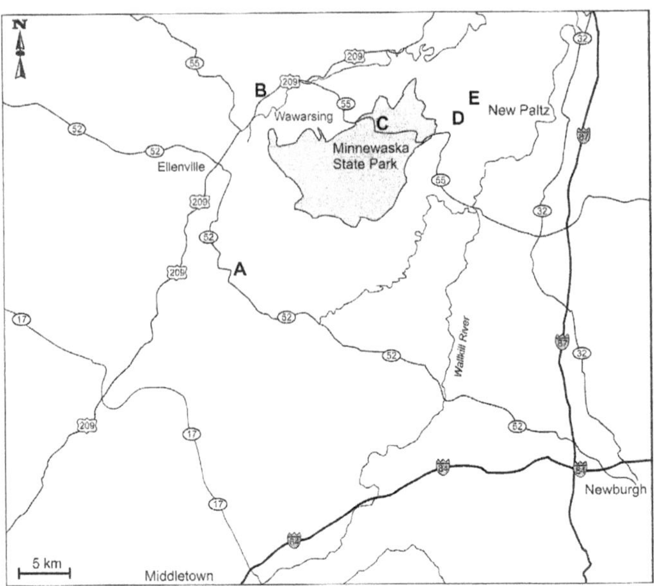

FIGURE 1. Locality map of the mid-Hudson Valley. A) Sam's Point; B) Onondaga Limestone in Wawarsing, NY; C) Mysterious "pods" in Minnewaska State Park; D) The "Shale Bank"; E) Eagle Cliff Road just to the west of Lake Mohonk.

Mohonk workers quarry and crush the shale in order to provide good footing and slow down erosion on the carriage roads. The outcrop consists of 14 m (45 ft) of predominantly dark gray shales and siltstones interbedded with fine-grained graywacke beds, occasional prominent pyrite layers, and disseminated sphalerite, chalcopyrite, and galena. Oscillation ripples can be observed on some bedding surfaces. Carbonaceous material occurs mostly as fine-grained patches throughout the matrix. The studied exposure illustrates a high degree of strain, mostly manifested by shiny quartz slickensided surfaces, small-scale cross-laminated and parallel-laminated strata, and ripple marks. Crinoid stems, some slightly disarticulated, and free columnals occur on different bedding surfaces, indicating a possible change in current regime. Scattered linear to sinusoidal horizontal burrow structures ranging in diameter from 0.5 to 3 cm are found on silty beds; some of the burrows are infilled with course quartz grains. Gutter casts are scattered throughout the section. The faunal constituents of this *Sowerbyella-Onniella* Community (Figure 4) include brachiopods (93%), crinoids (*Ectenocrinus*; 3%), bivalves (*Deceptrix?*; 3%), ostracodes (<1%), corals (<1%), trilobites (*Cryptolithus, Isotelus*; <1%), conulariids (<1%)

and unidentified burrowers (<1%). Brachiopods are represented by a low-diversity assemblage of *Sowerbyella* and *Onniella* with occasional *Dalmanella.* The fauna can be classified into distinct trophic groups: (1) high-level suspensions feeders (crinoids, corals); (2) low-level suspension feeders (brachiopods, bivalves); (3) animals that collect food from the sediment surface (ostracodes, trilobites); and (4) animals that feed within the sediment (burrowers). This partition of feeding niches leads to a reduced competitive trophic structure and therefore increased community stability. The community appears twice in the section, separated by about 4.9 m (16 ft) and thus appears to be stable. The depositional environment of this fauna may have been in relatively deep water as evidenced by gutter casts, at the bottom of which are shell accumulations, as well as a lack of bioturbation in the sediments. However, the moderately diverse fauna and oscillation ripples indicate a shallower environment. In addition, there are auriculate nuculoid bivalves that were shallow infaunal and deposit-feeding organisms. The sediments were probably not deposited at depths of greater than 2000 m, as suggested by McBride (1962), in the Martinsburg Formation at the Great Valley. Lehman and Pope (1990) noted that Bretsky (1970) divided the Reedsville and Martinsburg formations of the central Appalachians into three paleocommunities, one of which was the *Sowerbyella-Onniella* Community. This community, according to Bretsky (1970), corresponds to a deep-water depositional environment. The fauna at the "Shale Bank" very closely resembles the *Sowerbyella-Onniella* communities of Lehman and Pope at Swatara Gap and Bretsky's community of the central Appalachians, although the *Sowerbyella-Onniella* Community along the Shawangunk Ridge is much less diverse than the other two and was most likely deposited in shallower water. Bretsky never defined the absolute depth of his communities (Bretsky, 1970; Lehman and Pope, 1990).

"PODS" OF UNKNOWN ORIGIN

Within the Shawangunk Formation, we note occurrences of what we term "pods" that are domal in cross section with a flat base, the bottoms of which often contain quartz pebbles. The matrix is dark gray and consists of interstitial quartz ranging in size from fine (0.25-0.125 mm) to coarse (1.00-0.5 mm) sand and muscovite mica (Figure 5). The pods are aligned along bedding and, in many cases, crossbedded foresets, and range in size from 0.2 to 7 cm in diameter (at the base) and 0.2 to 6 cm in height. The rock that hosts

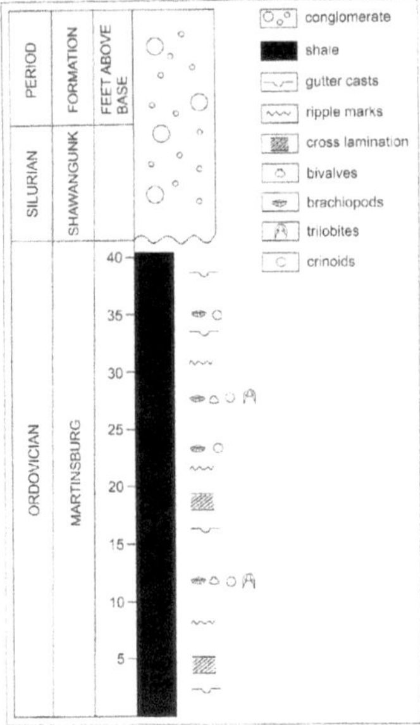

FIGURE 2. Columnar stratigraphic section of the Martinsburg and Shawangunk formations in the "Shale Bank" at Mohonk. The Martinsburg Formation, estimated to be between 3048-3658 m (10,000-12,000 ft) thick in the mid-Hudson Valley, is highly deformed. Based on field relations, the "Shale Bank" is considered to be near the top of the Martinsburg, but due to repetition of beds in the area, it is difficult to precisely determine exactly how close to the top of the formation it lies.

FIGURE 3. The steeply dipping Ordovician Martinsburg Formation in the "Shale Bank" on the Shawangunk Ridge.

the pods is composed of pressure solution welded quartz grains consistent with the quartzitic nature of the formation. There is no evidence of stratification. The origin of the pods is under investigation (Feldman et al., 2009), and possibilities include microbial mounds, sponges, or mud balls. Pods are similar in outline to some thrombolites in that the internal texture is non-laminated; however, there is no indication of microbial activity such as clotting. Though the pods may represent algal mats, their association with braided fluvial deposit renders this interpretation doubtful. It is possible that they are the remains of sponges, since the braided streams drained into a marine basin and the resulting tidal incursions may have supported sponges in an estuarine

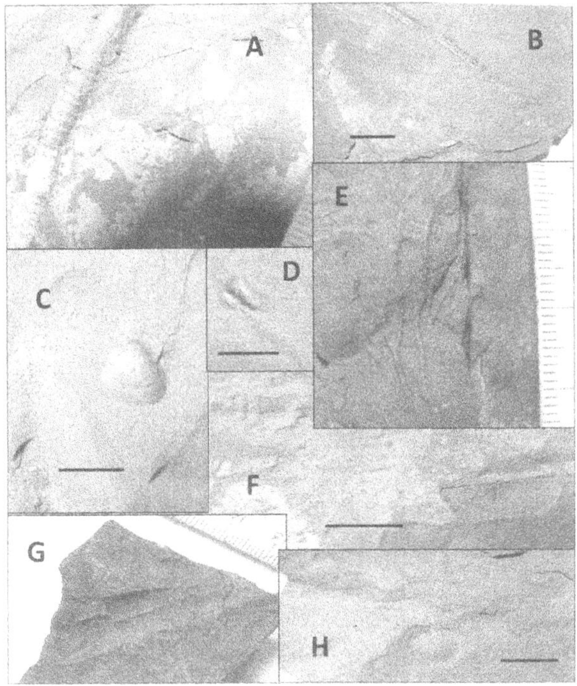

FIGURE 4. Faunal constituents of the *Sowerbyella-Onniella* Community in the "Shale Bank." A) articulated crinoid stem *(Ectenocrinus?)* indicating little if any current activity; B) slightly disarticulated *Ectenocrinus?* stem (scale bar = 1 cm); C) an auriculate nuculoid bivalve *(Deceptrix?)* typical of a shallow water environment (scale bar = 1 cm); D) unidentified ostracode adjacent to crinoid stem (scale bar = 1 cm); E) glyptocrinid partially buried under shale; F) the brachiopod *Sowerbyella* sp., ventral valve interior (scale bar = 1 cm); G) *Conularia* sp., a conulariid common in the Ordovician of the United States; H) interior mold of ventral valve of the brachiopod *Dalmanella* sp. (scale bar = 1 cm).

setting. Most sponges live in a marine environment. Support for this scenario is the occurrence of the trace fossil (ichnofossil) *Arthrophycus* recently discovered (by HRF) on the ridge in the Shawangunk Formation. Rindsberg and Martin (2003) proposed that a rhandophorid trilobite could have been the organism responsible for *Arthrophycus* burrows in Alabama, and they proposed that *Cryptolithus* may have been the tracemaker of *Arthrophycus.* The domed shape of the pods tends to preclude a mud-ball origin. At first glance, the long axes of the pods and gray matrix seem to be aligned along bedding planes, but upon closer inspection, they actually follow cleavage planes and may have a pressure solution origin. Further study is necessary to arrive at a definitive conclusion as to the mode of formation of the pods.

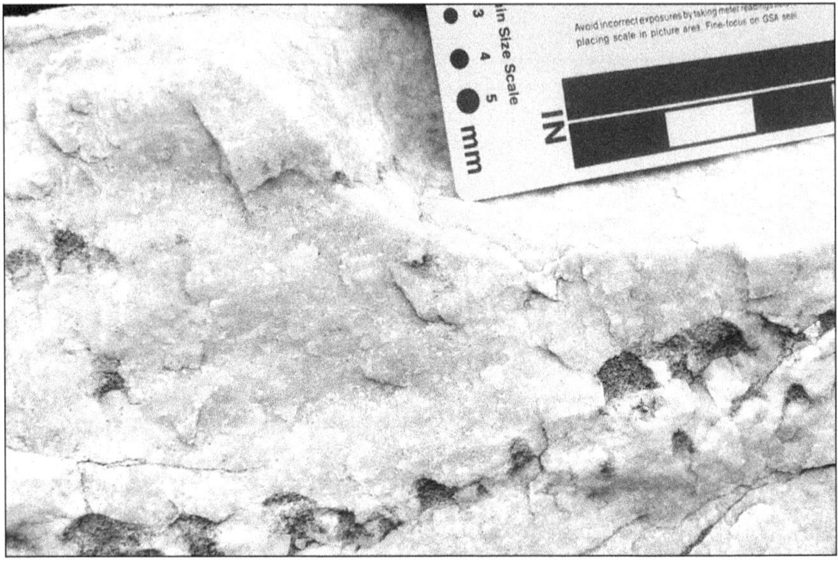

FIGURE 5. "Pods" aligned along bedding near the top of the Middle Silurian Shawangunk Formation near Minnewaska State Park.

SULFIDE MINERALIZATION

Lead-zinc sulfide deposits formed near Ellenville, New York, are well known and were mined from the latter half of the nineteenth century into the early years of the twentieth century (Gray, 1961; Ingham, 1940). However, a previously unrecognized metal sulfide deposit has been discovered by Smoliga in the Shawangunk Formation along Eagle Cliff Carriage Way (Figure 6). This small deposit consists of the Fe-sulfide phases pyrite and marcasite, lesser amounts of the Cu-Fe-sulfide chalcopyrite, and trace amounts of anglesite (Pb-sulfate). A larger area of Fe-oxide, hydroxides, and Fe-sulfate minerals, hematite, goethite, and jarosite, which form as weathering products of Fe-sulfides, has also been observed on Eagle Cliff. This finding would suggest that the sulfide deposit may be more extensive than what is seen in surface outcrop. Moreover, jarosite deposited along fracture surfaces has been observed on Sky Top ridge east of Eagle Cliff, suggesting sulfide occurrences in this area.

The metal sulfides appear to be associated with faulting and may have been remobilized from sulfides occurring in the underlying Martinsburg Formation. The relationship of the sulfide deposits found at Mohonk to the Ellenville deposits is unknown at this time; further study is underway.

FIGURE 6. Sulphide mineral vein, opaque in thin section, along Eagle Cliff Carriage Way in the Shawangunk Formation on the Ridge.

ONONDAGA LIMESTONE

An interesting outcrop of the Onondaga Limestone can be found in back of Wendel's Asphalt on the west side of Route 209 about 0.8 km (0.5 mi) north of Vernoy Kill. The limestone crops out in an abandoned quarry with in the Port Jervis Trough and is 15.5 m (51 ft) thick at this locality. The Onondaga passes downward into the underlying Schoharie Formation via a limy grit. Above the transitional zone, which is 0.6 m (2 ft) thick, there is another 7.6 m (25 ft) of the Edgecliff Member consisting of medium-grained to medium-bedded limestone, transitional with the lower Moorehouse Member of the Onondaga Formation. The contact between the top of the Edgecliff and the base of the Moorehouse is defined by the presence of large crinoid columnals and the coral *Amplexiphyllum.* This unit of the Moorehouse Member is overlain by a medium-bedded, light-gray limestone unit 2.7 m (9 ft) thick. The topmost unit of the Moorehouse consists of 5.2 m (17 ft) of very fine-grained, massive, medium-gray limestone with discontinuous seams of chert, trilobite fragments, and the brachiopod *Levenea subcarinata.* At the Eastern New York Correctional Facility, there is an excellent exposure of the Martinsburg-Shawangunk contact, about 6.1 m (20 ft) long. The angular discordance between the two formations is 4 degrees. The Shawangunk dips 22 degrees to the northwest in the northwest limb of the Ellenville arch. Mullions are prominent on the basal Shawangunk surface, and a shear fabric can be observed in both the Martinsburg and Shawangunk formations.

SUMMARY

The geology of the Shawangunk Ridge and environs is varied and often complex. Here we present an overview of a geologically important segment of the Appalachian Orogeny, specifically dealing with aspects of the geology on the Mohonk Preserve in the area around New Paltz. The Martinsburg Formation, which crops out on the Shawangunk Ridge, has yielded a diverse marine fauna of brachiopods, crinoids, bivalves, ostracodes, corals, trilobites, and conulariids. This moderate diversity suggests that feeding niches were partitioned, resulting in a reduced competitive trophic structure and therefore increased community stability. We described a new metal sulfide deposit in the Shawangunk Formation along Eagle Cliff Road on the west side of Lake Mohonk. This deposit consists of the Fe-sulfide phases pyrite and marcasite, lesser amounts of the Cu-Fe-sulfide chalcopyrite, and trace amounts of anglesite (Pb-sulfate). The Middle Devonian Onondaga Limestone that crops out on the west side of the Shawangunk Ridge within the Port Jervis Trough contains large crinoid columnals, the coral *Amplexiphyllum,* trilobite fragments, and the brachiopod *Levenea subcarinata.* The Onondaga in this area, near Wawarsing, is part of a carbonate ramp that thickens as it dips toward the south and the Port Jervis region.

ACKNOWLEDGMENTS

We thank John Thompson of the Mohonk Preserve for inviting us to present an oral version of this paper at the Northeast Natural History conference held in Albany, New York in the spring of 2011. The authors gratefully acknowledge John H. Puffer, Rutgers University, and two anonymous reviewers for critically reviewing the manuscript and providing suggestions for improvement. Many thanks to Mena Schemm-Gregory, University of Coimbra, for drafting Figures 1 and 2.

REFERENCES

Bretsky, P. W. 1970. Ordovician ecology of the central Appalachians. *Peabody Museum of Natural History, Yale University, Bulletin* 34: 1-150.

Epstein, J. B. 1989. Regional stratigraphy of Silurian rocks and an enigmatic Ordovician diamictite, southeastern New York. In A. P. Schultz (ed.), *Appalachian Basin Symposium: Program and Extended Abstracts. U. S. Geological Survey Circular Report C* 1028.

Epstein, J. B. 1993. Stratigraphy of Silurian rocks in Shawangunk Mountain, southeastern New York, including a historical review of nomenclature. *U. S. Geological Survey Bulletin* 1839: Ll- L40.

Epstein, J. E., and Lyttle, P. T. 1987. Structure and stratigraphy above, below, and within the Taconic unconformity, southeastern New York. In R. H. Waines (ed.), *New York State Geological Association, 59th Annual Meeting, Kingston, New York*, November 6-8, 1987. New Paltz, New York, State University of New York, College at New Paltz, Fieldtrip Guidebook, C1-C78.

Feldman, H. R., and Thompson, J. 2008. Top of the Gunks. *Natural History Magazine* 117: 36-38.

Feldman, H. R., Smoliga, J., Wilson, M. A., Schemm-Gregory, M., and Starr, J. 2009. Mysterious "pods" in the Middle Silurian Shawangunk Formation, mid-Hudson Valley, New York. *Geological Society of America Northeastern Section Meeting, Portland, Maine* 41: 88.

Gray, C. 1961. Zinc and lead deposits of Shawangunk Mountains, New York. *New York Academy of Sciences* 23: 315-331.

Ingham, A. I. 1940. The zinc and lead deposits of Shawangunk Mountain, New York. *Economic Geology* 35: 751-760.

Lehman, D., and Pope, J. K. 1990. Upper Ordovician tempestites from Swatara Gap, Pennsylvania: Depositional processes affecting the sediments and paleoecology of the fossil faunas. *Palaios* 4: 553-564.

Lindemann, R. H., and Feldman, H. R. 1987. Paleogeography and brachiopod paleoecology of the Onondaga Limestone in eastern New York. *New York State Geological Association, Guidebook for Fieldtrips, 59th Annual Meeting* 59: 1-30.

McBride, E. F. 1962. Flysch and associated beds of the Martinsburg Formation (Ordovician), central Appalachians. *Journal of Sedimentary Petrology* 32: 39-91.

Rindberg, A. K., and Martin, A. J. 2003. *Arthrophycus* in the Silurian of Alabama (SA) and the problem of compound trace fossils. *Palaeogeography, Palaeoclimatology, Palaeoecology* 192: 187-219.

Chapter Two

Level-bottom Brachiopod Communities in the Middle Devonian of New York

ABSTRACT

Depositional environments of the Onondaga Limestone from central to southeastern New York are found to be normal, subtidal marine, due to the absence of characteristic supratidal or intertidal sedimentary features and the presence of a typical, diverse, marine level-bottom community framework. Post-mortem transport has not been extensive, as evidenced by low articulation ratios, lack of abraded valves, and complete ontogenetic gradations within species, which precludes large scale winnowing. Sedimentation rates appear to have been greatest in eastern New York, where the Onondaga Limestone reaches a thickness almost three times that of the strata in central New York. Shaly beds in the central area represent periods of cessation of carbonate deposition rather than an influx of clastic material. Deposition terminated with the onset of deeper water characterized by a westerly advance of terrigenous sedimentation (the Marcellus Shale of the Hamilton Group). Nine brachiopod communities can be recognized in the Onondaga Limestone. There is a strong correlation between sediment-substrate and community type, rellecting the sedimentologic control of brachiopod community distribution. Sandy facies, cherty limestones and coral biostromes and bioherms are associated with inner-neritic deposition in Edgecliff time; argillaceous lime muds and lime sands are characteristic of mid-neritic deposition in Nedrow to Moorehouse time; and highly argillaceous lime muds are associated with outer-neritic deposition in Seneca time.

Although the Devonian strata of New York State have been studied by many workers (Anderson, 1967; Sulton et al., 1970; Walker and Laporte, 1970; Thayer, 1972, 1974; Bowen et al., 1974) the invertebrate communities of the Onondaga Limestone have not been analyzed with regard to community structure, nor has the brachiopod fauna been described adequately since the days of James Hall. The above cited studies have dealt mainly with Middle to Upper Devonian rocks; little attention has been paid to the brachiopod communities of the upper Lower to lower Middle Devonian formations, Data on Middle Devonian communities are not as plentiful as would appear from the above cited papers, as is evident from Boucot's publication (1975: 225).

Although there are scattered comments in the literature concerning communities, not enough information has accumulated to write a world-wide summary. For example, in the Appohimchi Subprovince, neither the Onondaga nor the Hamilton groups have been subjected to the necessary scrutiny, although a mass of information allows certain major categories to be roughly defined, particularly for the Hamilton. These have been well-known for some time and may approximate Benthic Assemblages of even actual communities.

Since the Onondaga Limestone represents, for the most part, a low-energy environment with little post-mortem transport, local abrupt changes in the physical characteristics of the fauna, such as diversity, density, and general composition, reflect the original distribution of species in the living communities. Since fine-grained sediments are the most favorable for the preservation of organisms but usually contain the least number of preservable species (Speden 1966), one might argue that the fossil communities of the Onondaga that are associated with fine-grained substrata may not be representative of the living communities. However, this does not apply to the hard-shell component (e.g. brachiopods) which would, in the absence of significant post-mortem transport or destruction due to biopredation or diagen esis, be closely representative of the living community.

Field studies for this project were undertaken during the summers of 1976 and 1977 in three separate areas: (1) the central area (Syracuse region); (2) the Cherry Valley area (Cherry Valley, New York); and (3) the eastern area (mid-Hudson Valley to Wawarsing, New York). Although about 100 outcrops were visited, only 30 were studied in detail due mainly to the availability and accessibility of fossils. When a silicified fauna was found to be present bulk samples were collected and processed by immersion in dilute HC1. Fragile specimens were soaked in sodium sulfate solution, dried, and brushed with alvar in order to prevent chipping.

Systematic descriptions of the brachiopods as well as locality data can be found in Feldman (1985). All of the material is in repository at the American Museum of Natural History, Department of Invertebrates. Abbreviations used in this paper are as follows: AMNH, American Museum of Natural

History; AMNH Loc., American Museum of Natural History Locality Number; USNM, United States National Museum.

STRATIGRAPHIC FRAMEWORK

Oliver (1954, 1956) has thoroughly reviewed the stratigraphy of the Onondaga Limestone in New York State and the reader is referred to his papers for an in-depth discussion of the stratigraphy of the formation. In the area dealt with in this study (Figure 1), four members of the Onondaga Limestone are recognized: Edgecliff, Nedrow, Moorehouse, and Seneca. The stratigraphy generally follows Oliver's (1956) scheme but there are some minor modifications which are incorporated into the discussion below.

Thickness: Although no complete sections of the Onondaga Limestone are exposed from which the exact thickness can be measured, several localities permit fairly accurate estimates of thickness (Figure 2). At the type locality of the Nedrow Member in the Onondaga Indian Reservation Quarry near Nedrow, New York, more than 16.5 m of limestone from the Upper

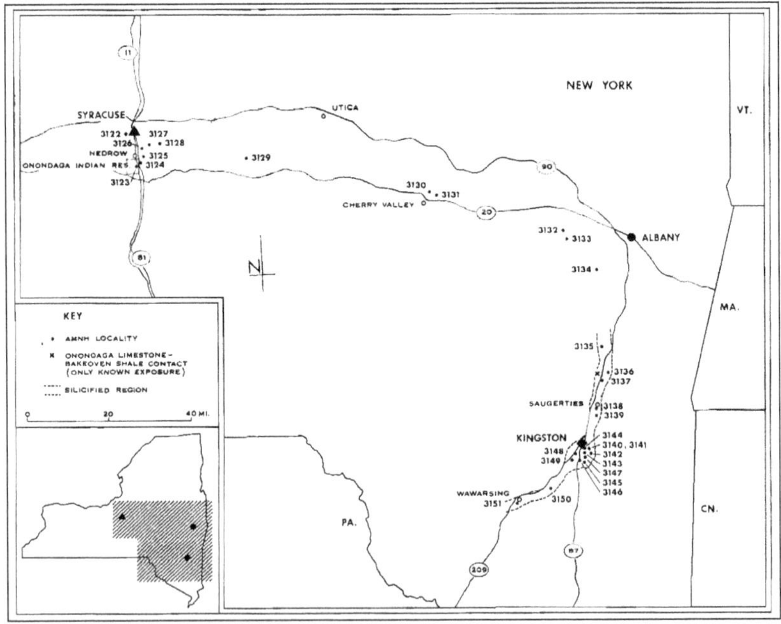

FIGURE 1. Index map of collecting localities and described sections in the Onondaga Limestone, central and southeastern New York.

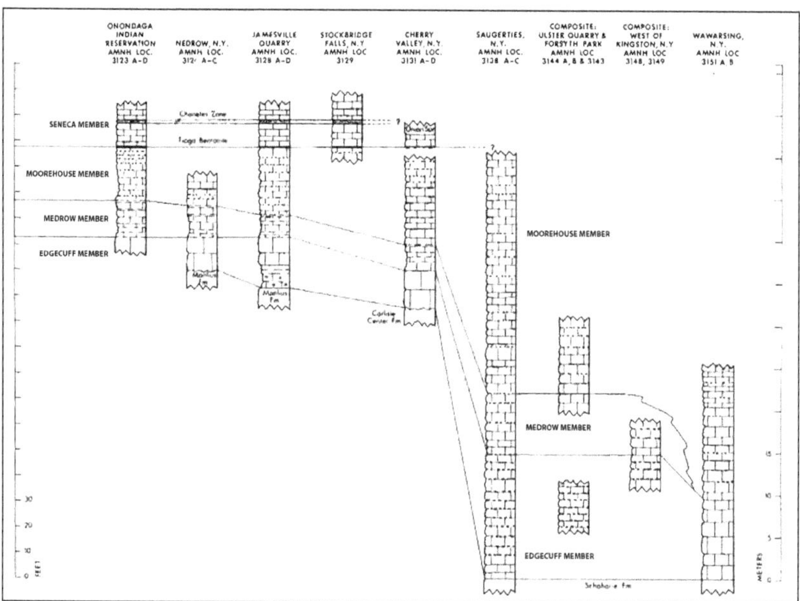

FIGURE 2. Measured stratigraphic sections of key outcrops on a west-east-southeast transect along the outcrop belt of the Onondaga Limestone in central and southeastern New York.

Edgecliff to the middle of the Seneca was measured. Chute and Brower (1964) have measured almost 18 m at the same locality. In the Jamesville Quarry at Jamesville, New York, recent blasting has exposed the most complete section known to date—21 m. This includes the Manlius-Edgecliff contact and more than 4 m of the Seneca Member, and compares well with Oliver's estimate of 21 m for the Syracuse area (1954: 624), which was arrived at by piecing together two separate sections. Further east, near Saugerties, New York, the Onondaga more than doubles in thickness to more than 49.6 m. The lower contact (Schoharie-Edgecliff) is exposed but the top of the Moorehouse is missing. This locality represents the most complete section known east of Cherry Valley.

EDGECLIFF MEMBER

The Edgecliff Member in the central area is a light gray, in places pink, coarse-grained limestone characterized by solitary rugose and tabulate corals which are often so common as to form biostromes (Oliver, 1956).

Toward the east, the Edgecliff undergoes a change. Although characteristic large crinoid columnals are still present, and remain so throughout the member in New York State, the lithology changes to a finer-grained rock with numerous light-weathering chert seams (in a crinoidal matrix) present in the upper three-fourths of the member. The top of these chert seams is considered to be the Edgecliff-Nedrow contact in the eastern area.

NEDROW MEMBER

In the central area, the Nedrow Member is easily recognizable by its typical recessed weathering caused by the highly argillaceous nature of the rock. Numerous shaly beds are present which often yield fairly well preserved fossils. Toward the east the Nedrow thickens and becomes less argillaceous, more massive, and coarse-grained. In the mid-Hudson Valley it is difficult to pick out the Nedrow as a distinct member due to the lithologic similarities between it and the Edgecliff and Moorehouse members. However, in the Saugerties area the Nedrow is considered to include all strata between the top of the light-weathering cherty seams of the Edgecliff and the first dark-weathering (gray to black) chert seam, which represents the base of the Moorehouse Member. At Wawarsing, the Nedrow is no longer recognizable.

Numbers in the left-hand portion of each box represent percentage of fauna while those in the right-hand portion represent absolute numbers identified and/or collected at each locality.

MOOREHOUSE MEMBER

In the central area, the Moorehouse Member is a medium gray, medium- to fine-grained limestone with abundant chert seams in the upper one-third. In the eastern area, the entire member is uniform throughout and no zonation based upon lithology is possible. Characteristic dark-weathering chert seams are present throughout the member at irregular intervals. At about 23-26 m above the Nedrow-Moorehouse contact, a silicified faunal association of platyceratid gastropods (*Platyceras dumosum*) and brachiopods (mainly *Atrypa "reticularis"*) is recognized in some places. The lithology of the Moorehouse changes to the southeast, becoming darker, finer-grained, and highly argillaceous. The fauna shows a very low diversity and is definitely atypical. Apparently, environmental conditions worsened due to poor circulation and low oxygen content. This view is supported by the low diversity and smaller sized fossils recovered from Moorehouse strata in southeastern New York.

SENECA MEMBER

The base of the Seneca Member is marked by a 0.15-0.20 m thick, greenish-gray to ochre colored (when fresh) ash layer of volcanic origin (Oliver, 1954) called the Tioga Bentonite. The bentonite weathers rapidly (to dull gray) and fresh samples are rare. East of Cherry Valley there are no known occurrences of the Seneca Member which, in the area under study, is a very highly argillaceous limestone, poorly fossiliferous except for a zone of chonetid brachiopods which are consistently found 3 m above the Tioga Bentonite ("*Hallinetes*" Zone or Zone J of Oliver, 1954; *Hallinetes* was formally known as *Chonetes*). Apparently, both the occurrence of the bentonite and a sharp transition to deeper water conditions paralleled a faunal break at the base of the Seneca. The upper contact is with the Marcellus Shale which is representative of a westerly advance of terrigenous sedimentation and relatively deep, poorly oxygenated water. The contact, gradational near Union Springs, New York, is sharply defined at AMNH Loc. 3131A. At some localities stylolites are common in the upper half of the member.

COMMUNITY ANALYSIS

In this study a community is defined as a recurrent association of taxa which was presumably controlled by a set of environmental factors such as: temperature, pressure, current and/or wave action, light penetration, nutrients, water chemistry, dissolved oxygen, and miscellaneous factors (biotic relationships such as symbiosis, mechanical wear, etc.). The communities are named after the dominant taxa. A total of 40 brachiopod species are recognized (Table 1) and, along with other faunal constituents, are listed in Tables 3 and 5-12. The taxa are also ranked according to abundance. Fragments of bryozoans, crinoids, and corals were noted but not tabulated in any quantitative fashion due to the difficulty of assessing their percentage composition within the community. Numbers of individual brachiopods were determined by the most abundant valve found. Relative abundance as

TABLE 1. Distribution of Brachiopods in the Onondaga Limestone. Numbers in the left-hand portion of each box represent percentage of fauna whereas those in the right-hand portion represent absolute numbers identified and/or collected at each locality.

BRACHIOPOD COMMUNITIES	LEPTAENA – MEGAKOZLOWSKIELLA							ATRYPA – COELOSPIRA-NUCLEOSPIRA				ATRYPA		
AMNH LOC. TAXA	3124A	3123B	3128B	3124B	3128C	3123C	3123D	3137	3138A	3135	3144A	3142	3138B	3131C
Atrypa "reticularis"	29|5	6|3	5|3	2|1	3|5	4|3	10|3	30|163	27|98	38|80	33|12	62|18	62|18	44|30
Megakozlowskiella raricosta	24|4	28|13	33|21	25|13	27|43	26|18	45|14	10|57	2|8	2|4	11|4			
Leptaena aff. "rhomboidalis"	24|4	15|7	25|16	27|14	35|57	35|24	35|11	<1|1	<1|1			3|1	7|2	
Orthotetacids indet.	12|2	9|4	6|4	6|3	5|8	7|5	3|1	<1|1		2|4	6|2	7|2		7|5
Pentamerella arata	6|1	6|3	8|5	4|2	4|6	6|4	6|2	4|21	2|9		6|2			
Levenea sp. A			3|2	10|5	2|3			<1|1	1|2			7|2	21|6	13|9
Dalejina sp. A	6|1	9|4		4|2	2|3	1|1		4|21	<1|1	4|8	6|2	3|1		
Athyris sp. A		11|5	3|2	2|1	3|4	3|2		1|6		1|3				
Coelospira Camilla			1|1		3|5	1|1		12|63	15|53	19|40	8|3	3|1	7|2	
Nucleospira aff. Ventricosa								12|63	16|59	5|11	8|3	3|1	3|1	
Pentagonia unisulcata								3|15	6|20		8|3	3|1		16|11
Megastrophia sp.		4|2		9|4	8|12	6|4			<1|2	1|3				9|6
Schizophoria cf. multistriata								2|12	6|21	1|2	3|1	7|2		
Pacificocoelia acutiplicata		4|2	16|10	2|1	9|14	4|3								
Acrospirifer duodenaria								10|54	7|25	8|16	8|3			
Cyrtina hamiltonensis								2|11	2|7	4|8	3|1			
Elytha fimbriata								2|10	1|3	2|4				
Gypidula sp.		2|1		2|1	1|1									
Stropheodonta cf. demissa				4|2				<1|1		1|3				
"Mucrospirifer" sp.								3|16	3|10	2|5				
Cupularostrum? sp. A								1|4	8|30	1|2				6|4
Rhipidomella?									<1|3					
"Chonetes" aff. lineata														
Stropheodontids indet.		2|1				1|1				3|7				
Ambocoelia sp.								<1|3		1|3				
Athyridaceans indet.								<1|1	<1|2					
Atribonium halli								<1|1		1|1				
Meristina cf. nasuta		4|2						<1|1						
Strophonella cf. punctulifera						1 |1								4|3
Cupularostrum? sp. B									1|5					
Cyrtina sp. A										1|2				
Dalejina sp. B								<1|2						
Levenea sp. B														
Trematospria? sp.				4|2										
Athyris sp. B									<1|2					
Amphigenia? sp.														
Charionoides aff. doris						1|1								
Eospiriferid? indet.										1|3				
Schuchertella sp.								<1|3						
Rhynchospirina sp.								<1|1						
total number of specimens	17	47	64	51	161	68	31	532	359	209	36	29	29	68

			ATRYPA-MEGAKOZLOWSKIELLA				LEVENEA COMM I			AMPHIGENIA	LEVENEA COMM II	PACIFICO-COELIA	CHONETES		
3141	3140	3139	3130	3131B	3133	3134	3138C	3143	3131D	3126	3151A	3125A	3123A	3128A	3129
40|17	52|12	58|48	20|12	25|20	32|21	33|22	23|3	9|2	27|3			8|14	1|1		
7|3	17|4	10|8	15|9	15|12	23|15	21|14						9|16	3|4		
2|1	9|2		7|4	1|1		5|3		9|2				10|18		<1|3	
		4|3	10|6	3|2	8|5	9|6						4|8		<1|2	
	4|1	1|1	3|2	4|3	6|4			9|2				3|6	1|1		
21|9		1|1	5|3	6|5		9|6	69|9	64|14	73|8		100|98				
5|2		4|3	2|1	3|2	5|3							3|6			
		2|3	2|1	3|2	3|2							3|5		<3|2	
2|1		9|7			5|3										
7|3	4|1	4|3	3|2	1|1	6|4										
7|3	13|3	5|4	20|12	5|4	3|2	5|3									
				3|2		5|3						1|2	2|3		
2|1			2|1	19|15	2|1	14|9									
					2|1							59|110			
			2|1	1|1											
2|1					3|2										
		1|1		3|2			8|1								
5|2			2|1		2|1										
			5|3	3|2	3|2										
			2|1	3|2											
														<1|1	
				1|1								1|1	93|142	99+|>4000	100|410
				3|2											
								9|2							
										100|12					
43	23	82	59	79	66	66	13	22	11	12	98	186	151	4008	410

TABLE 2. Brachiopod Communities of the Onondaga Limestone.

Community	Member(s) in which found	Geographic area	Lithology	Depositional environment
Atrypa-Coelospira-Nucleospira	Moorehouse	Eastern area	Moderately argillaceous	Mid-neritic
Atrypa-Megakozlowskiella	Moorehouse	Cherry Valley	Moderately argillaceous	Mid-neritic
Atrypa	Nedrow-Moorehouse	Cherry Valley and Eastern Area	Moderately argillaceous to non-argillaceous	Mid-neritic
Leptaena-Megakozlowskiella	Nedrow-Moorehouse	Central area	Moderately argillaceous to highly argillaceous	Mid-neritic
Pacificocoelia	Nedrow	Central area	Highly Argillaceous	Inner-neritic?
Levenea community I	Edgecliff	Eastern area	Cherty, slightly argillaceous	Inner-neritic
Levenea community II	Moorehouse	Eastern area	Highly argillaceous	Outer-neritic
Amphigenia?	Edgecliff	Central area	Sandstone, Limy matrix	Inner-neritic
"*Hallinetes*"	Seneca	Central area	Very highly argillaceous	Outer-neritic

noted in the field appears to be a function of three conditions: (1) matrix type, (2) weathering rate, and (3) presence or absence of a silicified fauna. A general pattern is perceptible which shows increasing abundance with deeper water conditions, except for the "*Hallinetes*" and *Levenea* Community II communities in which diversity decreases while dominance of single taxa increases. Peak abundance (of brachiopods) is attained in the *Atrypa-Coelospira-Nucleospira* Community and reflects conditions which seem to be most favorable to brachiopods: good circulation, adequate nutrient supply, normal salinity, normal temperature and pressure. Reduced abundances in some communities may, conversely, be due to the lack of favorable environmental conditions or paucity of available fossil material in the field due to preservational selectivity (silicified versus non-silicified fauna). The nine communities recognized in the Onondaga Limestone (Table 2) are discussed below along with comments on adaptive habits of the brachiopods, depositional environments, geographic ranges and lateral distribution.

ATRYPA-COELOSPIRA-NUCLEOSPIRA COMMUNITY

The *Atrypa-Coelospira-Nucleospira* Community is found in the mid-Hudson Valley in eastern New York and ranges from the area of Leeds to just south of Kingston. The limestones vary from mudstones to wackestones and the environment of deposition was probably mid-neritic with a moderately to highly argillaceous lime mud (or possibly lime sand) substratum. Diversity is greatest of all communities recognized (Table 3). A similarity is evident between this community and Copper's (1966) biotope of primitive, abundant Atrypidae in which variably sized atrypids occur with spiriferids and schizophorids in fine-grained sandstones, siltstones and shales with thin limestone interfingerings. Sedimentary structures are rare. A shallow environment, affected by currents and too muddy for coral growth, appears to have been indicated. The *Atrypa-Coelospira-Nucieospira* Community, however, has a diverse coral fauna possibly indicative of less mud.

Lenz's (1976) Lower Lochkovian *Howellella-Protathyris* Community possesses similar faunal elements in an offshore position: "*Schuchertella,*" *Atrypa, Schizophoria, Ambocoelia, Coelospira,* and *Nucleospira.* Major elements of his fauna are characterized by the following morphotypes: (1) broad, flat ("*Schuchertella*"); (2) relatively smooth spiriferid and atrypid genera (*Protathyris* and *Cryptatrypa*); (3) broad, unequally biconvex forms (*Schizophoria*); and (4) frilly forms (*Atrypa*). The *Atrypa-Coelospira-Nucleospira* Community possesses the same morphotypes:

(1) Orthotetacids, *Schuchertella* (broad, flat)
(2) *Nucleospira, Athyris* (smooth spiriferids)
(3) *Schizophoria* (unequally biconvex)
(4) *Atrypa* (with frills)

The greater diversity of the *Atrypa-Coelospira-Nucleospira* Community implies somewhat deeper water conditions, similar to Johnson's (1974) Lower Devonian leptocoelid-acrospiriferid biofacies from the Rabbit Hill Limestone, also argillaceous, and the *Trematospira* Zone of the McColley Canyon Limestone.

Low articulation ratios (Table 4) probably represent mild current activity. It is noteworthy that the atrypids, with articulation ratios of 0.94 and 1.14 at two localities (AMNH Loc. 3135, 3137) within this community,

Table 3. Faunal Constituents of the *Atrypa-Coelospira-Nucleospira* Community. AMNH Localities: 3135, 3137, 3138A, 3144A. vc = very common, c = common, r = rare, vr = very rare.

Taxa	Rank	Relative (%) abundance
Brachiopods *(N = 1130):*		
Atrypa "reticularis"	1	31
Coelospira camilla	2	14
Nucleospira aff. *ventricosa*	3	12
Acrospirifer duodenaria	4	8
Megakozlowskiella raricosta	5	6
Pentagonia unisulcata	6	3
Cupularostrum sp. A	7	3
Schizophoria cf. *multistriata*	7	3
Dalejina sp. A	8	2
Pentamerella arata	8	2
"Mucrospirifer" sp.	9	2
Cyrtina hamiltonensis	10	2
Elita fimbriata	11	1
Athyris sp. A	12	+
orthotetacids indet.	13	+
stropheodontids indet.	13	+
Ambocoelia sp.	14	+
Cupularostrum sp. B	15	+
Megastrophia sp.	16	+
Stropheodonia cf. *demissa*	16	+
Athyridaceans indet.	17	0.3
Athyris sp. B	17	0.3
eospiriferid? Indet.	17	0.3
Levenea sp. A	17	0.3
Schuchertella sp.	17	0.3
Atribonium halli	18	0.2
Cyrtina sp. A	18	0.2
Dalejina sp. B	18	0.2
Leptaena aff. *"rhomboidalis"*	18	0.2
Meristina cf. *nasuta*	19	0.1
Rhipidomella?	19	0.1
Rhynchospirina sp.	19	0.1

(Continued)

Table 3. (Continued)

Taxa	Rank	Relative (%) abundance
Gastropods:		
Platyceras dumosum		vc
Platyceras (Platystoma)		c
Platyceras sp.		c
Pseudophoracean indet.		r
Corals:		
Tabulates		
Favosites		c
Aulopora		c
Striatopora		c
Aulocystis (*Ceratopora*)		c
Rugosans:		
Acinophyllum		c
Heliophyllum		c
cf. *Amplexiphyllum*		c
Breviphrentis		c
"Heterophrentis"		c
cf. *Syringaxon*		c
cystimorph?		r
2 indet. Genera		r
Trilobites:		
Phacops?		r
indet. Fragments		r
Bryozoans:		
Dyoidophragma? sp. (Encrusted on *Schizophoria* cf. *multistriata,* external surface of pedicle valve)		vr
Trepostome fragments		r
Crinoids:		
Columnals		r
non-pinnulate inadunate ossicles		c

often display very stout articulation. Their shells disarticulate with some difficulty compared to most Silurian-Devonian brachiopods, since the teeth tend to interlock in some cases (A. J. Boucot, personal communication). The relatively high percentage of articulated *Atrypa* in high diversity collections makes this clear. *Atrypa* occurs in almost every collection, making the genus

TABLE 4. Articulation Ratios of Brachiopod Valves.

Taxa	AMNH Loc.	No. of articulated individuals	No. of pedical valves	No. of brachial valves	p.v./b.v.
"Mucrospirifer" sp.	3137	0	15	12	1.25
	3138A	2	88	4	2.00
Megakozlowskiella raricosta	3137	10	42	30	1.40
Coelospira camilla	3137	40	23	8	2.88
	3138A	119	50	33	1.51
Atrypa "reticularis"	3137	70	80	85	0.94
	3138A	23	43	34	1.72
	3135	14	25	22	1.14*
	3141	8	8	11	0.73
Acrospirifer duodenaria	3137	0	58	35	1.66
Nucleospira aff. *Ventricosa*	3137	11	33	26	1.27
	3138A	14	35	17	2.06
	3135	24	3	3	1.00
Pentagonia unisutcata	3137	2	16	13	1.23
	3138A	4	13	3	4.33
Schizophoria cf. *multistriata*	3137	2	7	6	1.17
	3138A	2	6	4	1.50

*There are several hundred articulated shells *in situ* at this locality.

very eurytopic. This is quite characteristic of its behavior in most of the Upper Llandovery-Frasnian. *Pacificocoelia* is another genus which is found at many localities and is almost as eurytopic as *Atrypa.*

The *Atrypa-Coelospira-Nucleospira* Community is characterized by many adaptive types, but only 13 genera of brachiopods comprise the bulk of the community; three of those *(Atrypa, Coelospira,* and *Nucleospira*) represent the trophic nucleus of low-level, epifaunal suspension feeders. Due to the predominance of nonpedunculate adult forms, a soft (muddy?) substrate can be inferred. Generally, brachiopods with pedicles live attached to firm substrates or possibly rocks, other small invertebrates, or benthic plants. The most dominant brachiopod in the community, *Atrypa,* lacked a functional pedicle in the ephebic and gerontic stages. In some neanic specimens a pedicle foramen is recognizable, leading to the assumption that the pedicle atrophied during ontogeny. Life position was probably influenced by the frilly border found on many specimens and thought to be present on many others, not as well preserved.

Coelospira possesses a small but distinct pedicle foramen, which leads to the question of attachment site on a soft substrate. No positive evidence was found, but small bits of organic material (shells, plants) could have served as sites of attachment considering the small size of the shells. Bretsky (1970) found *Zygospira,* a probable ancestor of *Coelospira* and morphologically similar, consistently associated with muds and lime muds, but did not address mode of attachment.

Nucleospira, a small, smooth brachiopod, displays no specific adaptation to a substrate. The small size and slightly globular form most likely aided it in resting on the sediment and enabled the shells to avoid sinking below the sediment-water interface.

Acrospirifer and *Megakozlowskiella* are medium-sized, strongly ribbed shells found in the *Amphigenia* Community of the Bois Blanc Formation and Camden Chert, and *Dicoelosia-Hedeina* Community (Boucot, 1975). *Acrospirifer* displays a rather alate morphology, similar to *Mucrospirifer,* which probably acted as a stabilizing feature on the sea floor. *Megakozlowskiella* is considerably larger and more globular than *Acrospirifer* and possesses higher and more prominent ribs. These coarsely ornamented brachiopods may indicate a position near wave base or possibly subjected to a certain degree of agitation. Evidence for this lies not only in their morphologic adaptations, but in accumulations of disarticulated shells and miscellaneous organic material in small lenses, parallel to bedding, especially prominent at silicified outcrops (e.g. AMNH Loc. 3138A) that have weathered extensively. The debris appears to have been washed and sorted by mild current or wave action.

All specimens of *Mucrospirifer* studied had an open delthyrium with no evidence of a stegidium or modifying plates. Thus, a pedicle was probably present throughout the life of the animal, although the size and degree of usefulness is questionable. Cowen (1968) correlated the decrease in function of the pedicle with an increase in the development of alae which acted to stabilize the shells on the substratum. *Mucrospirifer* might have possessed a rooted, branching pedicle in the manner of *Cryptopora gnomon* and *Chlidonophora chuni,* as pointed out by Bromley and Surlyk (1973), which are able to penetrate hard calcareous substrates.

Other less common but significant elements in the community, found only in the argillaceous units of the Onondaga Formation, are the large, somewhat globose brachiopods *Schizophoria* and *Pentagonia.* The gross external morphology of both, but especially *Schizophoria,* is similar to that of *Atrypa,* and consequently life position was most likely similar. However, no frilly borders were found on either *Schizophoria* or *Pentagonia.* Since the specimens studied lacked a pedicle foramen large enough to support a functional pedicle, it is inferred that *Schizophoria* and *Pentagonia* lived unattached on the substratum. The life position of *Pentagonia* was probably more toward 90 degrees, with the commissure directed normal to the sediment-water interface even though it might be argued that a more hydrodynamically stable position would be with the pedicle valve down. When the pedicle valve is directed down, however, the anterior commissure would be directed toward the sediment, a position highly unlikely for a ciliary suspension feeder. It is possible that the dorsal keel-like median ridge acted to stabilize the shell in the sediment and thus enable it to attain a life position somewhat less than 90 degrees but with the brachial valve down. Dutro (1971) has pointed out that a thick callosity in the posterior part of the pedicle valve undoubtedly acted in the stabilizing process. Shells in the collection also possess secondary shell material (at the base of the dental lamellae) in the pedicle valve. Dutro also noted (1971) that *Pentagonia* required a certain amount of current action in order to have survived and that no specimens of the genus have been recovered from muddy or silty rocks, supporting my contention that the depositional environment of the Onondaga Limestone was near wave base or subjected to mild current action, thus enabling *Pentagonia* to avoid clogging.

One of the most distinctive associations in the *Atrypa-Coelospira-Nucleospira* Community is that of the platyceratid gastropod (*Platyceras dumosum*) and articulated valves of the brachiopod *Atrypa "reticularis."* This phenomenon is especially evident in outcrop at AMNH Loc. 3135, where hundreds of spiny platyceratid gastropods (with spines essentially intact) occur over an extensive bedding plane, along with numerous *Atrypa* valves which appear to have accumulated in current-swept clusters. The close association of *Platyceras* and *Atrypa* cannot be explained at this time; it might be due to selective (diagenetic) preservational factors or to ecological ones. However, present in the Upper Moorehouse Member in the mid-Hudson Valley are numbers of crinoid columnals and non-pinnulate inadunate crinoid ossicles. No complete specimens have been recovered. In addition, there seem to be no camerate crinoids present in the eastern area above the Nedrow. The close association of the coprophagous gastropods and their (camerate) crinoid hosts has been recognized since their coincidental appearance in Black River and Trenton times, their Paleozoic expansion, and decline at the end of the Permian (Bowsher, 1955). Specimens from the Lower Mississippian of Iowa (Hampton Formation) are clearly attached to the anal opening of *Aorocrinus immalurus* (USNM No. S590). Another specimen of *Platyceras* is affixed to the anal opening of *Agaricocrinus iowaensis* (Osagean) from Indiana (USNM No. S4, 636) and shows the irregular aperture aligned with the outline of the crinoid. Specimens of *Platyceras dumosum* and *Platyceras* sp. that I collected were not found attached or with preserved apertures, thereby making the assumption of attachment somewhat tenuous. However, sufficient evidence has been presented in the literature (Bowsher, 1955) to postulate that the main food supply of the platyceratids was fecal matter excreted by the crinoids. At all localities where *Platyceras* was found, crinoid debris was also present. It must be assumed, therefore, that platyceratid gastropods present within the community did not live attached to the anal opening of the inadunate crinoids due mainly to the small size of the inadunates. Rather, the gastropods might have grazed along the bottom sediments and ingested the crinoidal fecal pellets which fell to the sea floor. No instances of gastropods attached to the anal opening of inadunate crinoids, to my knowledge, have been reported in the literature.

In conclusion, the presence of a normal marine level-bottom brachiopod fauna, crinoids, trilobites, gastropods, and especially 13 coral genera (tabulate and rugosan) supports the view that the environment of deposition was well oxygenated, albeit somewhat muddy.

ATRYPA-MEGAKOZLOWSKIELLA COMMUNITY

The *Atrypa-Megakozlowskiella* Community (Table 5) occurs midway between the central region (Syracuse area) and the mid-Hudson Valley in the mudstones and wackestones of the Moorehouse Member from Cherry Valley to Clarkesville, New York. The depositional environment was mid-neritic.

The adaptive habits of most of the occurring brachiopod genera have been discussed above and, since the lithologies are not radically different no further detailed discussion is needed. However, *Megakozlowskiella* is now a major faunal element and its life habits deserve some consideration. Externally it is a medium-sized to large shell, increasing in sphericity in later ontogenetic stages. Gerontic individuals are quite globose with the anterior margins almost incurved and abutting one another in a plane normal to the length axis of the shell. The animal most likely existed free on the ocean floor or rooted by a pedicle which extended through the open delthyrium.

TABLE 5. Faunal Constituents of the *Atrypa-Megakozlowskiella* Community. AMNH Localities: 3130, 3131B, 3133,3134. c = common, r = rare, vr = very rare.

Taxa	Rank	Relative (%) abundance
Brachiopods (*N* = 270):		
Atrypa "reticularis"	1	27.8
Megakozlowskiella raricosta	2	18.5
Schizophoria cf. *multistriata*	3	9.6
Pentagonia unisulcata	4	7.8
orthotetacids indet.	5	7.0
Levenea sp. A	6	5.1
Pentamerella arata	7	3.3
Leptaena aft. *"rhomboidalis"*	8	3.0
Nucleospira aff. *ventricosa*	9	2.6
Stropheodonta cf. *demissa*	9	2.6
Dalejina sp. A	10	2.2
Athyris sp. A	11	1.9
Megastrophia sp.	11	1.9
Coelospira camilla	12	1.1
"Mucrospirifer" sp.	12	1.1
Acrospirifer duodenaria	13	0.7
Hallinetes sp.	13	0.7
Cyrtina hamiltonensis	13	0.7
Elita fimbriata	13	0.7
Gypidula sp.	13	0.7
Pacificocoelia acutiplicata	14	0.7
Rhipidomella?	14	0.7
Other taxa present:		
Corals:		
Tabulates		
Aulopora		c
Favosites		r
Rugosans		
Amplexiphyllum?		r
"Heterophrentis"		c
Acinophyllum?		r
Breviphrentis		r

(Continued)

TABLE 5. (Continued)

Taxa	Rank	Relative (%) abundance
cf. Syringaxon		r
2 indet. Genera		r
Crinoids:		
non-pinnulate inadunate ossicles and columnals of indet. origin		c
Bryozoan fragments (Trepostome?)		r
Gastropods: *Tropidodiscus?*		vr
Trilobites: Indet. fragments		r

The delthyrium was probably closed during life, as there are indications on preserved specimens of narrow ridges bordering the delthyrium which would therefore have been closed off during life. The pedicle valve, with its greater convexity and deeper, more prominent ribs, was probably up in either case since the greater curvature of the external valve surface would have provided more hydrodynamic stability. The fact that there is less diversity here than in the *Atrypa-Coelospira-Nucleospira* Community might indicate a position closer to shore and therefore a nearer wave base, resulting in the need for a more suitable morphology. Brachial ribs could have served to anchor the brachiopod in the sediment. In addition, many neanic and gerontic individuals possess a large amount of secondary shell material deposited in the umbonal region, which weighted down the posterior of the shell and caused the anterior commissure to be raised at an angle above the sediment-water interface. In many unworn specimens, anterior frills and fine striae or capillae are present on the concentric growth lamellae. These may be the terminal ends of radial spines whose exact function is unknown at this time.

Minor, but significant, elements in the *Atrypa-Megakozlowskiella* Community are the numerous orthotetacid fragments (7.0%) collected and counted in the field. Possibly due to the nature of the shell (relatively thin), no complete valves were recovered. However, the morphology of these shells is such that they might have been able to swim. Stanley (1970) points out that the cross-sectional shape of *Placopecten magellanicus* is not so important in producing downward-directed swimming currents as in forming an efficient hydrofoil. When *Placopecten* glides through the water, the distance of water flow is longer over the more convex upper valve than beneath the less convex lower valve. The thinness of the valve and the hydrofoil shape seem to be key factors in enabling the bivalve to swim. Bowen et al. (1974) believe that the stropheodontid brachiopods *Stropheodonta* and *Nervostrophia,* found in the mid-shelf *Productella* Community of the Upper Devonian Sonyea Group, may have been able to swim somewhat like a *Pecten* by clapping their valves together. The following brachiopod genera of the *Atrypa-Megakozlowskiella* Community possess external morphologies that might have produced a hydrofoil given the right conditions: orthotetacids, *Stropheodonta, Megastrophia, Hallinetes,* and *Rhipidomella?.* The advantages of even a limited swimming ability on the part of brachiopods, especially those inhabiting soft sediments, are twofold: (1) the animal would be able to free itself from accumulating sediment by swimming a short distance into the water column and settling back again on top of the sediments, and (2) escape from predatory organisms.

ATRYPA COMMUNITY

The *Atrypa "reticularis"* Community (Table 6) occurs mainly in the mid-Hudson Valley but is represented as far west as Cherry Valley (AMNH Loc. 3131C). The fauna is found in moderately argillaceous mudstones and wackestones of Moorehouse and Nedrow age. Due to the mud content, the predominance of *Atrypa* (52.4%), and the fact that these atrypids are not smooth (Ager et al., 1976), a mid-neritic environment of deposition is postulated. The high degree of diversity in the community (18 brachiopod genera) precludes its isolation from other Onondaga fauna but does not explain the high percentage of *Atrypa* within the community. Copper (1966) claims that the strongest controls on atrypid distribution appear to have been bottom conditions (muds, silts) and trophic links (association with special coral groups). A strong similarity to Copper's (1966) European Eifel magnafacies and the Onondagan *Atrypa* Community is evident. Lithologically, the Eifel magnafacies is composed of calcareous shales, muddy limestones, and rare dolomites. The Nedrow and Moorehouse members of the Onondaga are essentially muddy limestones of varying degrees. The megafauna of the Eifel magnafacies is rich in rugose and tabulate corals, stromatoporoids, brachiopods, and crinoids (no ammonoids). Brachiopods are varied and multi specific in both environmental settings, with spiriferids,

TABLE 6. Faunal constituents of the *Atrypa* Community. AMNH Localities: 3131C, 3138B, 3139, 3140, 3141, 3142. c = common, r = rare.

Taxa	*Rank*	*Relative (%) abundance*
Brachiopods (*N* = 273):		
Atrypa "reticularis"	1	52.4
Levenea sp. A	2	9.9
Pentagonia unisulcata	3	8.1
Megakozlowskiella raricosta	4	5.5
Coelospira camilla	5	4.0
orthotetacids indet.	6	3.7
Nucleospira aff. *ventricosa*	7	3.3
Dalejina sp. A	8	2.2
Leptaena aff. *"rhomboidalis"*	8	2.2
Megastrophia sp.	8	2.2
Cupularostrum sp. A	9	1.5
Athyris sp. A	10	1.1
Schizophoria cf. *multistriata*	10	1.1
Strophonella cf. *punctulifera*	10	1.1
Gypidula sp.	11	0.7
Pentarnerella arata	11	0.7
Cyrtina hamiltonensis	12	0.4
Elita fimbriata	12	0.4
Other Taxa present:		
Corals:		
Tabulates		
Favosites		c
Aulopora		c
Rugosans		
Acinophyllum?		c
Heliophyllum?		c
cf. *Amplexiphyllum*		c
Breviphrentis		c
"Heterophrentis"		c
cf. *Syringaxon*		c
Trilobites:		
indet. fragments		r

rhynchonellids, athyrids, meristellids, and gypidulids present. Copper characterizes the depositional environment of his Eifel magnafacies as shallow, sheltered marine (mainly back reef, lagoonal?). However, there is no evidence of this being the case for the Onondaga. The amount of mud in the limestones of the *Atrypa* Community may be indicative of a position just below wave base, but accumulations of fragmented debris in parts of the outcrop area may also represent mild agitation (nearer wave base) or predation. Both salinity and temperature variations were low, and circulation was sufficient to keep the water well oxygenated. In any case, the amount of mud present was small enough to permit and proliferation of at least eight coral genera. Oliver (1956) states that the combined Nedrow-Moorehouse members are thinner and less pure in the central area than in the east, and that the water was shallower and rate of accumulation of mud relative to calcium carbonate was lower in the eastern part of New York.

LEPTAENA-MEGAKOZLOWSKIELLA COMMUNITY

The *Leptaena-Megakozlowskiella* Community (Table 7) occurs in central New York and is best observed in outcrops at the Onondaga Indian Reservation (Nedrow) and the Jamesville Quarry (AMNH Loc. 3123, 3128). The rocks are of Nedrow-Moorehouse age and range from mudstones to wackestones. The environment of deposition of the *Leptaena-Megakozlowskiella* Community was probably mid-neritic with a moderately to highly argillaceous substratum. The dominant element in the community, along with *Megakozlowskiella,* is *Leptaena* aff. *"rhomboidalis"* (29.9%). Although *Leptaena* is ubiquitous, in the Onondaga Limestone it is found predominantly in highly argillaceous rocks in which it is a major faunal element, but is noticeably rare in "purer" limestones. A possible explanation for this restricted occurrence may lie in the adaptive habits of the animal. The specimens collected were transversely subquadrate in outline and essentially concavo-convex with the pedicle valve strongly geniculate at the anterior and lateral commissures. No delthyrium was preserved. *Leptaena* might have possessed a rudimentary pedicle (as did many strophomenaceans with small apical foramens) which would have served as means of attachment in pre-neanic stages. However, even if a pedicle was present it would have been too thin and weak to support the entire shell and most likely would have

TABLE 7. Faunal Constituents of the *Leptaena-Megakozlowskiella* Community. AMNH Localities: 3123B, 3123C, 3123D, 3124A, 3124B, 3128B, 3128C.

Taxa	Rank	Relative (%) abundance
Brachiopods (*N* = 444):		
Leptaena aff. *"rhomboidalis"*	1	29.9
Megakozlowskiella raricosta	2	28.4
Pacificocoelia acutiplicata	3	6.8
orthotetacids indet.	4	6.1
Atrypa "reticularis"	5	5.2
Pentamerella arata	5	5.2
Megastrophia sp.	6	5.0
Athyris sp. A	7	3.2
Dalejina sp. A	8	2.5
Levenea sp. A	9	2.3
Coelospira camilla	10	1.6
Gypidula sp.	11	0.7
Meristina cf. *nasuta*	12	0.5
stropheodontids indet.	12	0.5
Stropheodonta cf. *demissa*	12	0.5
Trematospira? sp.	12	0.5
Charionoides aff. *doris*	13	0.22
Other taxa present:		
Corals:		
Tabulates		
Syringopora		c
Aulopora		c
Favosites		c
Rugosans		
Amplexiphyllum?		c
"Heterophrentis"		c
Acinophyllum		c
Gastropods:		
Straparollus		vr
Liospira?		vr
Ecculiomphalus		vr
Cephalopods:		
Foordites		vr
Trilobites:		
Phacops cristata		r
Odontocephalus		r
cf. *Dechenella*		r
dalmanitid fragments		r
Crinoids:		
Columnals (camerate)		r
Bryozoan fragments		r

TABLE 8. Faunal constituents of the *Pacificocoelia* Community. AMNH Locality: 3125A. r = rare.

Taxa	Rank	Relative (%) abundance
Brachiopods (*N* = 186):		
Pacificocoelia acutiplicata	1	59.1
Leptaena aff. "*rhomboidalis*"	2	9.7
Megakozlowskiella raricosta	3	8.6
Atrypa "*reticularis*"	4	7.5
orthotetacids indet.	5	4.3
Dalejina sp. A	6	3.2
Pentamerella arata	6	3.2
Athyris sp. A	7	2.7
Megastrophia sp.	8	1.1
Rhipidomella?	9	0.5
Other taxa present:		
Trilobite fragments		r

functioned solely to root or anchor the brachiopod to the substratum. In post-neanic stages the shell probably rested unattached on the ocean floor with the strongly geniculate anterior commissure raised above the sediment-water interface in order to prevent the infiltration of sediment into the body cavity. Accumulation of sediment on the concave brachial valve would have served to camouflage the animal and thereby conceal it from predators.

The suitability of *Megakozlowskiella* to a limy mud substrate was discussed above. However, it should be noted that there is a definite association between *Leptaena* and *Megakozlowskiella* on exposed bedding planes. The percentages of occurrence (*Leptaena* 29.9%; *Megakozlowskiella* 28.4%) in this community are remarkably close. These two brachiopod genera comprise a trophic nucleus of low-level suspension feeders, which, along with other minor elements in the community (e.g. *Pacificocoelia, Foordites*), and the high diversity (17 brachiopod genera), are indications of deposition in a mid-neritic environment.

The platyceratid gastropods and crinoidal debris that are so abundant in the *Atrypa-Coelospira-Nucleospira* Community seem to have been replaced here by the genera *Straparollus, Liospira?,* and aff. *Ecculiomphalus.* The platyceratids appear to be confined to the eastern area as Oliver (1956) reported. He found that *Platyceras dumosum* is common in the eastern Nedrow but rare elsewhere. Apparently the species preferred the environment represented by the eastern lithologies. The reason for the numerous platyceratids in the eastern Moorehouse Member (e.g. AMNH Loc. 3135) and their rarity in other members is not clear. One would think that the Edgecliff, especially in the central area with its characteristic large (up to 1 inch in diameter) crinoid columnals, would be a favorable environment for these gastropods, which enjoyed a symbiotic relationship with them. However, platyceratids were reported by Oliver (1956) as being mostly rare to very rare in the Edgecliff and I found none at all. One other factor may be that since many of the eastern faunas were silicified, the gastropods were preserved there and not in other areas.

PACIFICOCOELIA COMMUNITY

An anomalous *Pacificocoelia acutiplicata* Community (Table 8) is recognized within the Nedrow Member at only one outcrop: AMNH Loc. 3125A. The matrix is a mudstone in which the brachiopods are impressed on bedding planes and best collected by splitting large slabs parallel to the bedding. The shells are fairly flat with the pedicle valve gently convex. Ornamentation consists of about 10-12 plications with 2-3 concentric growth lines on larger specimens. Since no pedicle foramen was observed, an unattached existence in postneanic stages on the sea floor is inferred; no definite statement can be made regarding mode of life other than that.

A close *Leptaena-Megakozlowskiella* association (*Leptaena* 9.7%; *Megakozlowskiella* 8.6%) is evident here as well as in the *Leptaena-Megakozlowskiella* Community. One common factor shared by both communities is the high mud content of the rocks and consequently muddy substrata. (No corals were found in this community.) The diversity of the *Pacificocoelia* Community (10 brachiopod genera) is less than that of the *Leptaena-Megakozlowskiella* Community (17 brachiopod genera), possibly suggesting a position closer to shore since diversity generally decreases as shoreline is approached. Since only one outcrop was found in which this community is represented, further speculations are deferred until more material is at hand.

The positioning of the *Pacificocoelia* Community in a near shore environment (inner-neritic) is supported by Johnson's (1974) contention that an acrospiriferid-leptocoeliid biofacies, representative of shallower water, is characterized by abundant *Acrospirifer,* or ancestral large *Howellella* and *Pacificocoelia* or *Leptocoelina,* to the virtual exclusion of *Gypidula, Atrypa,* and *Schizophoria.* It is interesting to note that the *Pacificocoelia* Community bears no specimens of *Gypidula* or *Schizophoria,* although there is some *Atrypa* (7.5%) present. Johnson (1974: 810) warns, however, that taxa that are restricted to one biofacies in a single region may be limited by environmental factors that are not interregional. Such taxa may have broader or narrower biofacies distributions in other regions, or at other times. He claims that the genus *Pacificocoelia* is suspect in this way. Boucot (personal communication) mentions that *Pacificocoelia* is also dominant in the deep water Esopus Shale where environmental conditions are similar. This range of depth is not abnormal for Devonian leptocoeliids, which have far greater depth ranges in the Devonian than they do in the Silurian (Boucot, 1975).

LEVENEA COMMUNITY I

The fauna of this community (Table 9) is represented in mudstones and wackestones extending from Cherry Valley to Kingston. The lithologies of the Edgecliff in both areas are very similar, differing mainly in the vast amounts of light-weathering cherty seams in the mid-Hudson Valley. In the Cherry Valley region the Edgecliff is a medium to coarse-grained, massive to thick-bedded, light gray unit with little to no chert. In the central area the member is characterized by its large coral fauna, crinoid columnals, and little chert. However, in the east, the amount of chert increases drastically such that there appears to be an inverse relationship between the amount of chert present and the coral fauna. The extensive cherty eastern units are relatively poor in corals, while the noncherty central units are packed with corals. Thus, as Oliver (1956) states, it seems that either the conditions which produced the chert or the chert itself was a factor limiting coral growth. This might also be true of the brachiopods, which are poorly represented in the Edgecliff Member. Lithologically, the member is the "purest" within the Onondaga, indicative of clean, shallow (photic zone—upper?) water (inner-neritic) favorable to extensive coral proliferation. *Levenea* sp. A is the dominant faunal element in the community (67.4%). However, due to poor silicification and preservation, and the difficult nature of collecting in the eastern Edgecliff Member (N = 46),

TABLE 9. Faunal constituents of the *Levenea* Community I. AMNH Localities: 3131D, 3138C, 3143. r = rare.

Taxa	Rank	Relative (%) abundance
Brachiopods (*N* = 46):		
Levenea sp. A	1	67.4
Atrypa "reticularis"	2	17.4
Levenea sp. B	3	4.4
Leptaena aff. *"rhomboidalis"*	3	4.4
Pentamerella arata	3	4.4
Elita fimbriata	4	2.2
Other taxa present:		
Crinoids:		
camerate columnals		r

TABLE 10. Faunal constituents of the *Levenea* Community II. AMNH Locality: 3l51A, c = common.

Taxa	Rank	Relative (%) abundance
Brachiopods (*N* = 98):		
Levenea sp. A (depauperate fauna)	1	100
Other taxa present:		
Indet. trilobite fragments (depauperate?)		c

further discussion of the nature of the fauna within the community would be hazardous at best.

LEVENEA COMMUNITY II

This community (Table 10) occurs only at AMNH Loc. 3151A in mudstones in an abandoned quarry at Wawarsing, New York. The lithology differs from the lighter colored units in the mid-Hudson Valley. Here the rock is darker, finer grained, and muddier, with a sparse fauna consisting of trilobite fragments and *Levenea* sp. A. No other brachiopods were observed. Conditions seem to have become unfavorable due to an influx of terrigenous clastic material and transition to deeper water conditions. An outer-neritic environment is postulated. An identical lithology and fauna was observed near East Stroudsburg, Pennsylvania, in the Buttermilk Falls Limestone, a lateral equivalent of the Onondaga.

The *Levenea* Community II is regarded as separate from *Levenea* Community I for the following reasons: (1) the fauna is atypical, consisting solely of the species *Levenea* sp. A, and (2) the lithology differs radically and represents a distinct environment of deposition in a more offshore position.

AMPHIGENIA? SP. COMMUNITY

Twelve specimens of *Amphigenia?* sp. (Table 11) have been recovered from the basal Edgecliff of the central area (AMNH Loc. 3126) and serve as a tentative basis for naming this community. This sandy unit is very similar to the sand of the Bois Blanc Formation and may represent one of three conditions: (1) Reworked Oriskany sand, (2) reworked "Springvale" sand, or (3) unnamed sands of combined Oriskany and "Springvale" ages. There is no apparent means of differentiating these sands on the basis of lithology. The exact origin of the basal Edgecliff sand is not known at this time, but the general depositional environment is considered inner-neritic.

The presence of *Amphigenia?* sp. in a sandy facies is compatible with the robust morphology of the genus, which would enable it to survive a certain amount of tumbling on the bottom sediments. However, it is not compatible with Boucot's (1975) placement of the *Amphigenia?* Community in Benthic Assemblage 3 to 5. One resolution of this problem may be

TABLE 11. Faunal constituents of the *Amphigenia?* Community. AMNH Locality: 3126. c = common.

Taxa	rank	Relative abundance
Brachiopods (*N* = 12):		
Amphigenia? sp.	1	100
Other taxa present:		
Corals:		
Tabulates		
Favosites?		c
Rugosans		
Acinophyllum		c
Heliophyllum?		c
"Heterophrentis"		c

that the fragmental specimens of *Amphigenia?* sp. found in the Edgecliff (sandy facies) have been reworked and transported out of their original environment.

"*HALLINETES*" COMMUNITY

This community occurs only in the mudstones of the Seneca Member in the central area (AMNH Loc. 3123A, 3128A, 3129). It is a good marker horizon consistently found in a zone about 10 ft above the Tioga Bentonite. Due to the high mud content of the limestone and lack of wave-oriented shells, the environment of deposition is postulated as outer-neritic with an extremely soft substratum. Low diversity (Table 12), few coral genera, and a dark matrix are indicative of poor circulation and a low oxygen content. Linsley (1972) described a "*Chonetes*" facies from the "Pierceville" Quarry (Ludlowville Formation, Hamilton Group) which contains a much more diverse fauna and is dominated by "*Chonetes*" (76%). In the "*Hallinetes*" community of the Seneca Member "*Hallinetes*" is dominant (>99%) to the exclusion of bivalves, which are a significant element in the "Pierceville" facies. Based upon flume experiments, Linsley (1972) found that the most hydrodynamically stable position for chonetid shells is with the convex (pedicle) side up, since currents tend to move the shells and flip them over. The chonetids tended to orient themselves with the hinge line perpendicular to the current. In the "*Hallinetes*" Community of the Seneca no such orientation was recognized, suggesting an absence of strong currents. Also, there are no "clusters" of current-swept shells in a convex-up position. In 15 blocks of limestone, studied ratios of concave-up:convex-up shells indicate that brachiopods in life positions outnumber those presumably dead shells (convex-up) by an average of about 3:1, indicating a life assemblage (Table 13). No bimodal distribution was observed, suggesting that Linsley's (1972) notion that chonetids (of the "Pierceville") were seasonal breeders and that the population therefore should have consisted of adults plus juveniles that were growing up is not applicable here.

DISCUSSION

It should be noted that three communities within the Onondaga Limestone—*Levenea* Community I, *Levenea* Community II, *Amphigenia?*

TABLE 12. Faunal constituents of the *"Hallinetes"* Community. AMNH Localities: 3123A, 3128A, 3129. c = common, r = rare

Taxa	Rank	Relative (%) abundance
Brachiopods (N = 4569):		
"Hallinetes" aff. *lineatus*	1	99.7
Megakozlowskiella raricosta	2	<1
Leptaena aff. *"rhomboidalis"*	2	<1
Megastrophia sp.	2	<1
orthotetacids indet.	2	<1
Athyris sp. A	2	<1
Atrypa "reticularis"	2	<1
Pentamerella arata	2	<1
Other taxa present:		
Corals		
Rugosans		
"Heterophrentis"?		c
Amplexiphyllum?		c
Gastropods:		
Euomphalacean fragments		r
Trilobites:		
Phacops cristata		r
Odontocephalus		r
Crinoids:		
Camerate? crinoid columnals		r

Community—contain insufficient data for the purposes of drawing any conclusions regarding ancestral relationships other than stating, in general terms, that they appear to conform in structure to many Siluro-Devonian low-diversity communities typical of the Appohimchi Subprovince of Boucot (1975).

Four communities—*Leptaena-Megakozlowskiella, Atrypa-Coelospira-Nucleospira, Atrypa, Atrypa-Megakozlowskiella*—appear to be derived from Boucot's (1975) *Amphigenia* Community and share the following taxa: *Pentamerella, Leptaena, Strophonella, Atrypa, Coelospira, Pentagonia, Pacificocoelia, Meristina, Nucleospira, Acrospirifer, Mucrospirifer, Megakozlowskiella, Elita, Ambocoelia, Cyrtina,* and *Stropheodonta.* Since the *Amphigenia* Community is found in the Bois Blanc Formation, which directly underlies

TABLE 13. Ratios of living and dead shells in the *"Hallinetes"* Community.

Block No.	No. shells in concave-up position (living)	No. shells in convex-up position (dead)
1	34	17
2	40	16
3	35	5
4	76	17
5	11	7
6	28	8
7	22	12
*8	23 (top) 16 (bottom)	4 (top) 5 (bottom)
*9	4 (top) 21 (bottom)	5 (top) 9 (bottom)
10	17	6
11	14	4
12	15	7
13	14	9
*14	16 (top) 21 (bottom)	6 (top) 7 (bottom)
15	155	36
	Σ = 561 (76%)	Σ = 180 (24%)

*Both top and bottom of bedding plane were counted.

the Onondaga Limestone in western New York, a direct vertical and lateral (geographical) relationship is apparent. The presence of *Atrypa "reticularis"* within all of the highly diversified communities of the Onondaga Limestone is evidence for placement within a Benthic Assemblage 3 to 5 span (Boucot, 1975: 246). Other faunal elements such as *Amphigenia, Coelospira, Leptaena, Megakozlowskiella, Levenea, "Schuchertella," Nucleospira,* and *Stropheodonta* belong to communities ranging from Benthic Assemblage 3 to 5 and further serve to restrict the Onondagan communities to a position within that Benthic Assemblage span within the Appohimchi Subprovince of Early Devonian level-bottom communities.

The *"Hallinetes"* Community of the Seneca Member is a low-diversity community which seems to be derived from Boucot's (1975) Chonetid Community (Siegen-Ems). Its position is within a quiet-water, Benthic Assemblage 5.

The *Pacificocoelia* Community most likely originated from Boucot's (1975) *Coelospira-Leptocoelia* Community, typical of a position within Benthic Assemblage 4. Here, however, more data is needed for proper evaluation of community ancestry.

Boucot (1975) defined a Community Group as a member of communities that have homologous relations, although they may incorporate taxa from unrelated communities through the process of changing niche breadth, and also give off taxa to unrelated communities through the same process. Assignment to a Community Group is relatively straightforward in the case of only one Onondagan community. The *Pacificocoelia* Community fits well into the Leptocoelidae Community Group. The Devonian leptocoelids form parts of other communities and increase their environmental range after the Silurian. According to Boucot (1975) the leptocoelid communities have no set antecedents in the pre-Late Llandovery and may be derived from the Malvinokaffric Realm Silurian *Harringtonina* ("*Eocoelia*" *paraguayensis*) of the Early Llandovery.

The *Leptaena-Megakozlowskiella, Atrypa-Coelospira-Nucleospira, Atrypa,* and *Atrypa-Megakozlowskiella* communities, all apparently derived from the *Amphigenia* Community (Ems-Eifel) appear to fit into the *Striispirifer* Community Group, which is typically viewed as a high-diversity, quiet-water group with a Benthic Assemblage 3 range.

The "*Hallinetes*" Community cannot, at present, be assigned to any Community Group.

Based on gross visual estimates, field counts, and tabulations in the lab, a rough idea of relative abundance of various major invertebrate taxa within the brachiopod communities of the Onondaga Limestone may be reached. This abundance, or biovolume, is roughly equivalent to estimated biomass since actual biomass cannot be determined for fossil taxa (Walker, 1972). It is apparent, at least for the more diverse communities that most of Turpaeva's (1957) conclusions regarding structure and trophic relationships in Recent Arctic and Boreal communities apply here. All nine brachiopod communities are dominated by one trophic group whereas the next most dominant species belongs to another trophic group. In this case "species" is taken to mean major taxon since (1) communities considered are analyzed in terms of brachiopod species and, (2) biovolume calculations for colonial organisms (such as corals and bryozoans) and crinoids are imprecise. For example, the *Atrypa-Coelospira-Nucleospira* Community is dominated by low-level suspension feeders (brachiopods), with the next most dominant groups (in order of dominance) being: corals (high-level suspension feeders/ predators); gastropods (collectors?, browsers?, scavengers?); inadunate crinoids (passive high-level suspension feeders); trilobites (scavengers?, collectors?, predators?), and bryozoans (high-level suspension feeders). This trophic structure is found in all communities where more than one major taxon of invertebrates occur. Niche-partitioning results in non-competitive feeding and thus provides a measure of stability. In most of the communities studied, especially where significant numbers of specimens were collected, the dominant species utilized the food resources most fully by feeding at different trophic levels (Table 14).

Trophic structure is best studied among the silicified faunas of the mid-Hudson Valley, specifically limited to a narrow geographic belt extending from just north of Leeds, to around the area of Hurley, New York. Stratigraphically, most of the silicification is confined to the Upper Moorehouse Member. As one leaves Kingston, along Route 209, the silicification ceases suddenly after a few miles. Due to the sparse outcrops in that area the exact location is difficult to determine. There appears to be a sharp lithologic boundary between silicified and non-silicified zones such that the silicified zones are much less argillaceous and denser. Thus, there seems to be a restricted geographical area that has undergone silicification. The silicification may possibly be related to the tectonism of the mid-Hudson Valley, where faulting and jointing enabled percolating silica-rich solutions to infiltrate the limestones (e.g. AMNH Loc. 3137). A positive correlation between heavy jointing and silicification in that region is evident.

There appears to be a relationship between the amount of post-mortem transport and substratum type such that the coarser the sediments in which the community occurs the greater the amount of post-mortem transport. For example, the *Amphigenia?* Community, associated with a sandy substratum, is composed of shell fragments which display a great amount of abrasion and appear to be indicative of at least some transport. The *Atrypa-Coelospira-Nucleospira* Community, associated with moderately argillaceous sediments, contains shell fragments mostly free of any mechanical damage and species with low articulation ratios. Also, platyceratid gastropods have been recovered with their spines almost intact, indicative of a minimum amount of transport. The "*Hallinetes*" Community is associated with a very highly argillaceous substratum and represents a biocoenosis. Studies by Middlemiss

TABLE 14. Trophic levels of taxa within the brachiopod communities of the Onondaga Limestone listed according to relative abundance.

Major taxon	Trophic group	General definition	Food resource used
Brachiopods	Low-level suspension feeders	Filter food particles suspended in the water column near the sediment-water interface by means of a ciliary feeding mechanism (lophophore)	Dissolved organic nutrients, bacteria, miscellaneous particulate matter
Corals	High-level suspension feeders Predators?	Filter suspended food particles and/or jive prey captured (by nematocysts) without active pursuit	Small swimming or floating organisms
Gastropods	Collectors? Browsers? Scavengers?	Collect detritus from bottom sediments including dead and/or living plant material	Detritus from sediment including fecal pellets from inadunate crinoids, algae?, organically-rich sediment grains
Crinoids	Passive high-level suspension feeders	Food trapped from water currents well above sediment-water interface	Phytoplankton and zooplankton suspended and carried by currents in water column
Trilobites	Scavengers? Collectors? Predators?	Food actively pursued or collected from sediment	Detritus from sediment; Predation on worms or other soft-bodied invertebrates
Bryozoans	High-level suspension feeders	Suspended food particles filtered from water column at some distance above sediment-water interface	Dissolved organic nutrients and particulate matter
Cephalopods	Predators	Active capture of living prey	Macroorganisms of the marine benthos

(1962) and Cadée (1968) indicate that when macroinvertebrate remains are transported more than 5 km they become unidentifiable. The state of preservation of most of the invertebrates of the various Onondaga communities (lack of extensive disarticulation, little sorting) is good enough to preclude any post-mortem transport on the order of magnitude of even 1 km. Disarticulation of crinoid columnals and bryozoan fragments appears to be caused by water agitation and mild currents rather than extensive transport. In the case of the crinoids low sedimentation rates would account for the fact that no calices, arms, or whole stems have been recovered. Since the crinoids are so porous and their specific gravities low, any mild current, in the absence of rapid burial (high sedimentation rates) would tend to disarticulate them. Generally, high sedimentation rates favor complete preservation of crinoids (B. N. Haugh, personal communication). Observing geopetal infillings of some articulated brachiopods did not entirely clarify the extent of transport or the life orientation of various specimens. Most brachiopods, upon death, assume an orientation which is most stable hydrodynamically. This orientation is not necessarily the same as was assumed during life. For example, Ager (1963) points out that *Schizophoria,* typical of dome-shaped brachiopods, is most stable with the convex brachial valve up. However, this is not necessarily its life position. Field observations and geopetal infillings are of no help where postmortem transport had occurred since the shells might have accumulated sediment after death. This probably holds true for even slight transport or agitation on the bottom sediments.

The firmness of the substrate upon which the various brachiopod communities existed is subject to a certain amount of speculation. In general, brachiopods have been found associated with hard, firm, and soft substrates, with firm and soft predominating. Modes of attachment include: cemented (hard substrate), rooted (firm substrate), and burrowed (soft substrate). Although the predominance of epifaunal groups such as the articulate brachiopods and crinoids in many pure, fine-grained limestones may be considered by some (Johnson, 1964) to be indicative of a substrate of bare rock or stiff mud, no definite conclusions regarding the firmness of the substrates of the Onondaga, for the most part, may be made. In many instances, the presence of a lime mud substrate is suggested based upon morphologic adaptations of the brachiopod fauna (e.g. frills on *Atrypa*) and similar taxon-substrate associations (e.g. *Zygospira* of Bretsky [1970] on lime muds). No evidence for the existence of a hardground was observed as indicated by the absence of the following broad critieria for recognition of ancient hardgrounds: (1) encrustation and overhangs, (2) karst surfaces, (3) pyritic veneers, (4) indications of subaerial oxidation, (5) an encrusting epifauna (e.g. craniids), and (6) an irregular, pitted surface at the contact.

Based upon available data, the nine brachiopod communities of the Onondaga Limestone may be grouped into three main environments of deposition: inner-neritic; mid-neritic, and outer-neritic. This conclusion is based partly upon the diversity of the communities, which indicates a general trend (Figure 3), and partly upon changes in lithology. Generally speaking, diversity increases as distance from shoreline increases. The *Levenea* Community II and *"Hallinetes"* Community are considered to be outer-neritic mainly because of the extremely argillaceous nature of the limestones (mudstones). Although the diversity in these communities is relatively low, the mudstones are indicative of deeper-water deposition and the taxa are morphologically compatible with those faunas of Early Devonian level-bottom communities of the Appohimchi Subprovince (Boucot, 1975), which occur in a similar offshore position.

The *Leptaena-Megakozlowskiella, Atrypa-Coelospira-Nucleospira, Atrypa-Megakozlowskiella,* and *Atrypa* communities are considered mid-neritic, based partly on the moderately argillaceous limestones (mudstones-wackestones) in which they are found, and their relatively high species diversity. A rough gradient is postulated (see Figure 3) such that the number of brachiopod species and coral genera (the only two groups for which there is sufficient data for this plot) may be arranged in an onshore-offshore gradient. As the shoreline is approached, their diversity decreases. Further evidence for a mid-neritic environment is the presence of *Foordites* sp., a cephalopod, within the *Leptaena-Megakozlowskiella* Community. The Devonian ammonoids appear to have been most common in deeper waters and rare in near-shore facies (House, 1973).

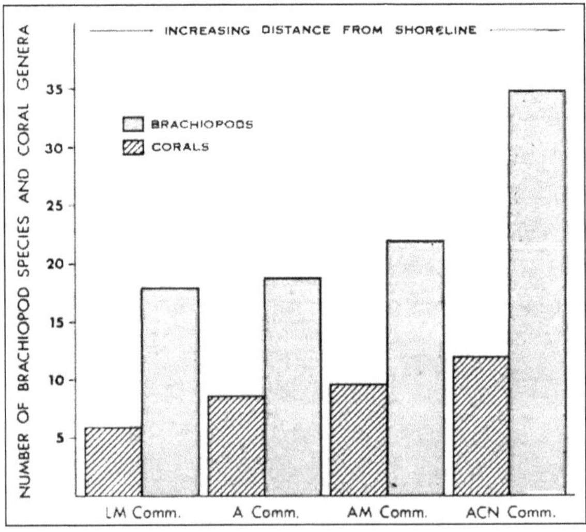

FIGURE 3. Hypothetical onshore-offshore relationships of Onondaga brachiopod communities based on relative diversity of brachiopod species and coral genera. LM Comm. = *Leptaena-Megakozlowskiella* Community; A Comm. = *Atrypa* Community; AM Comm. = *Atrypa-Megakozlowskiella* Community; ACN Comm. = *Atrypa-Coelospira-Nucleospira* Community.

The *Levenea* Community I is thought to represent inner-neritic conditions due to the low diversity (only six brachiopod species and occasional crinoid fragments) and relatively "pure" limestones (wackestones and packstones) indicative of cleaner, shallow water. Not enough data is at hand to make a definite determination at the present time.

The *Amphigenia?* Community is placed in an inner-neritic environment on the basis of the sandy matrix and the fact that the brachiopod valves present are abraded in a manner indicative of a near-shore facies with characteristic wave action.

The *Pacificocoelia* Community is tentatively placed in an inner-neritic position on the basis of its low diversity and high dominance by one species. Although in the Lower Devonian (Esopus Shale) *Pacificocoelia* occupies a deeper-water (outer-neritic?) position, its presence in the Onondaga Limestone may be indicative of a muddy basin within the inner-neritic zone.

The presence of a very uniform carbonate stratigraphy covering an extensive area and yielding almost homogeneous faunas suggests a region of negligible slope (Boucot, 1975). Thus, mixing of faunal elements from adjacent communities within the Onondaga is probably insignificant and therefore detailed analyses of community structures of this epeiric sea are useful for comparisons with similar communities and environments.

ACKNOWLEDGMENTS

I am especially indebted to Dr. Arthur J. Boucot of the Department of Geology, Oregon State University, for his critical review and valuable suggestions. The late Dr. W. A. Oliver, Jr., U. S. Geological Survey, Washington, D.C., provided much appreciated assistance in interpreting the stratigraphy of the Onondaga Limestone. Additional thanks are due to Drs. M. G. Bassett, R. L. Batten, R. S. Boardman, N. Eldredge, the late R. E. Grant, C. W. Harper, B. N. Haugh, M. R. House, R. K. Olsson, the late E. L. Yochelson, and Mr. S. Horenstein for helpful comments and discussion on various aspects of this paper. Partial support for this research was provided by grants from the Society of the Sigma Xi and the National Science Foundation, NSF Grant EAR 76-15402.

REFERENCES

Ager, D. V. 1963. *Principles of Paleoecology.* New York: McGraw-Hill.

Ager, D. V., Cossey, S. P. J., Mullin. P. R., and Walley, C. D. 1976. Brachiopod ecology in Mid-Paleozoic sediments near Khenifra, Morocco. *Palaeogeography, Palaeoclimatology, Palaeoecology* 20: 171-186.

Anderson, E. J. 1967. Paleoenvironments of the Coeymans Formation. PhD dissertation, Brown University, Providence.

Anderson, E. J. 1971. Environmental models for Paleozoic communities. *Lethaia* 4: 287-302.

Boucot, A. J. 1970. Practical taxonomy, zoogeography, paleoecology, paleogeography and stratigraphy for Silurian and Devonian brachiopods. *Proceedings of the North American Paleontological Convention* F: 566-611.

Boucot, A. J. 1975. *Evolution and Extinction Rate Controls.* Amsterdam: Elsevier.

Bowen, Z. P., Rhoads, D. C., and McAlester, A. 1974. Marine benthic communities in the Upper Devonian of New York. *Lethaia* 7: 93-120.

Bowsher, A. L. 1955. Origin and adaptation of platyceratid gastropods. *University of Kansas Paleontological Contributions Mollusca, Article* 5: 1-11.

Bretsky. P. W. 1969. Central Appalachian Late Ordovician communities. *Geological Society of America Bulletin* 80: 193-212.

Bretsky, P. W. 1970. Upper Ordovician ecology of the central Appalachians. *Peabody Museum of Natural History Bulletin 34,* Yale University.

Bromley. R. G., and Surlyk, F. 1973. Borings produced by brachiopod pedicles, fossil and Recent. *Lethaia* 6: 349-365.

Cadée, G. C. 1968. Molluscan biocoenoses and thanatocoenoses in the Ria de Arosa, Galicia, Spain. *Zoll. Verhandel uitgegeven door het Rijkomuseum van natuur lijke te Leiden* 95. Leiden: E. J. Brill.

Chute, N. E., and Brower, J. C. 1964. Stratigraphy and structure of Silurian and Devonian strata in the Syracuse area. *New York State Geological Association Guidebook, 36th Annual Meeting,* 90-101.

Copper, P. 1966. Ecological distribution of Devonian atrypid brachiopods. *Palaeogeography, Palaeoclimatology, Palaeoecology* 2: 245-266.

Cowen, R. 1968. A new type of delthyrial cover in the Devonian brachiopod *Mucrospirifer. Palaeontology 11*: 317-327.

Dutro, J. T., Jr. 1971. The brachiopod *Pentagonia* in the Devonian of eastern United States. *Smithsonian Contributions to Paleobiology* 3: 181-192.

Feldman, H.R. (1985). Brachiopods of the Onondaga Limestone in central and southeastern New York. *Bulletin, American Museum of Natural History,* 179: 289-377.

Hallam, A. (ed.) 1973. *Atlas of Paleobiogeography.* Amsterdam: Elsevier.

House, M. R. 1973. Devonian goniatites. In A. Hallam (ed.), *Atlas of Paleobiogeography.* Amsterdam: Elsevier.

Johnson, J. G. 1974. Early Devonian brachiopod biofacies of western and arctic North America. *Journal of Paleontology* 48: 809-819.

Johnson, R. G. 1964. The community approach to paleoecology. In J. Imbrie (ed.), *Approaches to Paleoecology*, 107-134. New York: Wiley.

Lenz, A. C. 1976. Lower Devonian brachiopod communities of the northern Canadian Cordillera. *Lethaia* 9: 19-28.

Lespérance, P. J., and Sheehan, P. M. 1975. Middle Gaspé limestone communities on the Forillon Peninsula. Quebec, Canada (Siegenian, Lower Devonian). *Palaeogeography, Palaeoclimatology, Palaeoecology* 17: 309-326.

Linsley, R. M. 1972. "Pierceville" Quarry (Bradley Brook Quarry) Paleoecology of the Ludlowville Formation, Hamilton Group. *New York State Geological Association Guidebook, 44th Annual Meeting*, 8-18.

Middlemiss, F. A. 1962. Brachiopod ecology and Lower Greensand Paleogeography. *Palaeontology* 5: 253-267.

Oliver, W. A., Jr. 1954. Stratigraphy of the Onondaga Limestone (Devonian) in central New York. *Geological Society of America Bulletin* 65: 621-652.

Oliver, W. A., Jr. 1956. Stratigraphy of the Onondaga Limestone in eastern New York. *Geological Society of America Bulletin* 67: 1441-1474.

Speden, I. G. 1966. Paleoecology and the study of fossil benthic assemblages and communities. *New Zealand Journal of Geology and Geophysics* 9: 408-423.

Stanley, S. M. 1970. Relation of shell form to life habits of the bivalvia (Mollusca). *Geological Society of America Memoir* 125.

Sutton, R. G., Bowen, Z. P., and McAlester, A. L. 1970. Marine shelf environments of the Upper Devonian Sonyea Group of New York. *Geological Society of America Bulletin* 81: 2975-2992.

Thayer, C. W. 1972. Marine paleoecology of the Upper Devonian Genesee Group of New York. PhD dissertation, Yale University, New Haven.

Thayer, C. W. 1974. Marine paleoecology in the Upper Devonian of New York. *Lethaia* 7: 121-156.

Turpaeva, E. P. 1957. Food interrelationships of dominant species in marine benthic biocoenoses. In B. N. Nikitkin (ed.), *Transa*, Institute of Oceanography, Marine Biology USSR Academy Sciences Press 20, 137-148. (Published in the United States by the American Institute of Biological Science, Washington, D.C.).

Walker, K. R. 1972. Trophic analysis: A method for studying the function of ancient communities. *Journal of Paleontology* 46: 82-93.

Walker, K. R., and Laporte, L. F. 1970. Congruent fossil communities from Ordovician and Devonian carbonates of New York. *Journal of Paleontology* 44: 928-944.

Wallace, P. 1972. Populations and paleoenvironments in the Devonian of the Cantabrian Cordillera, North Spain. *24th International Geological Congress, Section* 7: 121-129.

West, R. R. 1972. Relationship between community analysis and depositional environments: An example from the North American Carboniferous. *24th International Geological Congress, Section* 7: 130-146.

Ziegler, A. M., and Boucot, A. J. 1970. North American Silurian animal communities. *Geological Society of America Special Papers* 102: 95-106.

Ziegler, A. M., Cocks, L. R. M., and Bambach, R. K. 1968. The composition and structure of Lower Silurian marine communities. *Lethaia* 1: 1-27.

Chapter Three

Brachiopods of the Onondaga Limestone in Central and Southeastern New York

ABSTRACT

Thirty-nine species of brachiopods from central to southeastern New York are systematically described. This monograph is based upon 7,030 specimens collected from 30 localities and includes silicified as well as nonsilicified faunas. The following members of the Onondaga Limestone were sampled: Edgecliff, Nedrow, Moorehouse, and Seneca. The Onondaga thickens considerably toward the east with the greatest net change occurring in the Moorehouse Member (19 ft near Syracuse to 92 ft at Saugerties). The Seneca Member disappears just east of Cherry Valley and is not found in any strata of the mid-Hudson Valley. Crinoid columnals up to about 1 inch in diameter are characteristic of the Edgecliff Member across the state but become less abundant in the east. Brachiopod diversity is greatest in the Moorehouse Member and least in the Edgecliff Member. Of the 26 species of brachiopods in the underlying Bois Blanc Formation in western New York, nine occur in the Onondaga and show no evolutionary change. Morphologic variability in the Onondaga faunas was greater laterally than vertically. For the species studied here, the high degree of stasis in the Bois Blanc-Onondaga (Emsian-Eifelian) time interval may support a punctuational model as an evolutionary mode. During Eifelian time, the faunas found in central and southeastern New York belonged to the Appohimchi Subprovince of the Eastern Americas Realm and were part of the larger suite extending westward across the continent. The Onondagan brachiopods reviewed here were provincial in character with the following genera endemic to the Appohimchi Subprovince: *Charionoides, "Pacificocoelia,"* and *Pentagonia. Atribonium halli* and *Discomyorthis?* sp. are the only species described herein not previously reported from Onondaga strata. No new species or genera were erected.

INTRODUCTION

My purpose in this monograph is to systematically describe the brachiopods of the Onondaga Limestone of central and southeastern New York. By so doing I provide data which can enhance our understanding of brachiopod lineages from the Lower Devonian (e.g., Helderberg) through Middle

Devonian (e.g., Hamilton) strata, as well as aid in paleobiogeographic studies of Eifelian age faunas of New York and adjacent areas. Also, these data will be useful in resolving some of the problems inherent in correlating the Onondaga Limestone with the Detroit River Group (Anderdon Limestone, Lucas Dolomite, Amherstburg Dolomite, Sylvania Sandstone) of the Michigan Basin, southwestern Ontario, and central Ohio.

The Onondaga Limestone in New York State crops out from Buffalo, eastward to the Helderbergs, then southward toward Kingston and Port Jervis. I visited more than 100 outcrops from the Syracuse area to Wawarsing (Figure 1), but I studied only 30 in detail. Collecting in the Onondaga is difficult because of the density of the rock, especially in the central part of the state (excluding the shaly beds of the Nedrow Member), but occasional silicified and partly silicified faunas made this taxonomic study possible and provided internal morphologies much needed for specific identifications and comparisons. Silicification appears to increase in direct proportion to the occurrence of jointing and faulting, being most noticeable in the mid-Hudson Valley near Leeds and Saugerties. North and southwest along the outcrop belt, silicification decreases in intensity and occurrence. Silicified faunas were recovered in Upper Moorehouse strata along the New York State Thruway, at the Saugerties interchange, and at Leeds near route 23B and the Catskill Creek. They also occur sporadically throughout the entire section in the mid-Hudson Valley region. Degree of silicification varies greatly within the formation.

Large limestone blocks were retrieved and submersed in vats of commercial grade hydrochloric acid (muriatic acid). After etching, the silicified fossils were washed in a sodium sulfate solution, dried, and sorted biologically. An acid-proof resin (alvar) was gently brushed on the more fragile shells to facilitate handling and prevent breakage.

Stratigraphic (Oliver, 1954, 1956) and paleocommunity (Feldman, 1980; Lindemann and Feldman, 1981) studies relating to non-reefal aspects of the Onondaga Limestone have resulted in a fairly complete understanding of Onondaga stratigraphy across the state. The area from southwest of Kingston to Port Jervis, however, is in need of more detailed analysis. Outcrops in this area are few, and complete stratigraphic sections have yet to be recognized.

PREVIOUS WORK

Simeon De Witt, Surveyor-General of the State of New York, provided the first published illustration of a New York Devonian fossil in 1807 (Wells,

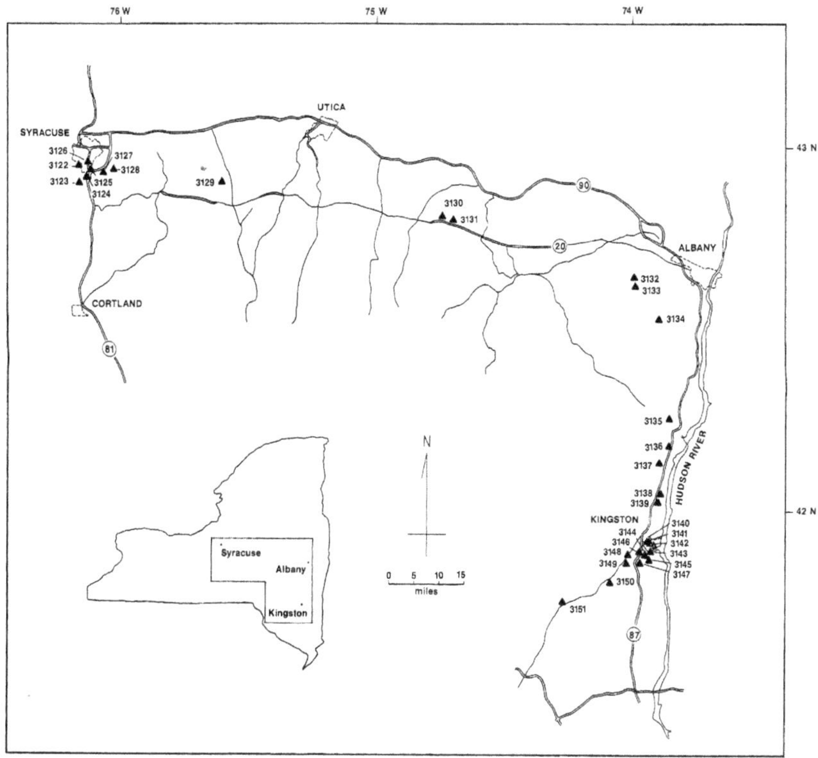

FIGURE 1. Index map of collecting localities and described sections in the Onondaga Limestone, central and southeastern New York.

1963: 14). The fossil was illustrated by a woodcut showing a gyroconic nautiloid from the Onondaga Limestone. Eaton (1832: table 4) made the first real attempt to analyze the fossils of the New York Devonian and mentioned 12 Onondagan species of a total of 41 listed, although none were brachiopods. Wells (1963: 63) points out that the end of these pioneer, "amateur," days of New York Devonian geology came in 1836 when Governor Marcy signed a bill establishing a geological survey of the state. Detailed studies of the New York Devonian began in earnest in the summer of 1836, when the geologists of the new State Survey (Conrad, Emmons, Hall, Mather, and Vanuxem) began fieldwork.

The first major published description of the Onondaga Limestone, including representative brachiopods, was by Vanuxem (1842), in which he described the geographic extent (e.g., outcrop belt) of the formation in New York State's Third Geological District. In his Fourth Annual Report, Vanuxem

(1840: 378) noted that the Onondaga Limestone "abounds in Testaceous fossils, the most characteristic ones have to yet be named." He also referred to the "Corniferous Limerock" of Prof. Eaton, and mentioned the occurrence of *Cyrtoceras* (*Foordites?*), found in the Nedrow Member (Feldman, 1980). Vanuxem (1840) also briefly described the Seneca Limestone, the "characteristic fossil of the Seneca being *Strophomena lineata* (*Hallinetes lineatus*)." Hall (1840: 454) listed *Atrypa wilsoni, Strophomena rugosa,* and *Leptaena indenta* from the Onondaga Limestone of the Fourth District but also stated that the list represents only "a few of the most common and characteristic fossils. The list will be greatly increased when we have succeeded in ascertaining the names of species" (453). Vanuxem noted (1842: 132) that the Onondaga Limestone "rarely exceeds ten or fourteen feet in thickness" and that it is "readily recognizable by its light grey color, crystalline structure, toughness, and its organic remains, which are very numerous." He also mentioned the occurrence of "smooth encrinal stems" ranging from 1/2 to 1 inch in diameter. In addition, Vanuxem (1842: 133) referred to an "elongated *Pentamerus*" found in the formation. From the above it appears as though Vanuxem was describing the Edgecliff Member of the Onondaga Limestone. The thickness of the Edgecliff in central New York ranges from 8 ft near Syracuse to 25 ft at Chittenango Falls (Oliver, 1956: 1446), and 12 ft at Nedrow (AMNH Loc. 3124, present report) to 19 ft at Jamesville (including the "sandy facies" AMNH Loc. 3128, present report). The member also typically contains thick crinoid stems and *Amphigenia* sp. in the central New York area (Oliver, 1954; Feldman, 1980). Three brachiopods were illustrated by Vanuxem (1842: 132): (1) *Pentamerus elongata* (*Amphigenia*); (2) *Consimilar hipparionix* (probably = *Rhipidomella*); (3) undulated delthyris (*Megakozlowskiella*). *Amphigenia?* sp. Has been found in the Edgecliff (Feldman, 1980); *Rhipidomella* was reported from the Edgecliff (Oliver, 1956); and *Megakozlowskiella* was found in the lowermost Nedrow (present report, AMNH Loc. 3123, just above the Edgecliff-Nedrow contact near Syracuse). Vanuxem (1842) referred to the "Corniferous Limestone" in which he included the "Seneca Limestone." He described this formation as achieving its maximum thickness of 60 to 80 ft, in Cherry Valley. This appears to include the Nedrow, Moorehouse; and Seneca members (measured in this report to a total thickness of 64.5 to 74.5 ft, AMNH Loc. 3131). Vanuxem (1842: 139) illustrated an "undulated cyrtoceras" from the Corniferous Limestone, which strongly resembles *Foordites* sp. found in Nedrow strata of the *Leptaena-Megakozlowskiella* Community (Feldman, 1980: 37). Vanuxem (1842: 141) indirectly referred to the *Chonetes* Zone when he mentioned *S. lineata,* a "small fossil so abundant in the upper part of the rock, which has heretofore been known by the name of Seneca Limestone," and directly (144): "The reason for considering it to be a district rock was . . . finding . . . *S. lineata* in great abundance." Vanuxem mentioned further that the Corniferous Limestone possessed parallel nodules of flint (chert) "upon which is the Seneca Limestone." Thus, he differentiated the Seneca Limestone from the underlying Corniferous Limestone and considered the Seneca as being the terminal part of the Corniferous Limestone.

Hall (1843), in his review of the geology of the Fourth District in New York (comprising the following counties: Wayne, Monroe, Orleans, Niagara, Seneca, Ontario, Yates, Livingston, Genesee, Erie, the western part of Tompkins, Chemung, Steuben, Allegany, Cattaraugus, and Chautauqua) described the lithology and faunas of the Onondaga and Corniferous limestones. Hall stated (156) that the formation is much more "developed in the First and Third districts where it contains a greater number of fossils and as a distinct mass is more persistent." Hall's (1843) stratigraphical nomenclature appears to conform to that of Vanuxem (1842). He mentioned the occurrence of *Atrypa* and *Delthyris,* but stated (1843: 160) that the "characteristic fossils of the Onondaga Limestone in the Third District, figured by Mr. Vanuxem on page 132 of his Report, cannot be considered as typical of this rock in the Fourth District. The *Pentamerus elongatus* occurs at Vienna, but I have not seen it elsewhere in the district." Hall described in that publication a more extensive brachiopod fauna from the Corniferous Limestone, including: *Atrypa scitula, A. prisca, Paracyclas elliptica, Strophomena acutiradiata, S. crenistria, S. undulata, S. lineata* (*Chonetes lineata*), *Delthyris duodenaria, Orthis lenticularis,* and *Orthis lentiformis.*

PURPOSE OF INVESTIGATION

The present report consists of systematic descriptions of brachiopods recovered from the Onondaga Limestone of central and southeastern New York State. All genera described here were previously reported or described from New York except for *Atribonium halli,* which is known from the Detroit River Group, an Onondaga correlative and *Discomyorthis?* sp. The Onondaga brachiopod fauna has been previously described by other workers (Eaton,

1832; Vanuxem, 1839, 1842; Hall, 1840, 1843, etc.) but the illustrations in all cases are lithographic plates and figures from woodcuts rather than photographs. This is the first paper describing the Onondaga brachiopod fauna using photographs. Also, most of the specimens described herein are silicified, whereas past workers' material was not. Silicified faunas enable a more detailed study of morphology since covered structures are exposed after etching.

The purposes of the present study are to: (1) systematically describe and characterize the large and significant brachiopod fauna of the Onondaga Limestone, which has not been redescribed since the days of James Hall; (2) provide data which can be utilized to evaluate lineages and evolutionary change (stasis or lack thereof) between Lower Devonian Helderberg and Schoharie age faunas and those of the Hamilton; (3) enable further paleo biogeographic implications to be drawn regarding Eifelian age faunas of New York, especially with regard to the westerly advance of these faunas in eastern North America; (4) help pinpoint and resolve some of the problems faced in correlating the Detroit River Group with the Onondaga Limestone.

TAXONOMIC NOTE

The taxon *Chonetes* is mentioned in the introduction in which the history of the Onondaga Formation is briefly discussed. However, subsequently the genus *Chonetes* is referred to by its correct taxonomic name *Hallinetes* (Racheboeuf and Feldman, 1990).

STRATIGRAPHIC CONSIDERATIONS

Four members of the Onondaga Limestone are recognized in the study areas (Figure 2): Edgecliff, Nedrow, Moorehouse, and Seneca. Detailed descriptions of these members may be found in Oliver (1954, 1956), but some additional comments are provided below.

EDGECLIFF MEMBER

In southeastern New York, the Edgecliff Member is finer-grained than in the central part of the state and is typified by abundant light-weathering chert seams often in a crinoidal matrix. Large crinoid columnals are

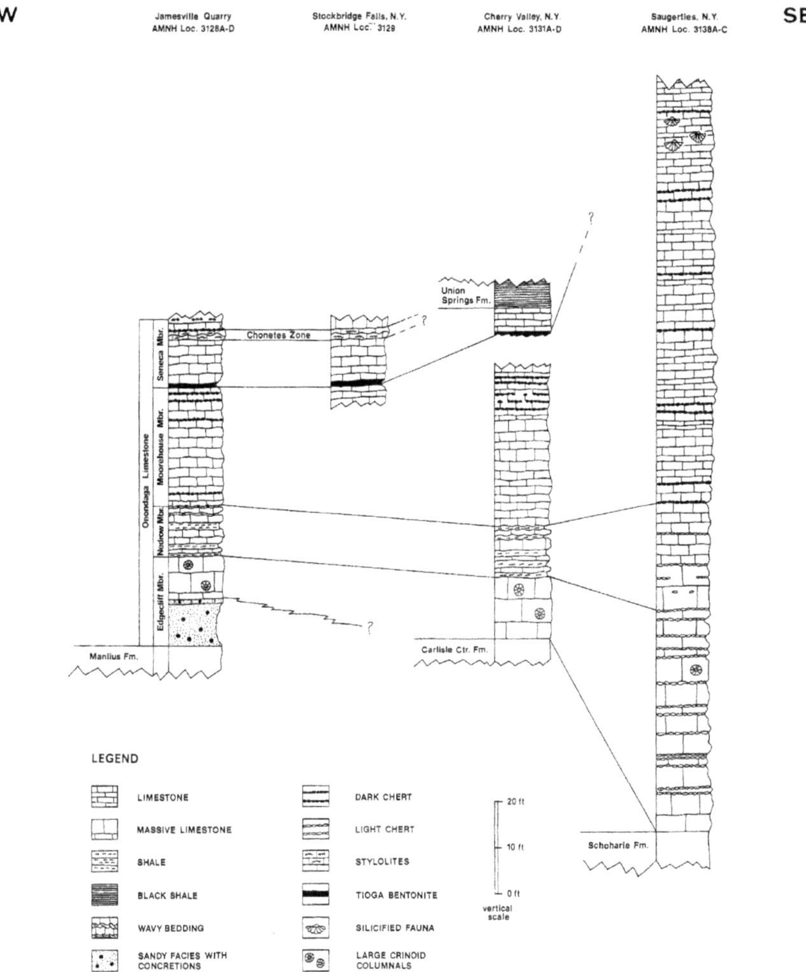

FIGURE 2. Measured stratigraphic sections of key outcrops on a west-east-southeast transect along the outcrop belt of the Onondaga Limestone in central and southeastern New York.

occasionally present, although they are more numerous in the Syracuse area. The Edgecliff varies in thickness from 12 to 19 ft in the Syracuse area, depending on the presence or absence of the basal (quartzose) sandy zone (see Oliver, 1966: 35, for a discussion of the age of the basal sandy zone), to a maximum of 48 ft near Saugerties. Brachiopods are relatively scarce in the Edgecliff and collecting is difficult due to the massive nature of the rock, its density and lack of joints, and slow weathering of the matrix. The following brachiopods have been collected from the Edgecliff for this study:

Amphigenia? sp.
Atrypa "reticularis"
Elytha fimbriata
Leptaena aff. *"rhomboidalis"*
Levenea aff. *subcarinata*
Levenea sp. A
Pentamerella arata

NEDROW MEMBER

Characterized by shaly beds, especially in the central part of the state, which, when weathered, result in recessed ledges along which fossils may be easily collected. In the east the Nedrow becomes thicker-bedded, less argillaceous, and coarser-grained. Collecting is more difficult in the eastern part of the outcrop belt. The following brachiopods have been collected from the Nedrow for this study:

Athyris sp. A
Atrypa "reticularis"
Charionoides aff. *doris*
Coelospira camilla
Costistrophonella cf. *punctulifera*
Cupularostrum sp. A
Dalejina aff. *alsa*
Gypidula sp.
Leptaena aff. *"rhomboidalis"*
Levenea aff. *subcarinata*
Megakozlowskiella raricosta
Megastrophia sp.
Nucleospira aff. *ventricosa*
Orthotetacid indet.
"Pacificocoelia" acutiplicata
Pentagonia unisulcata
Pentamerella arata
Stropheodontid indet.
Strophodonta demissa
Trematospira? sp.

MOOREHOUSE MEMBER

The eastern part of the outcrop belt is typified by dark-weathering chert seams which occur sporadically throughout the member. In the Syracuse area the chert is a bit lighter in color and concentrated more in the upper one-third of the member. Toward the southeast, the Moorehouse becomes noticeably darker, finer-grained, and highly argillaceous. The greatest change in thickness in the Onondaga Limestone occurs from Syracuse to the mid-Hudson Valley and is found within the Moorehouse Member.

Collecting is difficult in the uniformly bedded central area, but easier in the east where silicified faunas are accessible. The following brachiopods have been collected from the Moorehouse Member in this study:

Acrospirifer duodenaria
Ambocoelia sp.
Athyridacean indet.
Athyris sp. A
Athyris sp. B
Atribonium halli
Atrypa "reticularis"
Coelospira camilla
Cupularostrum sp. A
Cupularostrum sp. B
Cyrtina hamiltonensis
Cyrtina sp. A
Dalejina aff. *alsa*
Dalejina sp. A
Discomyorthis? sp.
Elytha fimbriata
Eospiriferid? indet.
Gypidula sp.
Hallinetes aff. *lineatus*
Leptaena aff. *"rhomboidalis"*
Levenea aff. *subcarinata*
Megakozlowskiella raricosta
Megastrophia sp.
Meristina nasuta
Mucrospirifer cf. *macra*
Nucleospira aff. *ventricosa*
Orthotetacid indet.
"Pacificocoelia" acutiplicata
Pentagonia unisulcata
Pentamerella arata
Rhynchospirina sp.
Schizophoria cf. *multistriata*
"Schuchertella" sp.
Stropheodontid indet.
Strophodonta demissa

SENECA MEMBER

The base of the Seneca Member in central New York is marked by the Tioga Bentonite, 10 ft above which occurs a zone of chonetid brachiopods (Zone J of Oliver, 1954). East of Cherry Valley, the Tioga Bentonite and Chonetid Zone have not yet been observed in the field. The Seneca Member is a highly argillaceous limestone in which fossils are not generally abundant (except for *Hallinetes* aff. *lineatus*). The following brachiopods have been collected from the Seneca Member in this study:

Atrypa "reticularis"
Athyris sp. A
Hallinetes aff. *lineatus*
Leptaena aff. *"rhomboidalis"*
Megakozlowskiella raricosta
Megastrophia sp.
Orthotetacid indet.
Pentamerella arata

BIOSTRATIGRAPHY

Boucot and Johnson (1968) report the occurrence of *Amphigenia elongata* from the Bois Blanc Formation of western New York, which directly underlies the Onondaga Limestone. Boucot (1959) mentions the occurrence of *A. preparva,* the earliest known species of the genus, from the Highland Mills Member of the Esopus Formation as well as from the Lower York River Sandstone in Gaspe. Grabau's (1906) faunal list for the Onondaga Limestone in eastern New York includes *A. elongata* as does Oliver's (1956) faunal list for the western part of New York. I have found poorly preserved fragments of *Amphigenia?* sp. in the basal sandy zone of the Edgecliff Member in the Syracuse area. The genus has not been recovered from any of the silicified rocks of the mid-Hudson Valley. Thus, additional collecting and study is indicated for refinement of Rensselaeriid zones in the eastern part of New York, specifically *Amphigenia* zonation in the post-Edgecliff rocks of the mid-Hudson Valley.

During the course of field studies, specimens of *Rhipidomella?* and *Cymostrophia?* have been observed in Moorehouse rocks in the vicinity of Saugerties, New York. The specimens were fragmentary, poorly preserved, nonsilicified, and impossible to extract. Attempts to crack out the occasional representatives of these genera resulted in destruction of the fossils. Additional work is needed to collect enough material for generic and specific assignment, although it may be stated with a fair amount of certainty that these forms occur rarely in the Onondaga.

The pre-Eifelian Bois Blanc Formation in western New York directly underlies the Onondaga Limestone. (See House, 1962 and Klapper, 1971 for further information on the Eifelian age of the Onondaga Limestone.) Oliver (1967) notes that the presence of key corals and the profusion of brachiopods is sufficient to separate the Bois Blanc from the coral-crinoid debris assemblage of the overlying Edgecliff Member of the Onondaga Limestone in western New York and the Niagara Peninsula of Ontario. Boucot and Johnson (1968) studied the brachiopods of the Bois Blanc Formation (USGS Loc. Nos. 4671-SD, 4672-SD) from western New York and provided a list of brachiopods from the Bois Blanc and its correlatives in eastern North America (Table 1). The following brachiopod species, occurring in the Bois Blanc, also occur in the Onondaga Limestone:

Acrospirifer duodenaria
Atrypa "reticularis"
Coelospira camilla
Dalejina aff. *alsa*

Megakozlowskiella raricosta
Meristina nasuta
"Muscrospirifer" cf. *macra*
Pentamerella arata
Strophodonta cf. *demissa*

The following brachiopod genera from the Bois Blanc also occur in the Onondaga Limestone:

Ambocoelia
Cupularostrum
Elytha
Leptaena
Nucleospira
"Schuchertella"

Amphigenia elongata from the Bois Blanc has been identified in this report as *Amphigenia?* in the Onondaga. Apparently there are nine species (possibly 16 species if the genera listed above prove to be conspecific with the Onondaga forms) that occur in the Onondaga. Three species range through the Seneca Member:

Atrypa "reticularis"
Megakozlowskiella raricosta
Pentamerella arata

The six remaining species are found through the Upper Moorehouse Member:

Acrospirifer duodenaria
Coelospira camilla
Dalejina aff. *alsa*
Meristina nasuta
"Mucrospirifer" cf. *macra*
Strophodonta cf. *demissa*

The above tabulations indicate a significant degree of stasis in terms of speciation (or lack thereof) in the Bois Blanc-Onondaga time interval. Although the present study is basically taxonomic and no attempt is made to specifically analyze evolutionary trends, the following generalizations are offered with regard to future evolutionary studies of Bois Blanc-Onondaga brachiopod faunas, as suggested by Gould and Eldredge (1977):

(1) The morphologic variability of the brachiopods is greater laterally than vertically.
(2) Of the 26 species of brachiopods in the Bois Blanc Formation, nine (34%) occur in the Onondaga and show no evolutionary change.

(3) The high degree of stasis present in the Bois Blanc-Onondaga (Emsian-Eifelian) interval may support Eldredge and Gould's (1972) idea of punctuated equilibrium rather than phyletic gradualism for the species described in this study.

ACKNOWLEDGMENTS

I thank A. J. Boucot of the Department of Geology, Oregon State University, for his extensive support, assistance, and critical review, without which this project would not have been possible. The late J. G. Johnson, of the same institution, deserves thanks for critically reading the manuscript and providing useful suggestions for improvement. I also thank the late W. A. Oliver, Jr., of the United States Geological Survey, Washington, D.C., for directing me to key outcrops in the study area, assisting in stratigraphic interpretation, and providing valuable comments and criticism during the research period. L. V. Rickard of the New York State Museum and Science Service, Albany, New York, deserves special thanks for spending time with me in the field and pointing out important subtleties of Devonian stratigraphy in the mid-Hudson Valley. I am particularly grateful to Niles Eldredge of the American Museum of Natural History for allowing me to use Museum facilities, especially the acid room, before I was appointed Research Associate. Fieldwork was partially supported by National Science Foundation grant EAR 76-15402 and a grant from the Society of the Sigma Xi.

ABBREVIATIONS

Institutions and Localities

AMNH — American Museum of Natural History, Department of Invertebrates
AMNH Loc. — American Museum of Natural History locality number
USNM — United States National Museum of Natural History, Department of Paleobiology, Smithsonian Institution
USNM Loc. — United States National Museum locality number
USGS — United States Geological Survey

Measurements

(L) — maximum length of shell
(W) — maximum width of shell

(T)	—	maximum thickness of shell
mm	—	millimeters
est.	—	estimated
b.v.	—	brachial (dorsal) valve
p.v.	—	pedicle (ventral) valve
def.	—	deformed
art.	—	articulated

SYNOPTIC CLASSIFICATION

Phylum Brachiopoda
Subphylum Rhynchonelliformea Williams and others, 1996
Class Rhynchonellata Williams and others, 1996
Order Orthida Schuchert and Cooper, 1932
Suborder Dalmanellidina Moore, 1952
Superfamily Dalmanelloidea Schuchert, 1913
Family Dalmanellidae Schuchert, 1913
Subfamily Isorthinae Schuchert and Cooper, 1931
Levenea aff. *subcarinata* (Hall, 1857) .... (183)1[1]

Geographic occurrence—AMNH Locs. 3124B, 3128B, C, 3130, 3131B-D, 3134, 3137, 3138A-C, 3139, 3141, 3142, 3143, 3151A.

Stratigraphic occurrence—Edgecliff, Nedrow, Moorehouse members.

Levenea species A .... (2)

Geographic occurrence—AMNH Loc. 3143.

Stratigraphic occurrence—Edgecliff Member.

Family Rhipidomellidae Schuchert, 1913
Subfamily Rhipidomellinae Schuchert, 1913
Dalejina aff. *alsa* (Hall, 1963) .... (61)

Geographic occurrence—AMNH Locs. 3121B-C, 3124A-B, 3128C, 3130, 3131B, 3133, 3135, 3137, 3138A, 3139, 3141, 3142, 3144A.

Stratigraphic occurrence—Nedrow, Moorehouse members.

Dalejina species A .... (2)

Geographic occurrence—AMNH Loc. 3137.

Stratigraphic occurrence—Moorehouse Member.

Discomyorthis? sp. .... (4)

Geographic occurrence—AMNH Locs. 3135, 3137.

[1] Numbers in parentheses are sample sizes.

Stratigraphic occurrence—Moorehouse Member.

Superfamily Enteletoidea Waagen, 1884

Family Schizophoriidae Schuchert and LeVene, 1929

Subfamily Schizophoriinae Schuchert and LeVene, 1929

Schizophoria cf. *multistriata* Hall (1859-1861) . . . . (65)

Geographic occurrence—AMNH Locs. 3130, 3131B, 3133, 3134, 3137, 3138A, 3141, 3142, 3144A.

Stratigraphic occurrence—Moorehouse Member.

Order Pentamerida Schuchert and Cooper, 1931

Suborder Pentameridina Schuchert and Cooper, 1931

Superfamily Clorindoidea Rzhonsnitskaia, 1956

Family Clorindidae Rzhonsnitskaia, 1956

Subfamily Pentamerellinae Sapelnikov, 1973

Pentamerella arata (Conrad, 1841) . . . . (75)

Geographic occurrence—AMNH Locs. 3123A-D, 3124A-B, 3125A, 3128B-C, 3130, 3131B, 3137, 3138A, 3139, 3140, 3143, 3144A.

Stratigraphic occurrence—Nedrow, Moorehouse members.

Order Strophomenida Öpik, 1934

Suborder Strophomenoidea

Superfamily Strophomenoidea King, 1846

Family Rafinesquinidae Schuchert, 1893

Subfamily Leptaeninae Hall and Clarke, 1894

Leptaena aff. *"rhomboidalis"* (Wilckens, 1769) . . . . (172)

Geographic occurrence—AMNH Locs. 3123B-D, 3124A-B, 3125A, 3128A-C, 3130, 3131B, 3134, 3137, 3138A-B, 3140, 3141, 3142, 3143.

Stratigraphic occurrence—Edgecliff, Nedrow, Moorehouse, Seneca members.

Order Orthotetida Waagen, 1884

Suborder Orthotetidina Waagen, 1884

Superfamily Orthotetoidea Waagen, 1884

Family Schuchertellidae Williams, 1953

Subfamily Schuchertellinae Williams, 1953

"Schuchertella" sp. . . . . (3)

Geographic occurrence—AMNH Loc. 3137.

Stratigraphic occurrence—Moorehouse Member.

Orthotetacids indet. . . . . (73)

Geographic occurrence—AMNH Locs. 3123B-D, 3124A-B, 3125A, 3128A-C, 3130, 3131B-C, 3133, 3134, 3135, 3137, 3139, 3142, 3144A.

Stratigraphic occurrence—Nedrow, Moorehouse, Seneca members.

Superfamily Strophomenoidea King, 1846

Family Strophonellidae Caster, 1939

Costistrophonella cf. *punctulifera* (Conrad, 1838) . . . . (4)

Geographic occurrence—AMNH Locs. 3123C, 3131C.

Stratigraphic occurrence—Nedrow Member.

Family Stropheodontidae Caster, 1939

Megastrophia sp. . . . . (43)

Geographic occurrence—AMNH Locs. 3123A-C, 3124B, 3125A, 3128C, 3131B-C, 3134, 3135, 3138A.

Stratigraphic occurrence—Nedrow, Moorehouse, Seneca members.

Strophodonta cf. *demissa* (Conrad, 1842) . . . . (13)

Geographic occurrence—AMNH Locs. 3124B, 3130, 3131B, 3133, 3135, 3137.

Stratigraphic occurrence—Nedrow, Moorehouse members.

Stropheodontid indet. . . . . (9)

Geographic occurrence—AMNH Locs. 3123B-C, 3135.

Stratigraphic occurrence—Nedrow, Moorehouse members.

Order Productida Sarytcheva and Sokolskaya, 1959

Suborder Chonetedina Muir-Wood, 1955

Superfamily Chonetoidea Bronn, 1862

Family Chonetidae Bronn, 1862

Subfamily Devonochonetinae Muir-Wood, 1962

Hallinetes aff. *lineatus* (Conrad, 1839) . . . . (4554)

Geographic occurrence—AMNH Locs. 3123A, 3128A, 3129, 3131B.

Stratigraphic occurrence—Moorehouse, Seneca members.

Order Rhynchonellida Kuhn, 1949

Superfamily Stenoscismatatoidea Oehlert, 1887 (1883)

Family Stenoscismatidae Oehlert, 1887 (1883)

Subfamily Stenoscismatinae Oehlert, 1887 (1883)

Atribonium halli (Fagerstrom, 1961) . . . . (2)

Geographic occurrence—AMNH Locs. 3135, 3137.

Stratigraphic occurrence—Moorehouse Member.

Superfamily Rhynchotrematoidea Schuchert, 1913

Family Trigonirhynchiidae Schmidt, 1965

Subfamily Trigonirhynchiinae Schmidt, 1965

Cupularostrum? species A . . . . (40)

Geographic occurrence—AMNH Locs. 3131C, 3135, 3137, 3138A.
Stratigraphic occurrence—Nedrow, Moorehouse members.

Cupularostrum? species B .... (5)

Geographic occurrence—AMNH Loc. 3138A.
Stratigraphic occurrence—Moorehouse Member.

Order Atrypida Rzhonsnitskaia, 1960
Suborder Atrypidina Moore, 1952
Superfamily Atrypoidea Gill, 1871
Family Atrypidae Gill, 1871
Subfamily Atrypinae Gill, 1871

Atrypa "reticularis" (Linnaeus, 1767) .... (617)

Geographic occurrence—AMNH Locs. 3122, 3123A-D, 3124A-C, 3125A-B, 3127, 3128C-D, 3139, 3131A-D, 3132, 3133, 3134, 3135, 3136, 3137, 3138A-C, 3139, 3140, 3141, 3142, 3143, 3144A-B, 3145, 3146, 3147, 3148, 3149, 3150, 3151B.
Stratigraphic occurrence—Edgec1iff, Nedrow, Moorehouse, Seneca members.

Superfamily Anoplothecoidea Schuchert, 1894
Family Anoplothecidae Schuchert, 1894
Subfamily Coelospirinae Hall and Clarke, 1895

Coelospira camilla Hall, 1867 .... (180)

Geographic occurrence—AMNH Locs. 3123C, 3128B-C, 3133, 3135, 3137, 3138A-B, 3139, 3141, 3142, 3144A.
Stratigraphic occurrence—Nedrow, Moorehouse members.

Family Leptocoeliidae Boucot and Gill, 1956

"Pacificocoelia" acutiplicata (Conrad, 1841) .... (141)

Geographic occurrence—AMNH Locs. 3123B-C, 3124B, 3125A, 3128B-C, 3133.
Stratigraphic occurrence—Nedrow, Moorehouse members.

Suborder Athyrididina Boucot, Johnson, and Staton, 1964
Superfamily Athyridoidea Davidson, 1881
Family Athyrididae Davidson 1881
Subfamily Athyridinae Davidson, 1881

Athyris species A .... (38)

Geographic occurrence—AMNH Locs. 3123B, C, 3124B, 3125A, 3128A-C, 3130, 3131B, 3133, 3135, 3137, 3139.
Stratigraphic occurrence—Nedrow, Moorehouse, Seneca members.

Athyris species B .... (2)

Geographic occurrence—AMNH Loc. 3138A.
Stratigraphic occurrence—Moorehouse Member.

Superfamily Meristelloidea Waagen, 1883

Family Meristellidae Waagen, 1883

Subfamily Meristellinae Waagen, 1883

Meristina cf. *nasuta* (Conrad, 1842) .... (3)

Geographic occurrence—AMNH Locs. 3123B, 3137.
Stratigraphic occurrence—Moorehouse Member.

Charionoides aff. *doris* (Hall, 1860) .... (1)

Geographic occurrence—AMNH Loc. 3123C.
Stratigraphic occurrence—Nedrow Member.

Pentagonia unisulcata (Conrad, 1841) .... (81)

Geographic occurrence—AMNH Locs. 3130, 3131B-C, 3133, 3134, 3137, 3138A, 3139, 3140, 3141, 3142, 3144A.
Stratigraphic occurrence—Nedrow, Moorehouse members.

Family Nucleospiridae Davidson, 1881

Nucleospira aff. *ventricosa* (Hall, 1857) .... (152)

Geographic occurrence—AMNH Locs. 3130, 3131B, 3133, 3135, 3137, 3138A-B, 3139, 3140, 3141, 3142, 3144A.
Stratigraphic occurrence—Nedrow, Moorehouse members.

Athyridacean indet .... (3)

Geographic occurrence—AMNH Locs. 3137, 3138A.
Stratigraphic occurrence—Moorehouse Member.

Order Retziidina Alvarez and Rong, 2002

Superfamily Rhynchospirinoidea Schuchert, 1929

Family Rhynchospirinidae Schuchert, 1929

Trematospira? sp. .... (2)

Geographic occurrence—AMNH Loc. 3124B
Stratigraphic occurrence—Nedrow Member

Family Rhynchospirinidae Schuchert, 1929

Rhynchospirina sp .... (1)

Geographic occurrence—AMNH Loc. 3137.
Stratigraphic occurrence—Moorehouse Member.

Order Spiriferida Waagen, 1883

Suborder Delthyridina Ivanova, 1972

Superfamily Delthyridoidea Phillips, 1841

Family Acrospiriferidae Termier and Termier, 1949

Subfamily Acrospiriferinae Termier and Termier, 1949

Acrospirifer duodenaria (Hall, 1843) .... (100)

Geographic occurrence—AMNH Locs. 3130, 3131B, 3135, 3137, 3138A, 3144A.

Stratigraphic occurrence—Moorehouse Member.

Family Mucrospiriferidae Boucot, 1959

Subfamily Mucrospiriferinae Boucot, 1959

"Mucrospirifer" cf. *macra* (Hall, 1857) .... (34)

Geographic occurrence—AMNH Locs. 3130, 3131B, 3135, 3137, 3138A.

Stratigraphic occurrence—Moorehouse Member.

Family Cyrtinopsidae Wedekind, 1926

Subfamily Cyrtinopsinae Wedekind, 1926

Megakozlowskiella raricosta (Conrad, 1842) .... (284)

Geographic occurrence—AMNH Locs. 3123A-D, 3124A-B, 3125A, 3128B-C, 3130, 3131B, 3133, 3134, 3135, 3137, 3138A, 3139, 3140, 3141, 3144A.

Stratigraphic occurrence—Nedrow, Moorehouse, Seneca members.

Superfamily Reticularioidea Waagen, 1883

Family Elythidae Frederiks, 1924

Subfamily Elithinae Frederiks, 1924

Elita fimbriata (Conrad, 1842) .... (21)

Geographic occurrence—AMNH Locs. 3131B, 3135, 3137, 3138A-C, 3139.

Stratigraphic occurrence—Edgecliff, Moorehouse members.

Eospiriferid? indet. .... (3)

Geographic occurrence—AMNH Loc. 3135.

Stratigraphic occurrence—Moorehouse Member.

Superfamily Ambocoelioidea George, 1931

Family Ambocoeliidae George, 1931

Subfamily Ambocoeliinae George, 1931

Ambocoelia sp. .... (6)

Geographic occurrence—AMNH Locs. 3135, 3137.

Stratigraphic occurrence—Moorehouse Member.

Order Spiriferidina Ivanova, 1972

Suborder Cyrtinidina Carter and Johnson, 1994

Superfamily Cyrtinoidea Frederiks, 1911

Family Cyrtinidae Frederiks, 1911

Cyrtina hamiltonensis (Hall, 1857) .... (30)

Geographic occurrence—AMNH Locs. 3133, 3135, 3137, 3138A, 3141, 3144A.

Stratigraphic occurrence—Moorehouse Member.

Cyrtina species A .... (2)

Geographic occurrence—AMNH Loc. 3135.

Stratigraphic occurrence—Moorehouse Member.

Order Terebratulida Waagen, 1883

Suborder Terebratulidina Waagen, 1883

Superfamily Stringocephaloidea King, 1850

Family Centronellidae Waagen, 1882

Subfamily Amphigeniidae Cloud, 1942

Amphigenia? sp .... (12)

Geographic occurrence—AMNH Loc. 3126.

Stratigraphic occurrence—Edgecliff Member.

Total Specimens Studied 7030

SYSTEMATIC PALEONTOLOGY

Phylum BRACHIOPODA

Order ORTHIDA Schuchert and Cooper, 1932

Suborder DALMANELLIDINA Moore, 1952

Superfamily DALMANELLOIDEA Schuchert, 1913

Family DALMANELLIDAE Schuchert, 1913

Subfamily ISORTHINAE Schuchert and Cooper, 1931

Genus *LEVENEA* Schuchert and Cooper, 1931

Type species: *Orthis subcarinata* Hall, 1857: 43.

Levenea aff. *subcarinata* (Hall, 1857) Figure 3.

Orthis subcarinata Hall, 1857: 43; 1859: 169, pl. 12, figs. 7-21.

Levenea subcarinata Schuchert and Cooper, 1932: 123, pl. 18, figs. 19-23, 25-32; Cooper, 1944: 353, pl. 138, figs. 21-23.

Exterior: The shells are small to medium sized, transversely suboval in outline, and ventribiconvex in lateral profile. The brachial valve bears a shallow, rounded sulcus which broadens anteriorly. Length is slightly greater than width. Maximum width is attained at or just anterior to midlength. The ventral interarea is short, slightly incurved, and apsacline. A triangular delthyrium is present which encloses an angle of approximately 60 degrees. On many specimens the delthyrium widens apically into a small, circular

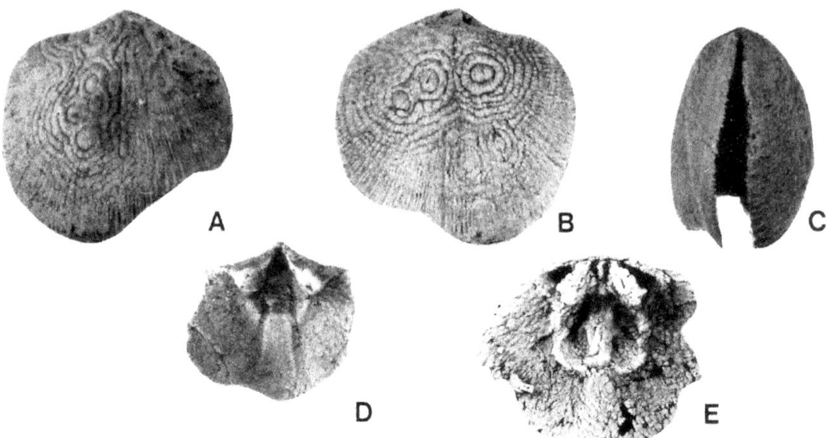

FIGURE 3. *Levenea* aff. *subcarinata* (Hall, 1957). A-C. AMNH Loc. 3134. Ventral, dorsal, and lateral views, AMNH 42756, x3. D. AMNH Loc. 3141. Pedicle valve interior, AMNH 39870, x2.5. E. AMNH Loc. 3141. Brachial valve interior, AMNH 39871, x2.5.

foramen. The dorsal interarea is poorly preserved. Ornamentation consists of indistinct, rounded, radial costellae which appear to increase in number anteriorly by bifurcation. Cross-sectional shape is unclear.

Pedicle valve interior: The hinge teeth are short, blunt, and subtriangular in cross section. They are supported by subpyramidal deposits of shell material in place of true dental lamellae. The muscle field is about one-third the valve length and subpentagonal in outline. The anterior rim of the muscle field is raised, separating it from the valve floor. The diductor tracks are represented by two longitudinal striae, which extend past the anterior rim of the muscle field almost to the anterior commissure. On smaller specimens no rim is present and the diductor tracks are not impressed.

Brachial valve interior: The sockets are short, deeply excavated, widely divergent, and slightly covered posteriorly by the dorsal interarea. The brachiophores project ventrally and slightly anteriorly. The cardinal process is set in a raised mound of secondary shell material within the notothyrial cavity and varies in shape from a simple longitudinal ridge to a bilobed process. Two specimens appear to be quadrilobed. A low, broad, rounded myophragm divides the adductor impressions medially. The adductor impressions are elongate suboval, fairly deeply impressed and bounded laterally and anteriorly by muscle-bounding ridges. No clear differentiation is evident between the anterior and posterior adductors.

Comparison: *Levenea* aff. *subcarinata* is smaller than Johnson's (1970: 77, pl. 2, figs. 8-18) *L. fagerholmi* and has a flatter pedicle valve. *Levenea navicula* (Johnson, 1970: 75, pl. 2, figs. 19-22; pl. 3, figs. 1-19) is more biconvex and has subparallel diductor tracks, whereas *L.* aff. *subcarinata* has diductor tracks which are somewhat divergent anteriorly. Johnson's (1970: 74, pl. 2, figs. 1-7) *Levenea* sp. A differs from *L.* aff. *subcarinata* in its stronger costation and dorsal interior, which has a more elongate muscle field.

The ventral muscle field of *Levenea* cf. *subcarinata* (Boucot, Gauri, and Southard, 1970: 8, pl. 2, figs. 3-5) has subparallel diductor tracks and is less divergent anteriorly, while the dorsal muscle field is slightly more elongate. The specimens are similar to *L.* aff. *subcarinata* in that they also possess sulcate brachial valves (although the Onondaga forms are very slightly sulcate), suboval outlines, and fine costae. In addition to specimens from the Kalkberg Limestone (USNM Loc. 11321), the New Scotland Formation (USNM Loc. 11317), and the Coeymans Limestone (USNM Loc. 11259), Boucot, Gauri, and Southard (1970: 9) briefly mentioned and described *Levenea* sp. from the Esopus Formation, Mountainville Member (USNM Loc. 11251), but due to a lack of available specimens were unable to assign a specific identification. Their specimens differ from *L.* aff. *subcarinata* in their subcircular outline and wider muscle fields. Helderberg age *Levenea* sp. from the Moose River Synclinorium (Boucot, 1973) resembles *L.* aff. *Subcarinata* in general outline, but the material is too fragmentary for further comparison.

Community occurrence: Feldman (1980) found this species in the *Leptaena-Megakozlowskiella, Levenea* Community I, and *Levenea* Community II communities.

Figured specimens: AMNH 3970, 39871, 42756.

Levenea sp. A, Figure 4.

Remarks: A single, well-preserved pedicle valve is available for study. It is concave with a subcircular outline slightly worn on the edges. The hinge line is short, interarea fairly high, apsacline, and moderately incurved. The delthyrium is triangular and encloses an angle of approximately 60 degrees. The beak is suberect. No fine ornament has been preserved.

A broad, flat, median ridge extends from deep in the umbonal cavity, increasing slightly in height until the anterior extremity of the muscle field is reached, then slopes off sharply until merging with the valve floor just short of midlength. The umbonal cavity, including the median ridge, is horizontally striated with what appear to be internal growth lamellae. Diductor

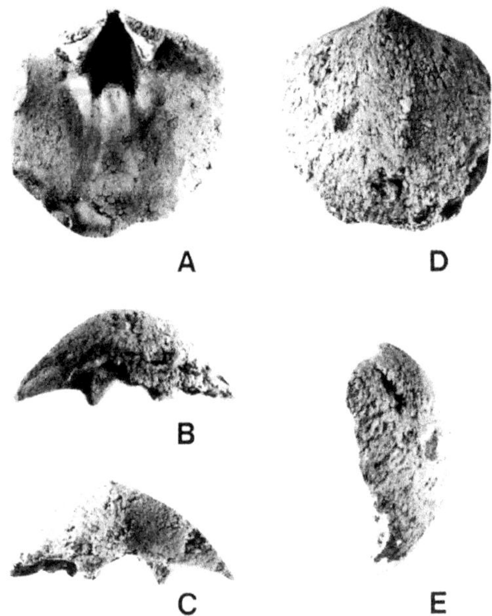

FIGURE 4. *Levenea* sp. A. A-E. AMNH Loc. 3143. Pedicle valve interior, lateral view, AMNH 39872, x1.5.

impressions are impressed upon the lateral margins of the umbonal cavity. The hinge teeth are large, pointed, and dorsally directed. They are supported by obsolescent dental lamellae.

This specimen of *Levenea* sp. A may be distinguished from *L.* aff. *subcarinata* by its larger size, larger, more pointed hinge teeth, and shorter muscle field.

Community occurrence: Feldman (1980) found this species in the *Levenea* Community I.

Figured specimen: AMNH 39872.

Family RHIPIDOMELLIDAE Schuchert, 1913
Subfamily RHIPIDOMELLINAE Schuchert, 1913
Genus *DALEJINA* Havlicek, 1953
Type species: *Dalejina hanusi* Havlicek, 1953: 5.

Dalejina aff. *alsa* (Hall, 1863), Figures 5-7, 8A-D.

Orthis alsus Hall, 1863: 33.

Rhipidomella alsa Hall, 1867: 36, pl. 4, figs. 2-7; Grabau, 1906: 181, fig. 95.

Dalejina alsa Boucot and Johnson, 1968: B7, pl. 1, figs. 11-27; Fagerstrom, 1971: pl. 1, figs. 1-2.

Exterior: The shells are transversely suboval to subcircular in outline, and ventribiconvex, with the pedicle valve slightly flatter and the brachial valve deeper, especially in the umbonal region. The hinge line is very short and straight in the apical area but becomes rounded as the lateral margins are approached. Maximum width (Table 1) is reached at or just anterior to midlength. The pedicle valve bears a slight median depression which rarely reaches the anterior commissure. The brachial valve often bears a corresponding median ridge. However, both of these features are variable. The anterior commissure is most often rectimarginate but can be slightly sulcate if the influence of the ventral median depression is strong enough. The ventral interarea is short, narrow, and apsacline. The dorsal interarea is anacline.

Radial ornamentation consists of numerous radial costellae which increase anteriorly both by intercalation and bifurcation. There are 18 to 20 costellae per 5 mm at the anterior commissure, near midline. The costellae are occasionally crossed by concentric growth lines near the anterior margins.

Pedicle valve interior: Short, somewhat obscure dental lamellae support the widely divergent, pointed, hinge teeth, which are subelliptical in cross-section. Shallow crural fossettes are present. On specimens whose apical region is preserved, there is usually a small pedicle callist. The umbonal cavity is fairly shallow and bounded laterally by the dental plates, which sometimes extend anterolaterally as poorly defined ridges encompassing the posterior one-half to one-third of the muscle field. The muscle impressions of small specimens are indistinct, but definitely nonflabellate.

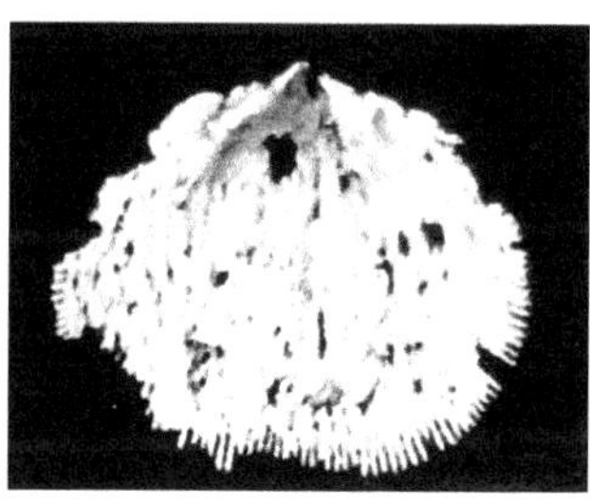

FIGURE 5. *Dalejina* aff. *alsa* (Hall, 1863). AMNH Loc. 3137. Pedicle valve interior, AMNH 42747, x2.

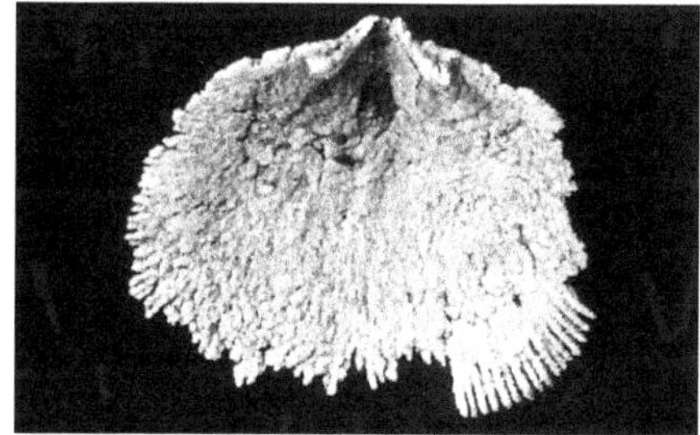

FIGURE 6. *Dalejina* aff. *alsa* (Hall, 1863). AMNH Loc. 3137. Pedicle valve interior, AMNH 42748, x3.

On larger individuals the diductor scars, although not well preserved, are flabellate and take up approximately one-third of the valve floor. No adductor impressions were preserved. A vague median ridge is present on some specimens. The muscle scars merge anteriorly with the valve floor such that no definite boundary between the two may be delineated. The valve floor is crenulated at the periphery by the impress of the medially grooved costellae.

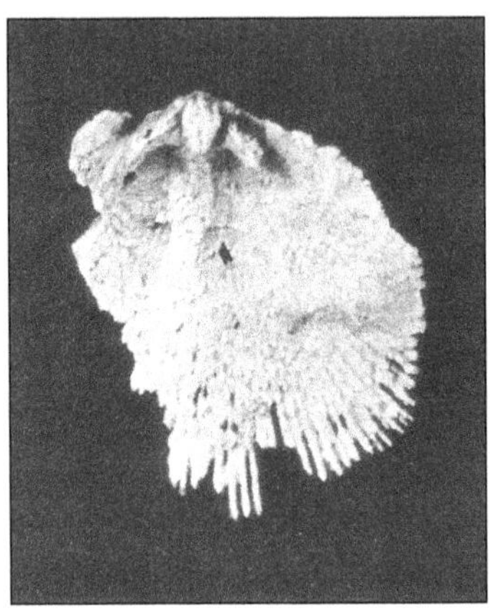

FIGURE 7. *Dalejina* aff. *alsa* (Hall, 1863). AMNH Loc. 3137. Brachial valve interior, AMNH 42749, x3.

Brachial valve interior: The sockets are fairly deep and broaden anterolaterally. The brachiophores attach directly to the valve floor and diverge from the beak at an angle of approximately 80 degrees. They are stout and subrectangular in cross-section. The medially located cardinal process consists of a variably-shaped protuberance, often poorly preserved, resting on a deposit of secondary shell material in the notothyrial cavity. A low, broad myophragm

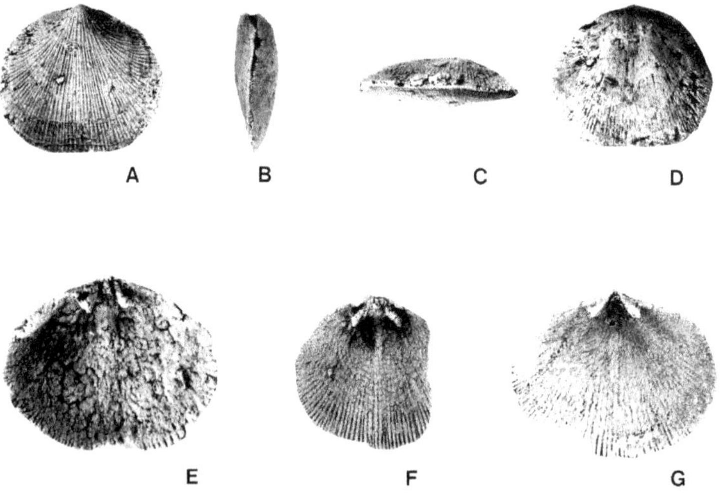

FIGURE 8. *Dalejina* aff. *alsa* (Hall, 1863). A-D. AMNH Loc. 3135. Ventral, lateral, anterior, and dorsal views, AMNH 39874, x2. *Discomyorthis?* sp. E. AMNH Loc. 3137. Brachial valve interior, AMNH 39876, x5. F. AMNH Loc. 3137. Brachial valve interior, AMNH 39873, x2. G. AMNH Loc. 3135. Pedicle valve interior, AMNH 39875, x2.

TABLE 1. Measurements (in mm) of Pedicle Valves of *Dalejina* aff. *alsa* (Hall, 1863).

AMNH Locality	(L)	(W)	(L)/(W)
3135	11.9	15.3	0.77
3137	5.8	6.6	0.88
3137	5.2	6.1	0.85
3137	10.3	12.5	0.82

extends from deep in the notothyrial cavity to about one-half to one-third the valve length. Muscle scars are generally faintly impressed but are more clearly defined as suboval pits on either side of the posterior portion of the myophragm (adductors?). The internal periphery is crenulated in the same manner as that of the pedicle interior.

Comparison: *Dalejina alsa* from the Bois Blanc Formation (Boucot and Johnson, 1968: B7, pl. 1, figs. 11-27) differs from *D.* aff. *alsa* in that the adductor scars are suboval with their anterior margins at approximately the center of the muscle field, with a thickening along their anterior edges that continues anteriorly as a myophragm (see pl. 1, figs. 11, 12 above). This is true only of larger shells. In small shells of both the Bois Blanc and Onondaga specimens the muscle field is not well impressed, and there is only a vague indication of flabellate muscle scars (especially in *D.* aff. *alsa*). The brachial interiors of *D.* aff. *alsa* (all silicified) are not preserved well enough to make meaningful comparisons with *D. alsa* other than to state that they appear to be *almost* identical.

Dalejina cf. *oblata* from the Green Pond Outlier in southeastern New York (Boucot Gauri, and Southard, 1970: 6, pl. I, figs. 7-13) may be differentiated from *D.* aff. *alsa* by the outline of the ventral adductors (narrower in *D.* cf. *oblata*).

Dalejina emarginata from the Keyser Limestone of Pennsylvania, Maryland, and West Virginia (Bowen, 1967: 21, pl. I, figs. 9-14) differs from *D.* aff. *alsa* in its more triangular ventral muscle field and the presence of feeble, radial myophragms within the diductor field and smaller size. The outlines of the adductors are remarkably similar to Boucot and Johnson's (1968) *D. alsa* (see Bowen, 1967: pl. 1, fig. 14). *Dalejina emarginata* spans from the Upper *Eccentricosta jerseyensis* Zone to the Lower *Meristella praenuntia* Zone of the Keyser. The Silurian-Devonian boundary is placed by Bowen (1967: 17) between these two faunal zones based on, in part, the occurrence of *Megakozlowskiella, Meristella,* and *Nanothyris,* genera known only in the Lower Devonian or younger strata in other areas of North America.

Boucot, Johnson, and Walmsley (1965), in a revision of the Rhipidomellidae, discuss the affinities of *Dalejina*, which is succeeded by *Rhipidomella* and *Aulacella* in the Middle Devonian. Johnson (1970: 80), in describing *Dalejina* sp. A, B, and C from central Nevada, mentions that the Nevadan forms appear to be distinct from the lineage that developed in the Appalachian Province and which is exemplified in the Helderbergian of the Appalachian Province by *Dalejina oblata.* Johnson further points out that the Appalachian lineage is characterized by an increase in size and degree of flabellation of the ventral muscle scar as pointed out by Amsden and Ventress (1963: 64, text-fig. 21).

Community occurrence: Feldman (1980) found this species in the *Leptaena-Megakozlowskiella, Atrypa-Coelospira-Nucleospira, Atrypa, Atrypa-Megakozlowskiella,* and *"Pacificocoelia"* communities.

Figured specimens: AMNH 42747, 42748, 42749.

Discomyorthis? sp., Figure 8E-G

Discussion: Two pedicle and two brachial valves of *Discomyorthis?* sp. were recovered from Upper Moorehouse strata in the mid-Hudson Valley. They differ from *Dalejina* aff. *alsa* in their more circular outline and larger ventral diductor scars. The pedicle valve bears a well-developed pedicle callist and short triangular hinge teeth. The internal periphery is well preserved and distinctly shows the medially grooved costellae. There is an indication of a low, broad myophragm. The brachial valve bears a triangular cardinal process and poorly developed brachiophores which diverge at an angle of about 40 degrees. The sockets are short and deeply excavated anteriorly. Indistinct linear muscle scars are present along with a suggestion of a broad myophragm. Although the internal periphery is poorly preserved, medially grooved costellae are evident. Aside from the morphological variances discussed above, the shells are identical with *Dalejina* aff. *alsa.* Further collecting and study is necessary in order to determine a specific assignment.

Community occurrence: Feldman (1980) found this species in the *Atrypa-Coelospira-Nucleospira* Community.

Figured specimens: AMNH 39873, 39875, 39876.

Superfamily ENTELETOIDEA Waagen, 1884

Family SCHIZOPHORIIDAE Schuchert and LeVene, 1929
Subfamily SCHIZOPHORIINAE Schuchert and LeVene, 1929
Genus *SCHIZOPHORIA* King, 1850
Type species: *Conchyliolithus Anomites resupinatus*
Martin, 1809: pl. 49, figs. 13, 14.

Schizophoria cf. *multistriata* Hall (1859-1861), Figures 9, 10

Conchyliolithus Anomites resupinatus Martin, 1809: pl. 49, figs. 13-14.

Schizophoria multistriata Grabau, 1906: 156, fig. 69; Goldring, 1935: 119, figs. 4IJ-K; 1943: 183, figs. 33N, O; Cooper, 1944: 357, pl. 140, figs. 10-11.

Exterior: The shells are medium-sized, suboval to subquadrate in outline, and unequally biconvex. The brachial valve is deeper and more uniformly convex. In younger specimens both valves become almost equally biconvex. The pedicle valve develops a broad, shallow sulcus on

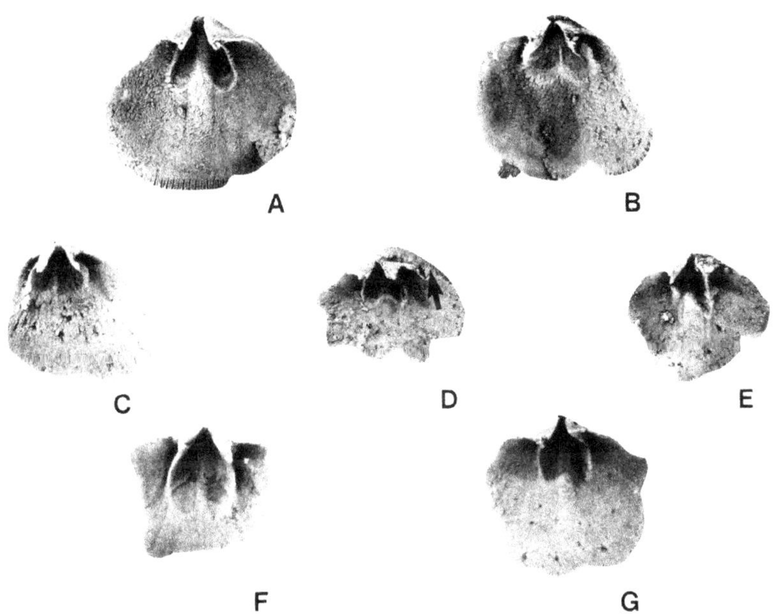

FIGURE 9. *Schizophoria* cf. *multistriata* Hall (1859-1861). A, B. AMNH Loc. 3138A Pedicle valve interior, brachial valve interior (articulated specimen), AMNH 39880a, 39880b, x1.5. C. AMNH Loc. 3135. Brachial valve interior, AMNH 42740, x2. D. AMNH Loc. 3138A Pedicle valve interior externally encrusted by the trepostome bryozoan *Dyoidophragma?* sp. (arrow), AMNH 42760, x1. E. AMNH Loc. 3146 Brachial valve interior, AMNH 42739, x1.2. F, G. AMNH Loc. 3138A. Brachial valve interior (articulated specimen), AMNH 42738a, 42738b, x2.

A

B

C

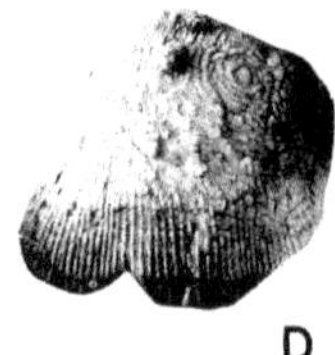
D

FIGURE 10. *Schizophoria* cf. *multistriata* Hall (1859-1861). A, D. AMNH Loc. 3138A. Pedicle and brachial exteriors (articulated specimen), AMNH 39880a, 39880b, x1.2. B. AMNH Loc. 3155. Brachial valve exterior, AMNH 42740, x2. C. AMNH Loc. 3146. Brachial valve exterior, AMNH 42739, x1.2.

ephebic specimens, whereas the brachial valve bears an even more indistinct fold. The hinge line is short and slightly rounded. Maximum width is attained at or just past midlength. The ventral interarea is apsacline, triangular, and fairly high in larger specimens and relatively narrow in younger ones. The dorsal interarea is narrower and ranges from orthocline to apsacline. The interareas of both valves are equal to about one-half the width.

Ornamentation consists of rounded to subangular radial costellae with broad, flat interspaces, which are usually preserved only at the anterior half of the valves. There are about 11 costellae in a 5 mm space near the anterior commissure at midline.

Pedicle valve interior: The hinge teeth are small and subrectangular in cross-section. They are supported by stout, anterolaterally diverging dental lamellae, which join the valve floor at about one-third the length. The dental lamellae form the lateral margins of the pyriform diductor impressions, which are bisected by a median, raised adductor platform that terminates slightly before the antenor extremity of the diductor scars is reached, thus giving a bilobed appearance to the entire muscle field. The delthyrium encloses an angle of about 45 degrees and opens into a small foramen apically. The anterior internal periphery is finely crenulated due to the impress of the costellae.

Brachial valve interior: Only one specimen in the collection possesses a preserved cardinal process. Located deep in the notothyrial cavity, the process is rectangular in outline and slightly worn. The posterior portion appears to be bilobed but not enough shell material is present to make additional conclusions regarding the morphology of the structure. The sockets are short, deep, curved and widely divergent. They are bounded anterolaterally by small fulcral plates. The interarea conceals the posterior portion of the socket. Attached to the posterior ends of the fulcral plates are the brachiophore bases which project ventrally and are concave at their extremities. On the end of each brachiophore there is a posterolaterally directed hook-like apophysis.

The muscle field is subcircular to suboval and bisected by a poorly defined myophragm which separates the adductor scars. In some instances the dorsal muscle field is pyriform in outline and appears to be bilobed, as does the ventral muscle field. The anterior rim of the field is often raised above the valve floor. The anterior internal periphery is finely crenulated due to the impress of the costellae.

Comparison: Imbrie (1959) described three species of *Schizophoria* from the Traverse Group of Michigan: *S. ferronensis* (364, pl. 48, figs. 1-7), *S. traversensis* (364, pl. 48, figs. 8-14), and *S. mescarina* (365, pl. 48, figs. 15-21). The ventral muscle field of *S. ferronensis* (also illustrated in Kesling and Chilman, 1975: pl. 24, figs. 12-14; pl. 64, figs. 1-5) is more lobate and shorter than that of *S.* cf. *multistriata.* In addition, *S.* cf. *multistriata* is subquadrate in outline, whereas *S. ferronensis* is smaller and subelliptical in outline. *Schizophoria traversensis* possesses a more pyriform ventral muscle field and is subelliptical in outline and *S. mescarina* may be distinguished from *S.* cf. *multistriata* by its elongate ventral muscle field.

Fragmentary specimens of *Schizophoria?* sp. ventral valves were described briefly by Boucot, Gauri, and Southard (1970: 91 , pl. 2, fig. 2a-b) from the Coeymans Limestone (USNM Loc. 11259) of the Green Pond Outlier, New York, which differ from *S.* cf. *multistriata* in their elongate muscle fields with subparallel lateral margins.

Johnson's (1970: pl. 8, figs. 1-12) *S. parafragilis* from the McMonnigal Limestone of the Great Basin may be differentiated from *S.* cf. *multistriata* mainly by its ventral muscle field, which is less bilobate anteriorly, more elongate, and subparallel to midline at the lateral margins.

Schizophoria cf. *paraprima* from the Lower Lochkovian and Pragian beds (sections 1, 2, 4) of Royal Creek, Yukon, described by Lenz (1977: 62, pl. 4, figs. 11-12, 15-37, 40, 42-44) show close affinities to Johnson's (1970) *S. parafragilis* of Nevada, but Lenz believes that the Yukon specimens are more closely related to *S. paraprima* (Johnson, Boucot, and Murphy, 1973). *Schizophoria* cf. *paraprima* differs from *S.* cf. *multistriata* in its narrower, more elongate ventral muscle field and ovate outline.

Community occurrence: Feldman (1980) found this species in the *Atrypa-Coelospira-Nucleospira, Atrypa,* and *Atrypa-Megakozlowskiella* communities.

Figured specimens: AMNH 39880a, 39880b, 42738a, 42739, 41740.

Order PENTAMERIDA Schuchert and Cooper, 1931
Suborder PENTAMERIDINA Schuchert and Cooper, 1931
Superfamily CLORINDOIDEA Rzhonsnitskaia, 1956
Family CLORINDIDAE, Rzhonsnitskaia, 1956
Subfamily PENTAMERELLINAE Sapelnikov, 1973
Genus *PENTAMERELLA* Hall, 1867
Type species: *Atrypa arata* Conrad, 1841: 55.

Pentamerella arata (Conrad, 1841), Figures 11A, B, D, 12A-D, 13.

Atrypa arata Conrad, 1841: 55.

Pentamerella arata Hall, 1867: 375, pl. 58, figs. 1-12; Kindle, 1901: pl. 615; Grabau, 1906: 184, fig. 100; Amsden, 1964: 233, pl. 40, figs. 9-14, text-fig. 5; Boucot and Johnson, 1968: B8, pl. 1, figs. 28-36.

Exterior: The shells are subglobose and broadly pyriform in outline with a strongly convex pedicle valve and a weakly convex brachial valve. The hinge line is short, curved, and narrow with no interarea evident. The maximum width is attained at or about midlength. The pedicle valve beak is short, strong, incurved, and appears not to be closely pressed against the brachial beak. The brachial beak is small, not as erect, and less incurved. A weak sulcus, becoming more distinct as the anterior commissure is approached, is present on the anterior half of the pedicle valve. A

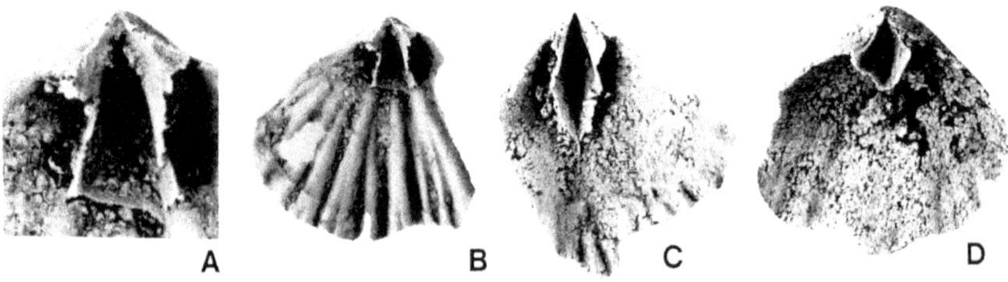

FIGURE 11. *Pentamerella arata* (Conrad, 1841). A, B. AMNH Loc. 3137. Pedicle valve interior, AMNH 39877, x4, x1.5. C. AMNH Loc. 3141. Pedicle valve interior, AMNH 39879, x3. D. AMNH Loc. 3137. Pedicle valve interior, AMNH 39878, x3. *Gypidula* sp.

corresponding fold on the single brachial valve in the collection is barely discernible. The anterior commissure is probably slightly sulcate (uniplicate?) and crenulate, although no specimens preserved this feature. Both valves are ornamented with numerous, rounded, bifurcating plications, which become narrower on the lateral slopes than near midline. The interspaces between plications are U-shaped and generally wider than the plications, which tend to become slightly V-shaped in cross-section on some specimens. Approximately five plications are located in the sulcus and six on the fold. The plications extend back to the beak in most instances, on both valves. Concentric growth lines, concentrated toward the anterior of the shells, are present on many of the pedicle exteriors studied.

Pedicle valve interior: The hinge teeth are small and V-shaped with the apex posteriorly directed. They attach directly to the lateral and anterior margins of the triangular delthyrium. A strong, elongate, deeply excavated spondylium is supported by a median septum such that the anterior nine-tenths of the spondylium projects out over the valve floor unsupported. The median septum is thin and bladelike and is confined to the posterior portion of the valve. In some specimens, the median septum extends past midlength. The valve floor is crenulated anteriorly due to the impress of the plications. No muscle scars were observed.

Brachial valve interior: The sockets are slight indentations, posterolaterally directed and well-worn. No cardinal process is present. A well-formed, triangular, spoon-shaped cruralium, shaped by the medial fusion of the outer plates, extends anteriorly and unites with the valve floor dorsally. The cruralium is the site for the attachment of the adductors. The notothyrial cavity is shallow and empty. The valve floor is strongly crenulated by the impress of the plications.

Discussion: Hall (1867: 375, pl. 58, figs. 1-21) describes *Pentamerella arata* from the Schoharie Grit and limestones of the Upper Helderberg Group in Albany and Schoharie counties, New York. Hall mentions additional occurrences without providing specific localities: Cherry Valley, Waterville, and Babcock's Hill in Oneida County; Lima in Ontario County; Caledonia in Livingston County; Leroy and Stafford in Genesee County; Clarence Hollow and 5 mi east of Buffalo in Erie County; Canada West and at the Falls of the Ohio. In Hall's (1867) illustrated specimens, only Figures 1-12 (pl. 58) resemble the Onondagan forms of *P. arata*. *Pentamerella* cf. *arata,* described from the Bois Blanc Formation in New York by Boucot and Johnson (1968: B8, pl. 1,

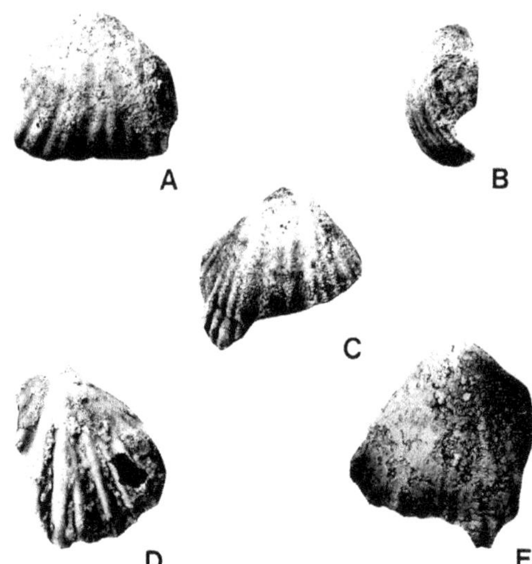

FIGURE 12. *Pentamerella arata* (Conrad, 1841). A, B. AMNH Loc. 3137. Pedicle valve exterior and lateral view, AMNH 42770, x1.5. C. AMNH Loc. 3135. Pedicle valve exterior, AMNH 42771, x1.5. D. AMNH Loc. 3137. Pedicle valve exterior, AMNH 39877, x1.5. *Gypidula* sp. E. AMNH Loc. 3141. Pedicle valve exterior, AMNH 39879, x3.

figs. 28-36, Boucot and Johnson Locality number 4672-SD), is identical with *P. arata* from the Onondaga except for a shallower and smaller spondylium.

Community occurrence: Feldman (1980) found this species in the *Leptaena-Megakozlowskiella, Atrypa-Coelospira-Nucleospira, Atrypa, Atrypa-Megakozlowskiella, Levenea* Community I, *"Pacificocoelia,"* and *Hallinetes* communities.

Figured specimens: AMNH 39877, 39878, 42770, 42771, 42772, 42773, 42774.

Genus *GYPIDULA* Hall, 1867

Gypidula sp., Figures 11C, 12E.

Remarks: Only two fragmentary brachial valves are available for study. They both possess brachial plates which are lyre-shaped in cross section, a spoon-shaped cruralium which unites with the valve dorsally, and a weak dorsal fold. Not enough material is at hand for a specific identification. Bowen (1967: 24, pl. 1, figs. 34-38; pl. 2, figs. 1-4) describes *Gypidula prognostica* from the Keyser Limestone (Bowen Localities 1, 3), which is similar to the Onondagan forms, but smaller. Boucot, Gauri, and Southard (1970: 10, pl. 4, figs. 2-6) report *G. coeymanensis* from the New Scotland and Coeymans formations (USNM Locs. 11270, 11259) of the Green Pond Outlier in southeastern New York. Their specimens approach the material recovered from the Onondaga in size and morphology, but at this point I am unable to determine any affinity due to partial silicification and poor preservation. Anderson and Makurath (1973) also report *G. coeymanensis* from the Coeymans Formation at Sharon Springs, New York. Their specimens, although not described, are illustrated and closely resemble *Gypidula* sp. described above.

Community occurrence: Feldman (1980) found this species in the *Leptaena-Megakozlowskiella, Atrypa,* and *Atrypa-Megakozlowsheila* communities.

Figured specimen: AMNH 39879.

Order STROPHOMENIDA Öpik, 1934
Superfamily STROPHOMENOIDEA King, 1846
Family RAFINESQUINIDAE Schuchert, 1893
Subfamily LEPTAENINAE Hall and Clarke, 1894
Genus *LEPTAENA* Dalman, 1828

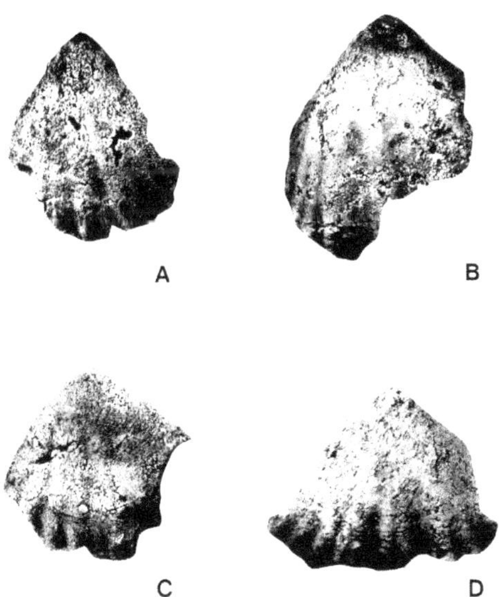

FIGURE 13. *Pentamerella arata* (Conrad, 1841). A. AMNH Loc. 3137. Pedicle valve exterior, AMNH 42772, x3. B. AMNH Loc. 3135. Pedicle valve exterior, AMNH 42773, x3. C. AMNH Loc. 3137. Pedicle valve exterior, AMNH 42774, x3. D. AMNH Loc. 3137. Pedicle valve exterior, AMNH 39878, x3.

Type Species: *Leptaena rugosa* Dalman, 1828: pl. 1, fig. 1.

Leptaena aff. *"rhomboidalis"* (Wilckens, 1769), Figures 14-16.

Leptaena rugosa Dalman, 1828: 93.

Leptaena aff. *" rhomboidalis"* Boucot, 1973: 20, pl. 6, figs. 12-16.

Exterior: The shells are medium to large and transversely subquadrate in outline, with the anterior margin often parallel to the hinge line. They range from concavoconvex to slightly biconvex with the pedicle valve strongly geniculate at the anterior and lateral commissure. The brachial valve is correspondingly geniculate within the pedicle trail. The hinge line is straight and is the place of maximum width. The pedicle interarea is linear, flat, and apsacline. No delthyrium was preserved, although there are indications of a delthyrial margin on one silicified specimen. The brachial interarea is linear and anacline.

FIGURE 14. *Leptaena* aff. *"rhomboidalis"* (Wilckens, 1769). AMNH Loc. 3128C. Impression of pedicle valve, AMNH 42758, x1.

Ornamentation consists of radial costellae (about 12 in a distance of 5 mm at midline, near the anterior margin and posterior to the point of geniculation) which extend past the point of geniculation and continue on the trail of the valves. Concentric rugae cross the costellae (six to 10 per valve) becoming larger and more pronounced anteriorly.

Pedicle valve interior: The hinge teeth were partially preserved on few

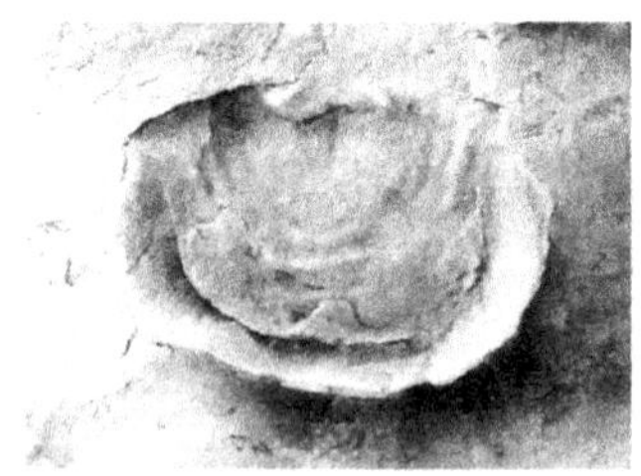

FIGURE 15. *Leptaena* aff. *"rhomboidalis"* (Wilckens, 1769). AMNH Loc. 3128C. Impression of pedicle valve, AMNH 39881, x1.

FIGURE 16. *Leptaena* aff. *"rhomboidalis"* (Wilckens, 1769). AMNH Loc. 3124A. Impression of pedicle valve interior, AMNH 39886, x1.5.

specimens. The teeth appear to be widely spaced, blunt, and stout, and anterolaterally directed. The muscle field is oval and bisected by a low, broad myophragm which terminates at the anterior margin of the muscle field. The myophragm bears a low, longitudinal ridge which ends just past the end of the myophragm, at about midlength. The myophragm and ridge are the sites for attachment of the adductors while the diductors are attached to the oval area bounded by a slightly raised, arcuate rim. The valve floor is crenulated due to the impress of the concentric rugae. Also, the impressions of the costellae are visible on thin-shelled specimens.

Brachial valve interior: The only brachial interior in the collection is lacking the cardinalia and muscle impressions. The interior is crenulated in an identical manner with that of the pedicle interior described above.

Comparison: The Onondagan specimens are similar to the ubiquitous *Leptaena "rhomboidalis"* in general morphology; however, material recovered is lacking in critical morphologic details, such that a definite specific assignment cannot be made at present. Bowen's (1967: 32, pl. 4, figs. 3-5) species from the Keyser Limestone of Pennsylvania and Maryland are also subquadrate in outline and about the same size. Rugae in both Onondaga and Keyser specimens weaken posteriorly. Specimens of *Leptaena* sp. from the Bois Blanc Formation described by Boucot and Johnson (1968: B8, pl. 2, figs. 1-6) are poorly preserved and serve only to record the presence of the genus in strata directly underlying the Onondaga in western New York (Boucot and Johnson Locality 4672-SD). Boucot's (1959: 752, pl. 96, figs. 1-2) fragmentary "wrinkled" forms of *L. "rhomboidalis"* from the Esopus Formation at Highland Mills, New York, appear to be closely related to the Onondagan forms but are too poorly preserved to make further significant comparison.

Community occurrence: Feldman (1980) found this species in the *Leptaena-Megakozlowskiella, Atrypa-Coelospira-Nucleospira, Atrypa, Atrypa-Megakozlowskiella, Levenea* Community I, *Pacificocoelia,* and *Hallinetes* communities.

Figured specimens: AMNH 39881, 39886, 42758.

Order ORTHOTETIDA Waagen, 1884
Suborder ORTHOTETIDINA Waagen, 1884
Superfamily ORTHOTETOIDEA Waagen 1884
Family SCHUCHERTELLIDAE Williams, 1953
Subfamily SCHUCHERTELLINAE Williams, 1953

Genus *SCHUCHERTELLA* Girty, 1904

"Schuchertella" sp., Figure 17.

Description: The taxon is represented in the collection by a single, fairly well-preserved ventral valve, damaged on one lateral extremity, along with miscellaneous shell fragments. The valve is medium-sized, transversely subelliptical in outline but subpyramidal in the umbonal region. The anterior of the valve is almost flat. The shell exterior is multicostellate; there are nine costellae in a space of 5 mm, measured along the anterior commissure at midline. Concentric growth lamellae are present originating at a distance of 10 mm from the beak, along midline. The valve is twisted apically. The interarea is apsacline, flat, and broadly triangular when viewed posteriorly. The delthyrium is covered by a convex pseudodeltidium. The hinge teeth are small, pointed, dorsally directed, and not supported by dental plates but act as extensions of longitudinal ridges lining the inner margins of the delthyrial cavity. The muscle scars are faintly impressed upon the floor of the valve and are defined by a circular field unequally divided by three slight ridges (probably sites of adductor attachment). There is no real median septum.

Remarks: *"Schuchertella"* sp. has coarser costae and a broader interarea than *"Schuchertella"* sp. A of Boucot and Johnson (1968, p. B9, pl. 2, figs. 17-30) from the Bois Blanc Formation in western New York (Boucot and Johnson Localities 4671-SD, 4672-SD). Boucot and Johnson (1968, p. B9, pl. 2, figs. 31-36) also describe *"Schuchertella"* sp. B from the Bois Blanc Formation (Boucot and Johnson Locality 4671-SD), which closely resembles *"Schuchertella"* sp. in that the costae are coarser and the growth lines are similarly spaced. However, since the Bois Blanc specimen is a dorsal valve, the internal morphologies are not readily comparable.

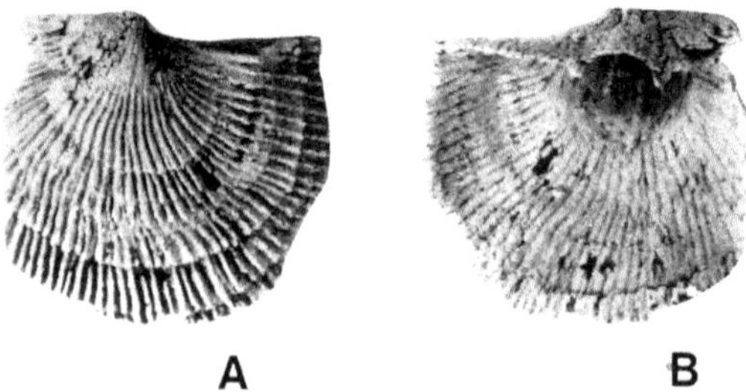

FIGURE 17. *"Schuchertella"* sp. A, B. AMNH Loc. 3137. Pedicle valve exterior, pedicle valve interior, AMNH 39882, x1.5.

Boucot, Gauri, and Southard (1970, p. 23, pl. 8, fig. 5a-c; pl. 8, figs. 6-8) describe *"Schuchertella?"* sp. A and sp. B from the Green Pond Outlier in southeastern New York. *"Schuchertella?"* sp. A (USNM Loc. 11259) is represented only by dorsal valves and is similar to *"Schuchertella"* sp. from the Onondaga in its relatively coarse costae. *"Schuchertella?"* sp. B (USNM Loc. 11245) more closely resembles *"Schuchertella"* sp. internally (see fig. 6b, rubber replica of ventral internal mold) in that the dental plates in the ventral interior are absent, the hinge teeth are small, and the interarea appears to be relatively broad.

Boucot (1973: 24, pl. 9, figs. 1-11) describes *"S." becraftensis* from the Moose River Synclinorium but cannot assign a definite generic name due to the lack of modern consideration of taxonomy and morphology of Silurian-Devonian orthotetaceans. The Maine shells are similar to *"Schuchertella"* sp. in costation and general internal morphology (compare pl. 9, fig. 11).

Community occurrence: Feldman (1980) found this species in the *Atrypa-Coelospira-Nucleospira* Community.

Figured specimen: AMNH 39882.

Orthotetacid indet., Figures 18-19.

Remarks: Several poorly preserved impressions of orthotetacid fragments are in the collection and many more have been observed in the field. The muscle fields appear to be small, and limited to the umbonal region. One cardinal process observed seemed to be an extension of a median partition. There is a

FIGURE 18. Orthotetacid indet. AMNH Loc. 3125A. Impression of pedicle valve interior, AMNH 39884, x2.

FIGURE 19. Orthotetacid indet. AMNH Loc. 3122. Impression of pedicle valve interior, AMNH 39885, x2.

FIGURE 20. *Costistrophonella* cf. *punctulifera* (Conrad, 1838). AMNH Loc. 3131C. Impression of pedicle valve with camerate crinoid scar AMNH 39883, x1.

well-developed hinge line but no preserved interareas. Ornamentation is finely costellate. Generic identification is not possible with the material at hand.

Community occurrence: Feldman (1980) found this species in the *Leptaena-Megakozlowskiella, Atrypa-Coelospira-Nucleospira, Atrypa, Atrypa-Megakozlowskiella, Pacificocoelia,* and *Hallinetes* communities.

Figured specimens: AMNH 39884, 39885.

Superfamily STROPHOMENOIDEA King, 1846
Family STROPHONELLIDAE Caster, 1939
Genus *COSTISTROPHONELLA* Harper and Boucot, 1978
Type species: *Strophomena semifasciata* Hall, 1863: 210.

Costistrophonella cf. *punctulifera* (Conrad, 1838), Figure 20.

Strophomena punctulifera Conrad, 1838: 117.

Strophodonta punctulifera Hall, 1859: 188, pl. 21, fig. 4; pl. 23, figs. 4, 5, 7e.

Strophonella cf. *punctulifera* Boucot, 1973: 23, pl. 8, figs. 14-18.

Strophonella punctulifera Hall and Clarke, 1892: 291, pl. 12, figs. 10-12; Schuchert et al., 1913: 323, pl. 59, figs. 8-10; Cooper, 1944: 339, pl. 130, figs. 19-20; Johnson, 1970: 113, pl. 20, figs. 1-6; pl. 21, figs. 13-14.

Costistrophonella punctulifera Harper and Boucot, 1978: 99, pl. 17, figs. 5-9; pl. 18, figs. 1-6.

Description: This species is represented in the collection by a single, poorly preserved pedicle valve exterior which is relatively large, wider than it is long, and subsemicircular in outline. Due to compression the valve is completely flattened. The hinge line is long and straight; no interarea is discernible. The ornamentation consists of numerous, subangular, radiating costae, which increase anteriorly both by bifurcation and intercalation. The interspaces are quite broad and shallow. There are 11 to 12 costae per 5 mm near the anterior commissure at midline. The costae are crossed by about seven fairly evenly-spaced concentric growth lines.

Remarks: *Costistrophonella* was erected as a new genus by Harper and Boucot (1978; see p. 99 for diagnosis). They report the geographic distribution of the genus in New York as follows. *Costistrophonella* cf. *punctulifera* occurs in Lower Helderberg rocks of Gedinnian age; *Costistrophonella* sp. has been reported from the New Scotland Formation, of Gedinnian age; and *C. ampla* has been found in the Schoharie Grit, of Emsian age. Until now *C. punctulifera* has not been found in rocks of Onondagan age in New York.

Community occurrence: Feldman (1980) found this species in the *Leptaena-Megakozlowskiella* and *Atrypa* communities.

Figured specimen: AMNH 39883.

Family STROPHEODONTIDAE Caster, 1939

Genus *MEGASTROPHIA* Caster, 1939

Subgenus *MEGASTROPHIA (MEGASTROPHIA)* Caster, 1939

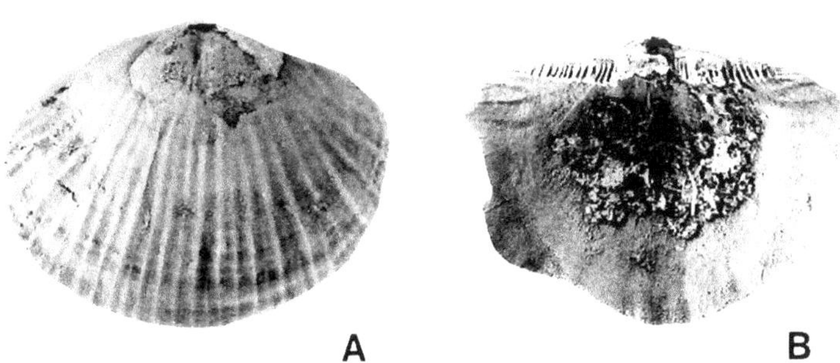

FIGURE 21. *Megastrophia* sp. A. AMNH Loc. 3135. View of exfoliated pedicle valve exterior, AMNH 39887, x1.5. B. AMNH Loc. 3138A. Pedicle valve interior, AMNH 39891, x2.5.

FIGURE 22. *Megastrophia* sp. AMNH Loc. 3123A. Pedicle valve exterior showing detail of ornament, AMNH 39888, x2.

Megastrophia sp., Figures 21, 22.

Description: The shells are medium to large, subsemicircular to transversely suboval in outline, somewhat alate, and concavoconvex in lateral profile. Maximum width is attained at the hinge line. No free brachial valves were recovered and the only articulated specimen is poorly preserved. The pedicle interarea is weakly apsacline and denticulate along its entire length. The delthyrium appears to be partially covered by a pseudodeltidium, but that general region is also not well preserved. The dorsal interarea is chipped off from the only articulated specimen in the collection. Its ventral umbo is exfoliated, exposing some of the muscle field, which appears to be identical with that of one other well-preserved (silicified) pedicle valve—subpyriform and bisected by a low myophragm which extends to about midlength. Suboval adductor impressions are located just ventral and anterior to the ventral process, which is bilobed and pustulose, and lie on either side of the posterior portion of the myophragm, which at that point is bladelike. Anteriorly, however, the myophragm becomes broader and flatter. The diductors were not impressed clearly, but there are vague indications of a broadly pyriform scar anterolateral to the adductor scars. On only one specimen is there a trace of the dorsal muscle field, and that is extensively exfoliated. However, there is a large, almost flabellate region in the center of which lies a narrow, suboval, longitudinally grooved scar. The ornament is finely parvicostellate, commonly with one or two secondary costellae between the primaries. Occasional specimens have as many as four secondary costellae associated with one larger, secondary costella.

Comparison: *Megastrophia* sp. is similar to *M. concava* (Hall) in its alate appearance and ornamentation. It is not as convex as *M. hemisphaerica* (Hall) and not as uniformly costellate. Harper and Boucot (1978: 20, pl. 39, fig. 16; pl. 40, figs. 1-4) report on the occurrence of *Megastrophia* (*Megastrophia*) Caster, 1939 from the Kanouse Sandstone (see Boucot, 1959: 752), the Schoharie Grit (see Hall, 1867), the Onondaga Limestone (see Hall, 1867: 91-92), and the Moscow Shale of the Hamilton Group (see Hall, 1867: 91).

Community occurrence: Feldman (1980) found this species in the *Leptaena-Megakozlowskiella, Atrypa-Coelospira-Nucleospira, Atrypa, Atrypa-Megakozlowskiella, Pacificocoelia,* and *Hallinetes* communities.

Figured specimens: AMNH 39887, 39888.

Genus *STROPHODONTA* Hall, 1850

Type species: *Strophomena demissa* Conrad, 1842: 258.

Strophodonta cf. *demissa* (Conrad, 1842), Figure 23.

Stropheodonta cf. *demissa* Boucot and Johnson, 1968: 89, pl. 1, figs. 7-16; Boucot, 1973: 21, pl. 6, figs. 17-19.

Remarks: Two moderately well-preserved, articulated specimens are available for study (Table 2) along with several fragmentary pieces which are so poorly preserved that it is impossible to determine whether they represent pedicle or brachial valves. One fragment bears what appears to be a well-worn, bilobed cardinal process. The shells are subcircular to shield-shaped in outline and concavoconvex in lateral profile. Both shells are wider

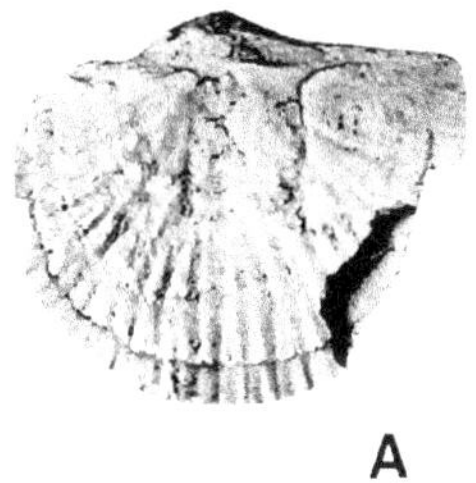

FIGURE 23. *Strophodonta* cf. *demissa* (Conrad, 1842). A, B. AMNH Loc. 3137. Dorsal and ventral views, AMNH 39890, x2. C. AMNH Loc. 3135. Ventral view, AMNH 42767, x2.

TABLE 2. Measurements (in mm) of Articulated Specimens of *Strophodonta* cf. *demissa* (Conrad, 1842).

AMNH Locality	(L)	(W)	(T)	No. of Costellae per 5 mm at Anterior Commisure
3137	13.0	15.5	3.5	8
3135	6.4	7.0 est.	1.5	5

than long. The point of maximum width is at the hinge line. The lateral margins are almost straight posteriorly, whereas the anterior margins are evenly rounded. All margins are crenulate and the anterior commissure is rectimarginate. The smaller specimen (neanic) bears coarser costellae, angular in cross section. On both shells the costellae are bifurcating with narrow interspaces, also angular in cross section.

Comparison: Specimens of *S. demissa* from the Kanouse Sandstone described by Boucot (1959: 751, pl. 95, figs. 1-4) are almost identical in outline. Costation in the Onondaga forms is generally coarser (especially AMNH 42767, fig. 18c). Boucot and Johnson's (1968: B9, pl. 2, figs. 7-16) *S.* cf. *demissa* from the Bois Blanc Formation (Boucot and Johnson Loc. 4672-SD) are almost identical in size, but due to poor preservation further comparison must be deferred.

Hall (1867) reports *S. demissa* from the Schoharie Grit, Onondaga Limestone, and Hamilton Group of New York. Hall also reports a continuation of the species in the Chemung Group where, found in arenaceous beds, it strongly resembles Schoharie specimens. *Stropheodonta demissa* from the Onondaga Limestone is reported by Hall (1867: 103) to be smaller than average for the species but are thick and robust with a high degree of convexity.

Community occurrence: Feldman (1980) found this species in the *Leptaena-Megakozlowskiella, Atrypa-Coelospira-Nucleospira,* and *Atrypa-Megakozlowskiella* communities.

Figured specimens: AMNH 39890, 42767.

Stropheodontid indet. Figure 24.

Remarks: This is represented by six fragmentary, silicified specimens which fit the general description of stropheodontids but are so poorly preserved that generic assignment is impossible. The shells are subcircular in outline, alate, and have a smooth exterior with irregularly spaced growth lines. In these respects they are similar to *Lissostrophia* (Harper

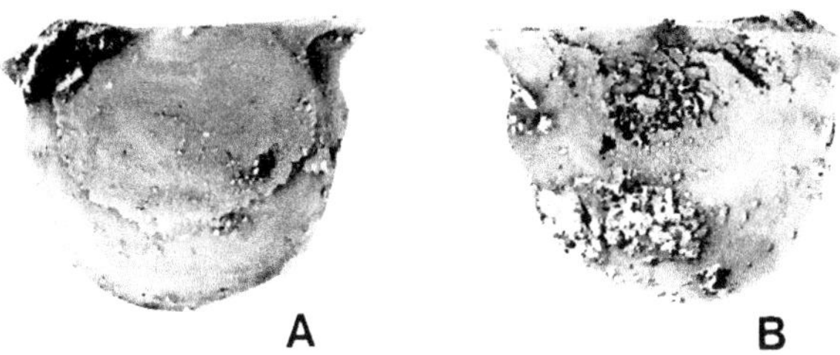

FIGURE 24. Stropheodontid indet. A, B. AMNH Loc. 3137. Pedicle valve exterior and pedicle valve interior, AMNH 39892, x3.5.

FIGURE 25. *Hallinetes* aff. *lineatus* (Conrad, 1839). A. AMNH Loc. 3138A. Ventral view, AMNH 42745, x2. B. AMNH Loc. 3129. Ventral view, AMNH 42744, x3.

FIGURE 26. *Hallinetes* aff. *lineatus* (Conrad, 1839). AMNH Loc. 3129. Impression of pedicle valve, AMNH 42755, x4.

and Boucot, 1978: 39, pl. 36, figs. 10-11, 14-16) but differ in their flatter umbo which does not project posteriorly to overhang the hinge line. The Onondaga shells are also similar to *Pholidostrophia (Mesopholidostrophia)* (Harper and Boucot, 1978: 41-42, pl. 50, figs. 13a, 23a) and *Pholidostrophia*

(Pholidostrophia) (Harper and Boucot, 1978: 42, pl. 50, fig. 7) in their smooth exteriors but differ in their smaller size. *Teichostrophia* (Harper and Boucot, 1978: 43-44, pl. 49, figs. 23, 25, 29) also has a smooth exterior but is subrectangular in outline.

Community occurrence: Feldman (1980) found this species in the *Leptaena-Megakozlowskiella* and *Atrypa-Nucleospira-Coelospira* communities.

Figured specimen: AMNH 39892.

Order PRODUCTIDA Sarytcheva and Sokolskaya, 1959
Suborder CHONETIDINA Muir-Wood, 1955
Superfamily CHONETOIDEA Bronn, 1862
Family CHONETIDAE Bronn, 1862
Subfamily DEVONOCHONETINAE, Muir-Wood, 1962
Genus *HALLINETES* Racheboeuf and Feldman, 1990

Chonetes aff. *lineatus* (Conrad, 1839), Figures 25-28.

Strophomena lineata Conrad, 1839: 64; Vanuxem, 1842: 139, fig. 5a; Hall, 1843: 175, fig. 8.

FIGURE 27. *Hallinetes* aff. *lineatus* (Conrad, 1839). AMNH Loc. 3129. View of poorly preserved slab of fine-grained, "muddy" limestone of the *"Hallinetes"* Zone (see Feldman, 1980: 29). Exfoliation and extensive weathering preclude positive generic identification. Note fairly well-preserved pedicle valve impression with median septum (arrow) and papillose internal surface, AMNH 42768, x2.5.

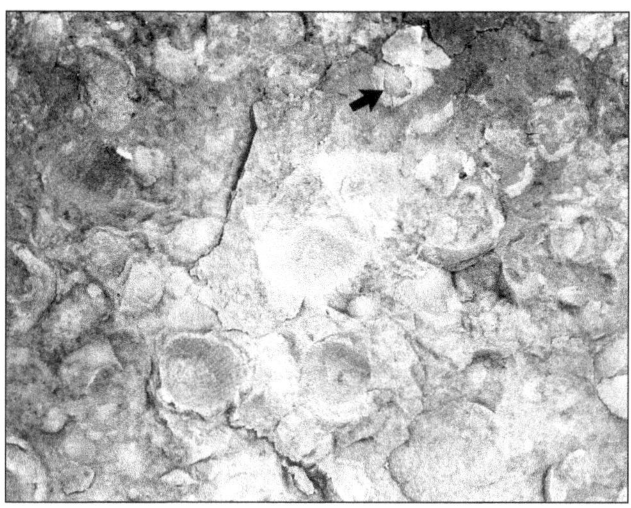

FIGURE 28. *Hallinetes* aff. *lineatus* (Conrad, 1839). View of same slab of poorly preserved limestone as illustrated in figure 24. In both figures note absence of identifiable brachial valve impressions and presence of exfoliated surfaces (arrow). The limestone at this locality contains numerous stylolites, AMNH 42768, x2.5.

Chonetes glabra Hall, 1857: 117, figs. 1-8.

Chonetes lineata Hall, 1867: 121, pl. 20, fig. 3; Hall and Clarke, 1892: pl. 16, fig. 34.

Remarks: Since the material at hand is poorly preserved and worn, and the few brachial interiors that were recovered were unsuitable for identification at the generic level, the specimens are questionably assigned to the genus *Hallinetes.*

Exterior: The shells are small, subsemicircular in outline, and concavo-convex in lateral profile. The interareas are very narrow. No delthyrial structures were preserved. Greatest width is attained at the hinge line or anterior to midlength. Spines were not observed. The valves are covered with fine capillae which increase anteriorly by bifurcation.

Pedicle valve interior: The pedicle valve interior bears a strong median septum often extending two-thirds the valve length, broadening at midlength, and tapering off anteriorly. The internal surface is papillose and impressed anteriorly by the external ornament.

Brachial valve interior: The only cardinal process observed was a faint longitudinal ridge at the posterior end of the valve. No alveolus was present, but one faint lateral groove (anderidium?) was evident. No other structures were preserved.

Comparison: *Hallinetes* aff. *lineatus* is smaller than *Chonetes scituta*, which has a less convex pedicle valve and is more finely capillate (AMNH 5164/1 from the Hamilton of Cumberland, Maryland). Cooper (1944: pl. 134, fig. 11) also illustrates a specimen of *C. scitula* but does not give the exact location. In all other respects the two species appear to be identical.

Grabau (1906) illustrates three species of *Hallinetes:* (1) *Chonetes coronatus* is larger than *Hallinetes* aff. *lineatus;* (2) *C. mucronatus* has coarser plications; and (3) *C. deflectus* resembles *Hallinetes* aff. *lineatus* in its striae and convexity.

Boucot (1959: 757, pl. 98, figs. 1-5) describes Chonetes sp. from the Esopus Formation at Highland Mills, New York, which differs from *Hallinetes* aff. *lineatus* in its larger size, more transverse outline (width is one and one-half times length), and straighter hinge line.

Discussion: Hall (1867) describes *Hallinetes lineatus* from the Onondaga (Corniferous) Limestone at Oneida Falls, between Jamesville and Manlius in Onondaga County and in the northern part of Seneca County in New York. Hall reports that the abundance of *C. lineata* decreases as one progresses east and west of the Jamesville-Oneida Falls area (that is, Onondaga County). The last easterly occurrence of the species (in the *Hallinetes* Zone of the Seneca Member) observed in the field is in Cherry Valley, New York, along route 20 (L. V. Rickard, personal commun.).

Community occurrence: Feldman (1980) found this species in the *Atrypa-Megakozlowskiella* and *Hallinetes* communities.

Figured specimens: AMNH 42744, 42745, 42755, 42768.

Order RHYNCHONELLIDA Kuhn, 1949

Superfamily STENOSCISMATACEA Oehlert, 1887 (1883)

Family STENOSCISMATIDAE Oehlert, 1887 (1883)

TABLE 3. Measurements (in mm) of *Atribonium halli* (Fagerstrom, 1961).

AMNH Locality	(L)	(W)	(T)	Number of Plications		Length of b.v.
				Sulcus	Fold	
3137	9.0	9.4	6.7	5	4	8.2
3135 (crushed)	9.2	10.6	–	5 est.	4 est.	8.1

FIGURE 29. *Atribonium halli* (Fagerstrom, 1961). AMNH Loc. 3137. Dorsal, ventral, anterior, posterior and lateral views, AMNH 39893, x3.

Subfamily STENOSCISMATINAE Oehlert, 1887 (1883)
Genus *ATRIBONIUM* Grant, 1965
Type species: *Stenoscisma halli* Fagerstrom, 1961: 29, pl. 9, figs. 48-51.

Atribonium halli (Fagerstrom, 1961), Figure 29.

Stenoscisma halli Fagerstrom, 1961: 29, pl. 9, figs. 48-51.

Stenoscisma rhomboidalis (Hall and Clarke) Fagerstrom, 1961: 29, pl. 9, figs. 45-47.

Atribonium halli (Fagerstrom) Grant, 1965: 52.

Remarks: The following description is based upon only two specimens, one of which was crushed. Although the complete specimen is silicified and well preserved, no adequate description of the internal morphology can be given here since the valves are articulated with no gape angle.

The shell is small (Table 3), rostrate, astrophic, impunctate, and subpentagonal in outline. When viewed in lateral profile, it is gently biconvex with a noticeable flexure in the pedicle valve at the beginning of the sulcus. The beak is short, rounded, and suberect. The commissure is uniplicate with a high-crested brachial fold and deep pedicle sulcus. The costae are weak and rounded, becoming indistinct as the beak is approached, disappearing on the brachial valve at about mid length. On the pedicle valve the costae disappear at about one-third the valve length. Two to three moderately strong costae are found on the flank. No growth lines are evident. Both valves are geniculate and butt against one another in a vertical plane at the anterior commissure, a generic character of the *Stenoscismatacea.* At the lateral and posterior margins, the valves also abut against each other showing no overlap. There are no indications of any incipient frills at the commissure. The beak is curved with short, blunt, beak ridges. There is a small pedicle foramen and triangular delthyrium constricted by what appear to be nearly conjunct deltidial plates.

Comparison: Grant (1965: 52) notes that *Atribonium halli* differs from all other species of the genus in having few (two or three) costae on the fold, and normally the same or a greater number on each flank. He distinguishes *A. halli* from Hall and Clarke's Indiana species (assigned to the genus *Coledium*; 1965: 97) in their smaller size, lower convexity, trigonal rather than ovate outline, weaker costae that begin farther anteriorly, and geniculation of each valve.

Atribonium halli is narrower and more convex than *A. simatum* and the height of its fold is similar to that of *A. cooperorum. Atribonium halli* is larger than *A. kernahani* and thinner than *A. rostratum* and *A. succiduum. Atribonium gregari* is larger, more globose, and lacks a flattened anterior. *Atribonium pauperum* differs in its strong, sharp plications which fold at both valves and the commissure (see Grant, 1965: pl. 3, fig. 3). *Atribonium pingue, A. simatum,* and *A. gregari* are the only known species of *Atribonium* that attain a maximum size significantly larger than the other species.

TABLE 4. Measurements (in mm) of Articulated Specimens of *Cupularostrum* sp. A and B.

AMNH Locality	(L)	(W)	(T)	Number of Plication		Sulcus Begins
				Fold	Sulcus	
Species A						
3131C	5.8	7.1	3.9	4	3	3.0
3135	5.1	5.8	2.8	6	5	2.9
3135	5.6	5..9	3.1	4	3	3.2
3137	6.4	6.5	3.6	5	4	3.1
3138A	4.8	4.2	2.5	4	3	2.5
3138A	4.9	4.7	2.7	4	3	3.0
3138A	4.8	4.5	2.7	5	4	
Species B						
3138A	4.0	3.1	2.2	–	–	–
3138A	3.7	3.4	2.3	–	–	–
3138A	4.0	3.1	2.4	–	–	–
3138A	4.1	3.6	2.1	–	–	–
3138A	4.5	4.4	2.2	–	–	–

Community occurrence: Feldman (1981) found this species in the *Atrypa-Coelospira-Nucleospira* Community.

Figured specimen: AMNH 39893.

Superfamily RHYNCHOTREMATOIDEA Schuchert, 1913
Family TRIGONIRHYNCHIIDAE Schmidt, 1965
Subfamily TRIGONIRHYNCHIINAE Schmidt, 1965
Genus *CUPULAROSTRUM* Sartenaer, 1961

Cupularostrum sp. A, Figure 30A-F

Exterior: The shells are small, equibiconvex, and subtrigonal to transversely suboval in outline. The posterolateral margins are straight, whereas the anterolateral margins are rounded. The pedicle beak is erect to slightly incurved, almost covering the more obscure brachial beak. The delthyrium is open and triangular with a small foramen located apically. No deltidial plates were observed. The point of maximum width is generally anterior to midlength. The pedicle valve bears a sulcus which originates at about midlength. The brachial valve bears a corresponding fold, considerably weaker than the sulcus, which becomes more prominent anteriorly, resulting in a uniplicate anterior commissure. Ornamentation consists of about 15 simple plicae, narrowly U-shaped in cross section. There are three plicae situated in the sulcus and four on the low fold (Table 4). The plicae are separated by U-shaped interspaces. On some specimens a simple, prominent growth line is present at or anterior to midlength.

Pedicle valve interior: Pedicle valve interiors are few and poorly preserved. There are two small, narrow, pointed hinge teeth supported posteriorly by thin, bladelike dental plates which lie close to the walls of the shallow umbonal cavity. The valve floor is crenulated due to the impress of the plications.

Brachial valve interior: The sockets are moderately shallow, widely divergent, and appear to be uncrenulated. The hinge plates are fairly wide and raised above a concave septalium, which is supported by a short, low, median septum and lacking a transverse plate. No cardinal process is present. The valve floor is strongly crenulated due to the impress of the plications.

Comparison: The brachial valves of *Cupularostrum* sp. A from the Onondaga Limestone are transversely oval, moderately deeply convex, and possess four plicae on the low fold. The interiors have a concave septalium

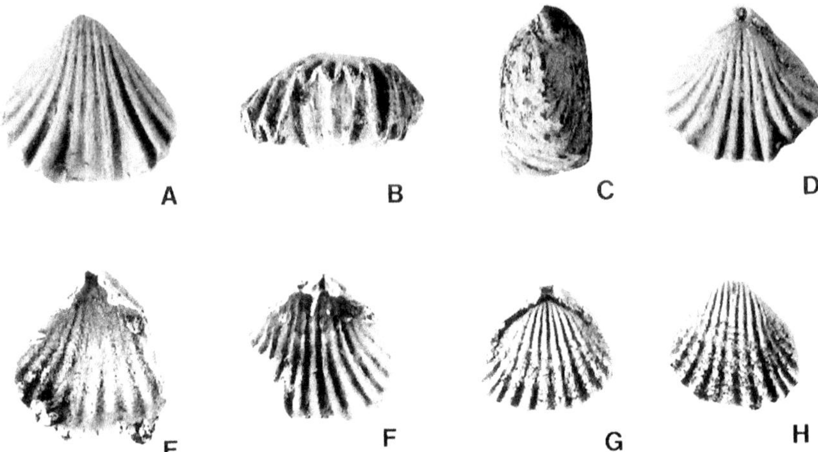

FIGURE 30. *Cupularostrum* sp. A. A-D. AMNH Loc. 3135. Ventral, anterior, lateral and dorsal views, AMNH 39894, x5. E. AMNH Loc. 3138A. Pedicle valve interior, AMNH 39895, x5. F. AMNH Loc. 3138A. Brachial valve interior, AMNH 39896, x5. *Cupularostrum* sp. B. G, H. AMNH Loc. 3138A. Dorsal and ventral views, AMNH 39897, x5.

supported by a low median septum. In these respects, the specimens are identical with the *Cupularostrum?* sp. from the Bois Blanc Formation (Boucot and Johnson, 1968: B11, pl. 3, figs. 21-29, Boucot and Johnson Loc. 4672-SD). Boucot, Gauri, and Southard (1970: 12, pl. 3, figs. 15-17) specimens from the Coeymans Limestone of the Green Pond Outlier assigned to *Cupularostrum?* sp. A are similar to the Onondaga forms, but do not possess a transverse plate extending across the septalium. *Cupularostrum macrocosta* (Boucot, 1973: 29, pl. 12, figs. 3-11) from the Moose River Synclinorium is larger (up to 3 cm in length) and has four to five plicae in the pedicle sulcus. The plications are finer in *C. macrocosta* than in the Onondaga and Coeymans species.

Community occurrence: Feldman (1980) found this species in the *Atrypa-Coelospira-Nucleospira* and *Atrypa* communities.

Figured specimens: AMNH 39894, 39895, 39896.

Cupularostrum sp. B, Figure 30G-H.

Remarks: These specimens are externally identical with those described above except for lack of a sulcus and fold (see Table 4). Specimens of *Cupularostrum* sp. A at the same ontogenetic stage possess a noticeable sulcus and fold. No free valves were recovered so that the internal structures cannot be compared; consequently specific designation must be deferred.

Community occurrence: Feldman (1980) found this species in the *Atrypa-Coelospira-Nucleospira* Community.

Figured specimen: AMNH 39897.

Order ATRYPIDA Rzhonsnitskaia, 1960
Suborder ATRYPIDINA Moore, 1952
Superfamily ATRYPOIDEA Gill, 1871
Family ATRYPIDAE Gill, 18 71
Subfamily ATRYPINAE Gill, 1871
Genus *ATRYPA* Dalman, 1828
Type species: *Anomia reticularis* Linnaeus, 1758: 702.

Atrypa "reticularis" (Linnaeus, 1767), Figure 31.

Anomia reticularis Linnaeus, 1758: 702.

Atrypa "reticularis" Boucot, 1959: 741, pl. 91, figs. 7-9; Boucot and Johnson, 1968: B12, pl. 3, figs. 30-49; Boucot, 1973: 36, pl. 15, figs. 1-6.

Exterior: The shells are medium to large and range from subcircular to elongate oval in outline. They are dorsibiconvex with the anterior end of some specimens becoming almost flat to concave at the margins. Maximum width occurs at about midlength on most individuals but tends to occur more toward the posterior than anterior to midlength. In lateral profile, the great convexity of the brachial valve in many individuals gives a disproportionately inflated appearance. The pedicle beak is suberect; the brachial beak is erect. There is a low, narrow, triangular interarea with a small pedicle foramen on neanic specimens. These features atrophy with age and are absent in the ephebic and gerontic stages. No delthyrium was observed. There is a variable pedicle sulcus and a broad brachial fold resulting in a slightly uniplicate anterior commissure. Young individuals are non-sulcate and rectimarginate.

FIGURE 31. *Atrypa "reticularis"* (Linnaeus, 1767) AMNH Loc. 3137. Dorsal view, AMNH 39889, x1.5.

The exterior is ornamented with well-rounded radial costellae, which increase in size and number anteriorly. The costellae are separated by U-shaped interspaces. On the pedicle valve the costellae appear to increase by bifurcation, while on the brachial valve they seem to increase by intercalation. Concentric growth lamellae cross the costellae, becoming more distinct and frilly anteriorly. Toward the anterior commissure the lamellae become more closely spaced. Posteriorly, however, they become obscured.

Pedicle valve interior: The large, stout hinge teeth are supported by short, narrow dental lamellae. The surface of the hinge teeth of less-worn specimens appear to be transversely corrugated in a manner similar to those of Boucot and Johnson (1968) which were grooved subparallel to the posterolateral margin of the valve and deeply crenulated to accept the median crenulated ridge within the sockets of the brachial valve. The muscle field is pyriform to subcircular, flabellate, and lies anterior to a broad, shallow, subtriangular delthyrial cavity. The adductor attachment site is represented by two raised areas near the base of the dental lamellae. The diductors attach to an anteriorly situated, longitudinally striated region often separated from the valve floor by a thick, raised, arcuate rim.

Brachial valve interior: The sockets are from moderately to widely divergent and are bounded by the valve margin posterolaterally and by curved hinge plates anteromedially. Extending the length of each socket and medially situated, there is a transversely grooved ridge which acommodates the grooved hinge teeth. On the inner margins of the sockets, short, thin crural bases are attached; these project anteroventrally. No cardinal process

FIGURE 32. *Coelospira camilla* Hall, 1867. A-D. AMNH Loc. 3135. Dorsal, ventral, anterior and lateral views, AMNH 39904, x3.5. E. AMNH Loc. 3138A. Pedicle valve interior, AMNH 39905, x3.5. F. AMNH Loc. 3135. Brachial valve interior, AMNH 39906, x3.5.

is present. However, the posterior portion of the small notothyrial cavity is often longitudinally striated for diductor attachment. The adductor scars are sub pyriform to suboval and are divided by either a single, thin, low myophragm or by a pair of myophragms. In both cases the posterior base of the myophragms is considerably thickened. When a single myophragm is present, the posterior base is pyramidal in cross section.

Discussion: The shells described above fit the general description of the genus *Atrypa* (*"reticularis" sensu lato*) in ornamentation, outline, and internal structure. Specimens of similar morphology have been found in the Kanouse Sandstone (Boucot, 1959), Bois Blanc Formation (Boucot and Johnson, 1968), and Tomhegan Formation (Boucot, 1973). Hall (1867) described *A. reticularis* from the Onondaga (Corniferous) Limestone (see Hall, 1867: pl. 51, figs. 11-13) of New York, Canada West, Ohio, Kentucky, Indiana, and Illinois. The occurrences of this species are so numerous that Hall (1867: 321) does not name them in detail but mentions that they are found in nearly every exposure of the Corniferous Limestone from Albany County to Black Rock, on the Niagara River.

Community occurrence: Feldman (1980) found this species in the *Leptaena-Megakozlowskiella, Atrypa-Coelospira-Nucleospira, Atrypa, Atrypa-Megakozlowskiella, Levenea* Community I, *Pacificocoelia,* and *Hallinetes* communities.

Figured specimen: AMNH 39889.

Superfamily ANOPLOTHECOIDEA Schuchert, 1894
Family ANOPLOTHECIDAE Schuchert, 1894
Subfamily COELOSPIRINAE Hall and Clarke, 1895
Genus *COELOSPIRA* Hall, 1863
Type species: *Leptocoelia concava* Hall, 1857: 107.

Coelospira camilla Hall, 1867, Figure 32.

Leptocoelia concava Hall, 1857: 107.

Coelospira camilla Hall, 1867: 329 (as *Coelospira concava*), pl. 52, figs. 13-19; Hall and Clarke, 1895, pl. 53, figs. 24-31; Boucot and Johnson, 1967: 1237, pl. 164, figs. 20-30; pl. 165, figs. 1-15; Boucot and Johnson, 1968: B13, pl. 4, figs. 1-25; Boucot, Gauri, and Southard, 1970: 17-18, pl. 5, figs. 17-19, 21-22.

Exterior: The shells are small, concavoconvex to planoconvex, and subcircular to suboval in outline. The hinge line is rounded in adults but relatively straight in small specimens and diverging at an angle of approximately 65 degrees. There is a small but distinct pedicle foramen equally well developed in juveniles and adults. The pedicle beak is incurved and the pedicle valve strongly convex in most specimens. No interarea is evident. The maximum width (figs. 33-34) is about one-third the valve length, occurring anterior to the pedicle foramen, in adults. In juveniles the maximum width is at midlength. The anterior and anterolateral commissures are well rounded, whereas the posterolateral commissure is straighter. In some specimens the brachial valve has a broad sulcus, which originates at the posterior extremity of the brachial valve and shallows out anteriorly.

The pedicle valve bears two medial plications (Table 5) commonly at least as large as the remaining radial plications on the flanks (numbering 11 on the average). The interspace between the two medial plications, in well-preserved specimens, bears two small ridges beginning at about midlength and extending to the anterior commissure. The interspaces between all plications on the flanks are U-shaped and are of approximately the same width as the plications, although the interspaces become broader than the plications as the lateral commissure is approached. There is some bifurcation on the lateral plications.

The brachial valve bears a medial plication which generally bifurcates at about one-third the valve length. The median interspace is either flat (most common condition) or bears a small median ridge. The plications become broader on the flanks and thinner toward the lateral commissure.

On many specimens, especially those in the ephebic to gerontic stage, several well-defined, concentric growth lines are concentrated toward the anterior commissure.

Pedicle valve interior: The hinge teeth are small, slightly concave and thin, attaching ventrally to obscure dental lamellae. The hinge teeth diverge anterolaterally at an angle of approximately 80 degrees. There are small, distinct crural fossettes on the medial side of each hinge tooth. A low, blunt myophragm, rectangular to rounded in cross section, divides the ventral muscle field and ends almost at midlength. Elliptical to sub oval diductor impressions are on either side of the myophragm. In some specimens the anterior boundary of the diductor impressions is well defined by a difference in elevation on the floor of the valve. In most cases, however, the muscle field grades into the valve floor without any noticeable change in elevation.

TABLE 5. Measurements (in mm) of Articulated Specimens of *Coelospira camilla* Hall, 1867.

3135	6.4	5.4	2.1	12	14
3135	5.1	5.2	1.6	13	14
3138A	3.8	3.8	1.5	–	15
3144A	5.3	5.2	1.4	11	13
3137	7.3	6.7	2.9	17	–
3137	3.5	3.4	1.1	–	–
3135	4.0	4.2	1.2	13	15
3135	7.4	6.7	2.6	10+	13
3137	6.6	6.1	2.4	14	12
3137	6.0	6.4	2.3	13	13
3137	5.0	5.0	1.7	11	15
3137	3.5	2.8	0.9	6	–
3135	2.9	3.2	0.9	10	11
3137	7.0	6.9	2.5	15	–
3135	6.0	6.9	1.5	–	–
3137	4.0	4.0	1.4	15	13
3137	4.4	4.5	1.5	17	14
3137	4.9	4.4	2.4	–	–
3135	6.5	7.0	2.4	17	14
3135	5.2	4.9	2.2	–	14
3135	3.1	3.2	1.0	12	12
3135	5.8	5.6	2.2	–	–
3138A	3.1	3.2	1.1	–	12
3138A	6.7	6.0	3.4	–	–
3138A	3.8	3.3	1.7	12	13
3138A	3.8	3.8	1.1	12	12
3138A	6.9	7.2	2.3	14	14
3138A	4.9	5.0	1.7	11	11
3138A	5.3	5.2	2.0	–	–
3138A	5.8	6.3	2.0	14	–
3138A	6.2	6.9	2.2	–	12
3138A	6.4	6.1	3.0	14	14

No adductor scars were evident. The valve floor at the anterior periphery is crenulated due to the impress of the plications.

Brachial valve interior: The sockets are deeply excavated and almost cylindrical in cross section, broadening slightly and shallowing out anterolaterally. Medial, strongly incurved socket plates form one border of the sockets adjacent to the cardinal process. The distal border is represented by a small ridge along the posterior shell margin. The cardinal process ranges from a simple moundlike protuberance to a quadrilobed form. A short, pointed, anteriorly tapering myophragm extends from the base of the cardinal process anteriorly and ends at about midlength to three-quarters

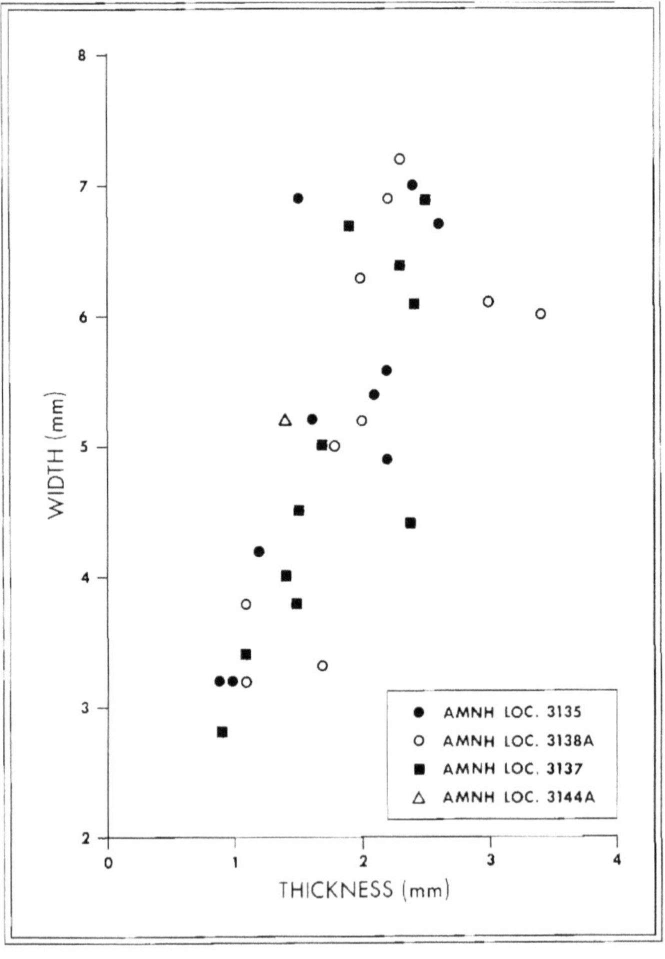

FIGURE 33. Width/thickness scattergram of *Coelospira camilla* Hall, 1867.

the length of the valve. In some specimens the plications are somewhat impressed upon the valve floor.

Comparison: Among the Devonian species of *Coelospira* the following comparisons may be made. These shells (Figure 35) are similar to *Coelospira* sp. of Boucot, Gauri, and Southard (1970), but differ in the degree of convexity of the brachial valve, tending toward the concavoconvex condition rather than planoconvex (less than 1 percent of *C. camilla* are planoconvex).

Coelospira concava of Boucot and Johnson (1967) differs from this species in the greater number of costae (commonly 14 to 17 on a pedicle valve). Also, the degree of concavity of the brachial valve is more pronounced in *C. concava*. *Coelospira camilla* of Boucot and Johnson (1967) differs from these specimens only in the greater prominence of the two medial costae on the pedicle valve. *Coleospira dichotoma* is generally larger, has more radiating costae, and has peripherally obsolescent bifurcating costae in large specimens.

Boucot and Johnson (1967: 1235, pl. 163, figs. 1-7, 10, 15-27) describe and list occurrences of *C. camilla*. The reader is referred to page 1235 for a

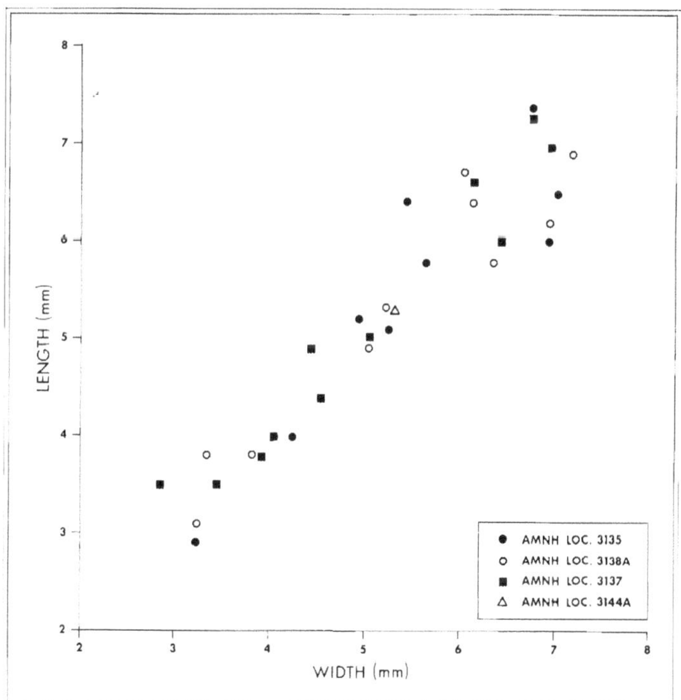

FIGURE 34. Length/width scattergram of *Coelospira camilla* Hall, 1867.

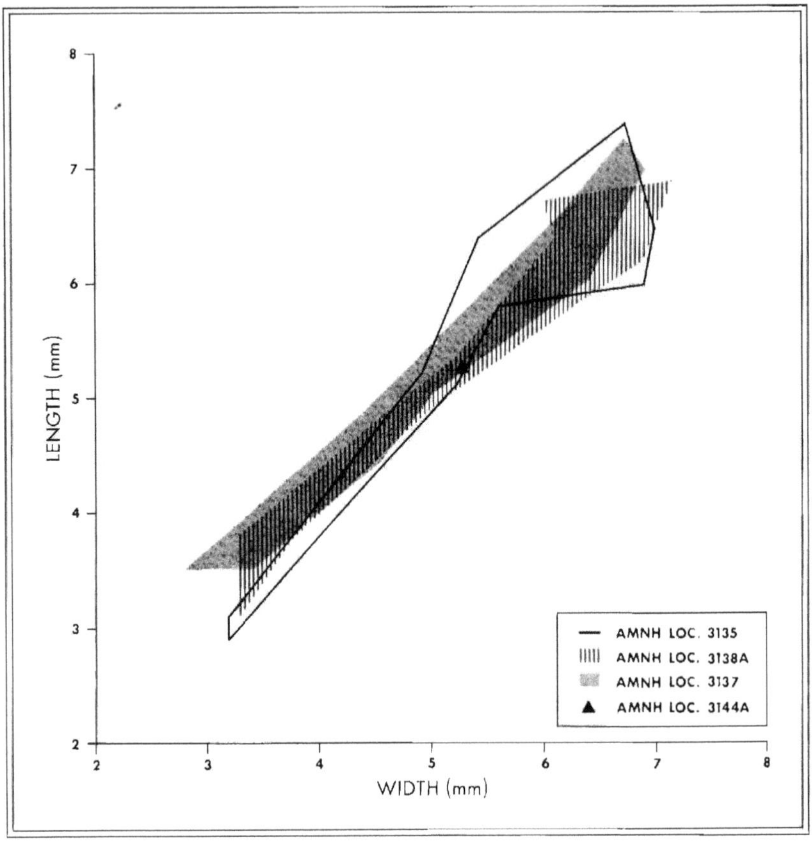

FIGURE 35. Comparative-size polygons for *Coelospira camilla* Hall, 1867, fitted visually.

FIGURE 36. *"Pacificocoelia" acutiplicata* (Conrad, 1841). AMNH Loc. 3128B. Lateral, dorsal, and ventral views, AMNH 42757, x5.

detailed locality list of the species' occurrence in Canada, Mexico, and the following states: New York, Quebec, New Brunswick, Oklahoma, Texas, and Nevada.

Community occurrence: Feldman (1980) found this species in the *Leptaena-Megakozlowskiella, Atrypa-Coelospira-Nucleospira, Atrypa,* and *Atrypa-Megakozlowskiella* communities.

Figured specimens: AMNH 39904, 39905, 39906.

Family LEPTOCOELIIDAE Boucot and Gill, 1956
Genus *PACIFICOCOELIA* Boucot, 1975
Type species: *Atrypa acutiplicata* Conrad, 1841: 54.

"Pacificocoelia" acutiplicata (Conrad, 1841), Figures 36, 37.

Anoplotheca acutiplicata Kindle, 1912: pl. 6, figs. 1-15.

Leptocoelia acutiplicata Hall, 1867: pl. 57, figs. 30-39.

Pacificocoelia acutiplicata Boucot, 1975: 361-362; Boucot and Rehmer, 1977: 1130, pl. 1, figs. 1-26.

Remarks: The bulk of material at hand is from the Nedrow Member, which is quite shaley. The specimens are not silicified and preservation is generally poor. Rarely are the shells found loose. Most of the collection consists of impressions and fragments imbedded in the matrix with few internal structures preserved. Since all free articulated specimens are deformed or somewhat crushed the true convexity of the shells in lateral profile is unknown. However, it appears as though the brachial valve is gently convex and the pedicle valve slightly more so. The shells are subcircular in outline with the length almost equal to the width. A weak pedicle sulcus is sometimes noticeable on larger specimens. No corresponding dorsal fold was observed. The hinge line is very short and becomes rounded anteriorly. No interareas were present. The entire apical region is poorly preserved. The anterior and lateral commissures are crenulate.

Ornamentation consists of 10 to 12 plications, more pronounced medially than on the flanks. In cross-section they are U-shaped, as are their interspaces. Compressed shells have V-shaped plications and interspaces. Distinct concentric growth lines (two or three per shell) occur on most large specimens.

Comparison: The author uses the name *"Pacificocoelia"* but notes that Koch (1981, and personal communication) has assigned *acutiplicata* to a new genus which differs from *Pacificocoelia* and *Leptocoelia* in its lack of a cardinal process. The Onondaga shells have a gently convex brachial valve, strong plications, and indications of a weakly impressed pedicle valve muscle field. The closest pacificocoelid to this species would probably be *P. nunezi texana*, which has a weakly impressed pedicle valve muscle field and three or four lateral plications on the pedicle valve. Other related forms with deeply impressed pedicle valve muscle fields include *P. biconvexa*, *P. nunezi nunezi*, and *P. murphyi*.

Community occurrence: Feldman (1980) found this species in the *Leptaena-Megakozlowskiella*, *Atrypa-Megakozlowskiella*, and *Pacificocoelia* communities.

Figured specimens: AMNH 39903, 42757.

Suborder ATHYRIDIDINA Boucot, Johnson, and Staton, 1964
Superfamily ATHYRIDOIDEA Davidson, 1881
Family ATHYRIDIDAE Davidson, 1881
Subfamily ATHYRIDINAE Davidson, 1881
Genus *ATHYRIS* M'Coy, 1844

Athyris sp. A, Figure 38A-J.

Exterior: The shells are small for the genus (about 9.9-13.2 mm wide) (Table 6), transversely suboval in outline, and subequally biconvex with the pedicle valve slightly deeper than the brachial valve. The convexity is accentuated in the umbonal region. The ventral beak is suberect, terminating in a small round foramen. The brachial beak is smaller and less noticeable. The pedicle valve bears a shallow sulcus with a corresponding low fold on the brachial valve, resulting in a weakly uniplicate anterior commissure. Some specimens are nonsulcate and rectimarginate. Ornamentation consists of fine concentric growth lines on both valves, which are quite lamellose in some individuals.

Pedicle valve interior: The hinge teeth are small, pointed, and dorsally directed. They are supported by weak dental plates diverging at an angle of approximately 45 degrees. The muscle scars are very faintly impressed and barely discernible. Possible adductor scars are just anterior to the base of the dental plates, forming a vague, oval impression.

Brachial valve interior: Since no free brachial valves were collected, a single articulated shell was dissected to provide the following description:

The sockets are short, U-shaped, and appear to deepen anteriorly. There is a broad, concave cardinal plate which rises ventrally in the apical region, forming a small lip. No muscle impressions were preserved.

Comparison: The specimens assigned to *Athyris* sp. A may be grouped as follows:

(1) Rectimarginate, nonlamellose
(2) Sulcate, nonlamellose
(3) Sulcate, lamellose.

All three show affinities to *Athyris cora* Hall but are significantly smaller. Hall (1867: 291-292, pl. 47, figs. 1-7) describes *A. cora* from the Hamilton at Delhi, New York, as well as a single specimen from the Chemung Group. *Athyris nuculoidea* from the St. Laurent Limestone (Cooper, 1945: 485, pl. 64, figs. 12-19) resembles *Athyris* sp. A in its lamellose ornamentation and small size. The Onondaga specimens are similar to *A. spiriferoides* (Eaton) only in the relatively divergent dental plates and muscle field; they differ significantly in size. Two specimens strongly resemble *A. lamellosa* (L'Eveille) with their narrow margins and strong concentric growth lamellae. However, their size and poor preservation preclude assignment to that species.

Community occurrence: Feldman (1980) found this species in the *Leptaena-Megakozlowskiella, Atrypa-Coelospira-Nucleospira, Atrypa, Atrypa-Megakozlowskiella, Pacificocoelia,* and *"Hallinetes"* communities.

Figured specimens: AMNH 39901, 39907, 39908, 39909, 42765.

Athyris sp. B, Figure 38K-L.

Remarks: Two pedicle valves are in the collection which may be differentiated from *Athyris* sp. A by their larger size (20.1 mm and 20.7 mm wide), subparallel dental plates, and narrow muscle field (see Table 6).

Community occurrence: Feldman (1980) found this species in the *Atrypa-Coelospira-Nucleospira* Community.

Figured specimen: AMNH 42766.

Superfamily MERISTELLOIDEA Waagen, 1883
Family MERISTELLIDAE Waagen, 1883
Subfamily MERISTELLINAE Waagen, 1883
Genus *MERISTINA* Hall, 1867

Type species: *Meristella maria* Hall, 1863: 212.

Meristina cf. *nasuta* (Conrad, 1842), Figure 39.

Atrypa nasuta Conrad, 1842: 265.

Meristella nasuta Boucot and Johnson, 1968: B13-B14, pl. 4, figs. 26-43.

Remarks: Two specimens of this species are in the collection. One is a silicified pedicle (?) valve collected from AMNH Loc. 3137 with almost no internal structures preserved, and the other is a nonsilicified, quite weathered, and exfoliated pedicle valve collected from AMNH Loc. 31238. The shells are convex, elongate, and suboval in outline with no noticeable interarea. Concentric growth lamellae are evident along the internal margins of the nonsilicified shell (AMNH 39900). Based on similar outline and external morphology to *Meristina nasuta* from the underlying Bois Blanc Formation (Boucot and Johnson, 1968: B13, pl. 4, figs. 26-36), these shells are assigned to the same genus.

Community occurrence: Feldman (1980) found this species in the *Leptaena-Megakozlowskiella* and *Atrypa-Coelospira-Nucleospira* communities.

Figured specimens: AMNH 39900, 42730.

Genus *CHARIONOIDES* Boucot, Johnson, and Staton, 1964

Type species: *Meristella doris* Hall, 1860: 84.

Charionoides aff. *doris* (Hall, 1860), Figure 40.

Meristella doris Hall, 1860: 84.

Charionoides doris Boucot, 1973: 64, pl. 20, figs. 14-22.

Remarks: A single, poorly preserved, convex pedicle valve was recovered with a fragment of the brachial valve attached at one lateral margin. The beak is relatively long and angular with a suggestion of a small pedicle foramen. The delthyrium was not preserved. The valve exterior is smooth with no indication of a pedicle sulcus.

FIGURE 37. *"Pacificocoelia" acutiplicata* (Conrad, 1841). AMNH Loc. 3128B. Ventral view, AMNH 39903, x3.

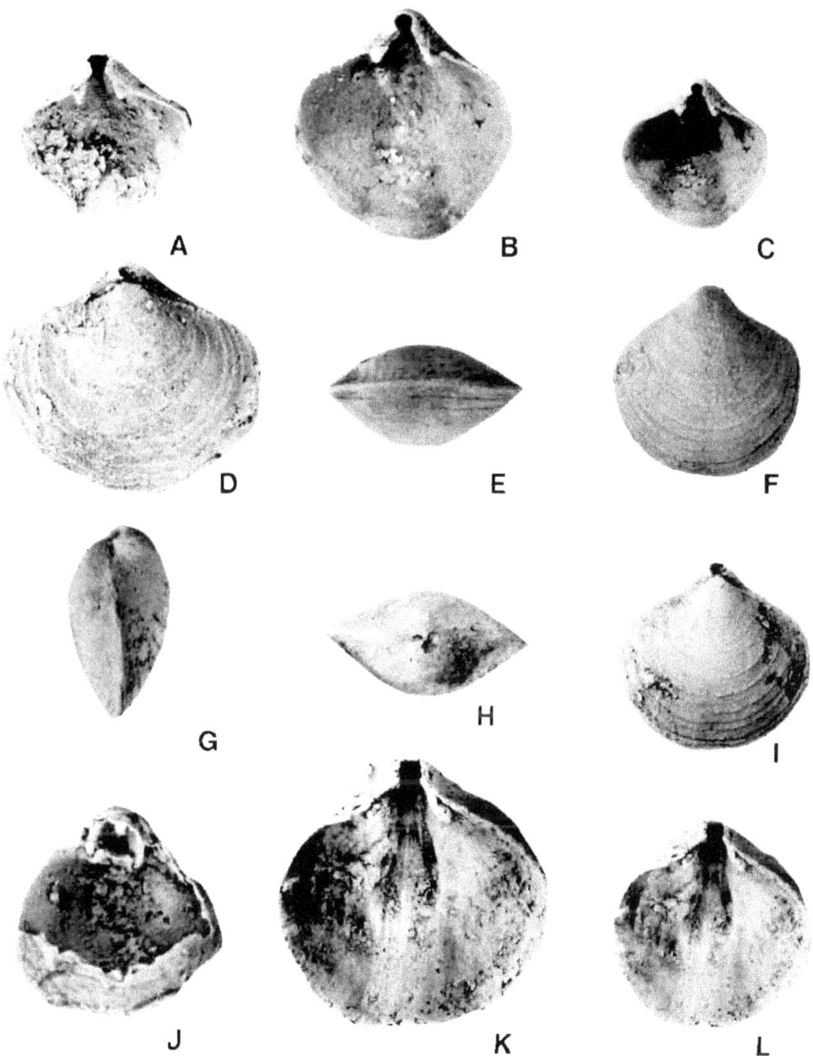

FIGURE 38. *Athyris* sp. A. A. AMNH Loc. 3135. Pedicle valve interior, AMNH 42766, x2.5. B, C. AMNH Loc. 3137. Pedicle valve interior, AMNH 39908, x3, x2. D. AMNH Loc. 3142. Dorsal view, AMNH 42765, x3.5. E-I. AMNH Loc. 3138A. Anterior, ventral, lateral, posterior, and dorsal views, AMNH 39907, x3. J. AMNH Loc. 3137. Articulated specimen with pedicle valve dissected exposing interior of umbonal region. Note the broad, concave cardinal plate rising ventrally and adjacent articulating hinge teeth and sockets, AMNH 39909, x3. *Athyris* sp. B. K, L. AMNH Loc. 3138A. Pedicle valve interior, AMNH 39901, x1.5.

TABLE 6. Measurements (in mm) of *Athyris* sp. A and B.

Specimen	AMNH Locality	Aticulated	p.v.	(L)	(W)	(T)
Sp. A	3137	X	–	11.7	12.2 est.	6.2
A	3137	X	–		11.5	7.5
A	3137	X	–	7.6	7.8	4.5
A	3137	–	X	11.6	11.7	–
A	3137	–	X	12.3	–	–
A	3135	–	X	11.4	12.4	–
Sp. A def	3123B	X	–	10.9	9.9	5.9
A	3138A	X	–	9.8	9.9	5.1
A	3138A	X	–	7.0	8.2	3.0
A	3144B	X	–	10.9	13.2	6.4
A	3136	X	–	8.3	9.4	5.5
Sp. B	3138A	–	X	16.4 est	20.1	–
Sp. B	3138A	–	X	20.3	20.7	–

[a] Dissected to expose cardinalia. *Note:* there are no free brachial valves in the collection.

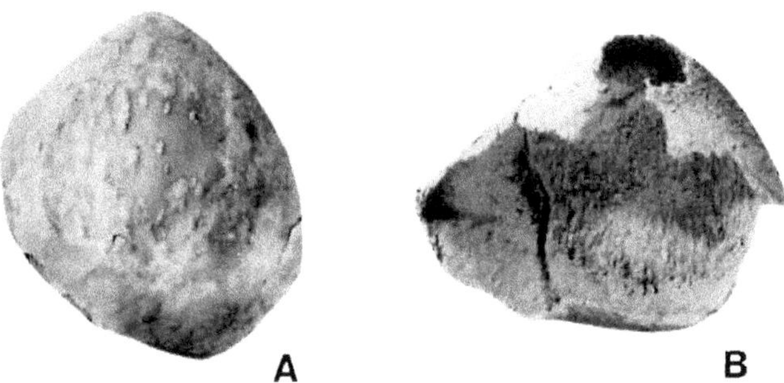

FIGURE 39. *Meristina* cf. *nasuta* (Conrad, 1842). A. AMNH Loc. 3123B. Ventral view, AMNH 39900, x1. B. AMNH Loc. 3137. Pedicle valve interior, silicilied and poorly preserved, AMNH 42730, x1.

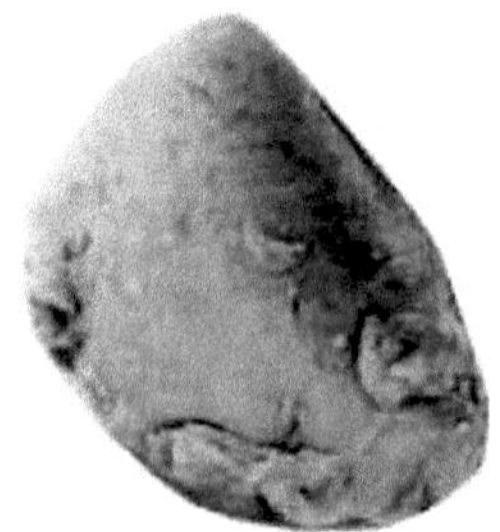

FIGURE 40. *Charionoides* aff. *doris* (Hall, 1860). AMNH Loc. 3123C. Ventral view, AMNH 39899, x2.

Fine concentric growth lines, evident only at the lateral margins of the pedicle valve and the brachial fragment, are present.

Hall (1867: 303, pl. 50, figs. 1-12) describes *C. doris* (*Meristella doris*) from the upper part of the Onondaga (Corniferous) Limestone near Williamsville (Erie County) and from the Schoharie Grit (Schoharie County), New York.

Community occurrence: Feldman (1980) found this species in the *Leptaena-Megakozlowskiella* Community.

Figured specimen: AMNH 39889.

Genus *PENTAGONIA* Cozzens, 1846

Type species: *Atrypa unisulcata* Conrad, 1841: 56.

Pentagonia unisulcata (Conrad, 1841), Figure 41.

Atrypa unisulcata Conrad, 1841: 56.

Atrypa uniangulata Hall, 1861: 101.

Meristella? unisulcata (Conrad), Hall, 1862: 158, pl. 2, figs. 17, 20-23 (not figs. 19, 24-25).

Meristella (*Pentagonia*) *unisulcata* (Conrad) Hall, 1867: 309, pl. 50, figs. 18-29 (not figs. 30-35).

Non *Meristella unisulcata* (Conrad) Nettleroth, 1889: 99, pl. 15, figs. 9-16.

Pentagonia unisulcata (Conrad) Stauffer, 1915: 104, 245, (not 160, 171, 175, 234); Dunbar, 1919: 87, 89; Goldring, 1935: 148, figs. 53B-D; Butts, 1940: 300-301, 304-305; 1941, pl. 115, figs. 17-21, 35; Cooper, 1944: 333, pl. 127, fig. 37; Oliver, 1954: 633-634, 638-640; 1956: 1452, 1456, 1462.

Non *Pentagonia unisulcata* (Conrad) Savage, 1930: 47, 50, 53, 62; 1931: 242, pl. 30, figs. 17-18.

Pentagonia unisulcata (Conrad) Dutro, 1971: 188.

Exterior: The shells are medium-sized (Table 7), astrophic, impunctate, pentagonal in outline in plain view, and when viewed posteriorly. The beak is suberect. The shells are dorsibiconvex with the greatest width attained between midlength and the anterior commissure, usually more toward the anterior margin. The brachial valve is cariniform due to the presence of a raised, rounded fold bearing a narrow, median groove which begins at the dorsal umbo and widens slightly anteriorly in some specimens forming two parallel to subparallel ridges extending almost half the

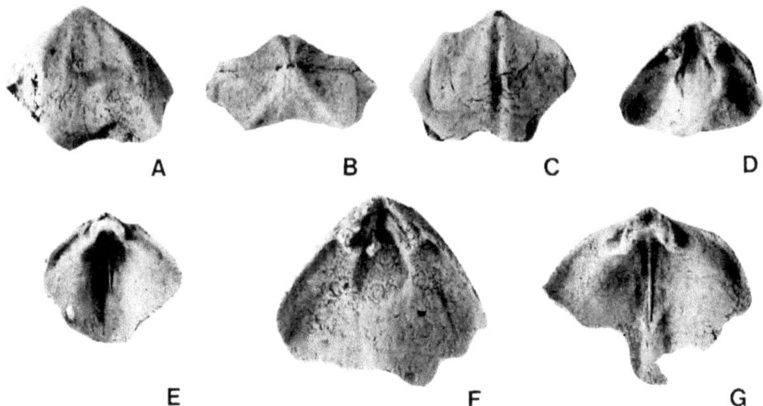

FIGURE 41. *Pentagonia unisulcata* (Conrad, 1841). A-C. AMNH Loc. 3138A. Ventral, posterior, and dorsal views, AMNH 42731, x1.5. D. AMNH Loc. 3138A. Pedicle valve interior, AMNH 42737a x1.3. E. AMNH Loc. 3138A. Brachial valve interior (articulated with AMNH 42737a), AMNH 42737b x1.3. F. AMNH Loc. 3138A. Pedicle valve interior, AMNH 39910a, x2. G. AMNH Loc. 3138A: Brachial valve interior (articulated with AMNH 39910a), AMNH 39910b, x2.

TABLE 7. Measurements (in mm) of Articulated Specimens of *Pentagonia unisulcata* (Conrad, 1841).

AMHN Locality	(L)	(W)	(T)
3138A	–	21.1	10.5
3138A[a]	–	14.6	–
3144A		17.1	8.5
3138A		18.8	9.7
3138A		20.9	10.3

[a] Badly damaged. *Note:* all specimens are incomplete to some extent.

length of the valve. The flanks are concave, dropping steeply adaxially, away from the sulcate fold.

The pedicle valve bears a relatively broad, shallow sulcus which widens considerably at the anterior. Two distinct ridges, defining the sulcus laterally, extend from the umbo across the posterolateral margins of the flanks to the anterolateral commissure. The anterior commissure is uniplicate.

The only ornamentation evident is vague concentric growth lines on the anterior portions of the shell.

Pedicle valve interior: The hinge teeth are short, blunt, and taper posteriorly. They are triangular in cross-section with the apex rounded and are supported by strong dental lamellae which extend one-fourth the valve length. Secondary shell material is deposited at the base of the dental lamellae and extends posteriorly to partially fill the umbonal cavity. The muscle field is broad, striate, and flabelliform, extending about one-third the valve length. A readily apparent difference in elevation, in the form of a low circular ridge, separates the anterior extremity of the muscle field from the valve floor in many specimens. In others the muscle field is gradationally absorbed into the floor of the shell. In most individuals there is an indication of a small pedicle foramen, although it is doubtful whether the ephebic form possessed a functional pedicle.

Brachial valve interior: The sockets are deep and cylindrical in cross-section. A prominent cardinal process dominates the posterior section of the brachial interior. The cardinal process is A-shaped when viewed from above and bears a deep pit on the distal end. The anterior end drops off steeply to the valve floor. When viewed laterally, the cardinal process appears to have a step, corresponding to the posterior depression (pit), parallel to the hinge axis. When viewed posteriorly the cardinal process appears to be bilobed.

A low, well-defined, median septum originates on the anterior face of the cardinal process, drops to the floor of the valve, and bisects the dorsal muscle field. The median septum extends approximately one-half to one-third the valve length. Two subparallel to parallel ridges define the medial aspect of the adductor scars which appear to be elliptical to oval in outline. These ridges extend for approximately half the length of the median septum.

Comparison: *Pentagonia unisulcata* may be differentiated from the stratigraphically younger *P. peersi* only by its lesser dimensions. Juveniles of *P. peersi* are identical in form with adults of *P. unisulcata. Pentagonia lenta,* ancestral to *P. unisulcata,* is smaller and subovate in plan view. The interiors, however, are identical. Dutro (1971: 182) reports that *Pentagonia* is well represented by *P. unisulcata* in Onondaga age rocks from New York to Virginia, but that species has not yet been reported from the western equivalents of the Onondaga in Ohio, Kentucky, Indiana, and Illinois. Boucot and Johnson (1968: B2) summarize the distribution of *Pentagonia* in rocks of Schoharie age. Conrad (1841) describes the genus *(Atrypa unisulcata)* from the Onondaga Limestone at Schoharie, New York.

Community occcurrence: Feldman (1980) found this species in the *Atrypa-Coelospira-Nucleospira, Atrypa,* and *Atrypa-Megakozlowskiella* communities.

Figured specimens: AMNH 39910a-b, 42731, 42737a-b.

Family NUCLEOSPIRIDAE Davidson, 1881
Genus *NUCLEOSPIRA* Hall, 1859
Type species: *Spirifer ventricosa* Hall, 1857: 57.

Nucleospira aff. *ventricosa* (Hall, 1857), Figure 42.

Spirifer ventricosa Hall, 1857: 57, not figs. 1 and 2.

Nucleospira ventricosa Hall, 1859: 220-221, pl. 14, fig. 1a-h; pl. 28B, figs. 2-9; Hall and Clarke, 1894: pl. 48, figs. 2-6, 18; Weller, 1903: 290, pl. 30, figs. 19-22; Schuchert, 1913: 430, pl. 73, figs. 10-12; Bowen, 1967: 37-38, pl. 5, figs. 16-17.

Nucleospira sp. Boucot and Johnson, 1968: H14, pl. 5, figs. 1-11.

Exterior: The shells are small (Table 8), transversely suboval in outline, and biconvex in profile with the pedicle valve slightly deeper than the brachial valve. The hinge line is curved. The brachial beak fits into the anterior end of the delthyrium which is partially covered by a concave pseudodeltidium in some specimens. Both beaks are erect. There is no interarea evident. The shell surface lacks radial ornamentation and is without a fold or sulcus. However, in some specimens the pedicle has an indistinct median depression. Concentric growth lamellae are present and noticeably concentrated toward the anterior of the valves. The anterior commissure is rectimarginate.

Pedicle valve interior: The hinge teeth are small, pointed, and hooked medially, and are supported by secondary shell material representing true dental lamellae, which are absent. Shallow sockets are located posterior to each tooth. The delthyrium is closed by a concave pseudodeltidium in some specimens; in others the pseudodeltidium is not preserved. In most specimens the delthyrium is open. A low, thin, median septum extends from the deep umbonal cavity to about seven-eighths the valve length. Often, the median septum is obscured by secondary shell material in the umbonal cavity and appears to originate at about midlength. The muscle field is weakly impressed but appears to be flabellate (diductor scars) with two longitudinal, suboval scars (adductors) bisected by the myophragm. The muscle field extends to just past midlength.

Brachial valve interior: The cardinal process is large and projects up from the valve floor such that when viewed from above the outline is triangular. The anterior edge is doubly concave, bisected by a ridge, giving the appearance of a lowercase "m." The ventral face of the cardinal process is scyphiform, whereas the apex is recurved posterodorsally. Well-developed, anterolateral diverging sockets are located at the base of the cardinal process. A low, bladelike median septum extends from the anterior vertical face, where it appears to be a longitudinal ridge, to just past midlength, tapering out anteriorly. Elongate, suboval adductor muscle impressions are bisected by the median septum. The diductors probably attached onto the vertical anterior face of the cardinal process since, when the valves are articulated, there is a very narrow space between the ventral face of the cardinal process and the floor of the delthyrial cavity. The spires, partially preserved on only one specimen, are laterally directed, becoming increasingly smaller

TABLE 8. Measurement (in mm) of Articulated Specimens of *Nucleospira* aff. *ventricosa* (Hall, 1857).

AMNH Locality	(L)	(W)	(T)
3138A	3.0	3.4	1.7
3138A	6.6	8.5	3.4
3135	9.1	10.5	4.7
3135	7.3	7.5	4.4
3135	4.9	5.5	3.2
3138A	10.6	11.7	7.0
3138A	4.2	4.8	2.5
3138A	4.6	4.6	2.8
3138A	4.9	5.4	3.2
3138A	7.6	8.7	4.6
3138A	9.2	10.6	5.5
3137	5.1	5.8	2.8
3137	5.8	6.6	3.2
3137	6.5	7.4	3.9
3137	6.5	7.0	3.8
3137	6.9	7.5	4.2
3137	8.0	7.5	4.1
3137	7.5	8.6	4.4
3137	9.7	11.1	5.1
3137	9.7	11.6	5.2

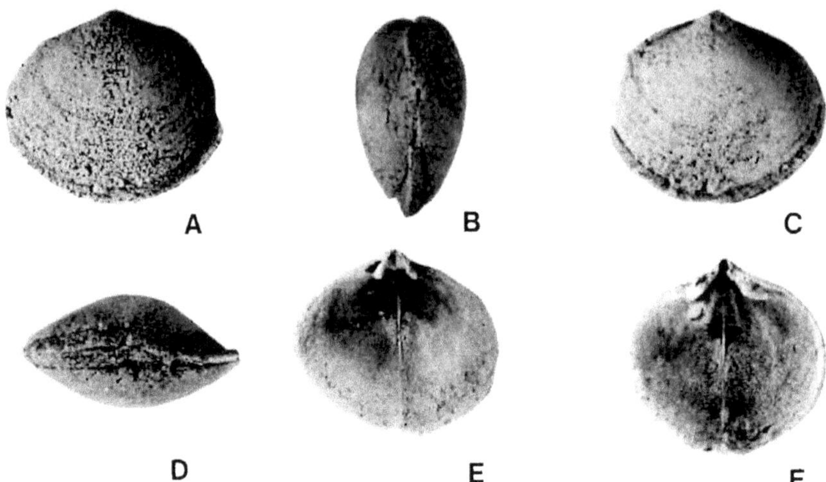

FIGURE 42. *NUCLEOSPIRA* AFF. *VENTRICOSA* (HALL, 1857). A-D. AMNH LOC. 3138A. DORSAL, LATERAL, VENTRAL, AND ANTERIOR VIEWS, AMNH 39914, X3. E. AMNH LOC. 3137. BRACHIAL VALVE INTERIOR, AMNH 39913, X3.5. F. AMNH LOC. 3138A. PEDICLE VALVE INTERIOR, AMNH 39912, X3.

with six turns on the side preserved. Not enough of the jugum is preserved for analysis.

Comparison: *Nucleospira* aff. *ventricosa* differs from *Nucleospira* sp. (Boucot, 1973, p. 64, pl. 20, figs. 23-27) of the Moose River Synclinorium in its transversely suboval outline. The pedicle interiors of both forms are similar, although the Moose River material is poorly preserved and precludes further comparison.

Nucleospira ventricosa from the Keyser Limestone (Bowen, 1967: 37, pl. 5, figs. 16-27) is similar in lateral profile but differs in the shape of the cardinal process (large and prominent in both forms). This difference may be due to intraspecific variation. Bowen (1967: 38) discusses variation in *Nucleospira* and concludes that most of the species have been based on variations in distinctness of the sulci, convexity of the valves, ratio of the length to width, presence of growth lines, and shell size; but the differences between species are nearly always vague. He places no species in synonymy because of the range in variation within *N. ventricosa.*

Nucleospira ventricosa from the New Scotland Formation and equivalents (Cooper, 1944: 331, pl. 127, figs. 8-9) has a cardinal process which differs from the Keyser and Onondaga shells in its subovate outline. *Nucleospira concinna* (Cooper, 1944: 331, pl. 127, figs. 5-7) more

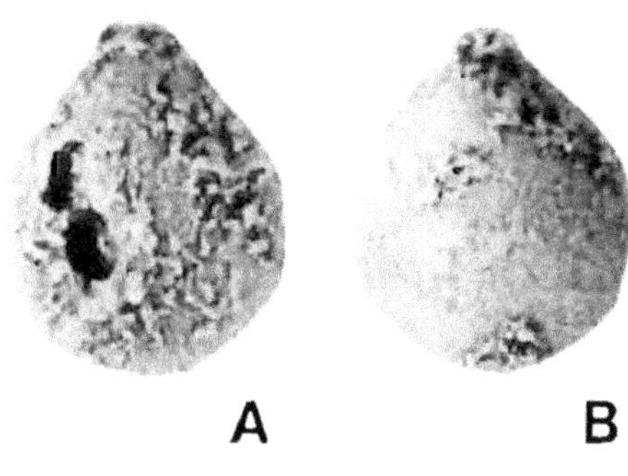

FIGURE 43. Athyridacean indet. A, B. AMNH Loc. 3138A. Ventral and dorsal views, AMNH 39902, x4.

FIGURE 44. *Trematospira?* sp. AMNH Loc. 3124B. Dorsal view of poorly preserved specimen, AMNH 39915, x1.

closely approaches *N.* aff. *ventricosa* in size but its interiors are not illustrated.

Nucleospira sp. from the Bois Blanc Formation (Boucot and Johnson, 1968: B14, pl. 5, figs. 1-11) most closely resembles *N.* aff. *ventricosa* in shell outline, size, and morphology of the cardinal process (see pl. 5, fig. 10). Since only fragmentary brachial valves are available no further comparisons can be made.

Community occurrence: Feldman (1980) found this species in the *Atrypa-Coelospira-Nucleospira, Atrypa,* and *Atrypa-Megakozlowskiella* communities.

Figured specimens: AMNH 39912, 19913, 19914.

Athyridacean indet., Figure 43.

Remarks: These shells were recovered from Upper Moorehouse strata in the southeastern part of New York State. The specimens probably belong to the superfamily Athyridacea, showing certain affinities to the Meristellidae. Due to the lack of material and absence of free valves with good internal morphology, these shells cannot be described or assigned at this time.

Community occurrence: Feldman (1980) found this species in the *Atrypa-Coelospira-Nucleospira* Community.

Figured specimen: AMNH 39902.

Suborder RETZIIDINA Alvarez and Rong, 2002

Superfamily RHYNCHOSPIRINOIDEA Schuchert, 1929

Family RHYNCHOSPIRINIDAE Schuchert, 1929

Genus *TREMATOSPIRA* Hall, 1859

Trematospira? sp., Figure 44.

Remarks: One specimen is available for study in the collection. The articulated shell is partly exposed at the ventral umbo, whereas approximately half of the dorsal valve is visible. The outline appears to be elongate, suboval, and moderately biconvex. The hinge line is curved and the interarea difficult to see, but appears to be extremely narrow. There are about 45 rounded, bifurcating costae on the dorsal exterior and about 19 on the ventral exterior. The peak is pointed and apically twisted.

Community occurrence: Feldman (1980) found this species in the *Leptaena-Megakozlowskiella* Community.

Figured specimen: AMNH 39915.

Family RHYNCHOSPIRINIDAE Schuchert and LeVene, 1929
Genus *RHYNCHOSPIRINA* Schuchert and LeVene, 1929

Rhynchospirina sp., Figure 45.

Remarks: This genus is represented by only one specimen, which is slightly damaged in the beak region. Since the beak region has been used to determine specific levels in rhynchospirinid taxonomy (Hall, 1857; Davidson, 1867; Schuchert, 1913; Kozlowski, 1929; Bowen, 1967), a specific assignment is not attempted here.

The shell is small (Table 9), pyriform in outline, and biconvex (subglobose) in lateral profile with an inflated brachial valve. Maximum width is attained at midlength. The pedicle beak is erect but broken at the apex and possibly incurved. The small pedicle foramen appears to be permesothyrid. The brachial beak is chipped off. The posterolateral margins are straight and diverge at an angle of about 90 degrees. The anterior margin is semicircular. The pedicle valve bears a weak sulcus but there is no corresponding fold visible on the brachial valve. The anterior commissure, however, is slightly uniplicate due to the influence of the sulcus. Ornamentation consists of eight subangular plications on the

TABLE 9. Measurements (in mm) of *Rhynchospirina* Species.

AMNH Locality	(L)	(W)	(T)
3137	6.0	5.8	3.9

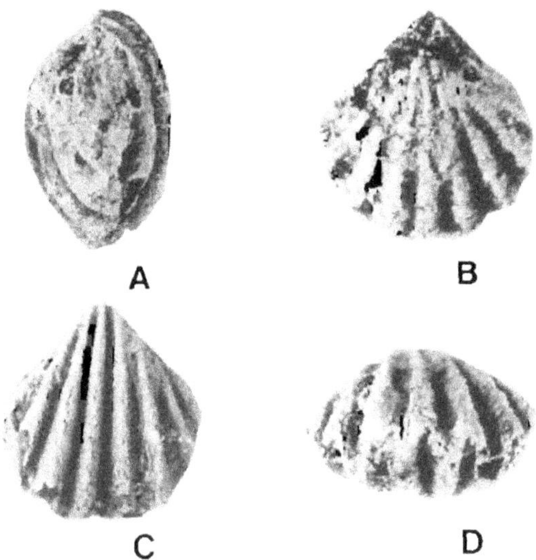

FIGURE 45. *Rhynchospirina* sp. A-D. AMNH Loc. 3135. Lateral, dorsal, ventral, and anterior views, AMNH 39918, x5.

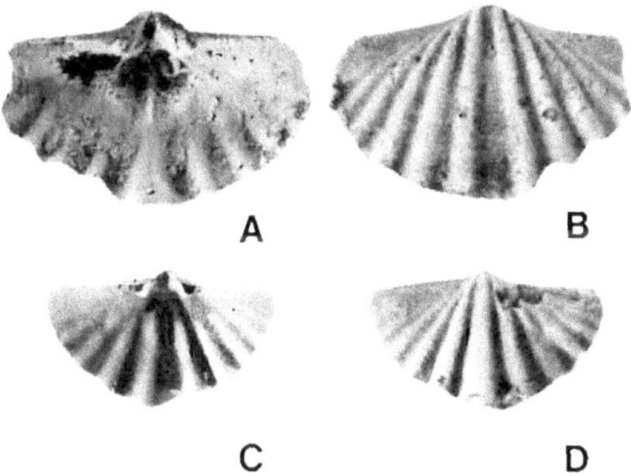

FIGURE 46. *Acrospirifer duodenaria* (Hall, 1843). A, B. AMNH Loc. 3135. Pedicle valve interior and exterior, AMNH 39916, x2, x1.5. C, D. AMNH Loc. 3137. Brachial valve interior and exterior, AMNH 39917, x2.

brachial valve and nine on the pedicle valve with subangular interspaces between the plications. As the anterior margins are approached, the plications become more distinct and well defined.

Discussion: Bowen (1967) describes two new species of *Rhynchospirina* from the Keyser Limestone of Pennsylvania and West Virginia. The genus has also been reported from the Decker Formation of New Jersey (Weller, 1903), the New Scotland Formation (New York), and the Oriskany Sandstone (Maryland) by Hall (1857).

Community occurrence: Feldman (1980) found this species in the *Atrypa-Coelospira-Nucleospira* Community.

Figured specimen: AMNH 39918.

Suborder DELTHYRIDINA Ivanova 1972

Superfamily DELTHYRIDOIDEA Phillips, 1841

Family ACROSPIRIFERIDAE Termier and Termier, 1949

Subfamily ACROSPIRIFERINAE Termier and Termier, 1949

Genus *ACROSPIRIFER* Helmbrecht and Wedekind, 1923

Type species: *Spirifer primaevus* Steininger, 1853,

by subsequent designation of Wedekind, 1926: 202.

Acrospirifer duodenaria (Hall, 1843), Figure 46.

Delthyris duodenaria Hall, 1843: 171, fig. 5.

Spirifer duodenaria Hall, 1867: 189, pl. 27, 28; Landes, Ehlers, and Stanley, 1945, pl. 12, fig. 4.

Hysterolites (Acrospirifer) worthenanus? Amsden, in Amsden and Ventress, 1963: 182, pl. 16, figs. 1-4, 6-8, 11-16, 5?, 9?, 10?

Acrospirifer duodenaria Boucot and Johnson, 1968: BI4-BI5, pl. 5, figs. 12-39.

Exterior: The shells are biconvex with a slightly more convex pedicle valve, and are transversely subelliptical in outline. The width, greatest at the hinge line, is commonly twice that of the length in most specimens studied, although that relationship is variable. The hinge line is long and straight. The pedicle interarea is long, low, narrow, moderately curved, and apsacline. There is a medial open delthyrium with no preserved deltidial plates. The pedicle beak is incurved toward the brachial umbo. The brachial interarea ranges from slightly orthocline to anacline (most common) and is exceedingly long and thin. The pedicle valve bears a narrow, triangular, moderately deep, noncostate sulcus originating in the umbonal region. The brachial valve bears a corresponding fold with a somewhat flattened crest.

On each pedicle flank there are five or six rounded plications becoming progressively narrower laterally. The interspaces are U-shaped, narrowing toward the bottom. The brachial flanks bear four or five plications, also with U-shaped interspaces. In some of the better-preserved specimens there are indications of fine, numerous, concentric growth lines. The anterior commissure is uniplicate. Fine, radial ornamentation was not preserved.

Pedicle valve interior: The hinge teeth are short, blunt, and somewhat subcircular in cross-section. The dental plates are short and almost obscured by the deposition of secondary shell material in the posterior portion of the umbonal cavity and along the inner margins of the delthyrium. In some specimens this secondary shell material results in a raised, platform-like mound located in the posterior region of the delthyrium. A low myophragm bisects the subcircular muscle field which consists of two suboval diductor impressions. There is no line of demarcation between the muscle field and the valve floor, which is crenulated anteriorly due to the impress of the plications.

Brachial valve interior: The dental sockets are short, deeply excavated, laterally directed, and subtriangular in outline. Short, incurved, crural bases attach to the inner margins of the sockets and merge into the socket floors. The hinge line overhangs the sockets on the posterior margin. The notothyrial cavity is empty and uncovered by chilidial plates. A vague, striated region represents the site of diductor attachment at the apex of the notothyrial cavity. Two adductor scars are separated by a low myophragrn extending to almost half the valve length. The muscle field is poorly impressed. The valve floor is crenulated due to the impress of the plications.

Comparison: *Acrospirifer murchisoni* from the Moose River Synclinorium (Boucot, 1973: 41, pl. 16, figs. 19-25) differs from *A. duodenaria* in its larger size and wide pedicle interarea. *Acrospirifer atlanticus*, also from the Moose River Synclinorium, is considerably larger and more alate. Boucot, Gauri, and Southard (1970: 14, pl. 4, figs. 22-26) describe *Acrospirifer?* sp., from the Esopus Formation to the Green Pond Outlier, which is a medium-sized shell considerably less transverse in outline than the Onondaga shells. *Acrospirifer* aff. *murchisoni* from the Great Basin (Johnson, 1970: 189, pl. 56, figs. 5-13; pl. 57, figs. 1-6) is less alate and more transversely suboval. Also, it generally has six plications on the pedicle valve flank. The Nevada shells usually have fewer plications (three to six on each pedicle valve flank).

In their discussion of *Acrospirifer duodenaria* from the Bois Blanc Formation, Boucot and Johnson (1968: B15) note that specimens of the species from the Kanouse Sandstone were assigned to *A. macrothyris* based on the belief that *A. duodenarius* represented the young of *A. macrothyris*. However, based upon their study of the Bois Blanc shells, Boucot and

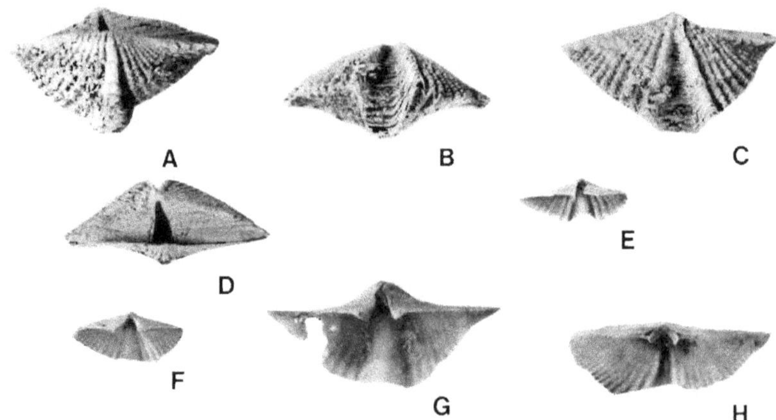

FIGURE 47. *Mucrospirifer* cf. *macra* (Hall, 1857). A-D. AMNH Loc. 3138A. Dorsal, anterior ventral, and posterior views, AMNH 39919, x1.5. E. AMNH Loc. 3137. Pedicle valve interior, AMNH 42767, x1. F. AMNH Loc. 3137. Pedicle valve interior, AMNH 42763, x1. G. AMNH Loc. 3137. Pedicle valve interior, AMNH 42761, x1. H. AMNH Loc. 3137. Brachial valve interior, AMNH 42762, x1.

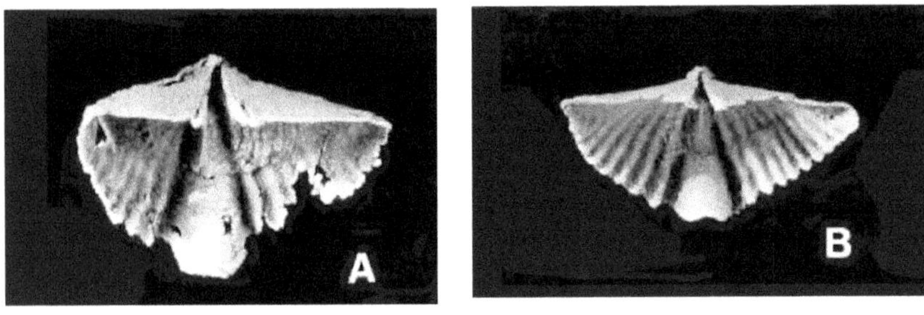

FIGURE 48. *Mucrospirifer* cf. *macra* (Hall, 1857). A. AMNH Loc. 3137. Pedicle valve interior, AMNH 42754, x2. AMNH Loc. 3138A. Pedicle valve interior, AMNH 42753, x2.

Johnson distinguish *A. macrothyris* by its much lower plications and narrow, shallow interspaces. The size difference between the two species is therefore real.

Hall (1867: 189-190) in describing *A. duodenaria* from the Schoharie Grit and Onondaga Limestone mentions that the species is known throughout "all the extent of the formation within the state," and occurs in western Canada and Ohio.

Community occurrence: Feldman (1980) found this species in the *Atrypa-Coelospira-Nucleospira* and *Atrypa-Megakozlowskiella* communities.

Figured specimens: AMNH 39916, 39917.

Family MUCROSPIRIFERIDAE Boucot, 1959

Subfamily MUCROSPIRIFERINAE Boucot, 1959

Genus *MUCROSPIRIFER* Grabau, 1931

Type species: *Delthyris mucronatus* Conrad, 1841: 54.

Mucrospirifer cf. *macra* (Hall, 1857), Figures 47-50.

Delthyris mucronatus Conrad, 1841: 54.

"Mucrospirifer" cf. *macra* Boucot and Johnson, 1968: B15, pl. 6, figs. 1-6; Boucot, 1973: 60, pl. 18, figs. 21-22.

Exterior: The shells range from small to large and are transversely subtrigonal to subsemicircular in outline and biconvex in lateral profile with the brachial valve a little flatter than the pedicle valve. The point of maximum width, commonly twice the length or more, is at the hinge line, which is straight. The relative widths of specimens studied is highly variable and often dependent upon the degree of alation. The ventral interarea is moderately high, long, slightly curved, and apsacline. The ventral beak, posterior to the interarea, is short and stubby, and ranges from nearly straight to suberect. An open, triangular delthyrium, enclosing an angle of about 20 to

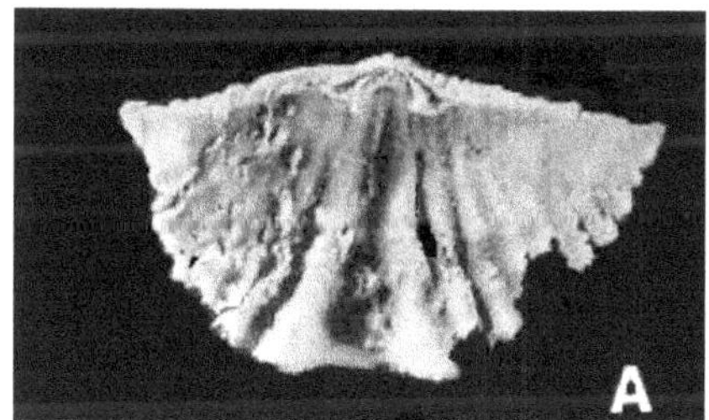

FIGURE 49. *Mucrospirifer* cf. *macra* (Hall, 1857). A. AMNH Loc. 3138A. Brachial valve interior, AMNH 42750, x2. B. AMNH Loc. 3137. Brachial valve interior, AMNH 42751, x2.

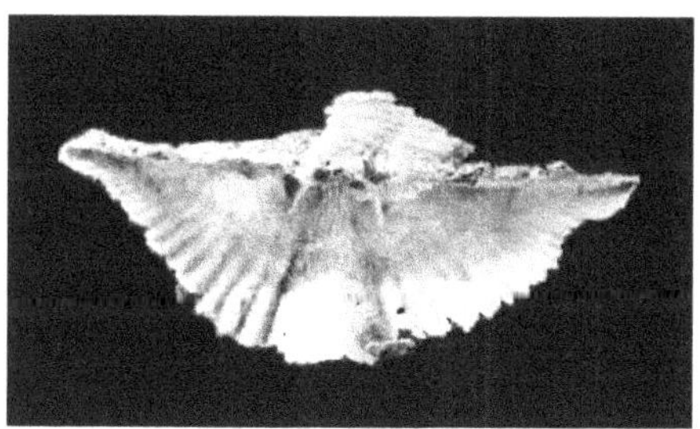

FIGURE 50. *Mucrospirifer* cf. *macra* (Hall, 1857). AMNH Loc. 3137. Pedicle valve interior, AMNH 42752, x1.5.

30 degrees, divides the interarea medially. No deltidial plates were observed. Faint, horizontal striations are present in the interarea of one specimen. The dorsal interarea is long, thin, ribbonlike, and apsacline. The brachial valve bears a high, medial fold flattened at the top and raised above the crests of the adjoining lateral plications. The pedicle valve bears a corresponding U-shaped sulcus. Both fold and sulcus originate in the umbonal region. The surface of the shells is covered by sharply defined plications (normally 9 to 15 on each flank) which range from U-shaped to subangular (almost chevron-like on some specimens) in cross section. Numerous, concentric, frilly growth lamellae are present, but no fine radial ornamentation was observed.

Pedicle valve interior: The hinge teeth are small, pointed, dorsally directed and somewhat crescentic in some specimens. They are supported by thin, relatively straight, prominent dental plates which extend to the floor of the valve. The dental plates diverge at an angle equal to that which is enclosed by the delthyrium (20 to 30 degrees). Muscle scars are not impressed and a short, thin myophragm may or may not be present. The valve floor is crenulated anteriorly due to the impress of the plications. The crenulations disappear toward the umbonal cavity due to deposition of secondary shell material.

Brachial valve interior: The sockets are long, thin, U-shaped, and widely divergent anterolaterally. The inner margin of the interarea overhangs the lateral edges of the sockets. The crural bases attach to the inner edges of the sockets and are concave in appearance, although their form is highly variable. Although muscle scars are not impressed, there is a thin, beadlike myophragm present on some specimens, which extends almost half the length of the valve. The only cardinal process observed is bilobed and rests apically on a thickened callus in the notothyrial cavity. Other specimens bear a thickened region, ventrally directed, in the notothyrial cavity, which is longitudinally striated and represents the site of diductor attachment. The valve floor is crenulated due to the impress of the plications.

Comparison: The brachial interior of the Onondaga shells is almost identical in its morphology to that of Boucot and Johnson's (1968: B15, pl. 6, fig. 4) *"Mucrospirifer"* cf. *macra* from the Bois Blanc Formation. Specimens from both the Onondaga and Bois Blanc have 15 rounded plications on each brachial flank (although there is some variation in Onondaga shells in which 9 to 12 plications are sometimes present). Boucot (1973: 60, pl. 18, figs. 21-22) describes *"M."* cf. *macra* from the Moose River Synclinorium based upon a single, poorly preserved pedicle valve internal mold. Although no meaningful comparison can be made from this material, it is noteworthy that the interarea is similar to that of some variants of "*M.*" cf. *macra* from the Onondaga Limestone.

Specimens of "*Mucrospirifer*" from the Middle Devonian Silica Formation (Kesling and Chilman, 1975) differ from "*Mucrospirifer*" cf. *macra* as follows: *M. prolificus* is larger, more alate, and has a wider interarea. *Mucrospirifer mucronatus* has a narrower pedicle interarea and *M. profundus* is more globose, less alate, and has a thicker shell. *Mucrospirifer grabaui* differs in the unusually long mucronate hinge extensions, resulting in an extreme alate morphology. The Onondaga shells lack a shallow median groove on the dorsal fold.

Community occurrence: Feldman (1980) found this species in the *Atrypa-Coelospira-Nucleospira* and *Atrypa-Megakozlowskiella* communities.

Figured specimens: AMNH 39919, 42751, 42752, 42753, 42754, 42761, 42762, 42763, 42764.

Family CYRTINOPSIDAE Wedekind, 1926
Subfamily CYRTINOPSINAE Wedekind, 1926
Genus *KOZLOWSKJELLINA* Boucot, 1958 (*KOZLOWSKIELLA* Boucot, 1957)
Subgenus *MEGAKOZLOWSKIELLA* Boucot, 1957
Type species: *Spirifer perlamellosus* Hall, 1857: 57.

Megakozlowskiella raricosta (Conrad, 1842), Figures 51-52.

Spirifer perlamellosus Hall, 1857: 57.

Delthyris raricosta Conrad, 1842: 262, pl. 14, fig. 18.

Spirifer raricosta Hall, 1867: 192, pl. 27, figs. 30-34; pl. 30, figs. 1-9.

Kozlowskiella (Megakozlowskiella) raricosta Boucot, 1957: pl. 3, figs. 18-19.

Megakozlowskiella cf. *raricosta* Johnson, 1970: 204, pl. 70, figs. 26-28.

Megakozlowskiella raricosta Boucot and Johnson, 1968: B16, pl. 6, figs. 7-15.

Exterior: The shells are subtransverse in outline, strophic, medium to large (up to 24.5 mm long and 33.6 mm wide) and ventribiconvex with a lessening convexity in both valves anteriorly. The hinge line is straight with the maximum width usually at the hinge line but occasionally slightly anterior to it. The pedicle interarea is moderately narrow and apsacline, but in some gerontic specimens tends to be almost catacline. Striae are present in the pedicle interarea which parallel the hinge line and extend the entire length of the interarea, becoming less distinct as the delthyrium is

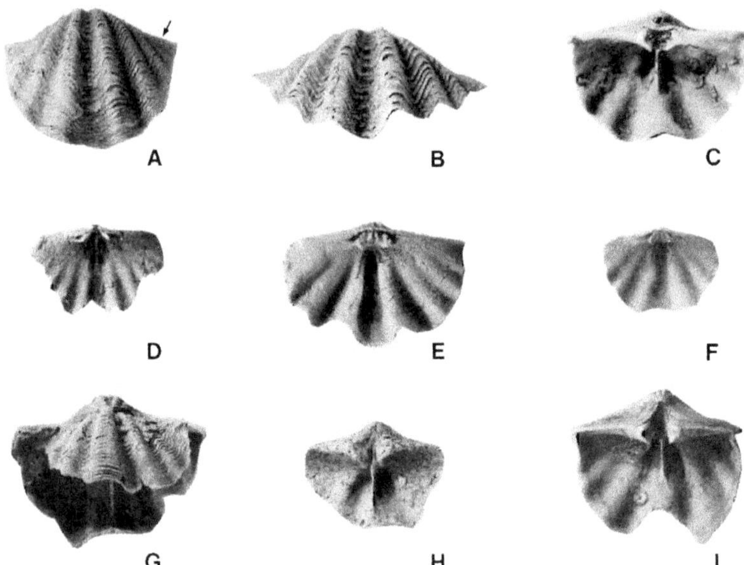

FIGURE 51. *Megakozlowskiella raricosta* (Conrad, 1842). A-C. AMNH Loc. 3137. Pedicle valve exterior (note arrow on fig. 46A denoting growth line along which an enlarged photograph was cropped in order to illustrate detail of external ornamentation), pedicle valve interior, AMNH 39920, x1.2, x2, x1.2. D. AMNH Loc. 3137. Brachial valve interior, AMNH 42735, x1.2. E. AMNH Loc. 3137. Brachial valve interior, AMNH 39922, x1.5. F. AMNH Loc. 3137. Brachial valve interior, AMNH 42736, x1.3. G. AMNH Loc. 3149. Dorsal view with anterior commissure dissected exposing well developed, high, thin median septum, AMNH 42759, x1. H. AMNH Loc. 3137. Severely weathered pedicle valve interior, AMNH 42734, x1.2. I. AMNH Loc. 3137. Pedicle valve interior, AMNH 42733, x1.3.

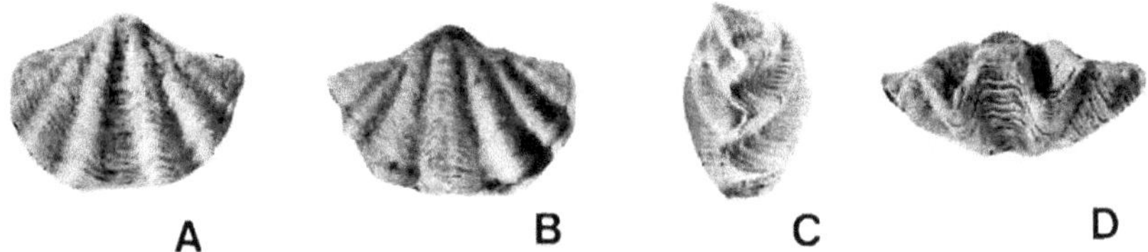

FIGURE 52. *Megakozlowskiella raricosta* (Conrad, 1842). A-D. AMNH Loc. 3135. Ventral, dorsal, lateral, and anterior views, AMNH 39921, x1.

approached. These striae appear to be extensions of the growth lines. The brachial interarea is extremely narrow and in articulated specimens is hidden by the ventral umbo.

There is a distinct, slightly flattened fold on the brachial valve and a corresponding deep, V-shaped sulcus on the pedicle valve. Both fold and sulcus originate deep in the umbonal region. There is usually a minimum of three plications on the flanks of both the pedicle and brachial valves. As the flanks are approached the plications become narrower and less prominent. Between the plications are deep U-shaped interspaces which, in rare cases, tend to be almost V-shaped.

The delthyrium includes an angle of approximately 60 degrees in most specimens. No deltidial plates were preserved although there are indications of narrow ridges on the sides of the delthyrium of some specimens indicating the probable presence of deltidial plates.

The anterior commissure is uniplicate with crenulations corresponding to the ends of the coarse plications on the flanks (two to five on each flank) (Table 10). Strong, concentric growth lamellae possess anterior frills (numbering 6 to 11 per 5 mm) in unworn specimens. Radial ornamentation consists of extremely fine striae, or capillae, noticeably more prominent at the anterior ends of the growth lamellae. These appear to be the terminal ends of radial spines and are especially well developed in the pedicle sulcus, where there are approximately 25 at the anterior commissure.

Pedicle valve interior: Small, pointed hinge teeth are attached to the dental lamellae along the margin of the hinge line. The dental lamellae are short, blunt, and directed medially at an angle of approximately 45 degrees. Two thin, posteriorly directed ridges extend from the base of the dental lamellae through the umbonal cavity toward the beak. In neanic and gerontic individuals, a large amount of secondary shell material is deposited in the umbonal region. A high, thin, well developed median septum extends from the anterior base of the umbonal cavity to a point approximately two-thirds to three-quarters of the valve length. The median septum never comes in contact with the base of the brachiophores but extends through the delthyrium to the beak. In older specimens, the anterior limit of the median septum is marked by a sudden convexity in the valve floor such that the convexity of the valve appears to increase greatly at that point. In one specimen the median septum increases in height at about midlength to a maximum of more than 50 percent of the distance to the floor of the brachial valve, and then drops off quickly to the pedicle valve floor. In all other specimens the median septum is relatively low.

TABLE 10. Measurements (in mm) of *Megakozlowskielia raricosta* (Conrad, 1842).

AMNH Locality	Valve Present	Length p.v.	Length b.v.	(W)	(T)	Number of plications p.v.	Number of plications b.v	Frills no./ 5mm	Length p.v./ width ratio	Length b.v./ width ratio
3149	art.	24.5	20.7	33.6	22.5	6	7	8	0.729	0.616
3135	art.	14.6	13.7	21.6	10.0	6	5	9	0.676	0.634
3137 def.	art.	23.7	19.9	26.7	18.0	6	–	7	0.888	0.745
3137	art.	16.2	–	23.1	14.6	8	7	11	0.701	–
3137	art.	17.0	12.8	20.4	11.5	6+	5+	6	0.833	0.627
3138A	p.v.	17.0		24.7	–	6	–	7	–	–
3135	p.v.	15.0		17.0	–	6	–	–	–	–
3135	p.v.	10.9		–	–	6	–	7	–	–
3137	p.v.	20.5		22.5	–	8	–	7	–	–
	p.v.	22.9		25.5	–	8	–	7	–	–
	p.v.	19.7		–	–	8	–	–	–	–
	p.v.	16.5		21.9	–	8	–	–	–	–
	p.v.	10.8		14.9	–	8	–	–	–	–
	p.v.	7.8		11.0	–	6	–	–	–	–
	p.v.	6.9		9.5	–	6	–	–	–	–
	p.v.	8.2		11.9	–	6	–	–	–	–
	b.v.		18.4	21.8	–	–	7	9	–	–
	b.v.		16.9	23.3	–	–	7	8	–	–
	b.v.		15.5	22.0	–	–	7	7	–	–
	b.v.		17.7	20.1	–	–	7	9	–	–
	b.v.		16.0	–	–	–	7	–	–	–
	b.v.		13.2	18.2	–	–	5	8	–	–
	b.v.		13.0	20.8	–	–	9	–	–	–
	b.v.		10.1	13.8	–	–	7	–	–	–
	b.v.		9.4	–	–	–	7	–	–	–
3138	art.	18.8	16.7	–	12.7	6	–	7	–	–
3139	p.v.	15.3	–	16.5	–	6	–	9	–	–

Note: the number of plications on the brachial valve includes the fold.

The muscle field is represented by two longitudinal pits (diductor impressions) adjacent to the median septum extending from the base of the dental lamellae to a point about one-third to one-half the valve length. Adductor and adjustor impressions are not discernible.

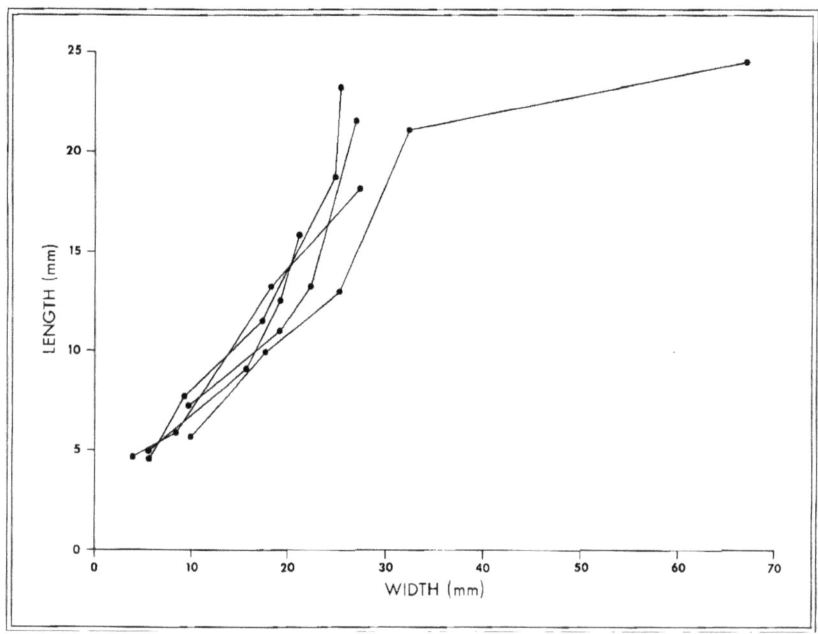

FIGURE 53. Length/width ontogenies of five pedicle valves of *Megakozlowskiella raricosta* (Conrad, 1842).

The crenulations of the external plicae are strong at the anterior periphery and extend posteriorly along the valve floor, becoming progressively weaker.

Brachial valve interior: The dental sockets are U-shaped, constricted posteriorly, and widen and shallow out anteriorly. The outer socket ridges are smooth and diverge laterally at an angle of from 20 to 35 degrees from their origin at the brachial interarea. The crural plates, small, slightly concave, and somewhat triangular, extend anteroventrally a short distance from the notothyrial platform.

The cardinal process is deeply striated in an anteroposterior direction. There is an indication of a bilobed cardinal process in some specimens studied, but due to erosion most display worn cardinalia. The cardinal processes are generally not well preserved. Anterior to the notothyrial cavity extends a low, medially situated, indistinct myophragm which is worn in most specimens. The myophragm generally extends about one-third to one-quarter the valve length, becoming progressively lower and thinner anteriorly. The myophragm divides the dorsal muscle field in which the sites of the adductors are represented by two pits, shallowing-out anteriorly and disappearing before midlength. The dorsal muscle field is not well preserved. The interior of the valve reflects the external plications and is strongly corrugated, especially anteriorly. The anterior commissure is crenulated.

Comparison: *Megakozlowskiella* may be differentiated from *Kozlowskiella* by its greater size (length, width, and thickness), and more convex brachial valve (Figure 53). Internally there is little to distinguish the two genera. The cardinalia are almost identical.

Kozlowskiella strawi Boucot (1957) is easily distinguished by its flatter brachial valve, less prominent median septum, and more alate appearance.

Kozlowskiella (M.) praenuntia (Swartz, 1929) has a flatter pedicle interarea, less incurved beak, and is generally smaller in size.

Plicocyrtina sp. differs from *Megakozlowskiella* in its more developed spondylium, weaker growth lamellae, and median plication within the pedicle sulcus.

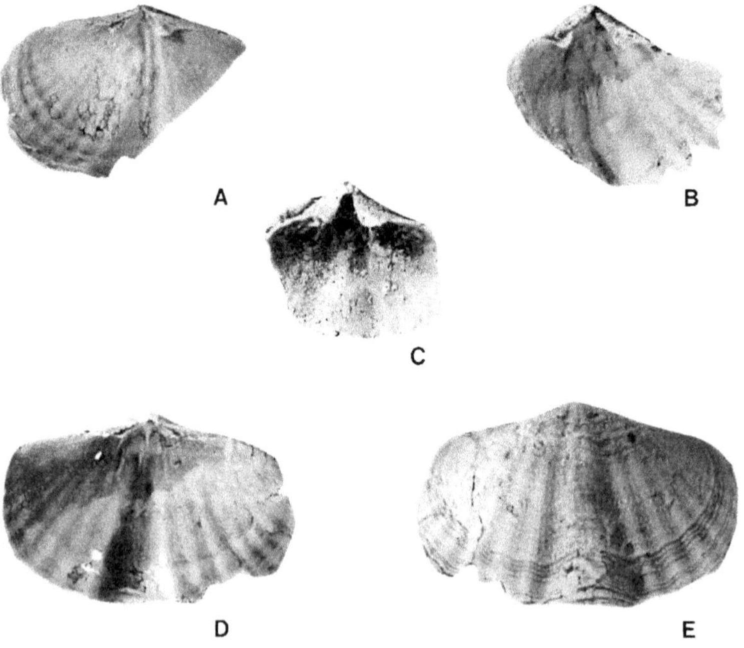

FIGURE 54. *Elytha fimbriata* (Conrad, 1842). A. AMNH Loc. 3138A. Brachial valve interior, AMNH 42741, x2. B. AMNH Loc. 3138A. Brachial valve interior, AMNH 42742, x2.5. C. AMNH Loc. 3138A. Pedicle valve interior, AMNH 42743, x2.5. D, E. AMNH Loc. 3139. Brachial valve interior and exterior, AMNH 39923, x2.

Megakozlowskiella magnapleura from the Great Basin, Nevada (Johnson, 1970: 202, pl. 71, figs. 1-19), has fewer plications and is more subquadrate in outline.

Hall (1867: 192, pl. 27, figs. 30-34; pl. 30, figs. 1-9) describes *M. raricosta* from the Schoharie Grit (Helderberg Mountains and Schoharie, New York), and from the Onondaga Limestone at Stafford, Caledonia, and Williamsville, all in western New York. Hall also reports occurrences in Columbus, Ohio, Falls of the Ohio, and Canada West.

Community occurrence: Feldman (1980) found this species in the *Leptaena-Megakozlowskiella, Atrypa-Coelospira-Nucleospira, Atrypa, Atrypa-Megakozlowskiella. "Pacificocoelia,"* and *Hallinetes* communities.

Figured specimens: AMNH 39920, 39921, 39922, 423733, 423734, 423735, 423736, 423759.

Superfamily RETICULARIOIDEA Waagen, 1883
Family ELYTHIDAE Frederiks, 1924
Subfamily ELYTHINAE Frederiks, 1924
Genus *ELITA* Frederiks, 1918
Type species: *Delthyris fimbriatus* Conrad, 1842: 263.

Elita fimbriata (Conrad, 1842), Figure 54.

Delthyris fimbriatus Conrad, 1842: 263.

Spirifer fimbriata Hall, 1867: 214, pl. 33, figs. 1-21.

Elytha sp. Boucot and Johnson, 1968: B18, pl. 7, figs. 1-5.

Elytha fimbriata Goldring, 1943: 236, fig. 43J; Cooper, 1944: 327, pl. 126, figs. 1-3.

Exterior: The shells are medium-sized (Table 11), biconvex in lateral profile, and transversely oval in outline. Both valves are approximately equal in depth with the brachial valve a little less convex. Maximum width is attained anterior to the hinge line, at about midlength. The hinge line is relatively short, about half the maximum width. The beak is short and erect. The ventral interarea is low, and ranges from slightly catacline to apsacline. The dorsal interarea is extremely low and narrow, and appears to be apsacline. The pedicle valve bears a distinct, moderately shallow, triangular sulcus which originates at the beak. The brachial valve bears a corresponding low, rounded fold.

The lateral slopes are covered by faint plications, broadly U-shaped in cross section, which disappear as the lateral margins are approached.

TABLE 11. Measurements (in mm) of Disarticulated Valves of *Elita fimbriata* (Conrad, 1842).

AMNH Locality	p.v.	b.v.	(L)	(W)	No. of plications per flank
3139	–	X	18.3	27.9	6
3139	–	X	15.2	–	7
3139	–	X	12.8	–	4
3137	X	–	7.9	9.5	3 est.
3137	X	–	11.2	15.0	5
3137	–	X	10.2	13.6	4 est.
3137	–	X	10.0	13.5 est.	5
3137	X	–	17.4	–	6
3137	X	–	11.2	12.3 est.	–
3137	X	–	11.5	13.4 est.	–

The plications are separated by equally indistinct U-shaped interspaces. Crossing the plications are concentric growth lamellae, more numerous anteriorly, which terminate in short, attenuated spines (5 per 1 mm in juveniles, but indistinct in larger specimens due to exfoliation). The rows of spines give the appearance of costellae when viewed without a lens. The anterior commissure is uniplicate.

Pedicle valve interior: The hinge teeth are short, pointed, and somewhat laterally directed. They are supported by slightly divergent dental lamellae, which extend to the floor of the valve and then anteriorly for several millimeters. There is an open delthyrium with no indication of deltidial plates or a pseudodeltidium. A long, low, narrow myophragm, extending from the apex of the delthyrial cavity through the faintly impressed muscle field, gradually merges with the ridge which represents the impress of the sulcus. The valve floor is faintly crenulated due to the impress of the plications.

Brachial valve interior: The sockets diverge, broaden anterolaterally, and are covered posteriorly by the overlapping hinge line. Broad crural plates, medially concave, partially adjoin the floor ofthe valve but do not unite into a septalium. In some specimens there is a moderate buildup of secondary shell material in the notothyrial cavity, but an insufficient amount to cause obsolescence of the crural plates. A short, low myophragm is present in adult specimens but not in juveniles. The muscle field is obscure and too faintly impressed to determine the position of the scars. The valve floor is crenulated by the impress of the plications, especially anteriorly. On some specimens the concentric growth lamellae are also impressed upon the floor of the valve, anteriorly.

Comparison: *Elytha* sp. described by Boucot and Johnson (1968: B18, pl. 7, figs. I-5) from the Bois Blanc Limestone is assigned here to *E. fimbriata* based upon its oval shape, short hinge line, length of the myophragm, and presence of dental lamellae. Although the material is fragmentary, the shells appear remarkably like those found in the overlying Onondaga Limestone (compare Figure 51C).

Hall (1867) described *E. fimbriata* from the Oriskany Sandstone at Saugerties, New York, and at Knox in Albany County. He also found it in the Schoharie Grit in Albany and Schoharie counties and in the Onondaga Limestone at Cherry Valley, Westmoreland (Oneida County), Onondaga Hollow (Onondaga County), Stafford (Genesee County), Williamsville, Clarence (Erie County), and in Canada West and Columbus, Ohio.

Community occurrence: Feldman (1980) found this species in the *Atrypa-Coelospira-Nucleospira, Atrypa, Atrypa-Megakozlowskiella,* and *Levenea* Community I communities.

Figured specimens: AMNH 39923, 42741, 42742, 42743.

Eospiriferid? indet., Figure 55.

Remarks: Because of paucity of material this shell cannot yet be assigned to a genus. However, the following observations can be made: the specimen is a single, mediumsized, convex pedicle valve, moderately transverse, plicate, and covered by fine radiating striae. There is a well-developed sulcus which originates in the beak region. Internally, the hinge teeth are supported by strongly arched dental plates which reach to the floor of the valve. The delthyrium is triangular with slightly corrugated edges indicating the possible presence of deltidial plates or a pseudodeltidium.

The general morphology fits Boucot's (1963) diagnosis of the superfamily Eospiriferinae. *Eospirifer* differs in its unplicated flanks, whereas *Striispirifer* is similar in its unplicated sulcus, laterally bordered by numerous striae. The radial ornamentation on the Onondaga material is very poorly preserved.

Savage (1974: 34) notes that within the eospiriferids, there are several intermediate stages of plication in addition to the unplicated flanks of *Eospirifer* at one extreme, and the broadly rounded plications immediately bordering the fold and sulcus of *Macropleura* at the other. Shells of *E. parahentius* from the Maradana Shale in New South Wales possess weak lateral plications that are variably developed. In the Onondaga specimen the plications are stronger and less broad.

Community occurrence: Feldman (1980) found this species in the *Atrypa-Coelospira-Nucleospira* Community.

Figured specimen: AMNH 39927.

Superfamily AMBOCOELIOIDEA George, 1931
Family AMBOCOELIIDAE George, 1931
Subfamily AMBOCOELIINAE George, 1931
Genus *AMBOCOELIA* Hall, 1860

Ambocoelia sp., Figure 56.

Exterior: The shells are small (the largest is 4.7 mm in length), ventribiconvex, and subcircular in outline. The pedicle valve is smooth and bears a weak sulcus which originates at the umbo and extends to the anterior commissure. The beak is incurved and the pedicle interarea apsacline. The hinge line is straight and the delthyrium open. The greatest width is attained anterior to the hinge line at about midlength. The brachial valve is slightly convex, with no ornamentation. The anterior commissure is rectimarginate to uniplicate to slightly intraplicate.

Pedicle valve interior: Short hinge teeth are located at the anterior end of thin fossetted tracks which form the inner boundaries of the delthyrium. No dental lamellae are present. A weak, low myophragm originates at the posterior end of the umbonal cavity and terminates at about midlength. Longitudinal muscle scars are weakly impressed on either side of the myophragm.

Brachial valve interior: No brachial interiors are in the collection.

Comparison: The shells are assigned to *Ambocoelia* sp. based on the presence of a median pedicle valve furrow and lack of plications on the flanks. *Metaplasia* has a sulcus on the brachial valve and a fold on the pedicle valve. *Plicoplasia* has a biplicate fold (see Boucot, Gauri, and Southard, 1970: 15) and a strong plication on the brachial sulcus.

Ambocoelia sp. from the Bois Blanc Formation (Boucot and Johnson, 1968: B18, pl. 7, figs. 6-10) has a shorter delthyrium and a less incurved, stubbier beak.

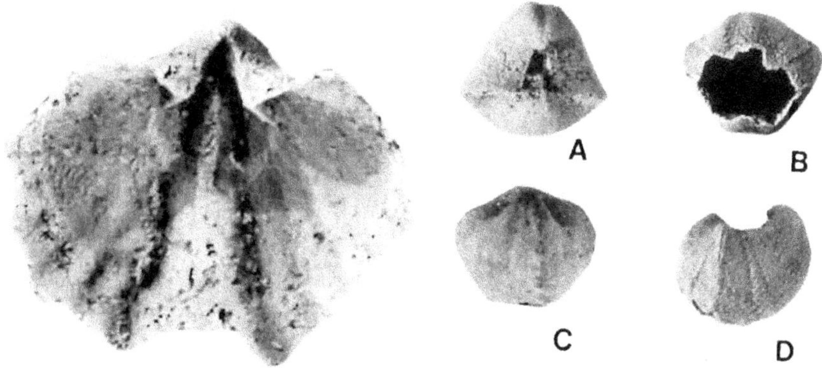

FIGURE 55. Eospiriferid? AMNH Loc. 3135. Pedicle valve interior, AMNH 39927, x2.5.

FIGURE 56. *Ambocoelia* sp. A-D. AMNH Loc. 3135. Posterior, anterior, ventral, and lateral views, AMNH 39924, x5.

Community occurrence: Feldman (1980) found this species in the *Atrypa-Coelospira-Nucleospira* Community.

Figured specimen: AMNH 39924.

Order SPIRIFERIDINA Ivanova, 1972
Suborder CYRTINIDINA Carter and Johson, 1994
Superfamily CYRTINOIDEA Frederiks, 1911 [*nom. transl.* Johnson, 1966, ex. Cyrtininae Frederiks, 1912]
Family CYRTINIDAE Frederiks, 1911
Genus *CYRTINA* Davidson, 1858
Type species: *Cyrtia hamiltonensis* Hall, 1857: 166.

Cyrtina hamiltonensis (Hall, 1857), Figures 57A-D, F, 58.

Cyrtia hamiltonensis Hall, 1857: 166.

Cyrtina hamiltonensis Hall, 1867: 268, pl. 27, figs. 1-4; pl. 44, figs. 26-33, 38-52.

Exterior: The shells are small (Table 12), hemipyramidal in outline, wider than long, with a straight hinge line anterior to which maximum width is usually attained. In some specimens, maximum width is at the hinge line. The ventral interarea is high, smooth, and catacline to slightly apsacline. The beak curves in at the apex of the pedicle valve, forming a concavity in the ventral aspect of the interarea. The delthyrium encloses an angle of about 20 to 30 degrees, but is variable. A convex pseudodeltidium covers the triangular delthyrium in about half of the specimens studied. In the others, the delthyrium is open, but longitudinal grooves are present along the margins of the delthyrial cavity representing the seat of attachment for the pseudodeltidium.

The pedicle valve bears a triangular, smooth sulcus, V-shaped in cross section, which originates at the apex of the umbo and extends and expands anteriorly until the anterior commissure is reached. Two to three rounded plications extend along the pedicle flanks. Concentric growth lamellae are present, mainly at the anterior portion of the valve. The anterior commissure is uniplicate.

The brachial valve bears a fold corresponding to the pedicle sulcus. Three to four lateral plications were observed on either side of the fold. The only ornamentation consists of concentric growth lamellae.

Pedicle valve interior: The small, blunt, hinge teeth astride the hinge margin are anterodorsal extensions of the elongate dental plates. The dental plates form a V-shaped spondylium bisected by a median septum running from the posterior extremity of the valve to the anterior end of the spondylium, where it drops precipitously to the valve floor and continues for approximately three-fourths the length of the valve. The spondylium bears a small, suboval, tichorhinum formed by struts extending from the abaxial slopes of the dental plates to the median septum. The tichorhinum probably functioned as a housing for the pedicle roots. Two longitudinal troughs, probably diductor supports, extend the length of the spondylium dorsal to the tichorhinum. The adductors could have attached to the keel-like median septum within the spondylium.

Brachial valve interior: There is only one free brachial valve in the collection so that an adequate description of the brachial interior is somewhat limited, especially since the single specimen is slightly damaged.

The cardinal process is short, blunt and transverse in outline. The posterior end is eroded but might have shown a triangular outline if present. Small, but well-excavated, anterolaterally diverging sockets lie on either side of the cardinal process. No muscle scars are evident. The interarea is narrow and apsacline to orthocline. The external plications are impressed upon the floor of the valve which is crenulated at the anterior commissure.

Comparison: *Cyrtina hamiltonensis* may be differentiated from the Middle Devonian *C. umbonata* (Cooper, 1944: 359, pl. 140, figs. 40-42) by its larger size and less incurved umbo. *Cyrtina* cf. *varia* (Johnson, 1970:

219-220, pl. 73, figs 1-14) has a shallower sulcus and more acute cardinal angles, and is less circular when viewed dorsally. *Cyrtina alpenensis* from the Middle Devonian Traverse Group of Michigan is considerably larger than *C. hamiltonensis* whereas the Mississippian *C. acutirostris* is smaller and more rectangular in outline.

Community occurrence: Feldman (1980) found this species in the *Atrypa-Coelospira-Nucleospira*, *Atrypa*, and *Atrypa-Megakozlowskiella* communities.

Figured specimens: AMNH 39925, 42775, 42776.

Cyrtina sp. A, Figure 57E, G-I

Remarks: Two articulated specimens of shells assigned to *Cyrtina* sp. A have been recovered. They are distinguished from *Cyrtina hamiltonensis* by

TABLE 12. Measurements (in mm) of Articulated Specimens of *Cyrtina hamiltonensis* (Hall, 1857).

				Number of Plication	
AMNH Locality	(L)	(W)	(T)	Dorsal	Ventral
3141	4.5	6.7	6.3	6	6
3138A	6.4	6.6 est.	6.5	–	6
3138A	7.2	8.1	5.5	6	6
3138A	4.6	5.0	4.0	6	–
3138A	5.4	6.6	5.2	4	6
3138A	3.6	4.5	3.8	6	6 est.
3137	9.8	11.7	11.7	8	10 est.
3137	3.0	4.3	3.2	4	6
3137	4.4	–	4.7	–	6 est.
3137	4.9	5.3	4.4	6	8
3137	5.8	8.0	5.7	4	6
3135	3.1	4.7	3.6	2	4
3135	4.1	6.5	4.7	4	6
3135	4.2	6.5	5.5	6	8
3135	5.4	7.8	6.8	4	6
3135	6.8	8.7	6.6	4	6
3135	5.5	8.0	7.0	–	8

[a] Excluding fold.

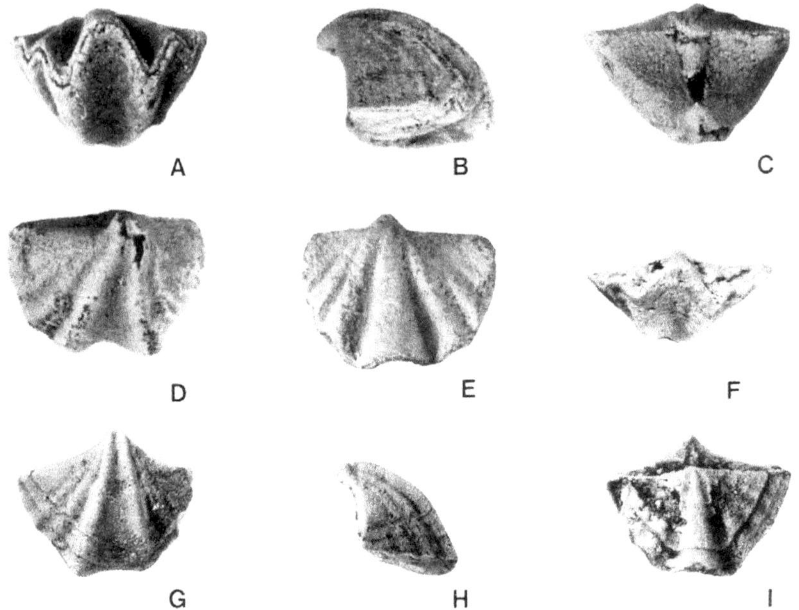

FIGURE 57. *Cyrtina hamiltonensis* (Hall, 1857). A-E. AMNH Loc. 3138A. Anterior, lateral, posterior, ventral and dorsal views, AMNH 39925, x4. *Cyrtina* sp. A. F-I. AMNH Loc. 3135. Anterior, ventral, lateral and dorsal views, AMNH 39926, x2.

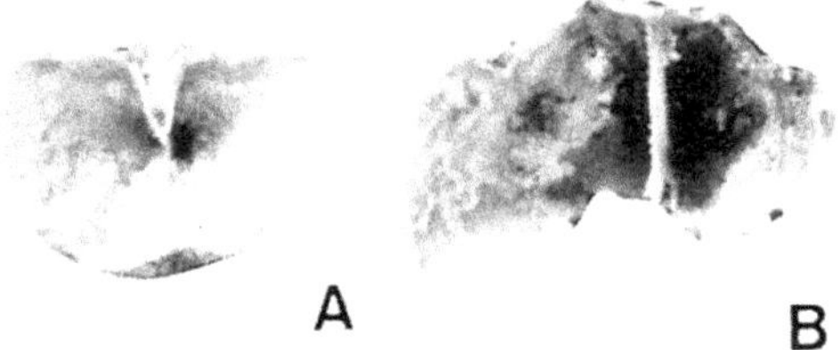

FIGURE 58. *Cyrtina hamiltonensis* (Hall, 1857). A. AMNH Loc. 3137. Pedicle valve interior, AMNH 42775, x4. B. AMNH Loc. 3137. Pedicle valve interior, AMNH 42776, x4.

FIGURE 59. *Amphigenia*? sp. AMNH Loc. 3126. Pedicle valve interior of poorly preserved specimen from the basal sandy (note matrix) zone of the Edgecliff Member, Onondaga Limestone, near Nedrow, New York, AMNH 39928, x2.

their larger size and more robust appearance. More material is need for specific designation.

Community occurrence: Feldman (1980) found this species in the *Atrypa-Coelospira-Nucleospira* Community.

Figured specimen: AMNH 39926.

Order TEREBRATULIDA Waagen, 1883
Suborder TEREBRATULIDINA Waagen, 1883
Superfamily STRINGOCEPHALOIDEA King, 1850
Family CENTRONELLIDAE Waagen, 1882
Subfamily AMPHIGENIIDAE Cloud, 1942
Genus *AMPHIGENIA* Hall, 1867

Amphigenia? sp., Figure 59.

Remarks: This species is represented by several spondylose fragments, suboval in outline and deeply convex, consistently found in the sandy basal Onondaga in the central region. The species seems to fit Boucot and Johnson's (1968) description of the unequally biconvex forms geographically limited to New York and the northern Appalachians (*A. elongata* and *A. parva*) rather than the subequally biconvex *A. curta* (Meek and Worthen) and *A. chickasawensis.* The spondylium is morphologically similar to the rhomboidal spondylium of *A. elongata* (Boucot and Johnson, 1968, pl. 8, figs. 1, 3).

Community occurrence: Feldman (1980) found this species in the *Amphigenia?* Community.

Figured specimen: AMNH 39928.

APPENDIX 1

The brachiopod faunas collected at key AMNH localities along the west-southeast transect of the outcrop belt of the Onondaga Limestone in central and southeastern New York are presented here in the form of histograms. The taxa at a specific locality are listed horizontally, whereas the percentage each represents at the given locality is plotted vertically. Taxa representing less than one percent of the fauna are not plotted. The total number of specimens collected at each locality is listed in Table 13. Based on the above data, several distributional patterns are evident.

Hallinetes aff. *lineatus* appears explosively in the *Hallinetes* Zone 10 ft above the Tioga Bentonite, from Syracuse to Cherry Valley (AMNH Locs. 3123A, 3128A, 3129) and occurs in constant density across the outcrop belt as far as Cherry Valley. At one locality (AMNH Loc. 3131B) in Cherry Valley the species was recovered from Upper Moorehouse rocks. The only record of its appearance in the east is in Upper Moorehouse strata in the mid-Hudson Valley (AMNH Loc. 3137).

"Pacificocoelia" acutiplicata occurs mainly in the central part of New York, except for its appearance as a minor faunal constituent in Moorehouse rocks (AMNH Loc. 3133 at Thompson's Lake). The species is also present in Upper Moorehouse rocks in the mid-Hudson Valley (AMNH Loc. 3137) but comprises less than one percent of the fauna. Toward the west abundance increases markedly, peaking at Nedrow, New York (AMNH Loc. 3125A) where it dominates the entire fauna (more than 60 percent).

The fauna of the *Atrypa-Coelospira-Nucleospira* Community found in the mid-Hudson Valley (especially at AMNH Locs. 3135, 3137) shows the greatest density of all Onondaga faunas studied in this report. In general, the fauna is recognizable in Upper Moorehouse rocks in the eastern part of the outcrop belt, but as one progresses west certain faunal elements are no longer present. For example, in Cherry Valley (AMNH Loc. 3131B) the

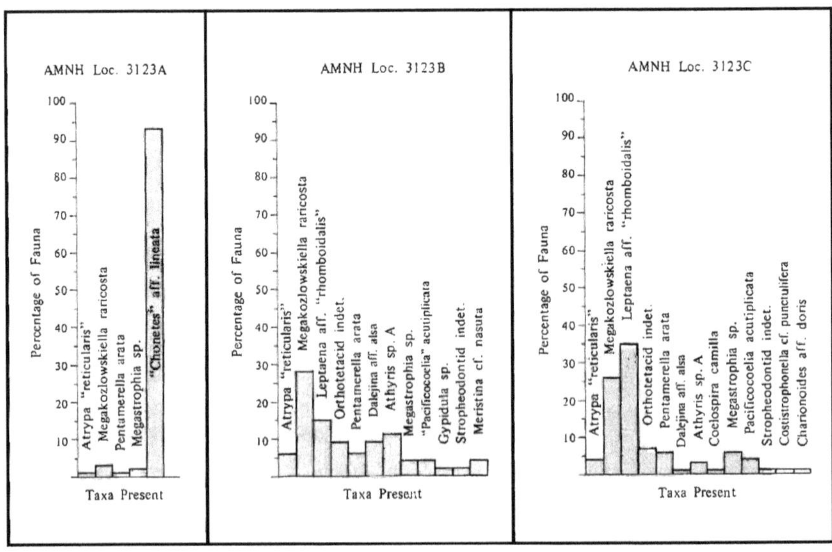

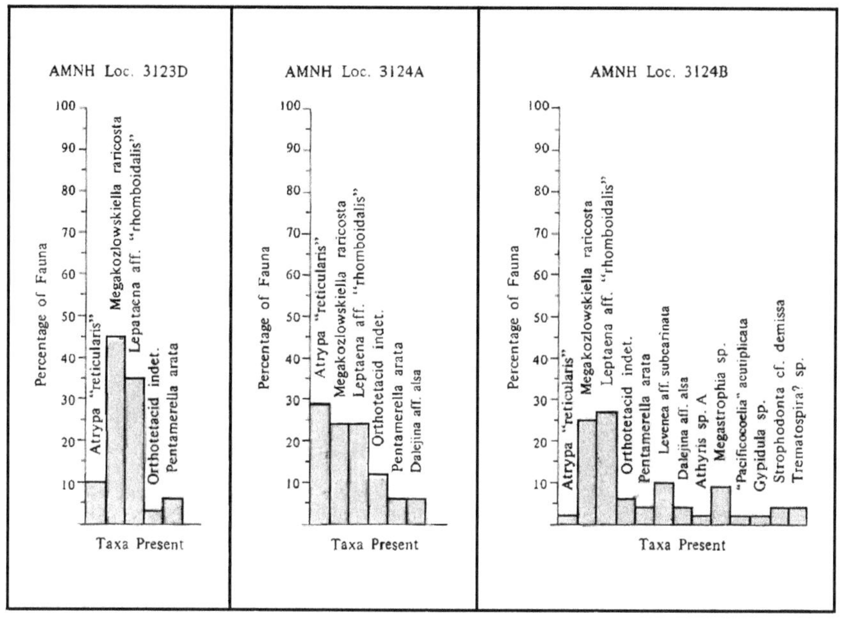
AMNH Loc. 3123D
Percentage of Fauna
Taxa Present
Atrypa "reticularis"
Megakozlowskiella raricosta
Lepataena aff. "rhomboidalis"
Orthotetacid indet.
Pentamerella arata
AMNH Loc. 3124A
Percentage of Fauna
Taxa Present
Atrypa "reticularis"
Megakozlowskiella raricosta
Leptaena aff. "rhomboidalis"
Orthotetacid indet.
Pentamerella arata
Dalejina aff. alsa
AMNH Loc. 3124B
Percentage of Fauna
Taxa Present
Atrypa "reticularis"
Megakozlowskiella raricosta
Leptaena aff. "rhomboidalis"
Orthotetacid indet.
Pentamerella arata
Levenea aff. subcarinata
Dalejina aff. alsa
Athyris sp. A
Megastrophia sp.
"Pacificocoelia" acutiplicata
Gypidula sp.
Strophodonta cf. demissa
Trematospira? sp.

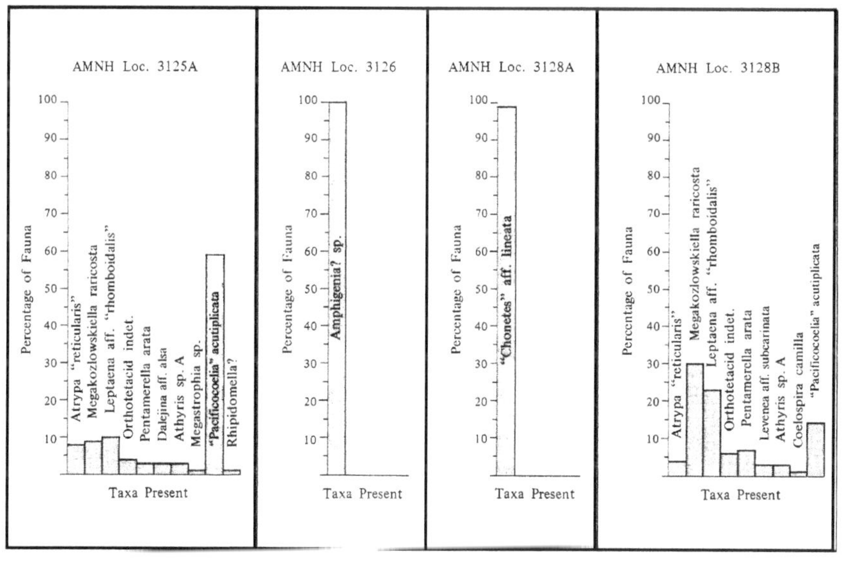
AMNH Loc. 3125A
Percentage of Fauna
Taxa Present
Atrypa "reticularis"
Megakozlowskiella raricosta
Leptaena aff. "rhomboidalis"
Orthotetacid indet.
Pentamerella arata
Dalejina aff. alsa
Athyris sp. A
Megastrophia sp.
"Pacificocoelia" acutiplicata
Rhipidomella?
AMNH Loc. 3126
Percentage of Fauna
Taxa Present
Amphigenia? sp.
AMNH Loc. 3128A
Percentage of Fauna
Taxa Present
"Chonetes" aff. lineata
AMNH Loc. 3128B
Percentage of Fauna
Taxa Present
Atrypa "reticularis"
Megakozlowskiella raricosta
Leptaena aff. "rhomboidalis"
Orthotetacid indet.
Pentamerella arata
Levenea aff. subcarinata
Athyris sp. A
Coelospira camilla
"Pacificocoelia" acutiplicata

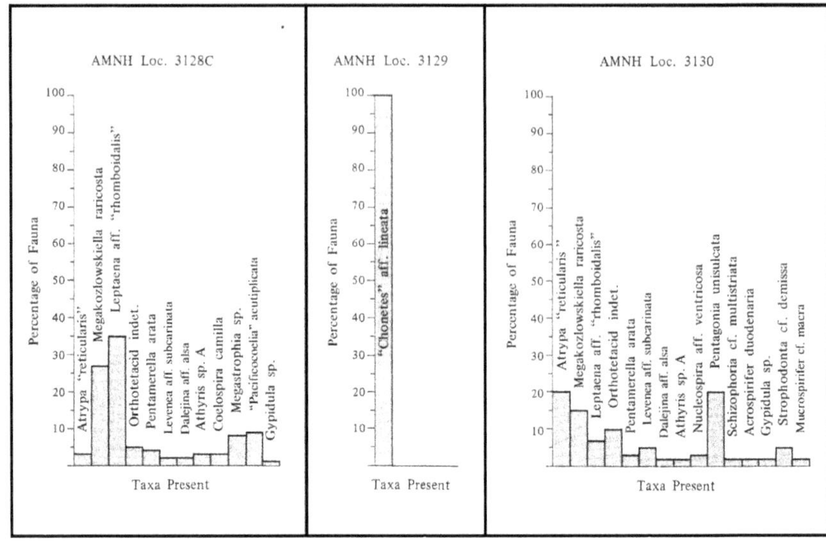
AMNH Loc. 3128C
Percentage of Fauna
Taxa Present
Atrypa "reticularis"
Megakozlowskiella raricosta
Leptaena aff. "rhomboidalis"
Orthotetacid indet.
Pentamerella arata
Levenea aff. subcarinata
Dalejina aff. alsa
Athyris sp. A
Coelospira camilla
Megastrophia sp.
"Pacificocoelia" acutiplicata
Gypidula sp.
AMNH Loc. 3129
Percentage of Fauna
Taxa Present
"Chonetes" aff. lineata
AMNH Loc. 3130
Percentage of Fauna
Taxa Present
Atrypa "reticularis"
Megakozlowskiella raricosta
Leptaena aff. "rhomboidalis"
Orthotetacid indet.
Pentamerella arata
Levenea aff. subcarinata
Dalejina aff. alsa
Athyris sp. A
Nucleospira aff. ventricosa
Pentagonia unisulcata
Schizophoria cf. multistriata
Acrospirifer duodenaria
Gypidula sp.
Strophodonta cf. demissa
Mucrospirifer cf. macra

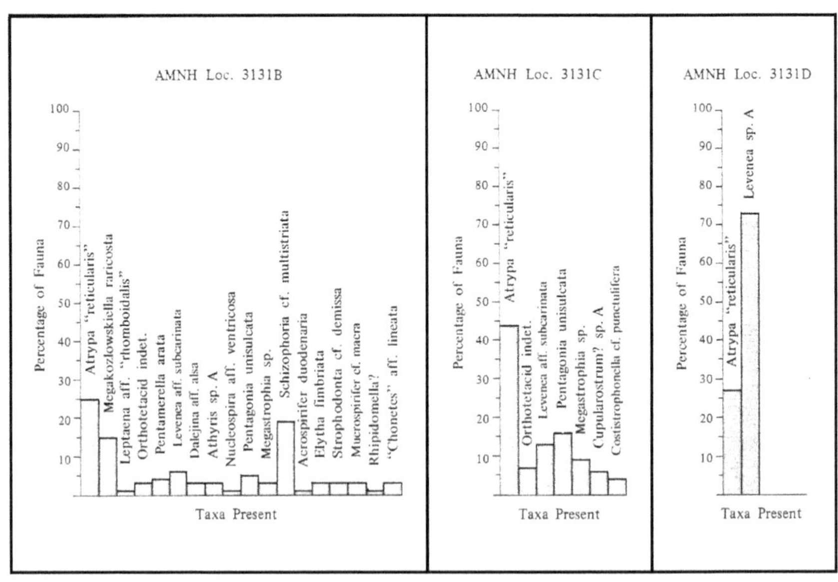
AMNH Loc. 3131B
Percentage of Fauna
Taxa Present
Atrypa "reticularis"
Megakozlowskiella raricosta
Leptaena aff. "rhomboidalis"
Orthotetacid indet.
Pentamerella arata
Levenea aff. subcarinata
Dalejina aff. alsa
Athyris sp. A
Nucleospira aff. ventricosa
Pentagonia unisulcata
Megastrophia sp.
Schizophoria cf. multistriata
Acrospirifer duodenaria
Elytha fimbriata
Strophodonta cf. demissa
Mucrospirifer cf. macra
Rhipidomella?
"Chonetes" aff. lineata
AMNH Loc. 3131C
Percentage of Fauna
Taxa Present
Atrypa "reticularis"
Orthotetacid indet.
Levenea aff. subcarinata
Pentagonia unisulcata
Megastrophia sp.
Cupularostrum? sp. A
Costistrophonella cf. punctulifera
AMNH Loc. 3131D
Percentage of Fauna
Taxa Present
Atrypa "reticularis"
Levenea sp. A

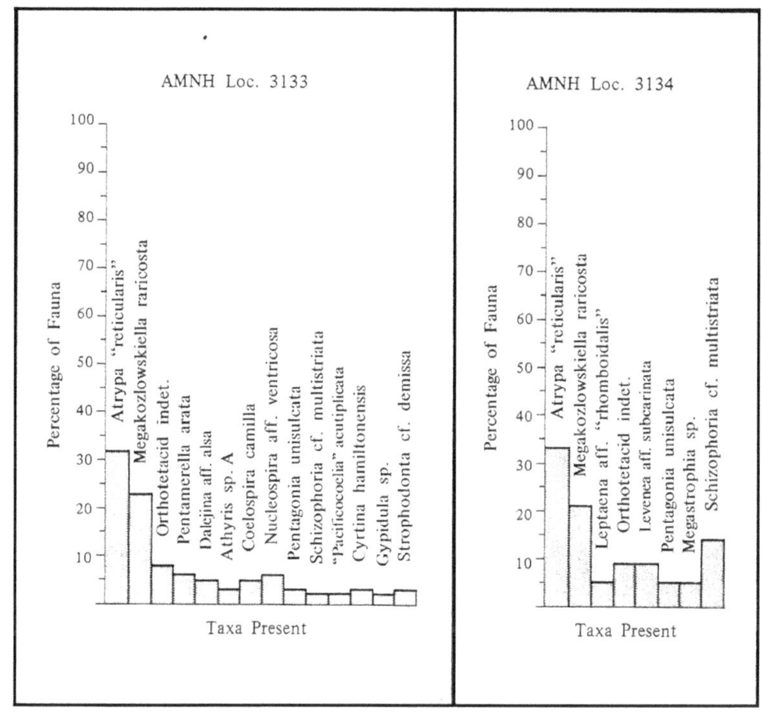
AMNH Loc. 3133
Percentage of Fauna
Taxa Present
Atrypa "reticularis"
Megakozlowskiella raricosta
Orthotetacid indet.
Pentamerella arata
Dalejina aff. alsa
Athyris sp. A
Coelospira camilla
Nucleospira aff. ventricosa
Pentagonia unisulcata
Schizophoria cf. multistriata
"Pacificocoelia" acutiplicata
Cyrtina hamiltonensis
Gypidula sp.
Strophodonta cf. demissa
AMNH Loc. 3134
Percentage of Fauna
Taxa Present
Atrypa "reticularis"
Megakozlowskiella raricosta
Leptaena aff. "rhomboidalis"
Orthotetacid indet.
Levenea aff. subcarinata
Pentagonia unisulcata
Megastrophia sp.
Schizophoria cf. multistriata

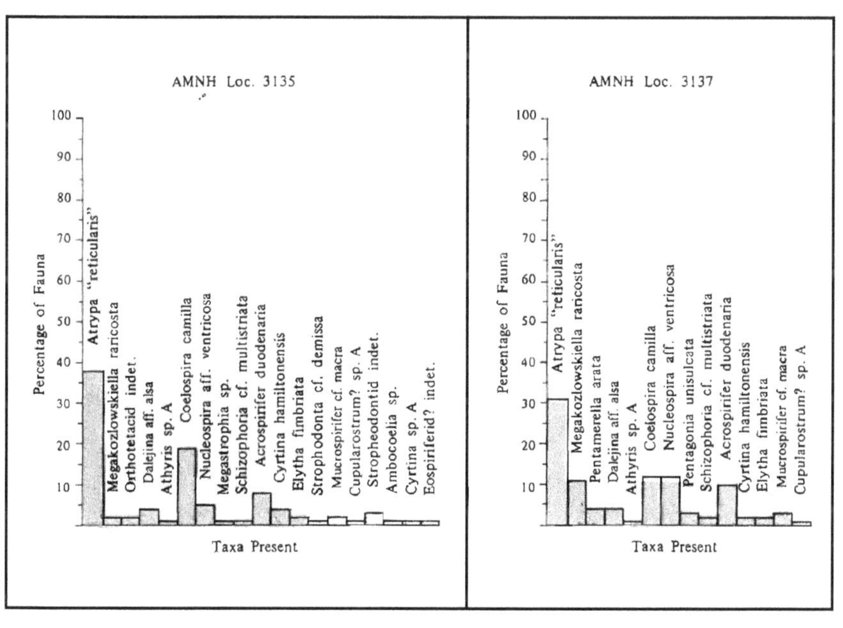
AMNH Loc. 3135
Percentage of Fauna
Taxa Present
Atrypa "reticularis"
Megakozlowskiella raricosta
Orthotetacid indet.
Dalejina aff. alsa
Athyris sp. A
Coelospira camilla
Nucleospira aff. ventricosa
Megastrophia sp.
Schizophoria cf. multistriata
Acrospirifer duodenaria
Cyrtina hamiltonensis
Elytha fimbriata
Strophodonta cf. demissa
Mucrospirifer cf. macra
Cupularostrum? sp. A
Stropheodontid indet.
Ambocoelia sp.
Cyrtina sp. A
Eospiriferid? indet.
AMNH Loc. 3137
Percentage of Fauna
Taxa Present
Atrypa "reticularis"
Megakozlowskiella raricosta
Pentamerella arata
Dalejina aff. alsa
Athyris sp. A
Coelospira camilla
Nucleospira aff. ventricosa
Pentagonia unisulcata
Schizophoria cf. multistriata
Acrospirifer duodenaria
Cyrtina hamiltonensis
Elytha fimbriata
Mucrospirifer cf. macra
Cupularostrum? sp. A

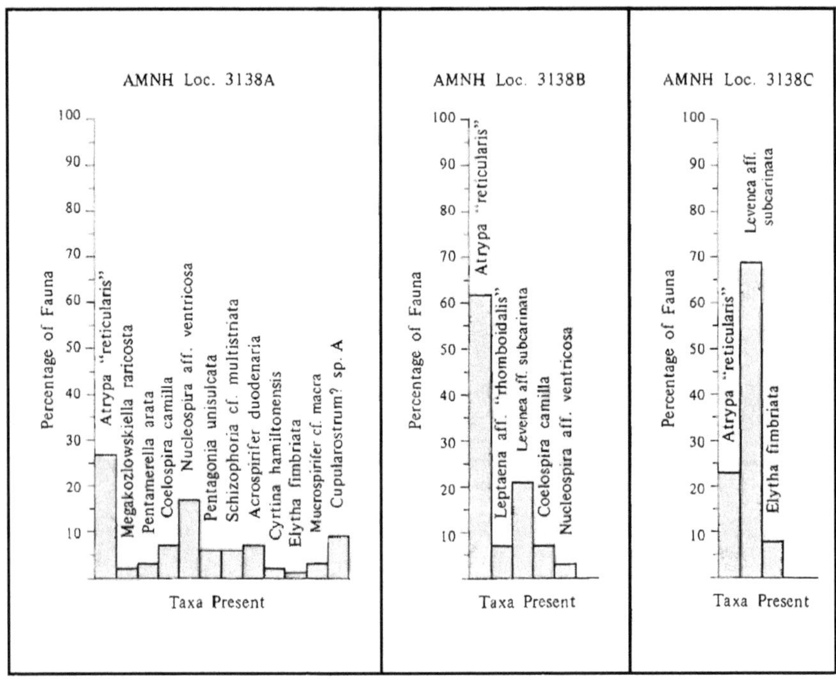
AMNH Loc. 3138A
Percentage of Fauna
Atrypa "reticularis"
Megakozlowskiella raricosta
Pentamerella arata
Coelospira camilla
Nucleospira aff. ventricosa
Pentagonia unisulcata
Schizophoria cf. multistriata
Acrospirifer duodenaria
Cyrtina hamiltonensis
Elytha fimbriata
Mucrospirifer cf. macra
Cupularostrum? sp. A
Taxa Present
AMNH Loc. 3138B
Percentage of Fauna
Atrypa "reticularis"
Leptaena aff. "rhomboidalis"
Levenea aff. subcarinata
Coelospira camilla
Nucleospira aff. ventricosa
Taxa Present
AMNH Loc. 3138C
Percentage of Fauna
Atrypa "reticularis"
Levenea aff. subcarinata
Elytha fimbriata
Taxa Present

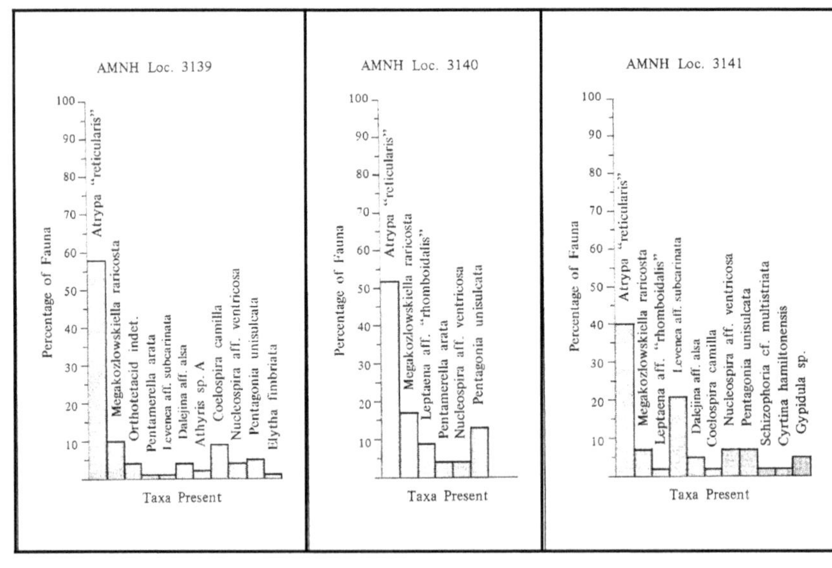
AMNH Loc. 3139
Percentage of Fauna
Atrypa "reticularis"
Megakozlowskiella raricosta
Orthotetacid indet.
Pentamerella arata
Levenea aff. subcarinata
Dalejina aff. alsa
Athyris sp. A
Coelospira camilla
Nucleospira aff. ventricosa
Pentagonia unisulcata
Elytha fimbriata
Taxa Present
AMNH Loc. 3140
Percentage of Fauna
Atrypa "reticularis"
Megakozlowskiella raricosta
Leptaena aff. "rhomboidalis"
Pentamerella arata
Nucleospira aff. ventricosa
Pentagonia unisulcata
Taxa Present
AMNH Loc. 3141
Percentage of Fauna
Atrypa "reticularis"
Megakozlowskiella raricosta
Leptaena aff. "rhomboidalis"
Levenea aff. subcarinata
Dalejina aff. alsa
Coelospira camilla
Nucleospira aff. ventricosa
Pentagonia unisulcata
Schizophoria cf. multistriata
Cyrtina hamiltonensis
Gypidula sp.
Taxa Present

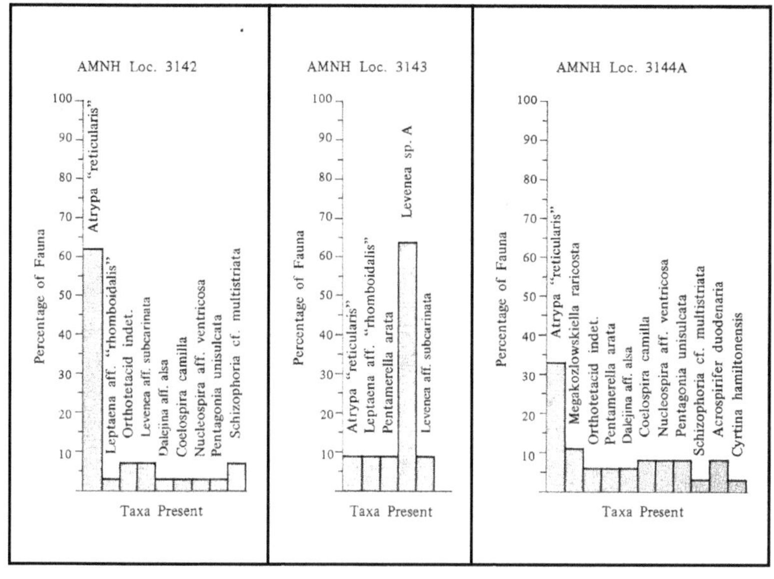
AMNH Loc. 3142
Percentage of Fauna
Atrypa "reticularis"
Leptaena aff. "rhomboidalis"
Orthotetacid indet.
Levenea aff. subcarinata
Dalejina aff. alsa
Coelospira camilla
Nucleospira aff. ventricosa
Pentagonia unisulcata
Schizophoria cf. multistriata
Taxa Present
AMNH Loc. 3143
Percentage of Fauna
Atrypa "reticularis"
Leptaena aff. "rhomboidalis"
Pentamerella arata
Levenea sp. A
Levenea aff. subcarinata
Taxa Present
AMNH Loc. 3144A
Percentage of Fauna
Atrypa "reticularis"
Megakozlowskiella raricosta
Orthotetacid indet.
Pentamerella arata
Dalejina aff. alsa
Coelospira camilla
Nucleospira aff. ventricosa
Pentagonia unisulcata
Schizophoria cf. multistriata
Acrospirifer duodenaria
Cyrtina hamiltonensis
Taxa Present

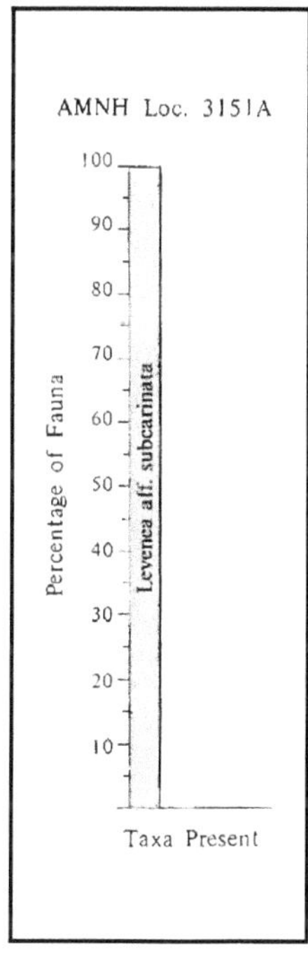
AMNH Loc. 3151A
Percentage of Fauna
Levenea aff. subcarinata
Taxa Present

following species are no longer present (the term "no longer present" in this discussion includes those species which appear as less than one percent of the fauna and should be read as "no longer present in significant amounts"): *Coelospira camilla, Cyrtina hamiltonensis, Cupularostrum* sp. A, strophe-odontid indet., *Ambocoelia* sp., and eospiriferid? indet. However, the following species not present in the east are present for the first time in Cherry Valley: *Leptaena* aff. *"rhomboidalis," Levenea* aff. *subcarinata, "Rhipidomella?,"* and *Hallinetes* aff. *lineatus.* Whereas the eastern faunas are dominated by *Atrypa "reticularis," Coelospira camilla, Nucleospira* aff. *ventricosa, Levenea* aff. *subcarinata, Pentaconia unisulcata,* and *Acrospirifer duodenaria,* the central faunas are generally dominated by *Leptaena "rhomboidalis"* and *Megakozlowskiella raricosta.*

APPENDIX 2: LOCALITY REGISTER

AMNH Loc. 3122, near Onondaga Hill, New York: A small exposure (1.5 ft) of the *Hallinetes* zone, Seneca Member, on the south side of Route 173, just west of Howlitt Hill Road. The rock is a very fine-grained, light gray limestone with no chert. Collecting is moderately good to poor.

AMNH Loc. 3123A-D, Nedrow, New York: This section is exposed in the Onondaga Indian Reservation Quarry just southwest of the junction of Route 11 and I-81. All four members of the Onondaga Formation are visible in outcrop at this locality, excluding the lower half (approximate) of the Edgecliff and the upper part of the Seneca. This is the type section of the Nedrow Member (Oliver, 1954). The Edgecliff Member is best exposed on the quarry floor at the southwest end of the quarry where the upper few feet are visible and accessible. Characteristic of the Edgecliff, especially in the central part of New York State, are the large crinoid columnals, up to 1 inch in diameter, weathered out on the quarry floor. The Nedrow-Edgecliff contact is best seen at the southern face of the quarry wall where there are shaly interbeds. The Moorehouse-Nedrow contact is difficult to pinpoint but can be placed with a high degree of consistency (in the central area) at the end of the series of shaly beds and at the beginning of a fine-grained, uniformly medium-bedded (8 to 10 inches thick) rock with characteristically straight contacts (as opposed to wavy in the Nedrow) between bedding planes. The Tioga Bentonite forms a reentrant (Seneca-Moorehouse contact) on the face of the quarry wall which is difficult to locate without climbing up the cliff and digging out the heavily weathered debris under which a fresh, platy, gray clayey layer (3 to 8 inches

TABLE 13. Distribution of Brachiopods in the Onondaga Limestone in Central and Southeastern New York.

AMNH Locality	Total Number of Brachiopods Collected
3123A	151
3123B	47
3132C	68
3123D	31
3124A	17
3124B	51
3125A	186
3126	12
3128A	4008
3128B	64
3128C	161
3129	410
3130	59
3131B	79
3131C	68
3131D	11
3133	66
3134	66
3135	209
3137	532
3138A	259
3138B	59
3138C	13
3139	82
3140	23
3141	43
3142	29
3143	22
3144A	36
3151A	98
	N = 7030

thick) is clearly visible. Although the top of the Seneca Member is eroded here, it is exposed across highway I-81 along the drainage ditch adjacent to the northbound lanes, where a small thrust fault cuts the top of the member along with the overlying Union Springs Black Shale, resulting in a zone of crumpling and jointing. Compare the description of the section below with those of Oliver (1954) and Chute and Brower (1964).

ONONDAGA LIMESTONE

Seneca Member (AMNH Loc. 3123A)

Unit	Unit Thickness (ft)	Interval above Base (ft)
4. Limestone, fine grained, medium bedded, medium gray.	4	12-16
3. Limestone, fine grained, medium bedded, medium gray, *Hallinetes* zone.	2	10-12
2. Limestone, fine grained, medium bedded, medium gray.	9.5	0.5-10
1. Tioga Bentonite: Deeply weathered reentrant from 3 to 8 inches thick.	0.5	0-0.5

Moorehouse Member (AMNH Loc. 3123B)

Unit		
2. Limestone, fine grained, medium bedded, medium gray, with abundant chert seams. Eumophalacean gastropods (rare) in outcrop.	7	12-19
1. Limestone, fine grained, medium bedded, medium gray with little chert.	12	0-12

Nedrow Member (AMNH Loc. 3123C)

Unit		
3. Limestone, shaly with interbedded shaly beds, medium to coarse-grained, medium to thin-bedded, medium to dark gray, crinoidal in places. Most limy beds are medium to coarse grained crinoidal limestone. The shaly beds are usually thin (1 to 4 inches thick). The uppermost shaly bed represents the Moorehouse-Nedrow contact.	23	2-25
2. Limestone, shaly, coarse-grained, medium to thin-bedded, with a crinoidal matrix. Large crinoid columnals seen in cross section, some of which are comparable in size with the columnals of the Edgecliff Member. In cross section, some of the columnals act as competent units with the softer shale flowing around them.	0.5-1	1-2
1. Limestone, coarse-grained, massive medium gray.	1	0-1

Edgecliff Member (AMNH Loc. 3132D)

Unit		
1. Limestone, medium to coarse-grained, crystalline, massive, light gray. Most of the quarry floor represents the top of the Edgecliff Member. Brachiopods on uppermost bedding plane (quarry floor): *Megakozlowskiella raricosta, Leptaena* aff. " *rhomboidalis*."	5	0-5

AMNH Loc. 3124A-C, Nedrow, New York: This section extends along Route II at the junction of highway I-81 and Route 11 just south of Nedrow, New York. As one progresses south along the outcrop the following members of the Manlius Formation, stratigraphically below the Onondaga Formation, are encountered: Clark Reservation Member, Jamesville Member, and Pools Brook Member. The contact between the Manlius and Onondaga formations is sharp and easily recognizable even though at this location the basal sandy zone of the Edgecliff Member is missing, resulting in a small disconformity. The exit ramp from the southbound lanes of I-81 cuts through the middle of the Nedrow Member, in which shaly beds weather out to form talus piles with fairly good fossil material. On the southeast side of the exit ramp the Nedrow Member is jointed; the joints are inclined and are discontinuous. The Nedrow-Edgecliff contact is placed at the first shaly bed, deeply weathered, and quite noticeable. The Moorehouse-Nedrow contact is placed at the top of the last two shaly beds, approximately 7 inches apart. The remainder of the Moorehouse and Seneca members is concealed.

ONONDAGA LIMESTONE

Seneca Member (AMNH Loc. 3123A)

Unit	Unit thickness (ft)	Interval above base (ft)
1. Limestone, medium to fine grained, medium bedded, medium gray with distinct chert seams. These chert seams app ear to correlate closely with identical seams in the same stratigraphic position in the Jamesville Quarry (Lower Moorehouse).	11.5	0-11.5

Nedrow Member (AMNH Loc. 3124B)

Unit		
4. Limestone, shaly, medium-grained, interbedded with shaly beds. Two distinct shaly beds approximately 7 inches apart, above which no shaly beds appear. The uppermost bed represents the Moorehouse-Nedrow contact.	2.5	10-12.5
3. Limestone, shaly, medium-grained, interbedded shaly beds, medium gray, with wavy contacts between bedding planes.	3.5	6.5-10
2. Limestone, medium-grained, massive, medium gray, at the top of which is a 2-inch thick shale bed.	2	2-4

1. Limestone, shaly, medium-grained, interbedded with numerous shaly beds. Weathers to a fissile talus pile. There is a sharp contact with the underlying Edgecliff Member at the lowermost (approximately 1 ft thick) deeply weathered shaly zone.	2	0-2

Edgecliff Member (AMNH Loc. 3124C)

Unit		
1. Limestone, medium to coarse-grained, massive, light gray crystalline, biostromal with typical Edgecliff lithology. Joined in places forming massive blocks. A small disconformity is evidenced by the missing lower-most sandy zone.	12	0-12

MANLIUS FORMATION

AMNH Loc. 3125A-B, Nedrow, New York: The Edgecliff and most of the Nedrow members are exposed in a large road cut on highway I-81 just south of AMNH Loc. 3126, approximately one-half mile north of Exit 16 (Nedrow), southbound lanes. The top few feet of the Nedrow Member are missing. At the base of the Edgecliff Member the basal sandy zone is accessible and well exposed. Fossil collecting is moderately good in the weathered shaly talus of the Nedrow and poor in the Edgecliff. However, this is not meant to imply that the Edgecliff is lacking in fossils; rather, that extraction from the dense, massive limestone is extremely difficult. Also, there are no silicified fossils in the area, so that collecting large blocks of limestone (with no exposed, weathered, bedding planes) would be relatively unproductive unless one is studying material that can be sectioned (e.g., corals).

ONONDAGA LIMESTONE

Nedrow Member (AMNH Loc. 3125A)

Unit	Unit Thickness (ft)	Interval above base (ft)
2. Limestone, shaly, medium to fine-grained, medium to thin-bedded, tan to gray interbedded with occasional chert nodules, which are often weathered out. Distinct, wavy contacts between the bedding planes of fissile shaly beds. Weathered shale forms talus pile where fossil collecting is moderately good.	7	3.5-10.5
1. Limestone shaly, medium grained, alternating light and dark bands of gray to tan beds. Poorly fossiliferous.	3.5	0-3.5

Edgecliff Member (AMNH Loc. 3125B)

Unit		
2. Limestone, fine to medium-grained, massive and jointed, light gray and dense. Typical Edgecliff lithology.	11.5	6-17.5
1. Sandstone, limy, coarse-grained, massive, jointed and friable in places, brown with *Amphigenia*? sp. and scattered phosphate nodules.	6	0-6

AMNH Loc. 3126, Nedrow, New York: Good exposures of the typical Edgecliff biostromal limestone and basal sandy zone with *Amphigenia*? sp. fragments occur on highway I-81, approximately 1.5 mi north of Exit 16 (Nedrow), southbound lanes, on the east side of the road.

ONONDAGA LIMESTONE

Edgecliff Member (AMNH Loc. 3125A)

Unit	Unit thickness (ft)	Interval above base (ft)
3. Limestone, medium to coarse-grained, massive, light gray crystalline biostromal with a noticeable abundance of crinoidal debris. Large crinoidal columnals.	10.5	10.5-15.5
2. Limestone, sandy, coarse-grained, massive, brownish, gradational with overlying unit.	4	1-5
1. Sandstone, limey, coarse-grained, friable and jointed in places. Dark brown with scattered phosphate nodules.	1	0-1

AMNH Loc. 3127, Clark Reservation State Park, Jamesville, New York: The Edgecliff Member outcrops at the lip of the abandoned waterfall, above a plunge basin, just west of Jamesville, New York. The rock is a water-worn, massive, light gray, crystalline limestone (approximately 14 ft thick) with basal sandy zone directly overlying the Manlius Formation, which, although exposed on the cliff wall, is extremely difficult to reach. As one proceeds to higher ground additional Onondaga outcrops become apparent, but exact stratigraphic placement is difficult.

AMNH Loc. 3128A-D, Jamesville, New York: This section is in the Allied Chemical Quarry (formerly Jamesville Quarry and Solvay Process Co.), #3 pit. This is the most complete section of the Onondaga Limestone in central

New York. It was inaccessible when Oliver (1954) did his work on the stratigraphy of the Onondaga but has since come under new management and is now open to visitors for study. Quarrying is quite active, resulting in few weathered surfaces. This is especially significant for the limy members (Edgecliff, Moorehouse, and to some degree, Seneca) since collecting in them is difficult unless specimens have had a chance to weather out. The Nedrow Member, with its shaly partings, provides talus piles of moderately weathered material from which some good specimens have been retrieved, although internal structures of the brachiopods are not generally visible. However, based upon external morphology, generic identifications are often possible. Bedding is horizontal and heavily jointed in many places. All members are accessible since the walls of the quarry are terraced. The Onondaga Limestone, the uppermost formation in the quarry, is missing only the top of the Seneca Member.

ONONDAGA LIMESTONE

Edgecliff Member (AMNH Loc. 3128A)

Unit	Unit Thickness (ft)	Interval above base (ft)
4. Limestone, muddy, fine-grained, medium to thick-bedded with some chert nodules. Weathered surface earth color; fresh surface dark gray. Heavily jointed at quarry wall. Fauna on uppermost bedding plane: *Athyris* sp., indet. brachiopod fragments, bryozoans, numerous horn corals, trilobite fragments (cf. *Odontocephalus*), euomphalacean gastropods.	2	12-14
3. Limestone, fine-grained, fresh surface dark gray, wavy contacts between bedding planes. Stylolites present (characteristic of the Seneca Member). At the top of this unit occurs a one inch thick chert band. *Hallinetes* Zone (Zone J of Oliver, 1954).	2	10-12
2. Limestone, medium to fine-grained, medium to thick-bedded, light gray to muddy, horizontally jointed.	9.5	0.5-10
1. Tioga Bentonite: Soft, gray clayey platy unit at the base of the Seneca Member. The Tioga Bentonite represents the Seneca-Moorehouse contact. Deeply weathered iron bearing minerals oxidized to a limonitic coating.	0.5	0-0.5
Moorehouse Member (AMNH Loc. 3128B)		
Unit		
3. Limestone, fine-grained, medium-bedded, light gray with distinct and continuous chert bands. Limonitic in places.	5	20-25

2. Limestone, fine-grained, medium to thick-bedded, light gray, no chert.	12	8-20
1. Limestone, fine-grained, medium-bedded, medium to light gray with some distinct chert bands. Fresh surface dark throughout all three units.	8	0-8
Nedrow Member (AMNH Loc. 3128C)		
1. Limestone, shaly, fine-grained, medium to thin-bedded, medium to dark gray with two distinct shaly beds 8 inches apart. The uppermost shaly bed represents the Moorehouse-Nedrow contact in the central New York area. On weathered surfaces the limestone is greenish and platy. Scattered chert nodules are present throughout the member, but are of no stratigraphic value. Good fossil collecting from weathered surfaces only. Brachiopods: *Leptaena* and *Megakozlowskiella* common.	3.5	7-10.5
Edgecliff Member (AMNH Loc. 3128D)		
Unit		
3. Limestone, medium to coarse-grained, massive crystalline, pinkish. Typical Edgecliff facies with large crinoid columnals up to 1 inch in diameter.	9.5	9.5-19
2. Limestone, sandy, coarse-grained, massive, tan, with small scattered phosphate nodules.	1	8.5-9.5
1. Sandstone, limey, coarse-grained, massive, dark brown with numerous large phosphate nodules.	8.5	0-8.5

AMNH Loc. 3129, Stockbridge Falls, New York: The section is located just west of Stockbridge Falls, New York, along Stockbridge Falls Road and Oneida Creek. Much of the outcrop is covered by debris, but about 10-15 ft of the Seneca Member is exposed. The Union Springs Black Shale outcrops just up the road and around the bend. The contact between the Seneca and the Union Springs is exposed on the hill at some small, isolated outcrops. The limestone is fine-grained, medium to thin-bedded, heavily weathered, and muddy on weathered surfaces but dark gray on fresh surfaces. The rock is jointed and blocky. The *Hallinetes* zone is exposed and collecting is moderately good. The Tioga Bentonite was not seen by the writer but is present (W. A. Oliver, Jr., personal communication). The Moorehouse Member is exposed in the creek (5+ ft).

AMNH Loc. 3130, Cherry Valley, New York: This section is 2.8 mi west of AMNH Loc. 3131 along the north side of Route 20. About 9 ft of the Moorehouse Member is exposed. The limestone is medium-grained, medium to dark gray, and medium-bedded with light chert. The fauna and lithology indicate stratigraphic placement within the (Upper?) Moorehouse: *Atrypa "reticularis,"* orthotetacid fragments, spiney platyceratids (correlates with Upper Moorehouse sections in the eastern part of the state), *Pentagonia unisulcata* (common), *Leptaena "rhomboidalis," Strophodonta* cf. *demissa,* and bryozoan fragments.

AMNH Loc. 3131A-D, Cherry Valley, New York: This large outcrop extends for approximately 250 yards along Route 20 on the west side of Cherry Valley. The section is almost complete, missing only the top of the Moorehouse and bottom of the Seneca members, which are actually concealed, but assumed to be present. The Tioga Bentonite was too deeply buried by talus debris to be seen by this writer, but its presence was verified by W. A. Oliver, Jr. and L. V. Rickard (personal communication). Both the lower contact (Carlisle Center Formation) and upper contact (Union Springs Black Shale) were observed.

ONONDAGA LIMESTONE

Edgecliff Member (AMNH Loc. 3131A)

Unit	Unit thickness (ft)	Interval above base (ft)
1. Limestone, shaly, very muddy, fine-grained, irregularly bedded, dark gray to brownish (mottled) in places. This appears to be equivalent to·Zone K of Oliver (1954). The rock becomes darker and muddier toward the top. Only rare fossil fragments were present, and they were too fragmentary for identification. The overlying Union Springs Black Shale contact is evident.	5.5	0-5.5
Moorehouse Member (AMNH Loc. 3131B) Unit		
2. Limestone, medium to fine-grained, medium-bedded, light gray, with numerous chert seams and very dense.	10	25-35
1. Limestone, medium to fine-grained, medium-bedded, light gray, dense, without chert.	25	0-25
Nedrow Member (AMNH Loc. 3131C) Unit		
2. Limestone, bracketed by two shaly beds approximately 1 ft apart. The uppermost shaly bed represents the contact with the overlying Moorehouse Member. The intervening limestone is medium-grained and medium gray.	1	9.5-10.5

1. Limestone, shaly, medium-grained, medium to thin-bedded, medium gray, with no chert. Interbedded shaly beds as in Jamesville Quarry and Nedrow, New York. Wavy contacts between the bedding planes. Typical Nedrow weathering (recessed). Fauna includes small, nonspiny platyceratids, *Levenea* sp., and *Pentagonia* sp.	9.5	0-9.5

AMNH Loc. 3132, Thompson's Lake, New York: A small outcrop (20 ft thick) of the Edgecliff Member is exposed on an east-west road just east of Route 157, north of Thompson's Lake. The limestone is massive and medium to dark gray, coarsely crystalline and biostromal (upper 12 ft). The lower 8 ft consists of thin-bedded, medium to coarse-grained, medium gray, fossiliferous limestone. The entire outcrop was packed with corals but the only brachiopod observed was *Atrypa "reticularis."*

AMNH Loc. 3133, Thompson's Lake, New York: Approximately 21 ft of the Moorehouse Member is exposed on the south shore of Thompson's Lake, on the property of the Thompson's Lake Hotel. Since the Marcellus Formation outcrops in a nearby stream bed (W. A. Oliver, Jr., personal communication), and based on faunal similarities, this section is placed in the upper Moorehouse. The limestone is dense, fine-grained, medium to thin-bedded, medium to light gray, and cherty. The chert has weathered out, giving the rock a pitted texture. The chert that is still present is dark weathering. The fauna includes: *Atrypa "reticularis," Megakozlowskiella raricosta* (very common), orthotetacid fragments, *Pentamerella* sp., *S. demissa*, *"Pacificocoelia"* sp.?, gypidulinids?, *Coelospira camilla*, spiny platyceratids (common 5 ft from base of unit), and crinoids (abundant in some beds in the middle of the unit).

AMNH Loc. 3134, Clarksville, New York: The section outcrops along Route 32, about 2 mi southeast of Clarksville, New York. The limestone is 22 ft thick and medium grained, medium bedded, light to medium gray, and coarsely crinoidal in places. There is some blue-gray chert, with dark weathering chert common in the upper 5 ft of the unit (seams). The fauna is typically Upper Moorehouse: *Atrypa "reticularis," Megakozlowskiella raricosta* (common), *Leptaena "rhomboidalis," Megastrophia*?, *Schizophoria* sp., *Levenea* sp., orthotetacid fragments, spiny platyceratids, and bryozoan fragments.

AMNH Loc. 3135, Leeds, New York: An extensive bedding plane near the top of the Onondaga Formation is exposed at the rear of Samantha's Inn on Route 23B in Leeds, New York. The outcrop runs parallel to Catskill Creek (east bank) and consists of more than 500 ft of medium gray, medium grained, somewhat cherty limestone dipping at a shallow angle to the west. On the surface of the rock, there are numerous platyceratid gastropods (spiny) and clusters of *Atrypa "reticularis"* along with many other silicified fossils typical of the Upper Moorehouse.

AMNH Loc. 3136, southwest of Catskill, New York: This roadcut is located along the New York State Thruway at the junction of the Thruway and Route 23A. A total of about 40 ft of the Edgecliff Member is exposed overlying the Schoharie Formation. The contact is gradational but is recognizable at the first buff layer toward the base of the Edgecliff. Slickensides are present and faulting has occurred within the Edgecliff to some degree. Thus, the exact thickness of the units, especially unit 2, is questionable. Lithologies correlate well with the strata at AMNH Localities 3138C, 3147, 3149.

ONONDAGA LIMESTONE

Edgecliff Member (AMNH Loc. 3128A)

Unit	Unit thickness (ft)	Interval above base (ft)
2. Limestone, medium grained, thick bedded to massive, light gray, with numerous light gray weathering chert seams. Crinoid columnals up to 1 inch in diameter.	35?	5-40
1. Limestone, medium grained, thick bedded to massive, light gray, with very little chert. The contact with the underlying Schoharie Formation is somewhat gradational but is placed at the first buff weathering unit.	5	0-5

AMNH Loc. 3137, Quatawichna-ach, on the Kaaterskill Stream, near Timmerman Hill, New York: This anticlinal exposure takes its name from the Indian "place where water all goes in a hole" (Chadwick, 1944), referring to the chert seams and massive joints that take the water underground as it passes through the limestone. The location is approximately 4.5 mi southwest of Catskill, New York, just east of Timmerman Hill, in the streambed. The stratigraphic position of the outcrop is placed at the top of the

Moorehouse Member since the Bakoven Shale, which directly overlies the Onondaga Formation in this area, outcrops just downstream. In addition, shale debris is evident in the outcrop area, although there is no indication of any shale exposures in the immediate vicinity. The Bakoven Shale is extremely soft and erodes easily, forming a line of weakness—the Bakoven Valley. The fauna at this locality is well silicified, for the most part. The silicification is not surficial but permeates the limestone.

ONONDAGA LIMESTONE

Moorehouse Member (AMNH Loc. 3137)

Unit	Unit thickness (ft)	Interval above base (ft)
6. Limestone, medium grained, medium bedded (4 to 8 inches thick), dense, with discontinuous dark weathering chert bands and sharp contacts between the bedding planes.	15	13-28
5. Limestone, medium grained, massive, light gray, extremely dense, with thin chert bands often forming nodules. Variable fractured with silicified fossils weathering out of the rock surface.	3	10-13
4. Dark chert band traceable as a marker throughout most of the outcrop.	0.5	9.5-10
3. Limestone, massive, light gray, fractured in places with discontinuous chert nodules and thin dark chert bands. Silicified fossil fragments observed in cross section.	2.5	7-9.5
2. Dark chert band traceable as a marker throughout most of the outcrop.	0.5	6.5-7
1. Limestone, medium grained, medium gray, dense, with numerous chert bands stringing out into dark nodules. Several extensive bedding planes are exposed upon which silicified fossil fragments partially weather out of the matrix. These consist mainly of corals, brachiopods, and trilobite fragments. *Atrypa* "*reticul*aris" and spiny platyceratid gastropods are common.	6.5	0-6.5

AMNH Loc. 3138A-C, Saugerties, New York, On the New York State Thruway at the Saugerties interchange: The most complete section of the Onondaga Formation in eastern New York is exposed in the roadcut made by the New York State Thruway. Although this area was tectonically active, resulting in numerous folds and faults, no evidence of significant faulting was found in the section measured. Slickensides in the strata bordering the southbound lanes, about 50 ft south of the measured section, tend to indicate small scale reverse faulting. An identical fauna occurs at 75 to 85 ft above the Nedrow-Moorehouse contact at two separate areas of this outcrop and at the few feet exposed at Leeds, New York (AMNH Loc. 3135). This fauna is also found at AMNH Loc. 3137 and at the Onondaga-Bakoven contact, thus providing strong evidence for the stratigraphic correlation of the various Moorehouse units in eastern New York.

ONONDAGA LIMESTONE

Moorehouse Member (AMNH Loc. 3138A)

Unit	Unit thickness (ft)	Interval above base (ft)
1. Limestone, medium grained, medium to thick bedded, irregularly bedded in places probably due to blasting the roadcut, medium to dark gray, dense in silicified zones. Dark weathering chert seams throughout the entire unit, which are characteristic of the Moorehouse Member in the east. Silicified faunas interspersed throughout but consistently found at 75-85 ft above the Nedrow-Moorehouse contact.	92+	0-92+
Nedrow Member (AMNH Loc. 3138B) Unit		
1. Limestone, medium grained, massive, light gray, with light gray weathering chert seams. The rock is very dense and the chert is sparse. The Moorehouse-Nedrow contact is placed at the first dark weathering chert seam. Some large crinoid columnals evident.	23	0-23
Edgecliff Member (AMNH Loc. 3138C) Unit		
2. Limestone, medium grained, massive, light gray, with numerous light gray weathering chert seams. The lithology here is identical with that at Forsyth Park (see below).	40.5	7.5-48
1. Limestone, medium grained, massive, light gray, with very little chert. The chert that is present is light gray. The Edgecliff-Schoharie contact is placed at the first buff-colored band of gritty limestone.	7.5	0-7.5

SCHOHARIE FORMATION

AMNH Loc. 3139, Saugerties, New York: This is a continuation of the large outcrop at AMNH Loc. 3138A on the New York State Thruway, near the Saugerties interchange. The stratigraphic position is Middle to Upper Moorehouse based upon lithology, fauna, and structure. The section is as follows:

ONONDAGA LIMESTONE

Moorehouse Member (AMNH Loc. 3128A)

Unit	Unit thickness (ft)	Interval above base (ft)
2. Dark chert band, discontinuous in places, varying in thickness from 6 to 12 inches, averaging about 10 inches.	0.5	8.85
1. Limestone, medium grained, medium bedded, light gray, dense, horizontally jointed with small (1 to 3 inches) dark chert bands. Contacts between the bedding planes vary from sharp to wavy.	6.5	0-1.5

AMNH Loc. 3140, Kingston, New York: The section outcrops at the junction of Hooker and Lincoln roads, on the east side of the street. Two-thirds of the exposed strata are Lower Moorehouse directly overlying the Nedrow Member, which consists of about 3.5 ft of massive, light gray, dense limestone with no discernible fossils in the outcrop and no silicified fauna within the matrix. Above lies 7 ft of dense, medium grained, medium gray Moorehouse Limestone with no noticeable chert, although it appears that the chert was incorporated into the matrix as darker bands within the limestone. Bedding is both massive and irregular. The section seems to be conformable with the Schoharie Formation, which outcrops due east at the bottom of the hill.

AMNH Loc. 3141, Kingston, New York: The outcrop is a few hundred feet south of AMNH Loc. 3139 on the west side of the road. Stratigraphically, the section appears to be Middle Moorehouse. There are dark weathering chert seams in a medium to dark gray, dense, medium grained limestone, which contains silicified fossils about 6 ft up on a ledge. Due to the dip of the strata the exact thickness of the outcrop is difficult to determine, but it approximates 10 ft.

AMNH Loc. 3142, rear of hospital in Kingston, New York: This section outcrops along the railroad tracks and is about 75 ft thick. The exposure represents the Middle to Upper Moorehouse and is laced with typically dark weathering chert seams. The limestone is medium grained, medium gray, dense, and correlates with the Middle to Upper Moorehouse rocks at AMNH Loc. 3138A. No upper contact was observed at the outcrop, although the top of the Onondaga Formation must have been close to the easternmost end of the section.

AMNH Loc. 3143, Forsyth Park and Zoo, Kingston, New York: Scattered outcrops of the Edgecliff Member occur throughout the park. The limestone is typical of the extremely cherty Edgecliff with numerous seams of light weathering chert in a medium to dark gray matrix. The rock is medium grained and up to about 20 ft thick (cumulative thickness, estimated) throughout the park.

AMNH Loc. 3144A-B, Ulster County Highway Department Quarry, Kingston, New York: The quarry is located on the west side of Kingston, and just west of Forsyth Park and Zoo. The strata here overlie the units at Forsyth Park (Edgecliff):

Moorehouse Member (AMNH Loc. 3128A)

Unit	Unit thickness (ft)	Interval above base (ft)
2. Limestone, medium to fine grained, medium bedded, medium to dark gray, with dark chert seams.	22	8-30
1. Limestone, medium grained, medium bedded, light to medium gray, with no chert.	8	0-8
Nedrow Member (AMNH Loc. 3144B) Unit		
1. Limestone, medium grained, medium bedded, medium to light gray, with light gray chert.	8	0-8

AMNH Loc. 3145, Kingston, New York: The Edgecliff Member of the Onondaga Limestone is exposed in a syncline about 100 yards north of the intersection of the Penn Central Railroad and West O'Reilly Street in Kingston. The Saugerties (30 ft), Aquetuck (44 ft), and Carlisle Center (143 ft) members of the Schoharie Formation directly underlie the Edgecliff Member of the Onondaga Limestone, which is approximately 26 ft thick at this locality. The contact between the Onondaga and Schoharie is gradational here and is difficult to place. However, based on lithologic criteria, I am placing the contact at the first dark band below the extensive light, cherty limestone, which typifies the Edgecliff Member in southeastern New York.

ONONDAGA LIMESTONE

Moorehouse Member (AMNH Loc. 3128A)

Unit	Unit thickness (ft)	Interval above base (ft)
2. Limestone, medium grained, massive, medium gray, with numerous light weathering chert seams throughout.	12	14-26
1. Limestone, medium grained, massive, medium to dark gray, with no chert. Contact gradational with underlying Schoharie Formation.	14	0-14

SCHOHARIE FORMATION

Saugerties Member
Unit

1. Sandstone, limy, brownish yellow to light gray, banded, medium grained.	30	187-217
Aquetuck Member Unit		
1. Sandstone, almost calcareous mudstone, brownish yellow to yellowish gray, with interbedded layers of nodular limestone and muddy lenses.	44	143-187
Carlisle Center Member Unit		
1. Mudstone, brownish yellow to light gray, calcareous gritty matrix, with an increasing limy content toward the top of the member.	143	0-143

AMNH Loc. 3146, near Kingston, New York: The section outcrops along the New York State Thruway 0.9 mi south of the Kingston interchange. The strata exposed represent the Nedrow (upper?) and correlate lithologically with the rocks at AMNH Loc. 3148 and 3138B. The total thickness here is 21 ft. The limestone is fine grained, massive, light gray, and jointed (probably due to the blasting when the thruway was constructed), with some light weathering chert.

AMNH Loc. 3147, near Kingston, New York: This section outcrops along the New York State Thruway approximately 1.25 mi south of the Kingston interchange, along the northbound lanes. Stratigraphically, the rocks are Edgecliff. The limestone is fine to medium grained with coarse crinoidal debris alternating with chert seams. Many of the chert seams have weathered out but there are still numerous seams and nodules in the matrix so that the lithology closely resembles that in Forsyth Park. Five ft from the base of the unit, there is a prominent lens of crinoidal material. Large crinoid columnals up to one inch thick are present along with *Atrypa "reticularis."* The total thickness is 22 ft.

AMNH Loc. 3148, near Hurley, New York: The outcrop is small, approximately 2.5 mi southwest of Kingston, along Route 209, on the northwest side of the road. Based on structural relationships and lithology, this section is placed in the Nedrow Member.

AMNH Loc. 3149, one-fourth of a mile southwest of AMNH Loc. 3148 on the southeast side of the road. The lithology is identical with that at AMNH Loc. 3147 (crinoidal with numerous light weathering chert seams). This section is stratigraphically lower than the units at AMNH Loc. 3148 (Figure 60).

ONONDAGA LIMESTONE

Nedrow Member (AMNH Loc. 3 I 28A)

Unit	Unit thickness (ft)	Interval above base (ft)
2. Limestone, fine grained, light gray, dense, massive, and crinoidal. Two distinct light weathering chert bands, discontinuous to nodular, are separated by about 2 ft (Figure 61). The fauna appears to be partly silicified with *Atrypa "reticularis"* fairly common.	4	9-13
1. Limestone, same lithology, massive, no chert, dense. The fresh surface is medium to dark gray.	9	0-9

AMNH Loc. 3150, near Accord, New York: The outcrop appears to be rather small in breadth when viewed from the road at approximately 3.3 mi northeast of Accord, New York, along Route 209, on the southeast side of the road. The exposure is about 15 ft thick, massive, cherty, medium grained, medium gray limestone, with crinoid columnals up to one inch in diameter. The front of the outcrop, along Route 209, is 6 ft thick and the only fossils noticeable on the exposed bedding planes are crinoid columnals. The rear of the outcrop is exfoliated. The chert is incorporated into the matrix and does not weather out to form nodules. There is no silicified fauna. Stratigraphically the section is either Edgecliff or Nedrow, but most likely Edgecliff, as the Nedrow cannot be lithologically differentiated this far southeast. Also, in Wawarsing, farther along Route 209, the Nedrow is no longer present.

AMNH Loc. 3151A, B, Wawarsing, New York: The Edgecliff and part of the Moorehouse members are exposed in an abandoned quarry approximately 100 ft north of Route 209, about 0.5 mi northeast of Vernooy Kill. The measurements are approximate due to the inaccessibility of the rocks and covered strata.

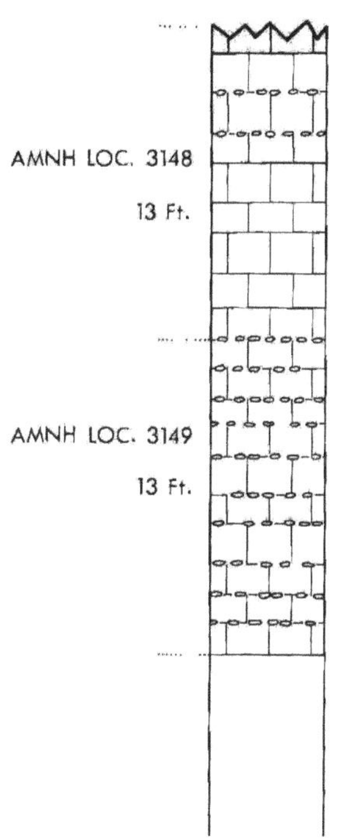

FIGURE 60. Composite stratigraphic section: AMNH Loc. 3148 and 3149. Note the even distribution of chert throughout the outcrop at AMNH Loc. 3149 compared with the relatively sparse amount of chert at AMNH Loc. 3148.

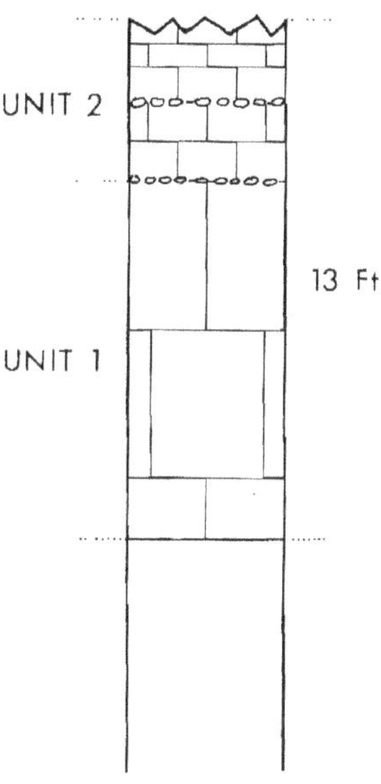

FIGURE 61. Detailed stratigraphic section of AMNH Loc. 3148, illustrating two distinct light weathering chert bands, the base of which represents the contact between units one and two at this outcrop. Silicification of the faunas seems to be more complete in unit two. There also appears to be a positive correlation between amount of chert present and silicification of fossils in the area of Kingston, New York. (See Feldman, 1980: 28 for index map delineating a silicified region.)

ONONDAGA LIMESTONE

Moorehouse Member (AMNH Loc. 3151 A)

Unit	Unit thickness (ft)	Interval above base (ft)
3. Limestone, very fine grained, massive, medium gray, small amounts of chert (discontinuous seams), with a maximum thickness of 1 inch. The chert seems to be incorporated into the matrix in a manner similar to AMNH Loc. 3150, resulting in a "banded" appearance. The upper half of the unit is medium gray while the lower half is light gray. Common here are trilobite fragments and *Levenea* sp. No silicified fauna is present.	17	34-51
2. Limestone, medium bedded, light gray. Beds are 4 to 10 inches thick. In the lower half of the unit the beds are up to 2 ft thick.	9	25-34
1. Limestone, massive, light gray. Some medium bedded subunits are 5 to 8 inches thick. Cherty in spots.	25	0-25
Edgecliff Member (AMNH Loc. 3151B) Unit		
2. Limestone, medium grained, medium bedded to massive, medium to dark gray, transitional with the Lower Moorehouse. The contact is defined on the basis of a darker color (Edgecliff) and the presence of large crinoid columnals. There does not appear to be any significant difference in the color of the Moorehouse from the Nedrow and Edgecliff members.	25	2-27
1. Siliceous, limy grit, transitional to the Schoharie Formation.	2	0-2

SCHOHARIE FORMATION

REFERENCES

Amsden, T. W. 1953. Some notes on the Pentameracea, including a description of one new subfamily. *Washington Academy of Science, Journal*, 43: 137-147.

Amsden, T. W. 1964. Brachial plate structure in the brachiopod Family Pentameridae. *Palaeontology* 7: 220-239.

Amsden, T. W., and Ventress, W. P. S. 1963. Early Devonian brachiopods of Oklahoma. *Oklahoma Geological Survey Bulletin* 94.

Anderson, E. J., and Makurath, J. H. 1973. Paleoecology of Appalachian gypidulinid brachiopods. *Palaeontology* 16: 381-390.

Boucot, A. J. 1957. Revision of some Silurian and Early Devonian spiriferid genera and erection of Kozlowskiellinae, new subfamily. Senck. *Lethea* 38: 311-334.

Boucot, A. J. 1958. Kozlowskiellina, new name for Kozlowskiella Boucot, 1957. *Journal* of *Paleontology* 32: 1030.

Boucot, A. J. 1959. Brachiopods of the Lower Devonian rocks at Highland Mills, New York. *Journal of Paleontology* 33: 727-769.

Boucot, A. J. 1963. The Eospiriferidae. *Palaeontology* 5: 682-711.

Boucot, A. J. 1973. Early Paleozoic brachiopods of the Moose River Synclinorium, Maine. *U. S. Geological Survey Professional Paper* 784, Washington, D.C.

Boucot, A. J. 1975. Evolution and extinction rate controls. *Developments in palaeontology and stratigraphy* 1. Amsterdam: Elsevier

Boucot, A. J., and Gill, E. D. 1956. *Australocoelia*, a new Lower Devonian brachiopod from South Africa, South America, and Australia. *Journal of Paleontology* 30: 1173-1178.

Boucot, A. J., and Johnson, J. G. 1967. Species and distribution of *Coelospira* (Brachiopoda). *Journal of Paleontology* 41: 1226-1241.

Boucot, A. J., and Johnson, J. G. 1968. Brachiopods of the Bois Blanc Formation in New York. *U. S. Geological Survey Professional Paper* 584B. Washington, D.C.

Boucot, A. J., and Rehmer, J. 1977. *Pacificocoelia acutiplicata* (Conrad, 1841) (Brachiopoda) from the Esopus Shale (Lower Devonian) of eastern New York. *Journal of Paleontology* 51: 1123-1132.

Boucot, A. J., Gauri, K. L., and Southard, J. 1970. Silurian and Lower Devonian brachiopods, structure and stratigraphy of the Green Pond Outlier in southeastern New York. *Palaeontographica* 135(A): 1-59.

Boucot, A. J., Johnson, J. G., and Staton, R. D. 1964. On some atrypoid, retzioid and athyridoid brachiopods. *Journal of Paleontology* 38: 805-822.

Boucot, A. J., Johnson, J. G., and Walmsley, V. G. 1965. Revision of the Rhipidomellidae (Brachiopoda) and the affinities of *Mendacella* and *Dalejina*. *Journal of Paleontology* 39: 331-340.

Bowen, Z. P. 1967. Brachiopods of the Keyser Limestone (Silurian-Devonian) of Maryland and adjacent areas. *Geological Society of America Memoir* 102.

Bronn, H. G. 1862. Die Klassen und Ordnungen der weichthiere (Malacozoa), vol. 3: 1-518. Leipzig and Heidelberg.

Buch, H. von. 1834. Uber Terebrateln. Abh. Dt. Akad. Wiss. Berlin, (1833): 21-144.

Butts, C. 1940. Geology of the Appalachian Valley in Virginia, part I. Geologic text and illustrations. *Virginia Geological Survey Bulletin* 52: 1-568.

Butts, C. 1941. Geology of the Appalachian Valley in Virginia, part II. Fossils, plates and explanations. *Virginia Geological Survey Bulletin* 52: 1-271.

Caster, K. E. 1939. A Devonian fauna from Colombia. Bulls. *American Paleontology* 24: 1-128.

Chadwick, G. H. 1944. Geology of the Catskill and Kaaterskill quadrangles. *New York State Museum Bulletin.*

Chute, N. E., and Brower, J. C. 1964. Stratigraphy and structure of Silurian and Devonian strata in the Syracuse area. *New York State Geological Association Guidebook, 36th Annual Meeting*, 90-101.

Clarke, J. M. 1908. Early Devonic history of New York and eastern North America. *New York State Museum Memoir* 9: 5-366.

Cloud, P. E., Jr. 1942. Terebratuloid Brachiopod of the Silurian and Devonian strata in the Syracuse area. *Geological Society of America Special Paper* 38.

Conrad, T. A. 1838. Report on the paleontological department of the survey. *New York State Geological Survey*, 2nd Annual Report, 107-119.

Conrad, T. A. 1839. Descriptions of new species of organic remains. *New York State Geological Survey, 3rd Annual Report*, 57-66.

Conrad, T. A. 1841. Fifth annual report on the paleontology of the State of New York. *New York State Geological Survey, 5th Annual Report*, 25-57.

Conrad, T. A. 1842. Observations on the Silurian and Devonian systems of the United States, with descriptions of new organic remains. *Academy of Natural Sciences Philadelphia Journal* 8: 228-280.

Cooper, G. A. 1944. Phylum Brachiopoda. In H. W. Schimer and R. R. Schrock, *Index fossils of North America*, 277-365. Cambridge, MA: The MIT Press.

Cooper, G. A.1945. New species of Brachiopods from the Devonian of Illinois and Missouri. *Journal of Paleontology* 19: 479-489.

Cozzens, I. 1846. Descriptions of three new fossils from the Falls of the Ohio. *New York Lyceum of Natural History Annals* 4: 157-159.

Dalman, J. W. 1828. Uppstallning och Beskrfning af di I Sverige funne Terebratuliter. *Kongliga Svenska Vetenskaps-akademiens Handlinger, Upssala and Stockholm* (1827): 85-155.

Davidson, T. 1851-1886. A monograph of the British fossil Brachiopod. *Palaeontographical Society of London Monograph*. 3: i-397.

Davidson, T. 1858. British Carboniferous Brachiopoda, pt. 5. *Palaeontographical Society of London Monograph* 2: 49-80.

Dunbar, C. O. 1919. Stratigraphy and correlation of the Devonian of western Tennessee. Tennessee State Geological Survey Bulletin 21: 1-127.

Dutro, J. T., Jr. 1971. The brachiopod *Pentagonia* in the Devonian of eastern United States. Smithsonian Contributions to Paleobiology. Paleozoic Perspectives: A paleontological tribute to G. Arthur Cooper. No. 3: 181-192.

Eaton, A. 1832. Geological text-book, for aiding the study of North American geology: Being a systematic arrangement of facts, collected by the author and his pupils. Albany: Webster and Skinners.

Eaton, A. 1840. Fourth annual report of the geological survey of the third district. *New York Geological Survey Annual Report* 4: 355-383.

Eldredge, N., and Gould, S. J. 1972. Punctuated equilibria: An alternative to phyletic gradualism. In T. J. M. Schopf (ed.), *Models in Paleobiology*, 81-115. San Francisco: Freeman, Cooper and Co.

Fagerstrom, J. A. 1961. The fauna of the Middle Devonian Formosa Reef Limestone of southwestern Ontario. *Journal of Paleontology* 35: 1-48.

Fagerstrom, J. A. 1966. Biostratigraphic significance of rhipidomellid brachiopods in the Detroit River Group (Devonian). *Journal of Paleontology* 40: 1236-1238.

Fagerstrom, J. A. 1971. Brachiopods of the Detroit River Group (Devonian) from southwestern Ontario and adjcent areas of Michigan and Ohio. *Geological Survey of Canada Bulletin* 204.

Feldman, H. R. 1980. Level-bottom brachiopod communities in the Middle Devonian of New York. *Lethaia* 13: 27-46.

Fischer de Waldheim, G. 1830. Oryctographie du Gouvernement de Moscou. 1st ed. ix + 26 pp., 60 pls. 1837. 2nd ed. 202 pp., Brachiopoda, pls. 20-26. Moscow.

Frederiks, G. 1912. Bemerkungen über einige oberpalaeozoische Fossilien von Krasnoujimsk. *Kazani Protocoly Obshchestva estestvoisputatelei* 69: 1-9.

Frederiks, G. 1918. Diagnoses generum et specierum novum. *Annuaire de la Societe Paleontologique de Russie* 2: 87.

George, T. N. 1931. *Ambocoelia* Hall and certain similar British Spiriferidae. *Geological Society of London, Quarterly Journal* 87: 30-61.

Gill, G. H. 1871. Arrangement of the families of Molluscs prepared for the Smithsonian Institution. *Smithsonian Miscellaneous Collection* 10 (227): 1-49.

Girty, G. H. 1904. New molluscan genera from the Carboniferous. *United States National Museum Proceedings* 27: 721-736.

Goldring, W. 1935. Geology of the Berne Quadrangle. *New York State Museum Bulletin* 303: 1-238.

Goldring, W. 1943. Geology of the Coxsackie quadrangle. *New York State Museum Bulletin* 332: 1-374.

Gould, S. J., and Eldredge, N. 1977. Punctuated equilibria: The tempo and mode of evolution reconsidered. *Paleobiology* 3: 115-151.

Grabau, A. W. 1906. Geology and paleontology of the Schoharie Valley. *New York State Museum Bulletin* 92.

Grabau, A. W. 1931, 1933. Devonian brachiopods of China; 1. Devonian brachiopods from Yunnan and other districts in South China. *Palaeontologia Sinica*, ser. 8, vol. 3: 1-752.

Grant, R. E. 1965. The brachiopod Superfamily Stenoscismatacea. *Smithsonian Miscellaneous Collection* 148 (2): 1-192.

Gray, J. E. 1840. Synopsis of the contents of the British Museum, 44th ed., London.

Hall, J. 1840. Fourth annual report of the geological survey of the fourth district. *Geological Survey of New York* 50: 389-455.

Hall, J. 1843. Geology of New York, Part 4, comprising of the survey of the fourth geological district. *Natural History of New York, Albany.*

Hall, J. 1850. On the Brachiopoda of the Silurian: Particularly the Leptaenida. *American Association for the Advancement of Science* 2: 347-351.

Hall, J. 1857. Descriptions of Paleozoic fossils. *New York State Cabinet 10th Annual Report,* Part C. Appendix, 41-186.

Hall, J. 1859, 1861. Paleontology of New York. *New York Geological Survey* 3: 1-152 (1859), pls. 1-120 (1861).

Hall, J. 1860. Contributions to Palaeontology. *New York State Cabinet of Natural History,* Annual report 13: 55-125.

Hall, J. 1862. Contributions to Palaeontology. *New York State Cabinet of Natural History,* Annual Report 15: 1-158.

Hall, J. 1863. Notice of some new species of fossils from a locality of the Niagara Group in Indiana; with a list of identified species from the same place. *Albany Institute Transactions* 4: 195-228.

Hall, J. 1867. Descriptions and figures of the fossil Brachiopoda of the Upper Helderberg, Hamilton, Portage and Chemung groups. *New York Geological Survey, Palaeontology* 4: 1-428.

Hall, J. 1894. Palaeontology. New York, 13th Annual Report of State Geologist 2: 601-943.

Hall, J., and Clarke, J. M. 1892. An introduction to the study of the genera of Paleozoic Brachiopoda, Part I. *New York Geological Survey, Palaeontology of New York* 8: 1-367.

Hall, J., and Clarke, J. M. 1894. An introduction to the study of the genera of Paleozoic Brachiopoda, Part II. *New York Geological Survey, Palaeontology of New York* 8; 1893, prepr., 1-317; 1894, pls. 21-84 (1895).

Harper, C. W., Jr., and Boucot, A. J. 1978. The Stropheodontacea. *Palaeontographica*, vol. 161 (A): 55-175; vol. 162 (A): 1-80.

Havlicek, V. 1953. O nekolika novych rameononozcich ceskeho a moravskeho stredniho devonu. *Ustředniho, Ustavu Geologickeho, Věstník* 28: 4-9.

Helmbrecht, W., and Wedekind, R. 1923. Versuch einer biostratigraphischen Gliederung der Siegener Schichten auf Grund von Rensselaerien und Spiriferen: Gluckauf, Berg. und Huttenmannische Zeits., Jahrg. 59 (41): 949-953.

House, M. R. 1962. Observations on the ammonoid succession of the North American Devonian. *Journal of. Paleontology* 36: 247-284.

Imbrie, J. 1959. Brachiopods of the Traverse Group (Devonian) of Michigan. *Museum Of Natural History, America Bulletin* 116: art. 4, 349-409.

Johnson, J. G. 1966. Middle Devonian Brachiopods from the Roberts Mountains, central Nevada. *Paleontology* 9: 152-181.

Johnson, J. G. 1970. Great Basin Lower Devonian Brachiopoda. *Geological Society of America Memoir* 121.

Johnson, M. E. 1977. Succession and replacement in the development of Silurian brachiopod populations. *Lethaia* 10: 83-93.

Kesling, R. V., and Chilman, R. B. 1975. Strata and megafossils of the Middle Devonian Silica Formation. *Papers on Paleontology*, no. 8. Friends of the University of Michigan Museum of Paleontology, Inc.

Kindle, E. M. 1901. The Devonian fossils and stratigraphy of Indiana. *Indiana Department of Geological Natural Resources, Annual Report* 25: 529-758; 773-775.

Kindle, E. M. 1912. The Onondaga fauna of the Allegheny region. *U. S. Geological Survey Bulletin* 508.

King, W. 1846. Remarks on certain genera belonging to the class Palliobranchiata. Annals and Magazine of Natural History London 18: 26-42; 83-94.

King, W. 1850. A monograph of the Permian fossils of England. Palaeontological Society of London Monograph 3: 1-258.

Klapper, G. 1971. Sequence within the conodont genus *Polygnathus* in the New York lower Middle Devonian. *Geologie et Palaeontologie* 5: 59-79.

Koch, W. F., II. 1981. Brachiopod community paleoecology, paleobiogeography, and depositional topography of the Devonian Onondaga Limestone and correlative strata in eastern North America. *Lethaia* 14: 83-103.

Kozlowski, R. 1929. Les Brachiopodes Gothlandiens de la Podolie Polonaise. *Palaeontologica Polonica* 1: 1-254.

Landes, K. K., Ehlers, G. M. and Stanley, G. M. 1945. Geology of the Mackinac Straits region and the subsurface geology of the northern southern peninsula. Michigan Geological Survey Division, Publication 44, Geological Series 37.

Lenz, A. C. 1977. Upper Silurian and Lower Devonian brachiopods of Royal Creek, Yukon, Canada. Part I: Orthoidea, Strophomenida, Pentamerida, Rhynchonellida. *Palaeontographica* 159 (A): 37-109.

Lindemann, R. H. 1980. Paleosynecology and paleoenvironments of the Onondaga Limestone in New York State. Unpublished PhD Thesis. Troy, New York, Rennselaer Polytechnic Institute.

Lindemann, R. H., and Feldman, H. R. 1981. Paleocommunities of the Onondaga Limestone (Middle Devonian) in central New York State. New York State Geological Association Guidebook for Field Trips in South-Central New York, 53rd Annual Meeting, State University of New York at Binghamton, 79-96.

Linnaeus, C. 1758. Systema Natura (10th ed.). vol. 1, Holmiae.

Martin, W. 1809. Petrificata derbiensia: Or figures and descriptions of petrifactions collected in Derbyshire. 28 pp., 52 pl. (Wigan.)

McLaren, D. J. 1965. Paleozoic Rhynchonellacea. In R. C. Moore (ed.), *Treatise on Invertebrate Paleontology*, Part H, Brachiopoda, H552-H597. Lawrence, KS: Geological Society of America and University of Kansas Press.

M'Coy, F. 1844. A synopsis of the characters of the Carboniferous limestone fossils of Ireland. Dublin.

Nettleroth, H. 1889. Kentucky fossil shells, a monograph of the fossil shells of the Silurian and Devonian rocks of Kentucky. *Kentucky Geological Survey.*

Oehlert, D. P. 1887. Brachiopodes. In P. H. Fischer, *Manuel de Conchyliologue et du Paleontologie Conchyliologieque au histoire naturelledes Mollusques vivants et fossiles*, 1189-1334. Paris: F. Savy.

Oliver, W. A., Jr. 1954. Stratigraphy of the Onondaga Limestone (Devonian) in central New York. *Geological Society of America Bulletin* 65: 621-652.

Oliver, W. A., Jr. 1956. Stratigraphy of the Onondaga Limestone in eastern New York. *Geological Society of America Bulletin* 67: 1441-1474.

Oliver, W. A., Jr. 1966. Bois Blanc and Onondaga formations in western New York and adjacent Ontario. *New York State Geological Association Guidebook, 38th Annual Meeting*, State University of New York at Buffalo, 32-43.

Oliver, W. A., Jr. 1967. Stratigraphy of the Bois Blanc Formation in New York. U. S. *Geological Survey Professional Paper* 584A. Washington, D.C.

Oliver, W. A., Jr. 1977. Biogeography of Late Silurian and Devonian rugose corals. *Paleogeography, Paleoclimatology, Paleoecology* 22: 85-135.

Phillips, J. 1841. *Figures and Descriptions of the Paleozoic Fossils of Cornwall, Devon, and West Somerset.* London.

Pitrat, C. W. 1965. Spiriferidina. In R. C. Moore (ed.), *Treatise on Invertebrate Paleontology*, Part H, Brachiopoda, H667-H728. Lawrence, KS: Geological Society of America and University of Kansas Press.

Racheboeuf, P. R., and Feldman, H. R. 1990. Chonetacean brachiopods of the "Pink *Chonetes*" Zone, Onondaga Limestone (Devonian, Eifelian), central New York. *American Museum Novitates* 2982: 1-16.

Sartenaer, P. 1961. Etude nouvelle, en deux parties, du genre *Camarotoechia* Hall et Clarke, 1893. Deuxième partie: *Cupularostrum recticostatum* n. gen., n. sp. *Institut royal des Sciences naturelles de Belgique Bulletin* 37: 1-15.

Savage, N. M. 1974. The brachiopods of the Lower Devonian Maradana Shale, New South Wales. *Palaeontographica* 146(A): 1-51.

Savage, T. E. 1930. The Devonian rocks of Kentucky. *Kentucky Geological Survey* 33: 1-161.

Savage, T. E. 1931. The Devonian fauna of Kentucky. In Paleontology of Kentucky. *Kentucky Geological Survey* 36: 217-246.

Schuchert, C. 1894. A revised classification of the spire-bearing Brachiopoda. *American Geologist* 13: 102-107.

Schuchert, C. 1913. Systematic paleontology. Lower Devonian, Brachiopoda. In Lower Devonian Volume (with T. P. Maynard). *Maryland Geological Survey*, 290-449.

Schuchert, C., and Cooper, G. A. 1932. Brachiopod genera of the suborders Orthoidea and Pentameroidea. *Peabody Museum of Natural History Memoir* 4, pt. I: 1-270.

Schuchert, C., and LeVene, C. M. 1929. Brachiopoda (Generum et Genotyporum Index et Bibliographia). Fossilium Catalogus 1, Animalia 42: 1-140.

Schuchert, C., Swartz, C. K., Maynard, T. P., Prosser, C. S., and Rowe, R. B. 1913. Lower Devonian. *Maryland Geological Survey*.

Schrock, R. R., and Twenhofel, W. H. 1953. *Principles of Invertebrate Paleontology*. New York: McGraw-Hill.

Stauffer, C. R. 1915. Stratigraphy of southwestern Ontario. *Geological Survey of Canada Memoir* 34.

Steininger, J. 1853. *Geognostisch Beschreibung der Eifel*. Trier: F. Lintz'schen.

Swartz, F. M. 1929. The Helderberg group of parts of West Virginia and Virginia. *U. S. Geological Survey Professional Paper* 158C: 27-69.

Vanuxem, L. 1839. Third annual report of the geological survey of the third district. *New York Geological Survey Annual Report* 3: 142-285.

Vanuxem, L. 1840. Fourth annual report of the geological survey of the third district. *New York Geological Survey Annual Report* 4: 355-383.

Vanuxem, L. 1842. Geology of New York, Part III, comprising the survey of the third geological district. Albany: W. and A. White and J. Visscher, Natural History of New York.

Waagen, W. H. 1882-1885. Salt Range fossils, Part 4 (2) Brachiopoda. *Palaeontologica Indica*, Memoir, ser. 13, vol. 1: 329-770.

Wedekind, R. 1926. Die Devonische Formation. In W. Saloman, *Grundzuge der Geologie* 2: 194-226. Stuttgart: E. Schweizerbastische.

Weller, S. 1903. The Paleozoic faunas. *New Jersey Geological Survey Report on Paleontology* 3: 1-462.

Wells, J. W. 1963. Early investigations of the Devonian System in New York, 1656-1836. Geological Society of America Special Paper 74: 1-74.

White, C. A. 1862. Description of new species of fossils from the Devonian and Carboniferous rocks of the Mississippi Valley. Boston Society of Natural History Proceedings 9: 8-33.

Wilckens, C. F. 1769. Nachricht von selten Versteinerungen, 78-79. Berlin.

Williams, A. 1953. North American and European stropheodontids: Their morphology and systematics. *Geological Society of America Memoir* 56.

Williams, A., and Wright, A. D. 1965. Orthida. In R. C. Moore (ed.), *Treatise on Invertebrate Paleontology*, Part H, Brachiopoda, H299-H359. Lawrence, KS: Geological Society of America and University of Kansas Press.

Ziegler, A. M. 1965. Silurian marine communities and their environmental significance. *Nature* 207: 270-272.

Ziegler, A. M., and Boucot, A. J. 1970. North American Silurian animal communities. In W. B. N. Berry and A. J. Boucot (eds.), *Geological Society of America Special Paper* 102: 95-106.

Chapter Four

Chonetacean Brachiopods of the "Pink *Hallinetes*" Zone, Onondaga Limestone (Devonian, Eifelian), Central New York

ABSTRACT

The "*Hallinetes*" Community (formerly "*Chonetes*") of the Seneca Member, Onondaga Limestone (Devonian, Eifelian), is redefined. The chonetacean brachiopods found therein belong to three different taxa: (1) *Hallinetes lineatus,* new genus, which is distinguished from other chonetacean genera by its disymmetrically arranged spines, which are perpendicular to the hinge line, and for which a neotype has been designated, (2) *Longispina mucronata* and (3) "*Eodevonaria*" *hemispherica*, for which lectotypes have been designated. *Hallinetes* is the most common chonetid found on the bedding planes (92%) while *Longispina* and "*Eodevonaria*" are much less common (5.4 and 2%, respectively). The *Hallinetes* Community of the "Pink *Hallinetes*" Zone corresponds to the assemblage within the mudstone matrix rather than to that on the bedding planes. It is a low-diversity community representative of a quiet water environment, probably in a Benthic Assemblage 4 position of Boucot (1975).

INTRODUCTION

The Onondaga Limestone in central New York (Figure 1) can be divided into four members (Oliver, 1954; Feldman, 1980, 1985): from base to top, the Edgecliff, Nedrow, Moorehouse, and Seneca. The total thickness of the formation in Jamesville, New York (AMNH Locality 3128) is 69.5 ft (Feldman, 1985). In the eastern and western part of the state, that is, away from the basinal axis, the total thickness approximately doubles. The "Pink *Hallinetes*" Zone (Oliver, 1954) in central New York is located in the Seneca Member, 10 ft above the base of a greenish-gray to ochre colored volcanic ash layer, the Tioga Bentonite (Figure 2) and is characterized in many localities by numerous stylolites. Overlying the Seneca in central New York is the Union Springs Black Shale, a correlative of the Marcellus Shale in the Cherry Valley

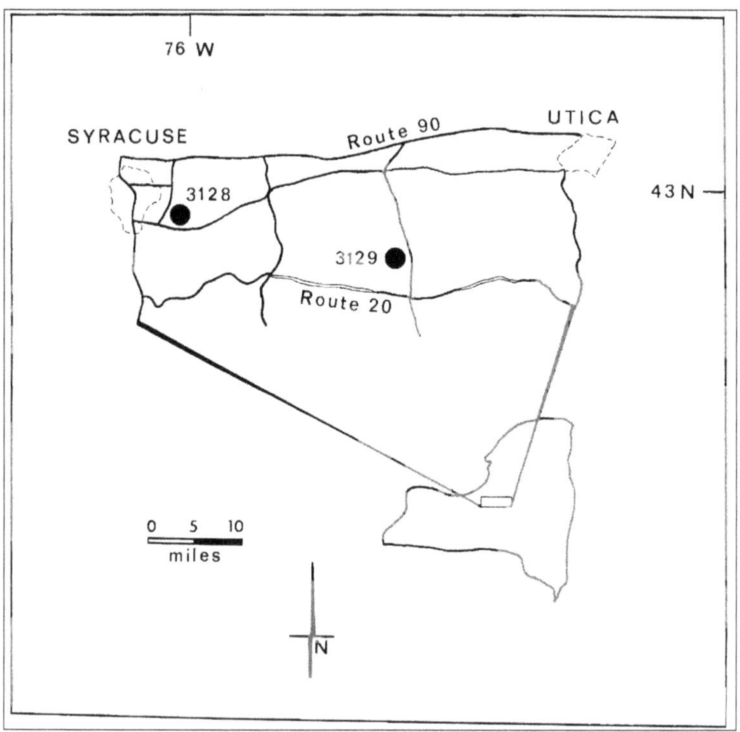

FIGURE 1. Index map of collecting localities in the "Pink *Hallinetes*" Zone, Onondaga Limestone, central New York. Numbers refer to AMNH Localities (see text for detail).

area, which represents a westerly advance of terrigenous sedimentation and relatively deep, poorly oxygenated water (Feldman, 1980).

Vanuxem (1839) recognized a nearly chert-free limestone unit with abundant "*Strophomena* (*Chonetes*)" *lineata* above the Corniferous Limestone, which he referred to as the Seneca Limestone. He considered the Seneca to be a separate unit, as is evidenced by his annual report (1839) and final report (1842). Workers have studied the biostratigraphy of the "Pink *Chonetes*" Zone (Oliver, 1954; Feldman and Lindemann, 1986), its community structure (Feldman, 1980: Lindemann and Feldman, 1981), the taxonomy of its brachiopod fauna (Feldman, 1985), and even reasons for the "pink" coloration of its dominant chonetid brachiopods (Zenger, 1967). Oliver (1954), in his classic study of the Onondaga Limestone in central New York, referred to the "Pink *Hallinetes*" Zone of the Seneca Member as Zone J.

He described it as a very thinly bedded limestone, about 4 ft thick and composed almost entirely of shells of "*H.*" *lineatus*, many of which are stained pink. Oliver (1954) noted that a bed of chert, similar to that of Zone I (8 ft thick), occurs at most localities about 6 inches below the top of Zone J in western New York. This chert bed, according to Oliver, includes the hyolithid *Coleolus crenatocinctum* and gastropod *Loxonema sicula.*

Oliver (1954: 641) found the following brachiopod fauna in the "Pink *Hallinetes*" Zone of western New York:

Camarotoechia tethys	*Athyris spiriferoides*
Atrypa "reticularis"	*"Schuchertella" pandora*
Atrypa spinosa	*Levenea lenticularis*
Coelospira camilla	*"Hallinetes" lineatus*
Atlanticocoelia acutiplicata	*Hallinetes mucronatus*
Elita fimbriata	*Pentagonia unisulcata*
Acrospirifer duodenarius	*Meristella nasuta*

He also noted the presence of the trilobite *Odontocephalus selenurus* and fish teeth.

In central New York, Oliver (1954: 635) found the following brachiopod fauna in the "Pink *Hallinetes*" Zone:

Coelospira camilla	*Levenea lenticularis*
Atrypa "reticularis"	*"Schuchertella" pandora*
Atrypa spinosa	*"Hallinetes" lineatus*
Chonostrophia reversa	*Meristella* sp. A

He also found the corals *Amplexiphyllum hamiltoniae* and *Heterophrentis* sp. B. Feldman (1985) found the following additional brachiopods in the "Pink *Chonetes*" Zone in central New York (AMNH locs. 3128A, 3129):

Megakozlowskiella raricosta	*Athyris* sp. A
Leptaena aff. *"rhomboidalis"*	*Megastrophia*
Pentamerella arata	orthotetacids indet.

Additional faunal constituents found by Feldman (1985) include the corals "*Heterophrentis*" and *Amplexiphyllum*, trilobites *Phacops cristata* and

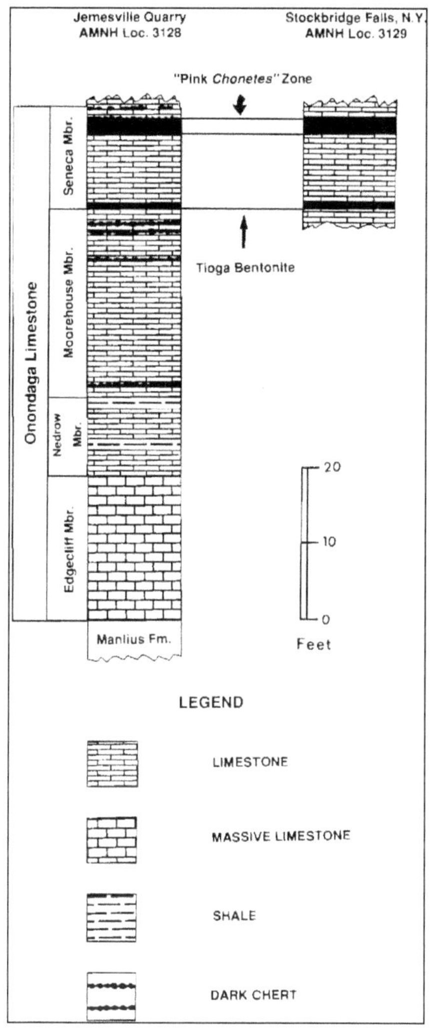

FIGURE 2. Measured stratigraphic sections of AMNH Loc. 3128 and 3129, Onondaga Limestone, central New York.

Odontocephalus, euomphalacean gastropod fragments, and camerate? crinoid columnals.

In this paper we revise the taxonomy of the chonetacean brachiopods found in the "Pink *Hallinetes*" Zone of the Onondaga Limestone in central New York and redefine the community based on new paleoecologic and taxonomic information.

ABBREVIATIONS

Institutions and Localities

AMNH	—	American Museum of Natural History, Department of Invertebrates
AMNH Loc.	—	American Museum of Natural History locality number
ANSP	—	Academy of Natural Sciences, Philadelphia
NYSM	—	New York State Museum, Albany
USNM	—	United States National Museum of Natural History, Department of Paleobiology, Smithsonian Institution

Measurements

(L)	—	maximum length of shell
(W)	—	maximum width of shell
(T)	—	maximum thickness of shell
mm	—	millimeters
est.	—	estimated
b.v.	—	brachial valve
p.v.	—	pedicle valve
b.p.	—	bedding plane
art.	—	articulated
ext.	—	exterior

SYSTEMATIC PALEONTOLOGY

Phylum BRACHIOPODA
Order PRODUCTIDA Sarytcheva and Sokolskaya, 1959
Suborder CHONETIDINA Muir-Wood, 1955
Superfamily CHONETOIDEA Bronn, 1862
Family CHONETIDAE Bronn, 1862
Subfamily DEVONOCHONETINAE, Muir-Wood, 1962
Genus *HALLINETES* Racheboeuf and Feldman, 1990

Hallinetes, new genus, Figures 3-5.

Type Species: *Strophomena lineata* Conrad, 1839.

Etymology: In honor of New York State paleontologist James Hall.

Diagniosis: Chonetid with vertical disymmetrically arranged hinge spines (4' -1, 2, 3 [see Table 2]). Brachial interior with a median septum not supporting the cardinal process. Alveolus well developed in juvenile brachial valves. Anderidia posteriorly fused with the elevated medial part of the inner cristae. No accessory septa.

Remarks: *Hallinetes* is provisionally assigned to the subfamily Devonochonetinae based on brachial interior morphology. The lack of accessory septa links the new genus to the "*Devonochonetes coronatus*" group (see Racheboeuf, 1981: 141; Racheboeuf and Branisa, 1985: 1439 for discussion). *Hallinetes* can easily be distinguished from all other chonetacean genera by its spines, which are perpendicular to the hinge line and disymmetrically arranged.

Morphology and arrangement of spines closely resemble those of the genus *Johnsonetes* Racheboeuf, 1987, from the Emsian Blue Fiord Formation of the Canadian Arctic Archipelago. However, *Hallinetes* lacks the median enlarged capilla and differs in its brachial interior, which is strongly pustulose with more divergent anderidia, and by the presence of a large alveolus.

Hallinetes resembles *Aseptonetes* Isaacson, 1977 from the Emsian Sicasica Formation of Bolivia, in its radial ornamentation as well as in spine morphology, but the latter possesses symmetrically arranged spines: both genera lack accessory septa.

The posteromedian part of the brachial interior of *Hallinetes* is somewhat similar to that of the genus *Saharonetes* Havlicek, 1984, from the Lower Carboniferous Ashkidah and Marar formations of Libya, as the inner cristae medially bend posteriorly toward the inner lobes of the cardinal process. However, in *Hallinetes* the inner cristae are much better developed, spines are oriented vertically instead of obliquely, and there are no accessory septa.

Although *Hallinetes* resembles *Johnsonetes, Aseptonetes,* and *Saharonetes*, there is no generic relationship evident between them, only morphological similarities.

Species questionably assigned to *Hallinetes,* new genus:

Strophomena setigera Hall, 1843.

Chonetes cf. *setigerus* (Hall), Kindle, 1912: 71, pl. 3, fig. 17.

Hallinetes lineatus (Conrad, 1839), Figures 3-5.

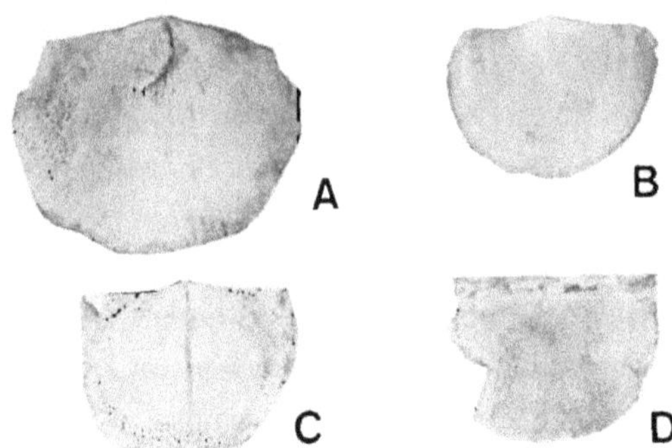

FIGURE 3. *Hallinetes lineatus* (Conrad, 1839). A. Pedicle valve, original of *Chonetes glabra* Hall, 1857, 1867, pl. 20, fig. 3D, AMNH 37235, x3. B. Pedicle valve, figured by Hall, 1867, pl. 20, fig. 3A, AMNH 37233a, x3. C. Pedicle interior, Hall's original, 1867, pl. 20, fig. 3E, AMNH 37233C, x3. D. Brachial exterior, Hall's original, 1867, pl. 20, fig. 3C, AMNH 37233B, x3.

Strophomena lineata Conrad, 1839: 64; Vanuxem, 1842: 139; Hall, 1843: 175, fig. 8.

Chonetes glabra Hall, 1867: 117, figs. 1-8.

Chonetes lineata Hall, 1867: 121, pl. 20, fig. 3; Hall and Clarke, 1892: pl. 16, fig, 34.

"*Chonetes*" *lineatus* Zenger, 1967: 161, fig. 1.

"*Chonetes*" aff. *lineata* Feldman, 1985: 321, figs. 26-28 (not 25).

Neotype: Unsuccessful attempts were made to locate the genotype of *Hallinetes* at various institutions (USNM, NYSM, ANSP, AMNH) for purposes of comparison. Since the genotype is apparently lost, we are herein designating a neotype, AMNH 37233A (fig. 3B herein).

Type locality: Unknown; according to Hall (1843: 175), this shell is abundant in Seneca County but rare farther west. Substitute type locality herein designated as AMNH Loc. 3128, Allied Chemical Quarry, Jamesville, New York.

Remarks: The species name *lineata* was first used by Conrad (1839), associated with the genus name *Strophomena.* James Hall was the first to establish the assignment of the species *lineata* to the genus *Chonetes* (Table 1), and gave the first complete description of the species. He considered Vanuxem as the author, and stated that the previously described species *Chonetes glabra* Hall, 1857 was a junior synonym of *Chonetes lineata* Vanuxem (Hall, 1867: 121).

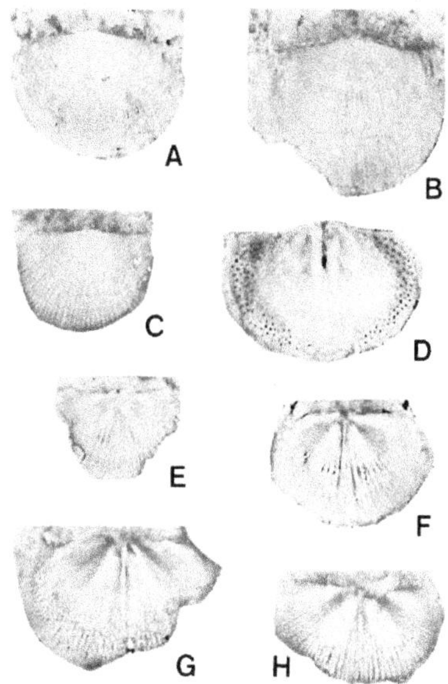

FIGURE 4. *Hallinetes lineatus* (Conrad, 1839). A. Pedicle valve, block 2, AMNH Loc. 3128, AMNH 43693, x3. B. Pedicle valve, AMNH Loc. 3128, AMNH 43694, x3. C. Latex mold of an immature pedicle valve, AMNH Loc. 3128, AMNH 43695, x3, D. Pedicle interior, AMNH 43695, x3. E. Latex cast of a damaged immature brachial interior, AMNH Loc. 3128, AMNH 43696, x3. F. Latex cast of a brachial interior, AMNH Loc. 3129, AMNH 43697, x3. G. Latex cast of a large, damaged brachial interior, AMNH Loc. 3128, AMNH 43698, x3. H. Brachial interior, latex, AMNH Loc. 3128, AMNH 43695, x3.

Exterior: Small shell with maximum width at hinge line in juvenile specimens but at midlength in adults. Maximum length about 9 mm; corresponding width about 11 mm; length/width ratio decreasing from 0.95 to 0.75 during ontogeny. Longitudinal profile weakly concavoconvex in small shells, becoming strongly arched in larger specimens; length/thickness decreasing from 7.5 to a mean value of 2.4 during ontogeny. Pedicle valve weakly flattened at the top. Umbo relatively small, narrow, strongly arched longitudinally. Ears small, triangular, often ill-defined and smooth. Ornamentation consists of low, rounded radial costellae with wider rounded interspaces; costellae increasing by intercalation on the pedicle valve and by bifurcation on the brachial valve; on the pedicle valve the first intercalation occurs about 1.2 mm from the beak. Two mm anterior of beak, costellae number 5 per mm; along the anterior margin they number 5-6 per mm. Costellae progressively narrower from beak to anterior margin as intercalations increase. Total

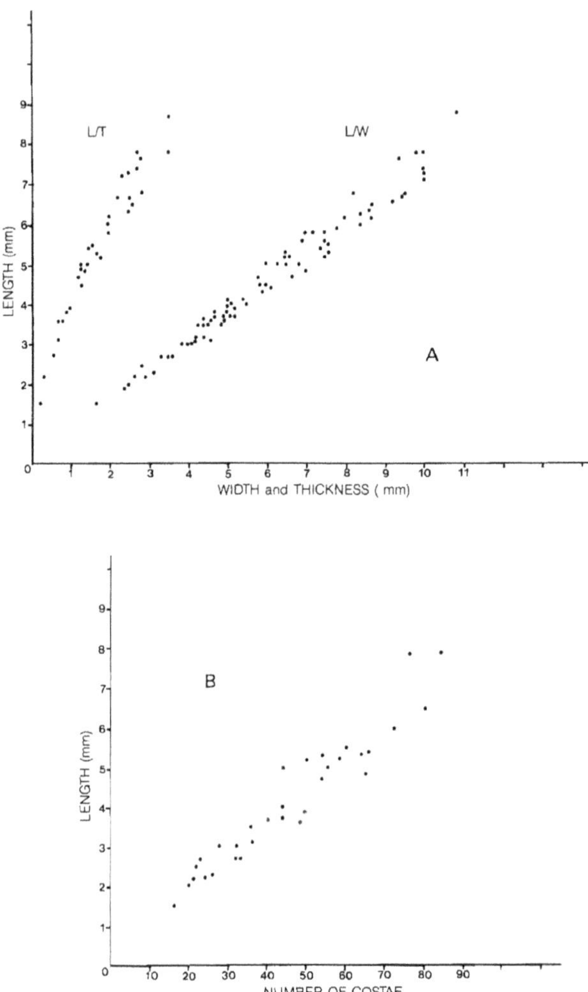

FIGURE 5. *Hallinetes lineatus* (Conrad, 1839). Scattergram showing: A. Length/width and length/thickness; and B. length/total number of costae.

number of costellae between 72 and 84 for shells which are more than 6.0 mm in length. Due to difficulties in preparing the material, the pseudodeltidium and chilidium have not been observed. Pedicle interarea concave and apsacline; brachial interarea almost linear and hypercline. In largest shells, three spines are inserted on the right side of the beak, only one on the left side. Spines are straight, perpendicular to hinge line, of orthomorph type. Their distribution is: 4, -1, 2, 3 (see Figure 4A) (Table 2).

Pedicle valve interior: Visceral cavity well defined. Stout myophragm extending anteriorly to midlength or two-thirds of the valve in largest specimens; myophragm narrows anteriorly, dividing a deeply impressed muscle field with subcircular adductor scars and subtriangular, anteriorly rounded diductor scars. Muscle field extending anteriorly up to one-third of the valve length. Vascular trunks originating at the anterior margin of adductors. Posterior ridges are relatively low, wide, and rounded, anteriorly divergent at about 100 degrees. Teeth are stout, transversely elongate, and subparallel to hinge line. Visceral cavity is deep, smooth, and well defined anterolaterally. Posterolateral parts of the valve are coarsely pustulose; anterolateral margins bear radially arranged endospines.

Brachial valve interior: Posteromedian part of valve deeply depressed between cardinal process, median septum, and proximal part of inner cristae. Median septum forms low, rounded ridge extending progressively toward anterior margin during growth; length about one-fourth of the valve length in small shells, reaching two-thirds the valve length in largest

TABLE 1. Measurements (in mm) of *Chonetes lineata* Hall, 1867 (originals of *Chonetes glabra* Hall, 1867).

Specimen	(L)	(W)	(T)	No. of costae per mm	Original specimen
AMNH 37235	8.2	–	4.1	4	Hall, 1867, pl, 20, fig. 3D
AMNH 37233a lectotype	5.8	80.	–	5	Hall, 1867, pl. 20, fig. 3A
AMNH 43707	5.8	8.0	2.2	5	
AMNH 43708	3.0	4.0	0.7	5	
AMNH 43709	6.8	8.5	2.8	5	
AMNH 43710*a*	5.8	8.2	–	–	
AMNH 43711	5.3	7.4	1.8	5	
AMNH 43712	3.4	4.6	0.8	5	
AMNH 43713	7.5	10.0	3.0	5	
AMNH 43714	6.5	8.3	2.6	5	

a Pedicle interior.

TABLE 2. Distribution of Hinge Spines on *Hallinetes lineatus*.

No. spines	4	1	2	3
Distance from beak (mean, mm)	3.63	0.32	0.94	2.46
No. observations	5	3	23	16

specimens. Outer cristae are reduced to very low and narrow linear ridges along the hinge line. Inner cristae are strongly developed as two stout rounded ridges almost parallel to hinge line, not fusing medially with the anterior of the cardinal process. Posterior edge of the inner cristae is deeply notched by dental sockets. Cardinal process internally bilobed, each lobe fusing anterolaterally with proximal part of inner cristae. Myophore quadrilobed. Anderidia anteriorly divergent at about 50-55 degrees, as straight, narrow, and elongated ridges fuse posteriorly with inner cristae. Adductor scars are deeply impressed in the valve floor of the largest shells. Brachial platform strongly convex, elevated above median septum, pustulose in juvenile specimens, smooth in adult shells. Posterolateral part of the valve between inner cristae and brachial platform is depressed. Inner surface is covered with relatively strong endospines, radially arranged, except for the inner cristae, muscle field, and brachial platform.

Material: Several hundred isolated valves and articulated shells, all from two localities in the "Pink *Hallinetes*" Zone, central New York.

Stratigraphic occurrence: The shells were collected from two localities, both within the Seneca Member, Onondaga Limestone.

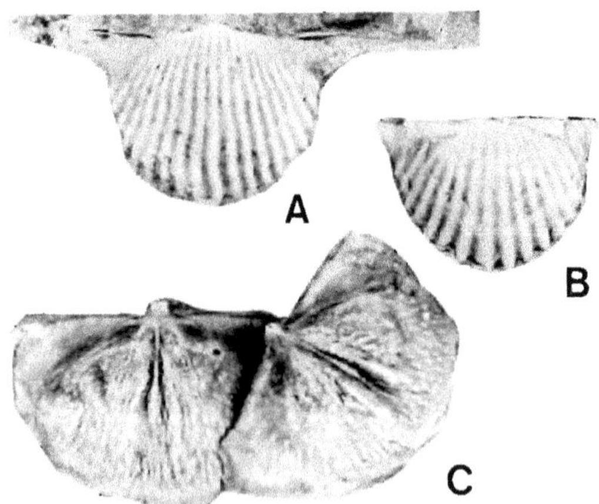

FIGURE 6. *Longispina mucronata* (Hall, 1843). A. Lectotype, pedicle valve with spines. Hall's original, 1867, pl. 21, fig. 1, AMNH 37246, x3. B. Brachial exterior, original of *Chonetes laticosta* Hall, 1857, figured in Hall, 1867, pl. 20, fig. 2, AMNH 37244, x3. C. Brachial interiors, original of *Chonetes scitula* Hall, 1857, figured in Hall, 1867, pl. 21, fig. 4F, AMNH 37254, x3.

The first (AMNH Loc. 3129A) is in the Jamesville Quarry #3 pit, which is the most complete section of the Onondaga in the central part of New York. Collecting in this quarry is difficult since it is very active. Not only is it hard to find weathered bedding surfaces, but the rocks are continually being blasted and removed, making it difficult to locate and recover fossil material. The Seneca Member here is 14 ft thick, with the Tioga Bentonite at the base. Ten feet above the Tioga is found the "Pink *Chonetes*" Zone, which consists of a fine-grained limestone with a fresh dark gray surface. There are wavy contacts between bedding planes and stylolites present.

The second (AMNH Loc. 3129) is located just west of Stockbridge Falls, New York, along Stockbridge Falls Road and Oneida Creek. Here, 10-15 ft of the Seneca Member is exposed, with an upper contact with the Union Springs Black Shale. The Seneca here is a fine-grained, jointed limestone, medium to thinly bedded, heavily weathered and muddy on weathered surfaces but dark gray on fresh surfaces.

Subfamily DEVONOCHONETINAE Muir-Wood, 1962
Genus LONGISPINA Cooper, 1942

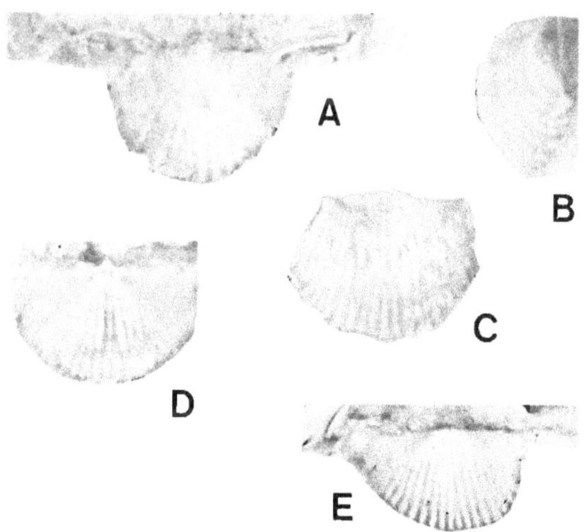

FIGURE 7. *Longispina mucronata* (Hall, 1843). A. Pedicle valve with spines, AMNH Loc. 3128, AMNH 43699, x5. B, C. Pedicle valve, lateral and ventral views, AMNH Loc. 3128, AMNH 43700, x3. D. Latex cast of a brachial valve, AMNH Loc. 3129, AMNH 43701, x3. E. Latex cast of an incomplete pedicle valve, AMNH Loc. 3128, AMNH 43706, x3.

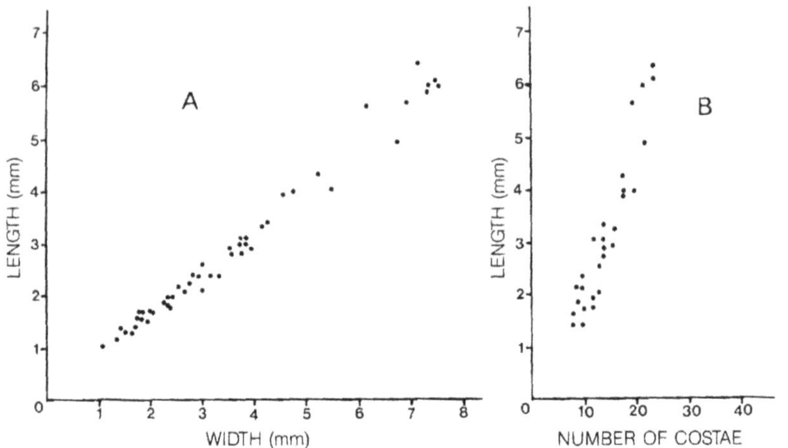

FIGURE 8. *Longispina mucronata* (Hall, 1843). Scattergram showing: A. Length/width; and B. Length/total number of costae for 55 pedicle valves.

Longispina mucronata (Hall, 1843), Figures 6-8, 12

Strophomena mucronata Hall, 1843: 180, fig. 3.

Chonetes laticosta Hall, 1857: 119.

Chonetes mucronata Hall, 1867: 124, pl. 20, fig. 1; pl. 21, fig. 1.

Chonetes mucronata Hall and Clarke, 1892: pl. 16, figs. 6, 7; 1894: pl. 20, fig. 3.

Type species: *Chonetes emmetensis* Winchell, 1866: 92.

Lectotype: Pedicle valve, AMNH 37246, figured in Hall, 1867 (pl. 21, fig. 1a-c) (Figure 6A herein; Table 3).

Type locality: Ontario County, New York State (more specific information unavailable; Hall [1857: 120] described the locality as a limestone a few miles southeast of Buffalo and in shales of the Hamilton Group on Canandaigua Lake). Substitute type locality herein designated as AMNH Loc. 3128, Jamesville Quarry, Jamesville, New York.

Exterior: Small shells, moderately concavoconvex with maximum width at hinge line. Anterior margin regularly rounded; lateral margins parallel or slightly divergent posteriorly. Maximum length about 6.0 mm; corresponding width about 7.5 mm. Length/width ratio between 0.75 and 0.8 mm. Umbo relatively small, weakly overlapping posterior edge of ventral interarea. Pedicle interarea concave and anacline; brachial interarea linear. Pseudodeltidium and chilidium not observed. Ornamentation consists of rounded, radial costae; along margins, costae number 12 to 24 during growth; costae are usually simple, progressively widening from umbo to margins. Costae are wider than interspaces on the pedicle valve while narrower than interspaces on the brachial valve. Along the anterior margin costae number 3 per 2 mm. Costae crossed by thin and regular concentric growth lines. According to Hall (1867: 125), some costae originate by bifurcation or by intercalation; this was not observed on the specimens from the Seneca Member. Three spines on each side of the beak of largest specimens; spines typical for the genus and symmetrically arranged.

Pedicle valve interior: The myophragm does not exceed one-third of the valve length. Muscle field deeply impressed on the valve floor with well-defined diductor scars. Adductors narrow, posteriorly situated. Vascular trunks strongly developed, anteriorly divergent from the anterior end of the myophragm. Visceral cavity well defined, peripheral margin of valve impressed by external radial costae. Teeth not observed on material available.

Brachial valve interior: Typical for the genus in every aspect of its morphology, with a low and narrow median septum supporting the cardinal process and extending anteriorly up to two-thirds the valve length; posteriorly elongated and narrow cardinal process; a pair of well-developed accessory septa and anderidia anteriorly divergent at 70 degrees.

Discussion: *Strophomena mucronata* Hall, 1857, is a typical representative of the genus *Longispina* Cooper, 1942, as attested by the shell shape and spine morphology, as well as interiors of both valves. The original of *Chonetes laticosta* Hall, 1857, figured in 1867 (pl. 20, fig. 2; AMNH 37244) is a brachial exterior or *L. mucronata,* as stated by Hall (1867: 125). Two dorsal interiors (AMNH 37254), one of which was figured by Hall (1867: pl. 21, fig. 20 as *Chonetes scitula,* are undoubtedly representatives of the genus *Longispina*; according to our own observations they must be assigned to the species *L. mucronata,* but *L. mucronata* and "*Chonetes*" *scitula* are distinct species.

TABLE 3. Measurements (in mm) of *Longispina mucronata* (Hall, 1843).

Specimens	(L)	(W)	Total no. of costae
AMNH 37246 lectotype, p.v.	6.6	8.5	20
AMNH 37247 p.v.	3.1	4.2	13
AMNH 37244 original *Chonetes laticosta,* p.v.	5.9	6.8	17

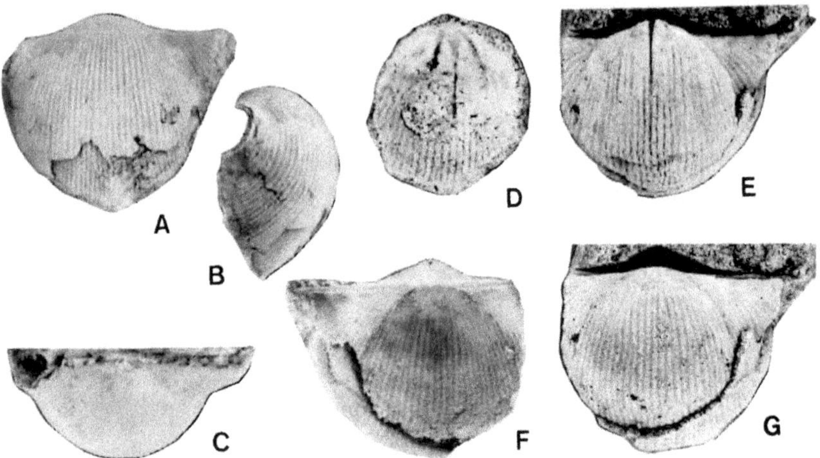

FIGURE 9. *"Eodevonaria" hemispherica* (Hall, 1857). A-C. Lectotype, pedicle valve in ventral (A), lateral (B), and posterior (C) views. Hall's original, 1857: 116, fig. 1; 1867, pl. 20, fig. 6b-d, AMNH 2825, x2. D-G. Decalcified articulated damaged shell, brachial interior (D), pedicle interior (E), latex cast of the brachial exterior (F), and interior (G), AMNH 37224, x2.

Material: A total of 105 specimens from the *Hallinetes* Community, mostly complete juvenile shells; only one brachial valve, and two pedicle valves were available for preparing the interiors; AMNH Locs. 3128, 3129.

Stratigraphic occurrence: *Longispina mucronata* is a common species of the Nedrow, Moorehouse, and Seneca members of the Onondaga Limestone; it is rare or very rare within the Edgecliff Member according to Oliver (1954, 1956). It is also found in the overlying Hamilton Group, specifically in the Marcellus Shale.

Family EODEVONARIIDAE Sokolskaya, 1960
Genus EODEVONARIA Breger, 1906

"Eodevonaria" hemispherica (Hall, 1857), Figures 9-12

Chonetes hemispherica Hall, 1857: 116; 1867: 118-119, pl. 20, fig. 6A-D.

Chonetes hemisphaerica Hall and Clarke, 1892: pl. 16, fig. 14.

Type species: *Chonetes arcuata* Hall, 1857.

Lectotype: Pedicle valve AMNH 2825 figured in 1857 (116, fig. 1), 1867 (pl. 20, fig. 6b-c), 1892 (pl. 16, fig. 14).

Type locality: Schoharie Grit (Hall, 1857: 116). Substitute type locality herein designated as AMNH Loc. 3128, Jamesville Quarry, Jamesville, New York.

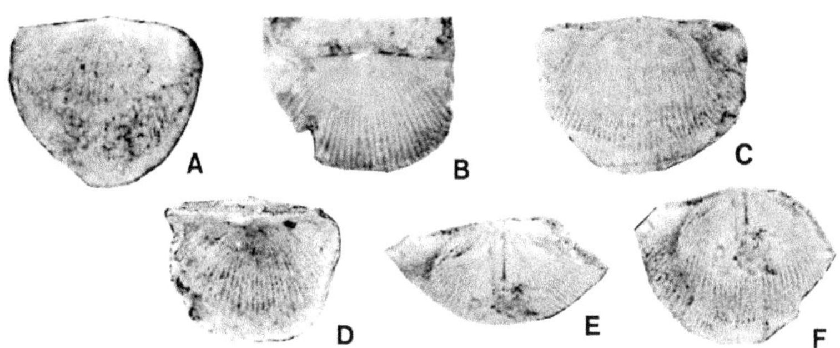

FIGURE 10. *"Eodevonaria" hemispherica* (Hall, 1857). A. Pedicle valve, latex cast, AMNH Loc. 3128, AMNH 43702, x3. B. Juvenile pedicle valve with spines, AMNH Loc. 3128, AMNH 43703, x3. C. Pedicle interior, AMNH Loc. 3128, AMNH 43704, x3. D. Articulated shell in dorsal view, AMNH Loc. 3129, AMNH 43705, x3. E, F. Damaged pedicle interior, posteroventral and ventral views, AMNH Loc. 3128. AMNH 43706, x3.

Exterior: Large shell, strongly concavoconvex with maximum width at hinge line. Maximum length is about 20 mm; length/width ratio about 0.66 mm; length/height ratio about 0.5 mm. Umbo is widely rounded, strongly arched dorsally, posteriorly overhanging hinge line by more than 2 mm in largest specimens. Ears are triangular, strongly convex, and well differentiated from the body of the pedicle valve. Ornamentation of low, rounded radial costae increases by bifurcation on the pedicle valve and by intercalation on the brachial valve. Along the anterior margin, costae number 3 to 4 per mm. At 2 mm anterior to the beak, the total number of costae is about 15, while at 5 mm from the beak the number increases to 35. Maximum number of costae is 70 (estimated) in largest shells. On the pedicle valve, first bifurcation occurs at 2.5 mm from the beak. Costae are separated by narrower spaces. Pseudodeltidium is reduced to a small triangular, convex plate at the apex of the delthyrium; chilidium not observed. Pedicle inter-area strongly concave and catacline, almost perpendicular to commissural plane. At least 6 spines on each side of beak, high angled and oblique at their base.

Pedicle valve interior: Only one specimen (AMNH 37224) was available for study. Muscle field is longitudinally divided by stout myophragm extending anteriorly up to midlength. Diductor scars are ill-defined, poorly impressed on the valve floor, wide, radially grooved, occupying the posterior half of the visceral cavity. Adductor scars relatively large, semioval in outline, almost semicircular; length is about 4 mm and corresponding width about 3 mm. Two narrow and deeply impressed vascular trunks (vascula media) originate anterior to adductor scars. Whole inner surface is impressed by external ribbing except for anterolateral margins of the valve, where endospines remain distinct, radially arranged. Hinge line strongly denticulate.

Brachial valve interior: Known from one incomplete internal mold only (AMNH 37224). A strong median septum extends anteriorly up to midlength; septum progressively widens posteriorly, developing a flattened triangular platform. Anderidia posteriorly fused with the platform, anteriorly diverging at 50 degrees. Cardinalia and other structures unknown. Inner surface covered with radially arranged distinct endospines.

Discussion: The species *hemispherica* (Table 4) undoubtedly belongs to the family Eodevonariidae as evidenced by its shape and denticulate hinge

TABLE 4. Measurements (in mm) of "*Eodevonaria*" *hemispherica* (Hall, 1857).

Specimen	(L)	(W)	(T)	Total no. of costae	No. of costae per specimen
AMNH 2825 lectotype, p.v.	19.7	+30	10	70 est.	3-4
AMNH 37224 b.v. ext.	18.5	24 est.	–	60 est.	3
AMNH 37225 p.v.	13.1	20 est.	7	64 est.	3
AMNH 37226 art.	12 est.	16+	–	–	3

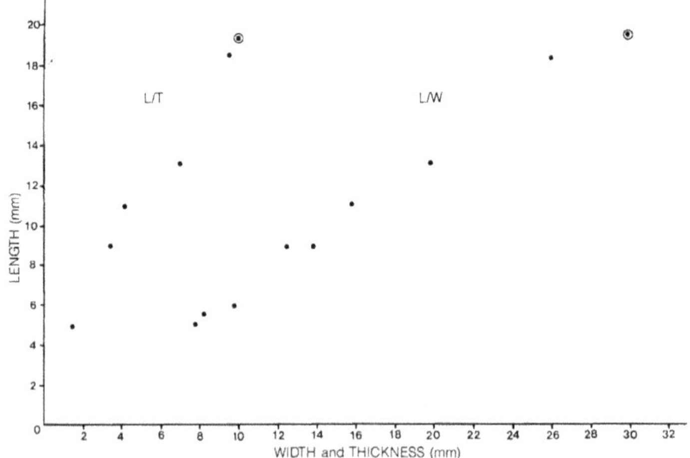

FIGURE 11. "*Eodevonaria*" *hemispherica* (Hall, 1857). Scattergram showing length/width and length/thickness. Circle denotes the lectotype.

line as well as other characteristics. However, the lack of complete brachial interiors precludes any generic assignment. For this reason *Chonetes hemispherica* Hall is here assigned to "*Eodevonaria*" according to recent recommendations (Racheboeuf, 1986).

The coarse radial ornamentation of this species is very similar to that of several taxa described from Central and South America. *Eodevonaria imperialis* (Caster), *E. subhemispherica* (Weisbord), and *E. inca* Isaacson show very close morphologies and similar ornamentation but more detailed information is needed for better comparisons. However, they would probably represent a group of globose, coarsely costate eodevonariids, distinct from *Chonetes arcuatua* Hall, 1857, type species of the genus *Eodevonaria* Breger.

Specimens from the "Pink *Hallinetes*" Zone are globose, coarsely costate eodevonariids which sporadically occur within this biostratigraphic zone. They are here assigned to the species "*Eodevonaria*" *hemispherica*. While mostly represented by fragmentary, isolated valves, this material has the characteristics of Hall's species. However, it differs from the type material in its smaller size; shell shape ornamentation and spines are similar. Specimens from the "Pink *Chonetes*" Zone are probably juveniles and the size difference does not preclude suggested specific assignment. The smaller size may also be due to environmental conditions.

Stratigraphic occurrence: According to Oliver (1956: 1452, table 1) *Chonetes hemisphericus?* is very rare in normal facies of the Edgecliff fauna and rare in the central facies of the Lower to Middle Moorehouse fauna in the Richfield Springs area (1956: 1462, table 3).

Material: Two articulated shells; 17 broken, or crushed pedicle valves.

The *Hallinetes* Community of the "Pink *Chonetes*" Zone

The "*Chonetes*" Community, which occurs in the mudstones of the Seneca Member, 10 ft above the Tioga Bentonite, was described by Feldman (1980), and was defined as a low-diversity assemblage in which "*Hallinetes*" aff. *lineatus* is highly dominant (>99%) with a high ratio of living shells (76%) according to the concave-up position of the specimens. While other faunal constituents are similar to those listed previously (Feldman, 1980: 40, table 12), the community needs redefinition, since the chonetacean brachiopods belong to three different taxa. Among these, *Hallinetes lineatus* is by far the most abundant species (92%); *Longispina mucronata* may be listed as common (5.4%) while "*Eodevonaria*" *hemispherica* is rare (2%).

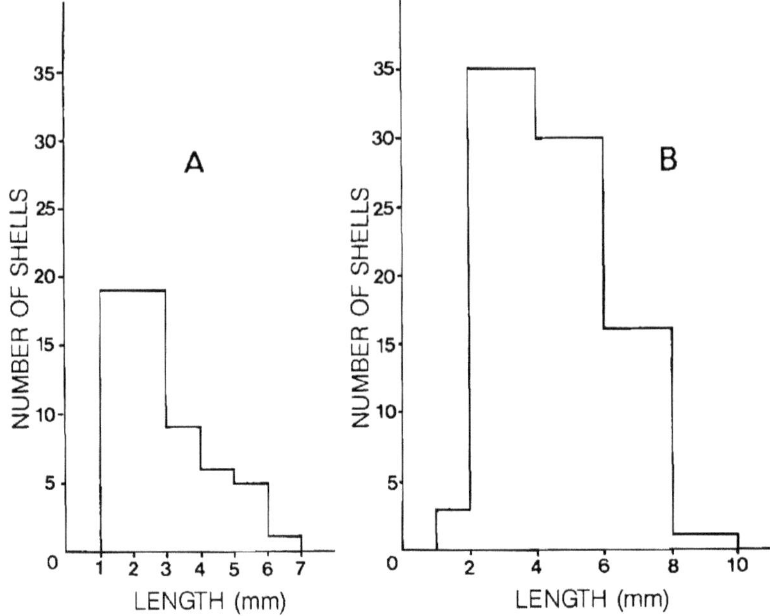

FIGURE 12. Histograms for 55 pedicle valves of *Longispina mucronata* (A) and 85 pedicle valves of *Hallinetes lineatus* (B) from the *Hallinetes* Community within the dark mudstone matrix of the "Pink *Chonetes*" Zone.

After tabulating the above results, based on a count of 907 specimens, it became apparent that the relative abundance of the taxa on bedding planes (both top and bottom) on one hand, and within the dark mudstone matrix on the other hand, reveals significant differences.

On the studied bedding planes, only very few specimens of *Longispina mucronata* (7) and "*Eodevonaria*" *hemispherica* (8) were counted, along with 630 isolated valves and articulated shells of *Hallinetes lineatus.* The specimens of *L. mucronata* occurring on the bedding planes are among the largest observed in the material studied.

Within the dark mudstone matrix, 41 small shells of *L. mucronata* and 11 of "*E.*" *hemispherica* have been counted together with 209 specimens of *Hallinetes lineatus*; most of the specimens are articulated shells and the shells of *L. mucronata* are very small to small juveniles.

Counting shows that the same chonetacean taxa occur on the bedding planes and within the mudstone matrix, but their relative abundance is different. Relative frequencies are, respectively, as follows: *Longispina*

TABLE 5. Counts of *Hallinetes lineatus* on Bedding Planes and within the Matrix.

Block. No.	Specimens	Concave-up position	Convex-up position	% articulated shells	% concave-up articulated shells
1	b.v.	66	22	8.3	7.0
Top b.p.	p.v.	89	42		
	art.	17	3		
	Total	172	67		
3	b.v.	10	6	11.0	10.0
Bottom b.p.	p.v.	29	7		
	art.	6	1		
	Total	45	14		
4	b.v.	15	12	9.1	2.7
Top b.p.	p.v.	45	27		
	art.	3	7		
	Total	63	46		
6	b.v.	6	15	8.2	2.3
Top b.p.	p.v.	25	32		
	art.	5	2		
	Total	36	49		
14	b.v.	9	11	6.6	6.6
Top b.p.	p.v.	30	20		
	art.	5	–		
	Total	44	31		
14	b.v.	15	6	6.7	3.1
Bottom b.p.	p.v.	25	14		
	art.	2	1		
	Total	42	21		
1	b.v.	8	14	2.1	2.1
Middle bed	p.v.	44	26		
	art.	2	–		
	Total	54	40		
4	b.v.	11	6	9.8	5.6
Middle bed	p.v.	21	26		
	art.	4	3		
	Total	36	35		

mucronata, 1 and 15%; "*Eodevonaria*" *hemispherica,* 1.2 and 4.2%; *Hallinetes lineatus,* 97 and 80%. The fact that only large shells lie on the bedding planes probably indicates that the tiny juvenile shells of *L. mucronata* have been winnowed away. This conforms with the isolated valves of *H. lineatus* found on the bedding surfaces, even though shells in a concave-up position are common (see Table 5). It is probable that the number of articulated shells in the concave-up position has been underestimated; in some cases the brachial valve of such oriented specimens appears to have been weathered off. In any case, the relative frequency of articulated shells in a concave-up position would not exceed twice the total number of specimens found.

Most of the shells lying on the bedding planes have broken spines while spines are nicely preserved, and always found, on shells encased within the matrix. This fact is further support for the interpretation that bedding plane assemblages are post mortem accumulations.

In conclusion, from the new observations of the chonetaceans from the "Pink *Hallinetes*" Zone, the true *Hallinetes* Community corresponds to the assemblage within the dark mudstone matrix rather than to the shells distributed on the bedding plane surfaces. The *Hallinetes* Community of the Seneca Member is a low-diversity, "highly dominated" (although not monospecific) community within a quiet water environment. This interpretation is supported by the coexistence of long-spined chonetaceans as well as by the reduced number of spines on the shells.

The Benthic Assemblage position (cf. Boucot, 1975) is more difficult to pinpoint, but in this case, a Benthic Assemblage 4 position is most probable, since it is below the lower limit for active photosynthesis and reef building activity (Yu et al., 1987: 6), as evidenced by a lack of reef building corals and scarcity of solitary corals (Feldman, 1980: 40). A normal marine environment is indicated by the presence of some bioturbation, which increases dramatically at the top of the formation. The fine-grained limestone matrix and dominance of *Hallinetes* are consistent with low (to intermittent) current activity and lower than normal oxygen, but not quite dysaerobic conditions, since trilobites, gastropods, and crinoids are present, although rare. The cause of relatively low diversity here is uncertain.

ACKNOWLEDGMENTS

We thank Drs. Arthur J. Boucot (Department of Zoology, Oregon State University), Paul Copper (Department of Geology, Laurentian University), Neil Landman (Department of Invertebrates) and David Grimaldi (Department of Entomology) (both of the American Museum of Natural History), for critically reading and commenting on the manuscript and offering valuable suggestions for improvement.

REFERENCES

Boucot, A. J. 1975. *Evolution and Extinction Rate Controls.* Developments in paleontology and stratigraphy I. Amsterdam: Elsevier.

Breger, C. L. 1906. On *Eodevonaria,* a new subgenus of *Chonetes. American Journal of Science* 21: 534-536.

Bronn, H. G. 1862. *Die Klassen und Ordnungen der weichthiere* (Malacozoa) 3: 1-518. Leipzig and Heidelberg.

Conrad, T. A. 1839. Descriptions of new species of organic remains. New York State Geological Survey Annual Report 3: 57-66.

Cooper, G. A. 1942. New genera of North American brachiopods. *Journal of Washington Academy of Sciences* 32: 228-235.

Feldman, H. R. 1980. Level-bottom brachiopod communities in the Middle Devonian of New York. *Lethaia* 13: 27-46.

Feldman, H. R. 1985. Brachiopods of the Onondaga Limestone in central and southeastern New York. *American Museum of Natural History Bulletin* 179 (3): 289-377.

Feldman, H. R., and Lindemann, R. H. 1986. Facies and fossils of the Onondaga Limestone in central New York. *New York State Geological Association Field Trip Guidebook, 58th Annual Meeting*, Cornell University, Ithaca, New York, 145-166.

Hall, J. 1843. *Geology of New York*, Part 4, comprising the survey of the fourth geological district. Albany: Natural History of New York.

Hall, J. 1857. Descriptions of Paleozoic fossils. N*ew York State Cabinet 10th Annual Report*, Part C. Appendix, 41-186.

Hall, J. 1867. Descriptions and figures of the fossil Brachiopoda of the Upper Helderberg, Hamilton, Portage and Chemung groups. *New York Geological Survey, Paleontology* 4: 1-428.

Hall, J., and Clarke, J. M. 1892. An introduction to the study of the genera of Paleozoic Brachiopoda, Part I. *New York Geological Survey, Paleontology of New York* 8: 1-367.

Kindle, E. M. 1912. The Onondaga fauna of the Allegheny region. *U. S. Geological Survey Bulletin* 508.

Lindemann, R. H., and Feldman, H. R. 1981. Paleocommunities of the Onondaga Limestone (Middle Devonian) in central New York State. *New York State Geological Association Guidebook for Field Trips in South-Central New York, 53rd Annual Meeting, State University of New York at Binghamton*, 79-96.

Muir-Wood, H. M. 1962. *On the Morphology and Classification of the Brachiopod Suborder Chonetoidea*. London: British Museum of Natural History Memoir.

Oliver, A., Jr. 1954. Stratigraphy of the Onondaga Limestone (Devonian) in central New York. *Geological Society of America Bulletin* 65: 621-652.

Oliver, A., Jr. 1956. Stratigraphy of the Onondaga Limestone in eastern New York. *Geological Society of America Bulletin* 67: 1441-1474.

Racheboeuf, P. R. 1976. Chonetacés (Brachiopodes) du Dévonien inférieur du Bassin de Laval (Massif Amoricain). *Palaeontographica* 152: 14-89.

Racheboeuf, P. R. 1981. Chonetacés (Brachiopodes) Siluriens et Dévoniens du sud-ouest de l'Europe. *Mémoire Société géologie minéralogie,* Bretagne 27: 1-294.

Racheboeuf, P. R. 1986. *Loreleiella* nov. gen., nouvel Eodevonariidé (Chonetacea, Brachiopoda) du Dévonien. *Geobios* 19: 641-646.

Racheboeuf, P. R. 1987. Upper and Lower and Lower Middle Devonian chonetacean brachiopods from Bathurst, Devon and Ellesmere Islands, Canadian Arctic Archipelago. *Geological Survey of Canada Bulletin* 375: 1-29.

Racheboeuf, P. R., and Branisa, L. 1985. New data on Silurian and Devonian chonetacean brachiopods from Bolivia. *Journal of Paleontology* 59: 1426-1450.

Sokolskaya, A. N. 1960. Superfamily Chonetacea. In Yu. Orlov (ed.), *Osnovy Paleontologii*, 221-223. Moscow: Akademiia Nauk SSSR. [vol. Bryozoa and Brachiopoda. In Russian]

Vanuxem, L. 1839. Third annual report of the geological survey of the third district. *New York Geological Survey Annual Report* 3: 142-285.

Vanuxem, L. 1842. *Geology of New York*, Part III, Comprising the survey of the third geological district. Albany: W. and A. White and J. Visscher, Natural History of New York.

Winchell, A. 1866. *The Grand Traverse Region.* A report on the geological and industrial resources in the Lower Peninsula of Michigan. Ann Arbor: Dr. Chase's Steam Printing House.

Yu, W., Boucot, A. J., Rong, J.-y., and Yang, X.-c. 1987. Community paleoecology as a geologic tool: The Chinese Ashgillian-Eifelian (latest Ordovician through early Middle Devonian) as an example. *Geological Society of America Special Paper* 211: 1-100.

Zenger, D. H. 1967. Coloration of the "Pink *Chonetes*" (brachiopod) of the Onondaga Limestone, New York. *Journal of Paleontology* 41: 161-166.

Chapter Five

Brachiopods of the Onondaga Formation, Moorehouse Member (Devonian, Eifelian), in the Genesee Valley, Western New York

ABSTRACT

In the Genesee Valley of western New York, the Moorehouse Member of the Onondaga Limestone contains 46 species of brachiopods, described herein. Of the 26 species of brachiopods in the underlying Bois Blanc Formation in western New York, 11 also occur in the Onondaga and show no evolutionary change between the Bois Blanc and the Onondaga Formation; that is, taxa found in the Bois Blanc-Onondaga interval indicate a high degree of stasis. During Onondaga time, there was a progressive increase in relative water depth throughout the basin, as indicated by a gradual northward shift of the carbonate facies as well as a northward migration of the overlying Hamilton Group. The Nedrow Member represents a minor transgressive cycle due to an influx of mud from the east, and the Moorehouse Member to Seneca Member interval represents a major transgression. Brachiopods in the Moorehouse Member include *Charionoides* and *Pentagonia,* genera endemic to the Appohimchi Subprovince. *Atribonium halli* and *Discomyorthis?* sp. are the only species herein not previously reported from Onondaga strata in western New York. Two new species are erected, *Athyris boucoti* and *A. leoni.* The most significant change in brachiopod faunas across the state during Moorehouse time is the increasing abundance of stropheodontids toward the west.

INTRODUCTION

As a continuation of my previous studies of Onondaga brachiopod taxonomy and paleoecology (Feldman, 1980, 1985; Feldman and Lindemann, 1986; Lindemann and Feldman, 1981, 1987; Racheboeuf and Feldman, 1990), I have sampled and described the brachiopod fauna in the Genesee Valley of western New York (Text-figure 1). The outcrop belt of the Onondaga Limestone in New York extends from Port Jervis (Tristate area) in the southeastern part of the state, northeastward to Kingston and the Helderbergs, and then westward toward Syracuse, Rochester, and Buffalo.

All of the brachiopods studied were recovered from the Moorehouse Member (see section on stratigraphic setting), a hard, dense limestone with little shale.

Two methods of collecting were employed. First, weathered fossils were collected from extensive, but rather rare, bedding surfaces in abandoned quarries, and second, blocks of limestone with silicified specimens were recovered, processed, and etched in an acid bath. Approximately 180 kg of limestone yielded 190 silicified shells. The brachiopod fauna of the Onondaga Limestone is particularly challenging to study because collecting is almost impossible unless bedding planes are accessible. Vertical roadcut faces are often not useful as faunal identification is difficult, and outcrops with well-silicified shells are not always exposed. As in the mid-Hudson Valley, silicification is sporadic and the degree of silicification ranges from poor to very good.

This study improves our understanding of brachiopod abundance and evolution in the Lower Middle Devonian, provides data for ecological and biogeographical studies, and improves correlations with more westerly limestone suites such as the Detroit River Group (Anderdon Limestone, Lucas Dolomite, Amherstburg Dolomite, Sylvania Sandstone) of the Michigan Basin, southwestern Ontario and north-central Ohio. Previous paleocommunity and stratigraphic studies of the non-reefal aspects of the Onondaga Limestone (Oliver, 1954, 1956) have resulted in a fairly good understanding of the formation in southeastern New York (except for the area between Ellenville and Port Jervis where outcrops and complete sections are sparse) and central New York (Syracuse). The area around Rochester (i.e. Genesee Valley), however, has not been studied in detail until now.

ACKNOWLEDGMENTS

I thank Arthur J. Boucot (Oregon State University, Corvallis, Oregon) for his continued assistance, encouragement, and support over the past decade, as my studies of the Onondaga Limestone have progressed from the mid-Hudson Valley, westward across New York State. He reviewed the manuscript and made numerous valuable suggestions for improvement. Touro College contributed $500.00 toward publication of this paper.

Carlton E. Brett, The University of Rochester (Rochester, New York), Gordon C. Baird, State University College at Fredonia (New York) and George

C. McIntosh, Rochester Museum and Science Center (Rochester, New York) spent valuable time with me in the field, and helped immensely in interpreting Onondaga stratigraphy in western New York State.

Paul Copper, Laurentian University, and J. Thomas Dutro, Jr., U.S. Geological Survey (Washington, D.C.), deserve thanks for critical comments on the manuscript. Ed Landing, New York State Museum and Science Service (Albany, New York), made the brachiopod collections in Albany available for study; I thank him for his hospitality. Fred Collier (U.S. National Museum of Natural History, Washington, D.C.), enabled me to examine the collections under his care and kindly loaned me specimens from the National collection in Washington.

Andrew Modell and Susan Klofak (both of the American Museum of Natural History, New York), deserve thanks for photographic work and specimen preparation, respectively. Laura Lynne Gallo, Sarah Lawrence College (Bronxville, New York) graciously assisted me in the field and during my examination of the collection in Albany, New York.

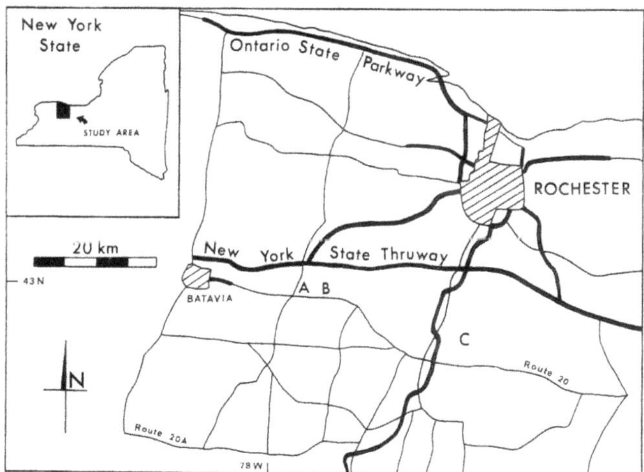

TEXT-FIGURE 1. Index map of collecting localities in the Onondaga Limestone (Moorehouse Member) Genesee Valley, western New York. A. AMNH Loc. 3152; a large exposure on the floor of the eastern part of an abandoned quarry, Lower Moorehouse Member. B. AMNH Loc. 3154; numerous exposures at the southwest comer of an active quarry operated by the Penfield Dolomite Company, Lower Moorehouse Member. C. AMNH Loc. 3153; small exposures along the southeast part of an active quarry operated by the General Crushed Stone Company; Upper Moorehouse Member (for exact locations see Appendix).

STRATIGRAPHIC SETTING

Five members of the Onondaga Limestone occur in the study area (Text-figure 2), the Edgecliff, Clarence, Nedrow, Moorehouse, and Seneca. A brief description of the lithologies of these members is provided here, but detailed stratigraphic descriptions may be found in Oliver (1954).

Edgecliff Member: In western New York the Edgecliff is a massive, light-gray, coarse, crystalline limestone about 5 m thick, packed with solitary rugose and tabulate corals which form biostromes in many places. Typical of the Edgecliff are large crinoid columnals and stems up to about 2.5 cm in diameter. Near Buffalo the Edgecliff undergoes a facies change with at least one large, lens-shaped, biohermal ("reef") structure, which contains extremely irregular bedding (Oliver, 1954: 635). Oliver (1954: 636) also reports the occurrence of small "micro-reefs" exposed in the biostrome at Clarence, 20 km east of Williamsville. In both western and eastern New York, bioherms occur in the Onondaga, representing favorable conditions for the growth of corals and crinoids. The Edgecliff is thought to be a moderately high energy, shallow-water, shelf carbonate which, in the western part of the state, becomes a crinoidal grainstone (Woodrow et al., 1989).

Clarence Member: The Clarence, 11.5 m thick, interfingers with the Edgecliff in places and essentially underlies the Nedrow Member in western New York. Some workers believe that the Clarence replaces the Nedrow in the western part of the state (Ozol, 1963). The extremely cherty Clarence (up to 75 percent by volume; April et al., 1984) consists of both vertically and horizontally coalescing chert nodules enclosing fine-grained lime mudstones (Selleck, 1985) and is easily recognized in outcrop. The rate of clastic deposition in the Clarence exceeded that of the Edgecliff, as indicated by the greater volume of clays. The Clarence Member represents slightly deeper water deposition than the Edgecliff; most of the silica was derived biogenically from dissolution and reprecipitation of sponge spicules and possibly radiolaria (Selleck, 1985). According to Woodrow et al. (1989), the Clarence may represent a facies belt enriched in siliceous organisms.

Nedrow Member: The Nedrow, 13 m thick, is a light- to dark-gray highly argillaceous limestone which forms recessed ledges when extensively weathered. Fresh cuts, as at Jamesville, New York (AMNH Loc. 3128), do not have the typical appearance of the Nedrow. In eastern New York, the Nedrow becomes less argillaceous and more difficult to recognize in outcrop, even when extensively weathered. The Nedrow is thought to represent deeper and muddier water deposition than the rest of the lower part of the formation and probably represents a minor transgressive cycle associated with an influx of clastics from the east (Woodrow et al., 1989).

Moorehouse Member: Typically uniformly bedded, the Moorehouse, 11 m thick, is a medium-gray, fine-grained micritic limestone with abundant dark-weathering chert. During Moorehouse time, there apparently were shallowing conditions similar to those that prevailed during Edgecliff-Clarence time (Kissling and Moshier, 1981). The top of the Moorehouse is marked by the occurrence of the Onondaga Indian Nation metabentonite (Tioga B metabentonite) which is not always present in the various quarries visited. Most of the brachiopods studied in the Onondaga in the last decade were recovered from the Moorehouse Member, due in part to accessibility of the fossils.

Seneca Member: The base of the Seneca Member (4 m thick in western New York) is marked by the "Tioga B" ash layer (about 15 cm thick). The Seneca is a medium- to dark-gray, light-weathering wackestone, sparsely fossiliferous, with occasional chert nodules throughout. The "Pink *Hallinetes* Zone" (Zone J of Oliver, 1954; formally "Pink *Chonetes* Zone"), 3 m above the ash layer and 1.8 m thick, is a thinly bedded limestone packed with chonetid brachiopods, many of which are stained pink. At most localities in western New York, a bed of chert can be found about 15 cm below the top of Zone J. The upper part of the Seneca is more argillaceous and distinctly darker in appearance, and is capped by a bone bed making the contact with the overlying basal Hamilton Group (Oatka Creek Shale) sharply defined in most places.

During Onondaga time, there was a general increase in water depth throughout the basin as indicated by a gradual northward shift of all carbonate facies, comprising the successive members and by northward migration with time of the Marcellus Shale (Kissling and Moshier, 1981). According to Woodrow et al. (1989), the Moorehouse-to-Seneca stratigraphic succession represents a major transgression, with the upper Moorehouse and lowermost Seneca aerobic facies passing upward into a dysaerobic, deeper water upper Seneca facies, which is finally succeeded by the minimally dysaerobic to anaerobic Union Springs environment.

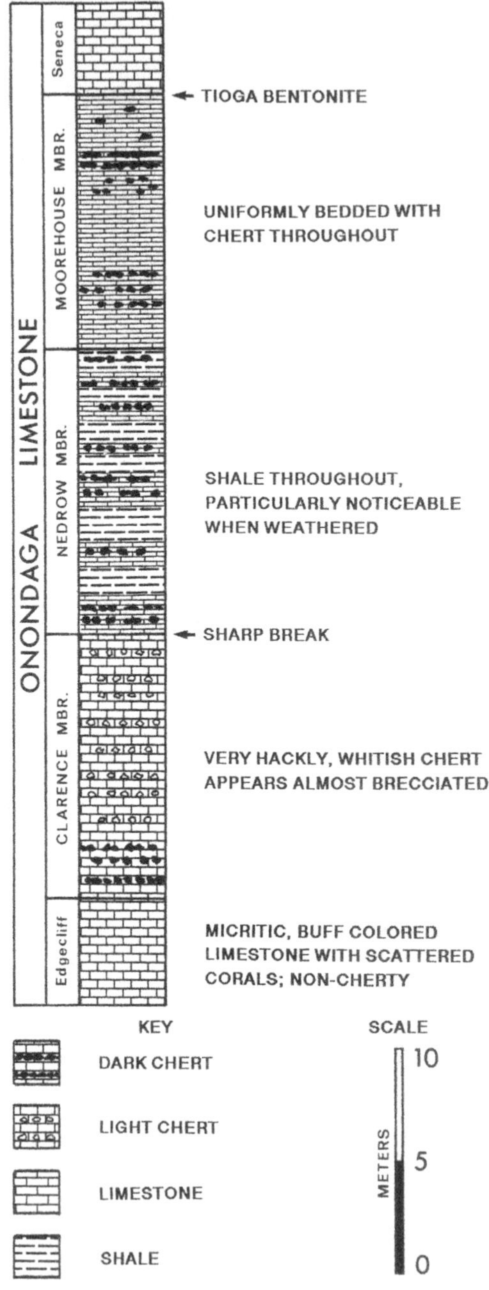

TEXT-FIGURE 2. Generalized stratigraphic column for the Onondaga Limestone in the Genesee Valley.

SYSTEMATIC PALEONTOLOGY

INTRODUCTION

Philosophical Considerations

One of the goals of this project is to help refine our understanding of the "invasion" of European taxa into eastern North America by providing data for a biogeographic study of the Early Middle Devonian faunas in New York. Species recognition is an integral part of this work. A biological definition of a species is: a group of populations that replace each other geographically or ecologically, and of which the neighboring ones intergrade or hybridize wherever they are in contact or are potentially capable of doing so (with one or more of the populations) in those cases where contact is prevented by geographical or ecological barriers (Mayr, 1976). In other words, the modern biological concept of a species is based on an organism's potential to interbreed and produce fertile offspring in natural conditions (Guensburg, 1984). The biological definition is untestable using fossils, and species have therefore been reported in this study based on morphology (i.e. morphospecies) in the sense of Hoover (1981). Although the concept of a species may and sometimes does differ among paleontologists depending on the group studied and the preservation of the particular collections on hand (Sohn, 1983), I consider, as do Cooper and Dutro (1982), species names merely conveniences for discussing clusters of morphologic variability within what are, for the most part, well established generic concepts. Taxonomic names are, as Hoover (1981) has noted, "handles" for discussion of a definite group of specimens, and their correspondence to genetic groups becomes increasingly vague as the temporal distance from the recent increases. I consider myself neither a "lumper" nor a "splitter." Of the two new species erected herein, one is based on 45 specimens, while the other is based on four specimens. As Raup and Stanley (1978) have so aptly stated, the fact that species discrimination depends largely on the experience of the person making the distinction has led to an informal definition of the species that is invoked with surprising frequency: "A species is a species if a competent specialist says it is." After studying the morphology of the new taxa, I have decided that they do indeed represent new forms and do not belong to any previously named species.

A Note on the Use of Open Nomenclature

The intent of the International Code of Zoological Nomenclature (Stoll et al., 1961) is to promote stability and universality in the scientific names of animals, and to insure that each name is unique and distinct. According to the code, all its provisions are subservient to these ends, and none restricts the freedom of taxonomic thought or action. Matthews (1973) noted that although the code sets a limit on its provisions, it does not in any way intend to impinge on the individual taxonomist's exercise of his or her judgement. Since the code provides no explicit guidelines for the use of "open nomenclature," I make the following brief comments regarding my use of it in this monograph. Additional insight into the question of open nomenclature may be found in Bengtson (1988), Kornicker (1979), Matthews (1973), and Richter (1943, 1948).

Open nomenclature is a device whereby an author expresses his or her judgement of his or her own material (Matthews, 1973). It is the procedure by which a taxonomist comments upon the identity of a specimen that cannot be readily or securely determined (Bengtson, 1988). The use of "cf." before a species-group name indicates a provisional determination for the species. Here, "cf." stands for *confer,* not *conformis,* and means "compare to" and not "compare with" (*sensu* Bengtson, 1988). I follow Bengtson's (1988) concept that the wording "compare to" expresses a possible identity, which is what most taxonomists have in mind when they use "cf.," whereas "compare with" rather implies a distinction. Further, "cf." is inserted between the genus and species name as in *Levenea* cf. *subcarinata,* and not *Levenea* cf. *L. subcarinata* as advocated by Lucas (1986). As noted by Bengtson (1988), the former expression conveys in an unambiguous way the message that the author considers the specimen in question to be "probably or possibly the species *subcarinata,* although there is not enough material to be sure, but if it is *subcarinata* it should be referred to the genus *Levenea."*

The use of "aff." indicates a new, undescribed taxon and relates it to a named taxon (*sensu* Bengtson, 1988). The use of "? sp." indicates an uncertain identification and that specific identification is not possible with the specimens at hand because of poorly preserved present or original material.

Terminology

The morphologic terms used herein follow the glossary in the brachiopod volume of the Treatise on Invertebrate Paleontology (Williams et al., 1997: H1-H3226).

ABBREVIATIONS OF REPOSITORY INSTITUTIONS AND LOCALITIES

Repository of all [unnumbered] specimens unless otherwise indicated:

AMNH — American Museum of Natural History, New York, New York.
AMNH Loc. — American Museum of Natural History locality number.
MCZ — Museum of Comparative Zoology, Harvard University, Cambridge Massachusetts.
NYSM — New York State Museum and Science Service. Albany, New York.
UMMP — University of Michigan Museum of Paleontology, Ann Arbor, Michigan.
USNM — United States National Museum of Natural History, Smithsonian Institution, Washington, D.C.

MEASUREMENT ABBREVIATIONS AND SUBSCRIPTS

In the tables of measurements: a.v. = articulated valves; p.v. = pedicle valve; b.v. = brachial valve; m = meters; cm = centimeters; mm = millimeters; kg = kilograms; L = maximum length; W = maximum width; T = maximum thickness. Several subscripts are used to qualify measurements. The subscript "d" indicates damage in that orientation and that the measurement was estimated to the nearest tenth of a millimeter. The subscript "c" indicates compression of the shell due to compaction, and the subscript "e" indicates exfoliation of the shell exterior. Measurement across a structure with bilateral symmetry, such as the hinge width, may be estimated, in cases of incompleteness, by doubling the half-measure (distance from symmetry plane to distal extremity). The subscript "h" indicates use of this procedure. The subscript "t" indicates that the measurement was estimated (e.g. number of plications on a flank).

SYSTEMATICS

Phylum BRACHIOPODA Dumeril, 1806
Subphylum RHYNCHONELLIFORMEA Williams and others, 1996
Class RHYNCHONELLATA Williams and others, 1996

Order ORTHIDA Schuchert and Cooper, 1932
Suborder DALMANELLIDINA Moore, 1952
Superfamily DALMANELOIDEA Schuchert, 1913
Family DALMANELLIDAE Schuchert, 1913
Subfamily ISORTHINAE Schuchert and Cooper, 1931
Genus LEVENEA Schuchert and Cooper, 1931
Type species: Orthis subcarinata Hall, 1857: 43.

Levenea cf. *subcarinata* (Hall, 1857), Plate 1, figures 1-4.

Orthis subcarinata Hall, 1857: 43; 1859: 169, pl. 12, figs. 7-21.

Levenea subcarinata Schuchert and Cooper, 1932: 123, pl. 18, figs. 19-23, 25-32; Cooper, 1944: 353, pl. 138, figs. 21-23.

Levenea aff. *subcarinata* Feldman, 1985: 304, fig. 3.

Description.—Exterior: Shell medium-sized (L = 17.0 mm, est.; W = 17.7 mm, est.), transversely suboval in outline, subcarinate. Ventral interarea short, incurved, and apsacline, with triangular delthyrium, which encloses an angle of 60 degrees. Rounded, radial costellae increase anteriorly by bifurcation; 25 costellae per 5 mm at midlength.

Pedicle valve interior: Hinge teeth strong, short, triangular in horizontal section, supported by very short dental lamellae. Muscle field subpentagonal, about 25 percent of valve length. Diductor tracks impressed longitudinally on valve floor.

Discussion: Levenea cf. *subcarinata* is identical to *L.* aff. *subcarinata* from the mid-Hudson Valley (Feldman, 1985: 304, fig. 3D). Apparently the species becomes rare toward the western part of the state; only one specimen was recovered in the Genesee Valley.

The general outline of the ventral muscle field of *Levenea* cf. *subcarinata* (Boucot et al., 1970: 8, pl. 2, figs. 3-5) is similar to the Genesee Valley shell.

The pedicle valves in Johnson's *Levenea fagerholmi* (1970: 77, pl. 2, figs. 8-18) are flatter and larger than in the Onondaga shell, while his *L. navicula* (75, pl. 2, figs. 19-22; pl. 3, figs. 1-19) shows greater biconvexity. Johnson's (1970: 74, pl. 2, figs. 1-7) *Levenea* sp. A may be differentiated from *L.* cf. *subcarinata* by its more prominent strong, rounded radial costellae.

Material: One pedicle valve.

Occurrence: AMNH Loc. 3152.

Family RHIPIDOMELLIDAE Schuchert, 1913
Subfamily RHIPIDOMELLINAE Schuchert, 1913
Genus DALEJINA Havlicek, 1953
Type species: Dalejina hanusi Havlicek, 1953: 5.

Dalejina cf. *alsa* (Hall, 1863), Plate 1, figures 5-12.

Orthis alsus Hall, 1863: 33.

Rhipidomella alsa Hall. 1867: 36, pl. 4, figs. 2-7; Grabau, 1906: 181, fig. 95.

Dalejina aff. *alsa* Feldman, 1985: 307, figs. 5-7, 8A-D.

Dalejina alsa Boucot and Johnson, 1968: B7, pl. 1, figs. 11-27; Fagerstrom, 1971: pl. 1, figs. 1-2.

Description.— Exterior: Shells range in size from small to medium-sized (Table 1), are dorsibiconvex with pedicle valve more arched posteriorly, and transversely suboval in outline; hinge line short and straight in apical area but becomes rounded as lateral margins approached; maximum width attained at or just anterior to midlength; in some adult shells, pedicle valve bears slight median depression while brachial valve bears corresponding ridge (these features are too indistinct to be called sulcus and fold); anterior commissure rectimarginate to very slightly sulcate. Ventral interarea apsacline, narrow, short; dorsal interarea anacline. Radial costellae increase by both intercalation and bifurcation; 12 to 15 costellae per 5 mm; costellae crossed by concentric growth lines near anterior margins.

Pedicle valve interior: Pointed hinge teeth expand anteriorly, widely divergent and supported by short dental lamellae. Shallow crural fossettes present. Umbonal cavity shallow, poorly defined, bounded laterally by dental lamellae. Muscle scars not clearly impressed. Small, low myophragm visible on silicified specimen. Valve floor crenulated at periphery by impress of costellae.

Brachial valve interior: Sockets deep, broaden anteriorly; brachiophores worn and diverge at angle of ninety degrees. Cardinal process single lobe resting on deposit of secondary shell material in notothyrial cavity. Neither myophragm nor muscle impressions preserved. Internal periphery crenulated by impress of costellae.

Discussion: Dalejina is distinguished from *Rhipidomella* by the medially grooved, flat crenulations found on the lateral and anterior margins of the shell. As most of the specimens in the collection are articulated, it is difficult

TABLE 1. Measurements (in mm) of *Dalejina* cf. *alsa* (Hall, 1863). See *Systematic Paleontology: Introduction: Measurement Abbreviations and Subscripts* for explanations.

AMNH loc.	(L)	(W)	(T)	Specimen type
3152	5.9	6.7	2.7	a.v.
3152	10.7	12.9	2.9	a.v.
3152	19.1	21.7	4.4	a.v.
3152	11.7	14.2	2.9	a.v.
3152	4.2	17.9	5.3	a.v.
3152	16.6	20.0	6.6	a.v.
3152	6.7	8.8	2.7	a.v.
3152	9.1	16.4	3.2	a.v.
3152	16.9	20.3	4.2	a.v.
3152	5.2	5.9	2.0	a.v.
3152	16.1	20.0	3.4	a.v.
3152	15.7	20.9	5.2	a.v.
3152	3.8	16.4	4.8	a.v.
3152	17.0	21.5	4.6	a.v.
3152	13.1	16.3	4.0	a.v.
3152	16.1	19.1	4.0	a.v.
3152	14.0	18.2	–	a.v.
3152	10.8	12.8	2.1	a.v.
3152	4.0	4.6	1.0	a.v.
3152	5.9	6.7	2.0	a.v.
3152	7.1	8.9	2.7	a.v.
3152	6.7	7.2	2.5	a.v.
3152	9.8	6.5	1.8	a.v.
3152	6.8	7.8	1.8	a.v.
3152	5.0	5.9	1.4	a.v.
3152	3.6	4.2	1.1	a.v.
3152	5.0	5.9	2.0	a.v.

to distinguish these shells from *Rhipidomella.* They are assigned to *Dalejina* for two reasons: (1) there are vague indications of flat, medially grooved internal crenulations on several shells that were ground and polished at the anterior commissures; and (2) the Genesee Valley shells are identical to *Dalejina* from the mid-Hudson Valley (Feldman, 1985: 306, figs. 5-8 A-D); no specimens of *Rhipidomella* were found there.

Based largely on external morphology, *Dalejina* cf. *alsa* from the Genesee Valley is identical to the mid-Hudson Valley shells, as noted above. *Dalejina alsa* from the underlying Bois Blanc Formation (Boucot and Johnson, 1968: B7, pl. 1, figs. 11-27) differs in that the brachiophores are more widely divergent in Boucot and Johnson (see pl. 1, fig. 23). Both, however, have similar cardinal processes extending from secondary shell material in the notothyrial cavity. Externally they are also very similar, although the Genesee Valley shells are more commonly transversely suboval rather than transversely subcircular in outline.

Material: Ninety-five articulated, six pedicle, three brachial valves.

Occurrence: AMNH Loc. 3152.

Genus DISCOMYORTHIS Johnson, 1970

Type species: Orthis musculosa Hall, 1857: 46.

Discomyorthis? sp., Plate 1, figures 13-15.

Discussion: These two shells are assigned to the genus *Discomyorthis* based on their larger size and large, flabellate (?) muscle field in one exfoliated shell. *Discomyorthis* from the Great Basin (Johnson, 1970: 84) is differentiated from *Dalejina* by its nearly planar pedicle valve and its unusually large ventral diductor scars. The only distinction between *Discomyorthis* and *Dalejina* is that the former is simply a large example of the latter. According to A. J. Boucot (oral communication, 1991), *Discomyorthis* is a large Eastern Americas Realm *Dalejina,* as contrasted with the much smaller *Dalejina* known from coeval beds elsewhere. Although the pedicle valves in the Genesee Valley shells are not planar, they are provisionally assigned to *Discomyorthis.*

Material: Two articulated valves.

Occurrence: AMNH Loc. 3152.

Superfamily ENTELETOIDEA Waagen, 1884

Family SCHIZOPHORIIDAE Schuchert and LeVene, 1929

Subfamily SCHIZOPHORIINAE Schuchert and LeVene, 1929

Genus SCHIZOPHORIA King, 1850

Type species: Conchyolithus anomites resupinatus Martin, 1809: pl. 49, figs. 13-14.

Schizophoria cf. *multistriata* Hall (1859-1861), Plate 1, figures 16-25.

Conchyolithus anomites resupinatus Martin, 1809: pl. 49, figs. 13, 14.

Schizophoria multistriata Grabau, 1906: 156, fig. 69; Goldring, 1935: 119, figs. 41J-K; 1943: 183, figs. 33n, o; Cooper, 1944: 357, pl. 140, figs. 10-11.

Schizophoria cf. *multistriata* Feldman, 1985: 309, figs. 9-10.

Description.— Exterior: Shells range from small to medium (Table 2), more subquadrate than suboval in outline and unequally biconvex; in most shells, the brachial valve is deeper and more uniformly convex; a broad, shallow sulcus developed on pedicle valve in ephebic specimens, although occasionally present on neanic shells; not all shells have a developed sulcus; slight fold may or may not oppose sulcus; hinge line short, varies from straight to slightly rounded; maximum width attained just past midlength; ventral interarea is apsacline, triangular, while dorsal interarea is narrower and may be apsacline or orthocline; rounded, radial costellae with interspaces that range from broad and flat to rounded and narrow; costellae inrease anteriorly by intercalation; about 17-19 costellae in a 5 mm space; older specimens have about 10 costellae in the same interval.

Pedicle valve interior: Hinge teeth poorly preserved in most specimens and variable in shape; in neanic forms they tend to be subpyriform in cross section, while in ephebic shells they are more ovate; teeth supported by thin, anterolaterally diverging dental lemallae, which join the valve floor at about one-fifth of the valve length; diductor scars bisected by median, raised adductor platform, which gives the muscle field a bilobed appearance; in one nonsilicified shell this platform carries five striae for its entire length; striae not preserved in silicified specimens; lateral margins of diductor field either are subparallel to adductor platform or diverge laterally; in either case the muscle field has a distinct bilobed appearance; delthyrial angle ranges from 40-50 degrees and delthyrium opens into a small foramen apically; costellae not impressed on internal periphery due to poor preservation.

Brachial valve interior: Neanic shell has a small, knoblike cardinal process deep in the notothyrial cavity; sockets deep, curved, bounded anterolaterally by poorly preserved fulcral plates; concave brachiophore bases, attached to posterior ends of fulcral plates, are directed ventrally and end in concave apophyses; muscle field is somewhat ovate and easily distinguished from pedicle valve muscle field; low myophragm divides adductor scars in ephebic shell, but is absent in neanic specimen; in larger shell the anterior rim of the muscle field is raised above the valve floor, but in smaller shells the muscle field merges with the valve floor anteriorly and no rim is

TABLE 2. Measurements (in mm) of *Schizophoria* cf. *multistriata* Hall (1859-1861). See *Systematic Paleontology: Introduction: Measurement Abbreviations and Subscripts* for explanations.

AMNH Loc.	(L)	(W)	(T)	Specimen type
3152	11.3	16.3	7.9	a.v.
3152	14.0	20.3	9.6	a.v.
3152	6.5	9.3	3.9	a.v.
3152	18.3	21.9	11.7	a.v.
3152	20.4	24.4	13.2	a.v.
3152	19.9	–	13.8	a.v.
3152	15.0	16.4	7.3	a.v.
3152	4.9	5.6	2.4	a.v.
3152	7.0	8.7	3.2	a.v.
3152	10.2	–	5.1	a.v.
3152	12.3	14.0	5.4	a.v.
3152	–	20.3	7.6	a.v.
3152	16.4	19.7	7.9	a.v.
3152	8.0	19.0	9.4	a.v.
3152	23.6	29.1	9.6	a.v.
3152	25.6 dh	32.4	14.7	a.v.
3152	19.8	24.4	11.8	a.v.
3152	16.1	23.0	12.2	a.v.
3152	18.4	23.4	9.4	a.v.
3152	16.2	19.0	7.0	a.v.
3153	15.8	–	10.7	a.v.
3153	16.9	–	8.0	a.v.
3153	17.6	–	9.6	a.v.
3153	21.6	–	11.4	a.v.
3153	19.7	–	8.1	a.v.

present; anterior internal periphery crenulated due to impress of costellae, but not well preserved due to etching.

Discussion: The Genesee Valley shells differ from *Schizophoria* cf. *multistriata* collected in southeastern New York (Feldman, 1985: 309, figs. 9-10) in that some specimens are quite globose, are more subquadrate in outline, and have more numerous costellae. As will be seen in the discussion below, variability in the ventral muscle field is such that this character cannot be used to define species.

Schizophoria parafragilis (Johnson, 1970: 91, pl. 8, figs. 1-12) from the McMonnigal Limestone of the Great Basin differs from the Onondaga forms in that its ventral muscle field is more elongate and more evenly bilobate. The same kind of elongate muscle field is found in *Schizophoria?* from the Coeymans Limestone, Green Pond Outlier, New York (Boucot et al., 1970: 91, pl. 2, figs. 2a-b).

Lenz (1977: 62, pl. 4, figs. 11-12, 15-37, 40, 42-44) described *Schizophoria* cf. *paraprima* and *Schizophoria* sp. from the Lower Lochkovian and Pragian strata of Royal Creek, Yukon, which are more ovate in outline than the Genesee Valley shells. The ventral muscle fields of both species from the Yukon are quite variable (note figs. 11, 21, and 38 for elongate scars and figs. 20, 24, 26, and 43 for more diverging, almost pyriform scars), as are the Onondaga specimens. Similar variability is noted in Johnson's *S. nevadaensis* (1970: pl. 9, fig. 12) which also has a pyriform muscle field.

Schizophoria traversensis from the Genshaw Formation, Traverse Group, Michigan (Imbrie, 1959: 364, pl. 48, figs. 8-14) has a similar ventral muscle field to those forms described above from Nevada and the Yukon in that it is also pyriform. It is also similar in its ornamentation, with 20 costellae per 5 mm at a distance of 15 mm from the pedicle beak. It differs, however, in its subelliptical outline. *S. ferronensis* from the Ferron Point Formation, Michigan (Imbrie, 1959: 364, pl. 48, figs. 1-7) is smaller and has a shorter and more lobate ventral muscle field than does *S.* cf. *multistriata.*

Material: Fifty articulated shells, thirty eight pedicle valves, four brachial valves.

Occurrence: AMNH Locs. 3152, 3153.

Order PENTAMERIDA Schuchert and Cooper, 1931
Suborder PENTAMERIDINA Schuchert and Cooper, 1931
Superfamily CLORINDOIDEA Rzhonsnitskaia, 1956
Family CLORINDIDAE Rzhonsnitskaia, 1956

Subfamily PENTAMERELLINAE Sapelnikov, 1973
Genus PENTAMERELLA Hall, 1867
Type species: Atrypa arata Conrad, 1841: 55.

Pentamerella arata (Conrad, 1841), Plate 1, figures 26-29; Plate 2, figures 1-3.

Atrypa arata Conrad, 1841: 55.

TABLE 3. Measurements (in mm) of *Pentameralla arata* (Conrad, 1841). See *Systematic Paleontology: Introduction: Measurement Abbreviations and Subscripts* for explanations.

AMNH loc.	(L)	(W)	(T)	Specimen type
3152	15.0	16.2	9.2	a.v.
3152	15.1	15.8	9.9	a.v.
3152	–	14.9	8.3	a.v.
3152	12.2	–	8.8	a.v.
3152	11.3	15.0	–	p.v.
3152	15.5	17.6	–	p.v.
3152	19.4	20.1	–	p.v.
3152	23.7	24.9	–	p.v.
3152	18.2	17.7	–	p.v.
3152	12.9	14.8	–	p.v.
3152	18.3	23.5	–	p.v.
3152	17.7	19.1	–	p.v.
3153	19.6	25.0	–	p.v.
3153	13.6	13.3 dh	–	p.v.

Pentamerella arata Hall, 1867: 375, pl. 58, figs. 1-12; Kindle, 1901: pl. 615; Grabau, 1906: 184, fig. 100; Amsden, 1964: 233, pl. 40, figs. 9-14, text-fig. 5; Boucot and Johnson, 1968: B8, pl. 1, figs. 28-36; Feldman, 1985: 312, figs. 11A-B, D, 12A-D, 13.

Description.—Exterior: Shells small to medium-sized (Table 3), subglobose, pyriform in outline, and ventribiconvex; pedicle valve bears shallow but distinct sulcus, while brachial valve bears corresponding fold; anterior commissure crenulate and sulcate; hinge line short, curved, with narrow pedicle interarea that decreases in height laterally; no interarea on brachial valve; small delthyrium present, which encloses an angle of 40 degrees; in young shells the delthyrium is open, however, there is probably no pedicle present, since the opening leads directly to the spondylium, which held diductor muscles (there would have been no room for a pedicle); maximum width attained just past midlength; pedicle beak short, slightly incurved; brachial beak smaller, erect; rounded plications increase anteriorly by bifurcation; interspaces V-shaped and about the same width as the plications; four plications found in the sulcus and five on the fold; concentric growth lines are present, more numerous anteriorly.

Pedicle valve interior: Hinge teeth small, triangular; short, deeply excavated spondylium supported by thin, bladelike median septum, which is confined to the posterior portion of the valve; valve floor crenulated anteriorly due to impress of the plications.

Brachial valve interior: Sockets shallow, well worn, poorly preserved; large, V-shaped (in cross section), deeply excavated, elongate cruralium extends anteriorly and unites with valve floor dorsally; adductor muscles attach directly to cruralium; notothyrial cavity small, shallow; valve floor crenulated due to impress of plications.

Discussion: Boucot and Johnson (1968: B8, pl. 1, figs. 28-36) describe three pedicle valves of *Pentamerella* cf. *arata* from the Bois Blanc Formation in western New York that are identical to the specimens recovered from the overlying Onondaga Limestone in the Genesee Valley. The latter specimens represent a larger sample and consequently display greater variation. For example, the Onondaga shells are not all rectimarginate, as are the Bois Blanc specimens, and when well preserved, they have a distinct pedicle interarea.

Hall (1867: 375, pl. 58, figs. 1-21) illustrated *Pentamerella arata* from the Schoharie Grit and limestones of the Upper Helderberg Group in Albany and Schoharie counties. The Genesee Valley shells very closely resemble Hall's specimen, illustrated in his Figures 5-8, which differs only in that the lateral commissure of Hall's shell is ventrally arched. Internally, Hall's brachial valve interior (Figure 17) is identical and his pedicle valve interior (Figure 18) differs in the greater length of the spondylium, a very variable character in *Pentamerella* (some of the Onondaga shells have an incomplete spondylium, which suggests a greater length).

Imbrie (1959: 370) described several new species of *Pentamerella* from the Traverse Group of Michigan, but did not illustrate any interiors. All show affinities to *P. arata* from the Genesee Valley, but some differences are noted: *P. pericosta* is more pyriform in outeline, *P. lingua* is more coarsely costate, *P. aftonensis* has a more inflated pedicle valve, while *P. tumida* is significantly larger.

Material: Seven articulated shells, fifty seven pedicle valves, four brachial valves.

Occurrence: AMNH Locs. 3152, 3154.

Order STROPHOMENIDA Öpik, 1934
Suborder STROPHOMENOIDEA Öpik, 1934
Superfamily STROPHOMENACEA King, 1846

Family RAPHINESQUINIDAE 1893
Subfamily LEPTAENINAE Hall and Clarke, 1894
Genus LEPTAENA Dalman, 1828
Type species: Leptaena rugosa Dalman, 1828: pl. 1, fig. 1.

Leptaena sp., Plate 2, figures 4-7.

Leptaena rugosa Dalman, 1828: 93.

Leptaena aff. *"rhomboidalis"* Boucot, 1973: 20, pl. 6, figs. 12-16; Feldman, 1985: 315, figs. 14-16.

Description.—Exterior: Shells typically thick (Table 4), transversely subrectangular to shield-shaped in outline, medium to large; maximum width in some shells at straight hinge line, but in others almost at anterior commissure; ears short, pointed; shells concavoconvex with pedicle valve strongly geniculate at anterior and lateral commissure; brachial valve correspondingly geniculate; trail very variable in length, ranging from just under one-half length in one specimen to more commonly one-fourth to one-third length of shell; pedicle interarea linear, flat, apsacline; delthyrium wide, 90 degrees, partially preserved in only two shells; no pseudodeltidium evident; brachial interarea flat, anacline, narrower than pedicle interarea; notothyrium and chilidium not preserved; 12 to 15 radial costellae in a distance of 5 mm, posterior to point of geniculation, and extending onto trail; six to 10 low, rounded rugae cover disc but are absent on trail.

TABLE 4. Measurements (in mm) of *Leptaena* sp. See *Systematic Paleontology: Introduction: Measurement Abbreviations and Subscripts* for explanations.

AMNH loc.	(L)	(W)	(T)	Specimen type
3152	16.3	17.6	3.3	a.v.
3152	17.6	19.5 dh	2.9	a.v.
3152	23.7	–	3.9	a.v.
3152	24.0	–	6.6	a.v.
3152	22.4	31.2	–	b.v.
3152	20.2	26.6	–	b.v.
3152	25.8	26.1	–	p.v.
3152	25.0	35.4	–	p.v.
3152	23.4	33.3	–	p.v.
3152	26.2	26.9	–	p.v.

Pedicle valve interior: Delthyrial cavity broadly triangular, floored by lamellose layers of shell material near apex; one short, stubby, anterolaterally directed hinge tooth rests on low dental lamella, fused to ridges bounding the muscle field; muscle field circular and bounded by low, continuous ridge; suggestion of low myophragm at anterior half of muscle field; adductor and diductor differentiation is not possible here, but it is likely that addcutors bisected by the myophragm and diductors positioned lateral to them; pustules present on valve floor.

Brachial valve interior: Shallow sockets, bounded by small socket plates, diverge anterolaterally; large, rounded, bifid cardinal process covered at base by very fine radiating striae; coarsely pustulose disc bounded by apophragm, which reaches maximum height anteriorly; apophragm separates disc from geniculate trail, which is also strongly pustulose; deeply impressed muscle field raised on platform above the valve floor; width one-third that of disc and slightly more than one-half its length; low myophragm bisects both posterior and anterior adductor scars; posterior scars subcircular while anterior scars uniform, narrow depressions anteriorly directed; anterior scars have no boundary ridges *sensu stricto,* rather they are sunk into muscle platform; posterior scars, however, have strong boundary ridges, especially laterally and anterolaterally.

Discussion: Assignment to *Leptaena "rhomboidalis"* cannot be made for the following reasons: (1) the species *"rhomboidalis"* as restricted by Kelly (1967: 591) is a Silurian species from northwestern Europe, i.e., Gotland and the United Kingdom (see also Bassett, 1974: 115-116), the youngest forms of which range up into the Leintwardinian; and (2) the morphology differs from *L. rhomboidalis* specifically in the brachial muscle field and more transverse outline of the Gotland shells.

Leptaena depressa (Bassett, 1974: 111, pl. 29, figs. 1-9; pl. 30, figs. 1-8) has weaker and lower rugae and a weaker brachial muscle platform. *L. depressa* also has a more quadrate or rhomboid shape (Kelly, 1967: pl. 98, figs. 4-9). Bowen's (1967: 32, pl. 4, figs. 3-5) *Leptaena* sp. cf. *L. "rhomboidalis"* from the Keyser Limestone of Pennsylvania and Maryland shows affinities to the Genesee Valley shells, but differs in its brachial muscle field. Boucot and Johnson's (1968: B8, pl. 2, figs. 1-6) *Leptaena* sp. from the Bois Blanc Formation in western New York has a similar pedicle muscle field. Additional collecting should shed light on the relationship of the two forms.

Material: Five articulated valves, six pedicle valves, four brachial valves.
Occurrence: AMNH Loc. 3152.

Order ORTHOTETIDA Waagen, 1884
Suborder ORTHOTETIDINA Waagen, 1884
Superfamily ORTHOTETOIDEA Waagen, 1884
Family SCHUCHERTELLIDAE Williams, 1953
Subfamily SCHUCHERTELLINAE Williams, 1953
Genus SCHUCHERTELLA Girty, 1904
Type species: Streptorhynchus lens White, 1862: 28.

"*Schuchertella*" sp., Plate 2, figure 8.

Description.—Exterior: Shells transversely subelliptical in outline with straight hinge line; pedicle valve planar with maximum width attained just past midlength; pedicle interarea flat, broadly triangular, apsacline, crossed by fine growth lines; delthyrium covered by convex pseudodeltidium; ornamentation consists of radial costellae, with eight in a space of 5 mm measured along the anterior commissure at midline; occasional weak, concentric growth lamellae cross costellae, which increase in number anteriorly by intercalation.

Pedicle valve interior: Hinge teeth short, low, triangular in cross section, not supported by dental lamellae; low myophragm separates bilobed, faintly impressed muscle field; costellae impressed along entire internal periphery of shell.

Discussion: "Schuchertella" sp. from the mid-Hudson Valley (Feldman, 1985: 316, fig. 17) is similar to the Genesee Valley shells, differing somewhat in the musculature of the pedicle interior. This is most likely due to variation or poor preservation. In any case, more material is needed for further meaningful comparisons.

"Schuchertella" sp. A of Boucot and Johnson (1968: B9, pl. 2, figs. 17-30) from the Bois Blanc Formation of western New York closely resembles the Genesee Valley specimens, again differing significantly only in the pedicle valve muscle field. Their *"Schuchertella"* sp. B (pl. 2, figs. 31-36) is more coarsely costate.

Bowen's (1967: 28, pl. 3, figs. 1-7) *Schuchertella prolifica* from the Keyser Limestone differs from the Onondaga shells in its more semielliptical outline and weakly costate internal surface.

Johnson (1970: 106-110, pls. 18-19) illustrated four species of *"Schuchertella"* from the Great Basin of Nevada, of which only "*S.*" sp. B (pl. 18, figs. 15-20) resembles "*S.*" sp. from the Onondaga Limestone of western New York.

Boucot et al. (1970: 23, pl. 8, figs. 5a-c, 6-8) described *Schuchertella?* sp A and B from the Green Pond Outlier in southeastern New York. "*S.*" sp. B (USNM Loc. 11245) resembles the Genesee Valley shells in that it has similar, although smaller, hinge teeth and a relatively broad interarea. "*S.*" sp. A (USNM Loc. 11259) is more coarsely costate, having angular costae.

Boucot (1973: 24, pl. 9, figs. 1-11) illustrated "*S.*" *becraftensis* from the Moose River Synclinorium, which resembles *"Schuchertella"* sp. from the Genesee Valley in its (narrower) bilobed pedicle muscle field.

Material: One articulated shell, one pedicle valve.

Occurrence: AMNH Loc. 3152.

Order STROPHOMENIDA Öpik, 1934
Superfamily STROPHOMENOIDEA King, 1846
Family LEPTOSTROPHIIDAE Caster, 1939
Genus PROTOLEPTOSTROPIA Caster, 1939
Type species: Strophomena blainvillei Billings,
1874: 28-29, figs. 1a-1b; pl. 3, fig. 1.

Protoleptostrophia? sp., Plate 2, figures 9-11.

Description: Maximum width probably at straight hinge line, but due to missing posterolateral margin this is not certain; shell wider than it is long (maximum width = 21.3 mm; maximum length = 18.4 mm), very gently convex and transversely subquadrate in outline; numerous costellae present, between which are interspaces of about the same width; costellae increase anteriorly by intercalation and are crossed by concentric, rugae-like growth lines.

Discussion: Differentiation between the genera *Protoleptostrophia* and *Leptostrophia* cannot be made on the basis of pedicle valve morphology alone, because there are no reliable criteria known for discriminating between pedicle valves of the two genera (Boucot, 1973). In order to distinguish the two genera, one must examine the brachial valve interior; *Protoleptostrophia* lacks socket plates and has a prominent chilidium. True

Leptostrophia is unknown anywhere in beds of Schoharie and Onondaga age (Boucot, oral communication, 1991).

The shells are similar to those described by Boucot et al. (1970: 21, pl. 7, figs. 9-12) from the Green Pond Outlier in southeastern New York, but adequate material from the Onondaga must be obtained.

Material: Four pedicle valves.

Ocurrence: AMNH Loc. 3152.

Family LEPTOSTROPHIIDAE Caster, 1939
Subfamily LEPTODONTELLINAE Williams, 1965
Genus BRACHYPRION Shaler, 1865
Type species: Strophomena leda Billings, 1860: 55, figs. 2-3.

"Brachyprion" cf. *mirabilis* (Johnson, 1970), Plate 2, figures 12-13.

Description: Shells available for study range in size from small to large (Table 5) and are moderately conavoconvex in lateral profile. Hinge line straight, long, denticulate; maximum width at hinge line; pedicle interarea narrow, concave outward, orthocline; no brachial interareas nor pedicle interiors preserved; ornamentation unequally parvicostellate; fine costellae develop anterior to umbo in some specimens, while in others the entire shell surface is parvicostellate; costellae superimpose upon and interrupt concentric rugae; on one shell (AMNH 44173), rugae tend to be somewhat zig-zag patterned in umbonal region.

Discussion: Harper and Boucot (1978: 127) noted that cymostrophid ornamentation is developed elsewhere in the Stropheodontids and, by itself, is not an indication of affinity. Because no muscle scars are preserved and no brachial valves are in the collection, it is difficult to identify these shells to the species or genus level; they are tentatively assigned to *"B." mirabilis*. *Cymostrophia* differs in its transverse outline and strongly concavoconvex

TABLE 5. Measurements (in mm) of *"Brachyprion"* cf. *mirabilis* (Johnson, 1970). See *Systematic Paleontology: Introduction: Systematic Paleontology: Abbreviations and Subscripts* for explanations.

AMNH loc.	(L)	(W)	Specimen type
3152	37.6	47.6	p.v.
3152	33.0	48.0	p.v.
3152	–	46.8	p.v.

geniculate profile with the trail longer than the central disk. *Shaleriella* is commonly smaller and has better developed zig-zag rugae interrupted by primary costellae. The Onondaga shells show the closest affinity to *"B." mirabilis* (Johnson, 1970: 115, pl. 22, figs. 1-12) in that both have parvicostellate ornamentation superimposed upon a concentric development of interrupted rugae. As noted by Johnson (1970: 115), *"B." mirabilis* from Nevada needs a new generic name, but the material available for study is insufficient to establish a new genus. The same holds true for the Onondaga shells.

Material: Five pedicle valves.

Occurrence: AMNH Loc. 3152.

Brachyprion? sp., Plate 2, figures 14-17

Description: Shells medium-sized to large (Table 6), moderately to strongly concavoconvex, apsacline to anacline, often wider than long, somewhat alate, generally poorly preserved except for some details of ornament on the pedicle valve exterior; hinge line denticulate; number and quality of denticles varies greatly (maximum of 25); sometimes vaguely preserved, in others hinge line incomplete; ornamentation unequally parvicostellate, with three to eight costellae between widely-spaced costae.

TABLE 6. Measurements (in mm) of *Brachyprion?* sp. See *Systematic Paleontology: Introduction: Measurement Abbreviations and Subscripts* for explanations.

AMNH loc.	(L)	(W)	Specimen type
3152	27.5	29.0	p.v.
3152	28.0	38.5	p.v.
3152	33.7	44.0 d	p.v.
3152	26.5	37.6 d	p.v.
3152	25.5 d	33.4	p.v.
3152	26.8	40.0 dh	p.v.
3152	19.9	34.0	p.v.
3152	19.6	38.8 d	p.v.
3152	14.0	22.0 dh	p.v.
3152	19.2	–	p.v.
3152	22.8	–	p.v.
3152	21.8	33.4 dh	p.v.
3152	23.7	–	p.v.

Discussion: These shells are provisionally assigned to *Brachyprion* (*Protomegastrophia*) because of their size and ornamentation. *Brachyprion (Brachyprion)* is usually smaller in size (less than 2 cm). *Brachyprion (Eomegastrophia)* is very similar in morphology, but differs in that it can have uniformly costate ornamentation (Sheehan, 1971: 240) and is restricted to the late Llandoverian through early Wenlockian (Harper and Boucot, 1978: 15) whereas *Brachyprion* (*Protomegastrophia*) ranges into the Gedinnian (Harper and Boucot, 1978: 18). The shells are not assigned to *Strophonella* because of their parvicostellate ornamentation. *Strophonella* also lacks widely spaced costae. As the internal morphology (particularly the muscle field and cardinal process) is unknown, positive generic and specific assignments cannot be made.

Material: Seventeen pedicle valves.

Occurrence: AMNH Loc. 3152.

Family STROPHODONTIDAE Caster, 1939
Genus MEGASTROPHIA Caster, 1939
Type species: Strophomena (*Strophodonta*) *concava* Hall, 1857: 140.

Megastrophia? sp., Plate 2, figure 18.

Description: Transversely oval muscle field with radial grooves; diductor scars narrow, elongate longitudinally, subelliptical in outline, and separated by very low myophragm; adductors larger, surround diductors anterolaterally, transversely oval in outline, and radially grooved; myophragm separating diductors extends through adductor field and becomes somewhat higher; shell almost identical to another from the Onondaga of the mid-Hudson Valley (Feldman, 1985: 318, fig. 21); muscle field very similar to Harper and Boucot's (1978: pl. 40, figs. 9a-b) *Megastrophia* (*Megastrophiella*) but has a more transverse outline; both have a low myophragm separating diductor and adductor muscles.

M*aterial:* One pedicle valve.

Occurrence: AMNH Loc. 3152.

Genus PLICOSTROPHEODONTA Sokolskaya, 1960
Type species: Orthis murchisoni D' Archiac and
de Verneuil, 1842: 371, pl. 36, fig. 2.

Plicostropheodonta? sp. Plate 2, figure 19.

Description: Shell moderately convex, wider than long, with interarea flat to very slightly concave outward, and orthocline to somewhat anacline; hinge denticulate for at least one-half the valve length but, due to poor preservation, lateral extremities of hinge show no evidence of denticles; delthyrium poorly preserved but open, and umbo projects posteriorly past hinge line; coarse, rounded costae separated by interspaces of commonly the same width, but occasionally two times as wide, superimposed on fine, uniform costellae; costae originate at beak and increase anteriorly by bifurcation and intercalation.

Discussion: Harper and Boucot (1978: 21, pl. 44, figs. 1-9; pl. 45, figs. 1a-b) described *Plicostropheodonta murchisoni*, which is larger and more coarsely plicate, from the Seifener Series, higher Middle Siegenian, Germany. The Onondaga specimen is tentatively assigned to *Plicostropheodonta* because of its relatively coarse costae superimposed on fine uniform costellae. This morphological feature, however, is likely due to evolutionary convergence.

Material: One pedicle valve.

Occurrence: AMNH Loc. 3152.

Genus STROPHODONTA Hall, 1850

Type species: Strophomena demissa Conrad, 1842: 258, pl. 14, fig. 14.

Strophodonta demissa (Conrad, 1842), Plate 2, figures 20-26; Plate 3, figures 1-5.

Stropheodonta cf. *demissa* Boucot and Johnson, 1968: B9, pl. 1. figs. 7-16; Boucot, 1973: 21, pl. 6, figs. 7-19; Feldman, 1985: 319, fig. 23.

Description.—Pedicle valve exterior: Shells weakly mucronate, semicircular to shield-shaped in outline, concavoconvex in lateral profile, slightly wider than long (Table 7); pedicle valve umbo extends noticeably past interarea; pedicle interarea orthocline to slightly apsacline: delthyrium open and broadly triangular; flat pseudodeltidium sometimes present; usually, pseudodeltidium and chilidium (if present) are not easily differentiated; coarse, uniform costae superimposed on finer uniform costellae, commonly increasing anteriorly by bifurcation and intercalation; interspace distance approximately equal to costae width posteriorly, but increases to two or (rarely) three times the width near anterior commissure; about eight to 10 costae per 5 mm.

Pedicle valve interior: Muscle field roughly suboval and apparently confined by low ridge anteriorly, although this may simply be a small depression in the valve floor; there is an indication of low platform a posterior end of muscle field, which probably supported adductors; ventral sockets are medium-sized and form a pair of pseudoteeth in only one specimen (AMNH 44185), which articulate with pseudosockets in brachial valve floor; shells endospinose on exfoliated surfaces.

Brachial valve interior: Cardinal process lobes are directed posteriorly and joined basally to form a U-shaped structure; attachment faces of lobes are non-striate and grooved medially; small sockets adjoin cardinal lobes laterally and diverge at angle of about 30 degrees from hinge line; muscle field extends anteriorly for about one-half the length of valve, consists of two sets of adductors elevated on a platform; a medial and posterolateral pair; medial scars extend from the middle of posterolateral scars, elongate and divided by low myophragm, laterally, surrounded by raised ridge; posterolateral scars each subelliptical to reniform, shorter in length and separated by low, round myophragm, which, in some cases, extends anteriorly and diverges, forming lateral ridges, which surround medial scars; posterolateral scars commonly more deeply impressed; indications of short breviseptum in some shells, along with slightly divergent (anteriorly) brachial ridges; breviseptum commonly continuous with myophragm; moderately high peripheral ridge present; external ornamentation only impressed on interior of endospinose valve floor at periphery.

Discussion: The brachial valve interior identical to Hall's (1867) specimen (AMNH 37208) but smaller. In the Onondaga shells, the pedicle valve is more convex than Hall's (1867) specimens (AMNH 5153C, D) but equal in convexity to AMNH 5153A (from the Hamilton of Genesee County, Darien, New York). In general, the Onondaga shells are smaller than Hall's types.

Strophodonta demissa may be differentiated from *Brachyprion (Brachyprion)* by its larger, broader cardinal process, in which the cardinal lobes are joined basally into a U-shaped structure, its raised brachial valve muscle field, pseudoteeth, and pseudosockets. Ornamentation is coarser in *S. demissa* and the Onondaga shells have relatively coarse uniform costellae imposed on finer uniform costae. *Brachyprion (Brachyprion)* from the Onondaga, as discussed above, has unequally parvicostellate ornamentation. These shells may show evolutionary convergence with *Plicostropheodonta* as the ornamentation is very similar and the brachial valve interiors are identical.

TABLE 7. Measurements (in mm) of *Strophodonta demissa* (Conrad 1842). See *Systematic Paleontology: Introduction: Measurement Abbreviations and Subscripts* for explanations.

AMNH loc.	(L)	(W)	(T)	Specimen type
3152	7.9	9.6	2.9	a.v.
3152	7.3	9.4	2.3	a.v.
3152	9.2	11.4	–	a.v.
3152	9.6	7.3	3.2	a.v.
3152	8.5	10.7	–	a.v.
3152	10.1	12.8	2.7	a.v.
3152	8.4	11.0	2.0	a.v.
3152	7.1	9.1	2.7	a.v.
3152	9.7	14.0	–	a.v.
3152	19.8	25.6	9.1	a.v.
3152	18.4	–	6.7	a.v.
3152	16.7	20.9	–	a.v.
3152	18.7	20.8	–	a.v.
3152	19.8	20.5	5.6	a.v.
3152	20.3	21.9	6.0	a.v.
3152	18.0	23.2 dh	5.0	a.v.
3152	15.2	16.9	2.3	a.v.
3152	21.4	24.0 d	–	a.v.
3152	20.1	21.2	7.0	a.v.
3152	19.9	21.9	6.7	a.v.
3152	18.3	20.6	5.1	a.v.
3153	19.1	–	6.6	a.v.
3153	18.4	20.9	5.0	a.v.
3153	15.9	17.4	5.0	a.v.
3153	16.9	17.1	4.8	a.v.

Strophodonta demissa differs from *Megastrophia* in having a less transverse muscle field in the pedicle valve; it is subelliptical to subtrigonal in the brachial valve of *Megastrophia.*

Material: Sixty articulated shells, twenty one pedicle valves, eight brachial valves.

Occurrence: AMNH Loc. 3152.

Family STROPHONELLIDAE Caster, 1939

Genus STROPHONELLA Harper and Boucot, 1978

TABLE 8. Measurements (in mm) of *Strophonella punctulifera* (Conrad, 1838). See *Systematic Paleontology: Introduction: Measurement Abbreviations and Subscripts* for explanations.

AMNH loc.	(L)	(W)	(T)	Specimen type
3152	14.7	24.3	–	a.v.
3152	20.3	32.8 dh	–	a.v.
3152	18.6	–	–	a.v.
3152	17.0	20.2 dh	–	a.v.
3152	18.4	25.5 dh	–	a.v.
3152	9.2	11.0	–	a.v.
3152	18.4	24.8 dh	–	a.v.
3152	–	34.0	–	a.v.
3152	13.0	22.2	–	a.v.

Type species: Strophomena punctulifera Conrad, 1838: 117.

Costistrophonella punctulifera (Conrad, 1838), Plate 3, figs. 6-11.

Strophomena punctulifera Conrad, 1838: 117.

Strophodonta punctulifera Hall, 1859: 188, pl. 21, fig. 4; pl. 23, figs. 4-5, 7e.

Strophonelia cf. *punctulifera* Boucot, 1973: 23, pl. 8, figs. 14-18; Johnson, 1970: 113-115, pl. 20, figs. 1-6; pl. 21, figs. 13-14.

Strophonella punctulifera Hall and Clarke, 1892: 291, pl. 12, figs. 10-12; Schuchert, 1913: 323, pl. 59, figs. 8-10; Cooper, 1944: 339, pl. 130, figs. 19-20; Johnson, 1970: 113, pl. 20, figs, 1-6; pl. 21, figs. 13-14.

Costistrophonella cf. *punctulifera* Feldman, 1985: 317, fig. 20.

Description.—Exterior: Brachial valve strongly convex in largest (Table 8) and most complete specimen (AMNH 44198); pedicle valve strongly concave; shells resupinate in lateral profile; hinge line at least partially denticulate, but denticles not well preserved; rounded and angular radial costellae of uniform width increase by bifurcation in some specimens and by intercalation in others; costellae separated by interspaces, which, in some shells, are same width as costellae, and in other shells range from one to three times their width.

Discussion: Strophonella punctulifera differs from *Strophonella (Strophonella)* in having costellae of uniform width that are separated by interspaces of one or two times their width. In *Strophonella (Strophonella)* the costellae are characteristically more widely spaced. *Strophonella punctulifera* from the Onondaga Limestone has identical ornamentation and muscle scars to that of *Costistrophonella* cf. *punctulifera* from the Great Basin of Nevada (Johnson, 1970: 113, pl. 20, figs. 2, 4). The cardinal lobes of the Nevada shells vary slightly in that they diverge from one another at a more moderate angle. *Costistrophonella* sp., a Helderberg equivalent (Harper and Boucot, 1978, pl. 17, figs. 9a-b) from the Gedinnian of Gaspe, Quebec, differs in its more angular costellae. The Onondaga brachial valve (AMNH 44198) differs from Harper and Boucot's (1978, pl. 18, fig. 1a) brachial interior and differs only in that it has a T-shaped platform which supports the cardinal lobes.

Strophonella punctulifera differs from *C. ampla* (Hall, 1867: 93-96, pl. 14, figs. 1a-i; Harper and Boucot, 1978: pl. 17, figs. 5-8; pl. 18, figs. 4-6) in its coarser and more angular costellae.

Material: Twenty articulated shells, two pedicle valves, eight brachial valves.

Occurrence: AMNH Loc. 3152.

Strophonella ampla (Hall, 1857), Plate 3, figures 12-14.

Type species. Strophomena ampla Hall, 1857: 111.

Strophomena (Strophodonta) ampla Hall, 1857: 111.

Strophodonta ampla Hall, 1867: 93-96, pl. 14, figs. 1a-i.

Strophonella ampla Hall and Clarke, 1892: pl. 12, figs. 13-15; Clarke, 1908: 197-198, pl. 37, fig. 12.

Description.—Exterior: Two specimens have been assigned to *Costistrophonella ampla* based on their distinctly finer costaellae.

Brachial valve interior: Cardinal process lobes low, separated basally, and diverge from each other at an angle of 45 degrees; each lobe divided by longitudinal groove and unstriated; small median groove separates cardinal lobes; low myophragm extends anteriorly from a T-shaped platform, splits anteriorly and separates adductor muscle field; muscles impressed more deeply posteriorly and merge imperceptibly with valve floor anteriorly; muscle scars longitudinally striated, somewhat dendritic, and roughly oval in outline; internal shell surface endospinose, especially posteriorly.

Discussion: Shells conform with the finer costellae typical of the species illustrate by Hall (1867: pl. 14, figs. 1a-i). Johnson's (1970: pl. 20, figs. 1-6; pl. 21, figs. 13-14) species *Costistrophonella* cf. *punctulifera* from the Great Basin of Nevada has more coarse and angular costellae.

Material: One pedicle valve, one brachial interior.

Occurrence: AMNH Loc. 3152.

Suborder CHONETIDINA Muir-Wood, 1955
Superfamily CHONETOIDEA Bronn, 1862
Family CHONETIDAE Bronn, 1862
Subfamily DEVONOCHONETINAE Muir-Wood, 1962
Genus LONGISPINA Cooper, 1942
Type species: Chonetes emmetensis Winchell, 1866: 92.

Longispina mucronata (Hall, 1843), Plate 3, figures 15-18.

Strophomena mucronata Hall, 1843: 180, fig. 3.

Chonetes laticosta Hall, 1857: 119.

Chonetes mucronata Hall, 1867: 124, pl. 20, fig. 1; pl. 21, fig. 1.

Chonetes mucronata Hall and Clarke, 1892: pl. 16, figs. 6-7; 1894: pl. 20, fig. 3.

"Chonetes" aff. *lineata* Hall, Feldman, 1985: 320, fig. 25A.

Description.—Exterior: All shells in collection are small (Table 9), subquadrate, concavoconvex; maximum width at hinge line; anterior commissure rounded, lateral commissures parallel; pedicle interarea concave and anacline; brachial interarea straight; remnant of single pseudodeltidium observed on one specimen but too poorly preserved to describe, other than to note that it includes an angle of approximately 60 degrees and is rather small; no chilidium observed; 12 to 15 rounded costae progressively widen from umbonal area toward margins; costae wider than interspaces on pedicle valve; brachial valve exterior too poorly preserved to comment on ornamentation; three costae per 2 mm along anterior commissure; no growth lines evident; neither intercalation nor bifurcation of costae observed, although Hall (1867: 125) noted that some costae originated by bifurcation and some by intercalation (in Hall's specimens there were up to 20 costae present); spines not observed.

Pedicle valve interior: One poorly preserved pedicle valve interior is available for study. The only morphological features of note are indications of two short hinge teeth, a large round muscle field, and impress of costae along interior margins of valve floor.

Discussion: Longispina emmetensis (Winchell, 1866) from the Traverse Group of Michigan (Imbrie, 1959: 397, pl. 64, figs. 23-26) differs from *L. mucronata* in that it has finer costae and is larger. Also, the costae of *L. emmetensis* increase very infrequently by both implantation and bifurcation.

TABLE 9. Measurements (in mm) of *Longispina mucronata* (Hall, 1843). See *Systematic Paleontology: Introduction: Measurement Abbreviations and Subscripts* for explanations.

AMNH loc.	(L)	(W)	(T)	Specimen type
3152	4.0	5.3	–	a.v.
3152	5.9	6.2	–	a.v.
3152	5.1	5.6 dh	–	a.v.
3152	5.0	5.8	1.0	a.v.
3152	4.0	4.6	–	a.v.
3152	6.0	7.0	–	a.v.
3152	5..4	–	1.9	a.v.
3152	3.8	5.1 dh	0.5	a.v.
3152	4.1	5.2	–	a.v.
3152	5.3	–	–	a.v.
3152	5.6	7.3	–	a.v.
3152	4.3	5.7	–	a.v.
3152	3.8	4.9 dh	–	a.v.

Material: One articulated shell, five pedicle valves.
Occurrence: AMNH Locs. 3152, 3154.

Family EODEVONARIIDAE Sokolskaya, 1960
Genus EODEVONARIA Breger, 1906
Type species: Chonetes arcuatus Hall, 1857: 76.

"*Eodevonaria*" cf. *hemispherica* (Hall, 1857), Plate 3, figures 19-21.

Chonetes hemispherica Hall, 1857: 116; 1867: 118, pl. 20, fig. 6: Hall and Clarke, 1892: pl. 16, fig, 14.

"Eodevonaria" hemispherica Racheboeuf and Feldman, 1990: 10-13, figs. 9-12.

Chonetes lineata Feldman, 1985: 320, fig. 25B.

Description.—Pedicle valve exterior: Shells large (Table 10), concavo-convex, with maximum width at hinge line; umbo strongly arched dorsally, overhangs hinge line; ears strongly convex, triangular, well differentiated from body of pedicle valve; pseudodeltidium small, triangular convex plate at apex of delthyrium; pedicle interarea strongly concave, catacline, and almost perpendicular to commissural plane; no spines preserved; ornamentation consists of low, rounded radial costae, separated by narrower interspaces, which increase by bifurcation; costae number about two to four per millimeter at anterior commissure; at 5 mm from beak there are approximately 45 costae, while largest shells have about 70 costae at anterior commissure.

Pedicle valve interior: One poorly preserved specimen studied in the field serves as the basis for this description. Faint myophragm divides muscle field to about midlength; diductor scars with radial striations found at posterior third of visceral cavity; large, oval adductor scars, almost semicircular in outline; hinge line strongly denticulate; internal periphery of anterior commissure grooved due to impress of costae; endospines found on anterolateral margins of valve, arranged in radial sequence.

Discussion: Due to the lack of internal brachial valve morphology, generic assignment is somewhat questionable, although the shape, denticulate hinge line, and other characters indicate assignment to the family Eodevonariiciae.

Racheboeuf and Feldman (1990) found the same species in the *Hallinetes* Zone (formerly "Pink" *Chonetes* Zone) in central New York (AMNH Locs. 3128, 3219). Both forms have rather coarse radial ornamentation very similar to shells from Central and South America. *Eodevonaria inca* (Isaacson, 1977), *E. imperialis* (Caster, 1939), and *E. subhemispherica* (Weisbord, 1926) have similar ornamentation and morphologies. Although they would

TABLE 10. Measurements (in mm) of *"Eodevonaria"* cf. *Hemispherica* (Hall, 1857). See *Systematic Paleontology: Introduction: Measurement Abbreviations and Subscripts* for explanations.

AMNH Loc.	(L)	(W)	Specimen type
3152	6.9	14.9	p.v.
3152	9.6	14.3	p.v.
3152	10.3	14.2	p.v.
3152	8.0	14.4 dh	p.v.
3152	8.6	–	p.v.
3152	9.8	13.1	p.v.
3152	9.6	13.4	p.v.
3152	8.1	7.8	p.v.
3152	8.0	13.4 dh	p.v.
3152	8.3	13.7	p.v.
3152	8.7	8.8	p.v.
3152	5.7	10.5 dh	p.v.

probably represent a distinct group of globose, coarsely costate Eodevonariids, more detailed information is necessary for better comparisons.

Material: Ten pedicle valves.

Occurrence: AMNH Locs. 3152, 3154.

Eodevonaria cf. *arcuata* (Hall, 1857), Plate 3, figure 22.

Chonetes arcuata Hall, 1857: 116-117; 1867: 119, pl. 20, figs. 7a-f.

Eodevonaria arcuata Williams and Breger, 1916: 52-54, pl. 3, figs, 6, 9, 11; Boucot and Harper, 1968: 153, pl. 27, figs. 1-7.

E. imperialis Caster, 1939: 122-126, pl. 7, figs. 11-14, 17-18; pl. 9, fig. 3.

E. imperialis var. *parva* Caster, 1939: 126-128, pl. 7, figs. 9-10, 15-16; pl. 9, figs. 4-7.

E. imperialis var. *transversa* Caster, 1939: 128-129, pl. 7, figs. 5-6; pl, 11, figs. 18-20.

E. sp. cf. *E. arcuata* Boucot, 1959: 756, pl. 97, figs. 11-16.

Description: One pedicle valve exterior along with the posterior section of the brachial valve was exposed from a block after etching in an acid bath. The specimen is silicified and the piece from which it protrudes is chert, making it impossible to further prepare the internal shell morphology. Shell large for Onondaga chonetids (L = 10.5 mm; W = 13 mm), strongly convex and transverse to possibly somewhat subcircular in outline; maximum width at hinge line; posterolateral margin of pedicle valve flattened; brachial valve not visible; pedicle interarea appears to be flat and orthocline, with remnants of small pseudodeltidium present, the convexity of which cannot be determined; very faint sulcus present on pedicle valve; about 17 costellae found in 5 mm interval at distance of 5 mm from beak; 24 denticles along entire length of hinge line; no spines evident.

Discussion: Eodevonaria cf. *arcuata* differs from *Eodevonaria* cf. *hemispherica* in that it is more finely costellate.

Eodevonaria arcuata intermedia (Amsden and Ventress, 1963) from the Sallisaw Formation (early Emsian), Oklahoma, is very similar to *Eodevonaria arcuata* from the Onondaga, but is not placed in synonomy for two reasons. The first is due to the observation of Boucot and Harper (1968: 156) that the number of costellae in the Oklahoma shells (as well as those from the Camden Chert of Tennessee) is too variable in collections of a single species from a single locality and is thus considered to be of questionable specific value. The second is that the Oklahoma shells lack a pedicle sulcus and the Onondaga specimen has a faint sulcus.

Eodevonaria acutiradiata, first described by Hall (1843: 171, fig. 3, as *Strophomena acutiradiata*) and presumed to be from the Onondaga Limestone of New York (see Amsden and Ventress, 1963: 168), has a shallower pedicle valve and is more finely costellate (12 to 13 per 5 mm).

Material: One pedicle valve.

Occurrence: AMNH Loc. 3153.

Order RHYNCHONELIDA Kuhn, 1949
Superfamily RHYNCHOTREMATOIDEA Schuchert, 1913
Family MACHAERARIIDAE Savage, 1996
Genus MACHAERARIA Cooper, 1955
Type species: Rhynchonella formosa Hall, 1857: 76, figs. 1-5.

Machaeraria sp., Plate 3, figures 23-27.

Description: Shell medium-sized (maximum length 14.5 mm, maximum width 19.2 mm, maximum thickness 10.1 mm), nonstrophic, transversely elliptical in outline, biconvex in lateral profile; brachial valve considerably deeper than pedicle valve; pedicle beak nearly straight, extending just posterior to brachial umbo; beak ridges small, with round, apically located permesothyridid foramen; delthyrium filled by incurved brachial beak, thereby obscuring any evidence of delthyrial plates; no interarea present; posterolateral margins almost straight, diverge at angle of slightly greater than ninety degrees; anterior commissure uniplicate; maximum width attained at about midlength; pedicle valve has a strong sulcus, which dies out posteriorly and disappears altogether at pedicle umbo, while brachial valve has a corresponding fold, which also becomes obsolescent posteriorly; radial costae angular in cross section, with 11 on pedicle flanks and five in sulcus; 10 costae on brachial flanks and six on fold; extremely fine, numerous (20 per mm), evenly spaced concentric growth lines present.

Discussion: Bowen's (1967) *Machaeraria whittingtoni* from the Keyser Limestone (MCZ 9502a-b) is very similar to *Machaeraria formosa* sp. from the Onondaga Limestone, but has three costae in the sulcus and four on the fold. *Machaeraria* from the Becraft Limestone of New York (Hall, 1859: pl. 35, figs. 60p, r) lacks a permesothyridid pedicle foramen. *Machaeraria formosa* seems to consist of two different species, one very slightly sulcate (AMNH 3398, 33401) and the other (AMNH 2481) sulcate and suboval in outline and similar to the Onondaga specimen, but having three costae in the sulcus and four on the fold. Boucot (1973: 35, pl. 14, figs. 14-21) described *M. mainensis* from the base of the Upper Silurian Hobbstown Formation that is similar to the Onondaga form, but again, as in *M. formosa,* it has only three costae in the sulcus and four on the fold. Also, in the Maine shells, as in some of Hall's (1857) types, the pedicle valve sulcus is prolonged into a tongue that abuts against the brachial valve fold; this feature is absent in *Machaeraria* sp. from the Onondaga Limestone. *M. carolina* (Hall), described by Boucot and Johnson (1968: B10, pl. 3, figs. 11-20), from the Bois Blanc Formation in western New York is more trigonal in outline, has a shallower sulcus and 24 costae on the best preserved pedicle valve, five of which are in the sulcus. Of all known specimens of the genus, these shells are the closest in morphology to the Onondaga specimen.

Until more material becomes available for study, especially of the internal morphology, specific assignment must be deferred. Johnson (1970: 142) noted that the types of *"Rhynchonella" carolina* Hall (1867: pl. 54) are actually *Machaeraria* and their probable Eifelian age makes *Carolina* the youngest known species of *Machaeraria. Machaeraria* sp. from the Onondaga Limestone is also Eifelian in age (Feldman, 1985: 293). The Onondaga form is decidedly distinct from the Silurian and Early Devonian species of *Macheraria* and may be transitional to *Callipleura* (type species *C. nobilis* Cooper, 1942), a Hamilton age genus.

Material: One articulated shell.

Occurrence: AMNH Loc. 3152.

Superfamily STENOSCISMATOIDEA Oehlert, 1887 (1883)
Family STENOSCISMATIDAE Oehlert, 1887 (1883)
Subfamily STENOSCISMATINAE Oehlert, 1887 (1883)
Genus ATRIBONIUM Grant, 1965
Type species: Stenoscisma halli Fagerstrom, 1961: 29, pl. 9, figs. 48-51.

Atribonium halli (Fagerstrom, 1961), Plate 4, figures 1-11.

Stenoscisma halli Fagerstrom, 1961: 29, pl. 9, figs. 48-51.

Stenoscisma rhomboidalis (Hall and Clarke) Fagerstrom, 1961: 29. pl. 9, figs. 45-47.

Atribonium halli (Fagerstrom) Grant, 1965: 52; Feldman, 1985: 313, fig. 29.

Description: Shells small, rostrate, nonstrophic, subpentagonal in outline, ventribiconvex with short, suberect beak and short, blunt beak ridges; pedicle foramen small; no deltidial plates preserved; anterior commissure uniplicate with prominent brachial fold and deep pedicle sulcus; costae weak, rounded, fading toward beak, disappearing on both valves just short of midlength; four costae on fold and three on sulcus of larger specimen, but none preserved on smaller one; three costae found on flanks; no visible growth lines evident; both valves are geniculate and butt against one another in a vertical plane at the anterior commissure (a generic character in the Stenoscismatacea); valves also butt against each other at lateral and posterior margins with no overlap; no evidence of incipient frills at commissure.

Discussion: The shell is almost identical to those collected from the mid-Hudson Valley (Feldman, 1985: 323, fig. 29) and differ in the number of costae on the sulcus and fold (not necessarily a significant difference when dealing with such a small sample). Grant (1965: 52) noted that *Atribonium halli* differs from all other species of the genus in having few (two or three) costae on the fold, and normally the same or a greater number on each flank. Since the Genesee Valley species so closely resembles the mid-Hudson Valley shells, the reader is referred to Feldman (1985: 324) for further comparisons that are applicable to both.

Material: Three articulated shells.

Occurrence: AMNH Loc. 3153.

Superfamily RHYNCHOTREMATOIDEA Schuchert, 1913

Family TRIGONIRHYNCHIIDAE Schmidt, 1965

Genus CUPULAROSTRUM Sartenaer, 1961

Type species: Cupularostrum recticostatum

Sartenaer, 1961: 6, pl. 1, figs. 1-7; pl. 2, figs. A-C.

TABLE 11. Measurements (in mm) of *Cupularostrum?* sp. See *Systematic Paleontology: Introduction: Measurement Abbreviations and Subscripts* for explanations.

AMNH loc.	(L)	(W)	(T)	Number of plications		Sulcus begins*
				fold	sulcus	
3152	13.9	13.6	6.0	6	5	3.4
3153	–	8.0	–	–	6	5.0

* Measured from the beak; both valves are articulated.

Cupularostrum? sp., Plate 4, figures 12-15.

Description: Two articulated specimens are available for study, one of which is embedded in a piece of chert, pedicle valve up, thus obscuring the anterior margin of the shell. Shells medium-sized for the genus (Table 11), subtrigonal in outline; posterolateral margins straight; pedicle beak slightly damaged but appears to be suberect; open, triangular delthyrium evident, with a small foramen located apically; no deltidial plates observed; maximum width attained anterior to midlength, almost at anterior commissure; pedicle valve bears sulcus that originates just anterior to umbo; brachial valve bears corresponding but weaker fold; 21 simple, sharply rounded to subangular plicae separated by V-shaped interspaces; six plicae in sulcus, but number on fold questionable.

Discussion: Boucot (1973: 30) noted that an examination of the cardinalia is crucial to assigning a species to the genus *Cupularostrum*, and that external form and ornamentation are not diagnostic. True *Cupularostrum* has a septalium commonly roofed over by a perforate hinge plate and crenulate dental sockets. *Cupularostrum macrocosta* from the Moose River Synclinorium, Maine (Boucot, 1973: 29, pl. 12, figs. 3-11) has four to five plicae in the sulcus of the pedicle valve and seven to nine on the flanks, for a total of 11 to 14. The overall size and outline is similar to the Genesee Valley specimens.

Cupularostrum sp. A from the mid-Hudson Valley (Feldman, 1985: 325, figs. 30A-D) is smaller and has only 15 plicae. In all other respects, regarding what can be determined from observing the exteriors, they are identical. They are also very similar to *Cupularostrum?* sp. from the underlying Bois Blanc Formation of western New York (Boucot and Johnson, 1968: B11, pl. 3, figs. 21-29) in outline and external morphology.

Material: Two articulated shells.

Occurrence: AMNH Locs. 3152, 3153.

Order ATRYPIDA Rzhonsnitskaya, 1960
Suborder Atrypidina Moore, 1952
Superfamily ATRYPOIDEA Gill, 1871
Family ATRYPIDAE Gill, 1871
Subfamily ATRYPINAE Gill, 1871
Genus ATRYPA Dalman, 1828
Type species: Anomia reticularis Linnaeus, 1758: 702.

Atrypa "reticularis" (Linnaeus, 1767), Plate 4, figures 16-24.

TABLE 12. Measurements (in mm) of *Atrypa "reticularis"* (Linnaeus, 1767). See *Systematic Paleontology: Introduction: Measurement Abbreviations and Subscripts* for explanations.

AMNH loc.	(L)	(W)	(T)	Specimen type
3152	14.4	14.7	8.7	a.v.
3152	18.6	19.6	9.7	a.v.
3152	23.7	23.0	12.1	a.v.
3152	22.2	24.7	14.3	a.v.
3152	17.6	17.4	12.0	a.v.
3152	15.9	17.0	9.0	a.v.
3152	20.0	20.6	11.6	a.v.
3152	19.7	21.9	11.9	a.v.
3152	23.5	22.5	12.7	a.v.
3152	22.9	23.7	9.8	a.v.
3152	22.8	21.1	15.6	a.v.
3152	14.7	16.8	8.9	a.v.
3152	21.0	22.4	12.5	a.v.
3152	14.9	16.0	6.6	a.v.
3152	21.6	22.4	11.5	a.v.
3152	16.9	16.0	8.5	a.v.
3152	10.2	10.3	5.0	a.v.
3152	19.1	17.4	10.7	a.v.
3152	19.8	21.9	11.7	a.v.
3152	20.2	20.3	10.0	a.v.
3153	18.9	22.0	10.8	a.v.
3153	16.7	15.8	9.5	a.v.
3153	13.4	–	6.2	a.v.
3153	15.7	13.8	4.9	a.v.
3153	20.6	24.4	9.3	a.v.

Anomia reticularis Linnaeus, 1758: 702.

Atrypa "reticularis" Boucot, 1959: 741, pl. 91, figs. 7-9; Boucot and Johnson, 1968: B12, pl. 3, figs. 30-49; Boucot, 1973: 36, pl. 15, figs. 1-6; Feldman, 1985: 326, fig. 31.

Description.—Exterior: Shells mostly medium-sized (Table 12), dorsibiconvex, subcircular to elongate, suboval in outline; maximum width attained at about midlength; pedicle beak suberect and overhangs brachial umbo, partially concealing erect brachial beak; pedicle valve bears sulcus and brachial valve bears corresponding fold; anterior commissure slightly uniplicate; rounded radial costellae increase anteriorly by bifurcation and intercalation; concentric growth lamellae interrupt costellae, becoming more numerous, distinct, and frilly anteriorly.

Pedicle valve interior: Large hinge teeth supported by short, narrow dental lamellae; muscle field flabellate, somewhat trigonal in outline and longitudinally striated, with relatively coarse, uniformly sized, slightly radiating low ridges (diductors), which abut anteriorly against a very low rim; adductors, located in posterior of muscle field, represented by two smooth indentations separated by low ridge (myophragm?); ovarian/gonadal depressions deeply pitted.

Brachial valve interior: Diverging sockets bounded posterolaterally by valve margin and anteromedially by curved hinge plates; socket plate transversely grooved; short, thin crural bases projecting anteroventrally, attached to inner margins of sockets; no cardinal process observed; adductor muscle scars faintly striated, almost fiabellate, subpyramidal, and divided by low myophragm, which is thick posteriorly, thins anteriorly, and bifurcates into two myophragms ending at anterior rim of muscle field; faintly pitted ovarian depression.

Material: One-hundred-eleven articulated valves, thirty-nine pedicle valves, thirty-five brachial valves.

Occurrence: AMNH Locs. 3152, 3153, 3154.

Superfamily ANOPLOTHECOIDEA Schuchert, 1894
Family ANOPLOTHECIDAE Schuchert, 1894
Subfamily COELOSPIRINAE Hall and Clarke, 1895
Genus COELOSPIRA Hall, 1863
Type species: Leptocoelia concava Hall, 1857: 107.

Coelospira camilla Hall, 1867, Plate 4, figures 25-38.

Leptocoelia concava Hall 1857: 107.

Coelospira camilla Hall, 1867: 329 (as *Coelospira concava),* pl. 52, figs. 13-19; Hall and Clarke, 1895: pl. 53, figs. 24-31; Boucot and Johnson, 1967: pl. 164, figs. 20-30; pl. 165, figs. 1-15: Boucot and Johnson, 1968: B13, pl. 4, figs. 1-25; Boucot et al. 1970: 17-18, pl. 5, figs. 17-19, 21-22; Feldman, 1985: 327, fig. 32.

Description.—Exterior: Shells small (Table 13), concavoconvex, subcircular in outline; hinge slightly rounded with small pedicle foramen in well preserved specimens; no interarea evident; pedicle beak incurved and pedicle valve strongly convex; maximum width attained at about one-third length; in some shells, brachial valve sulcate medially but planar anteriorly; ornamentation consists of 12 rounded radial plication with U-shaped interspaces of about same width, which become wider as anterolateral margins approached; plications rarely increase anteriorly by bifurcation.

Pedicle valve interior: Hinge teeth small, thin, generally poorly preserved and supported by obscure dental lamellae; crural fossette on medial side of each hinge tooth; diductor muscle scars bisected by low myophragm of varying thickness, which ends at about one-third valve length; in some shells, scars elongate, subelliptical and fairly deeply impressed and anterior boundary of diductor impressions are strongly defined by the difference in elevation on the valve floor; in others, scars are obscure and muscle impressions grade into the valve floor imperceptibly; adductor scars represented by subtriangular depression just past the anterior end of the myophragm, sandwiched between extremities of diverging diductor impressions, located almost in the exact center of the valve floor, which is faintly crenulated anteriorly due to impress of plications.

Brachial valve interior: Sockets diverge anterolatally, deeply excavated and bordered medially by incurved socket plates; cardinal process trilobed, possibly quadrilobed protuberance from the base of which extends short, low, often broad anteriorly tapering myophragm; myophragm begins as swelling before tapering to bisect elongate to suboval adductor muscle field that terminates in raised rim at midlength; valve floor impressed with plications along periphery.

Discussion. Adductor scars are not preserved in any specimens of *Coelospira camilla* from the mid-Hudson Valley (Feldman, 1985: 327, fig. 32); otherwise the shells are identical. The Genesee Valley specimens are almost identical to those collected from the Bois Blanc Limestone of western New York (Boucot and Johnson, 1968: B12, pl. 4, figs. 1-25), differing in their slightly finer plicae. Boucot et al. (1970: 17, pl. 5, figs. 17-19, 21-22) illustrate *Coelospira* sp., which differ in ornamentation from *C. Camilla* from the Green Pond Outlier in southeastern New York. The pedicle exterior has a medial fold bearing a thin lira medially that is bounded by two primary costae, each giving off a costella abaxially.

Laterally three primary costae are present, resulting in a total of 11 costae, costellae, and lirae. The brachial exterior bears a progressively thickening costa that supports a fine lira medially. Lateral to medial costae are present with secondary costellae, resulting in a total of 11 to 15 costae, costellae, and lira. The shells further differ in that there is a nonlobate cardinal process.

TABLE 13. Measurements (in mm) of *Coelospira camilla* Hall, 1867. See *Systematic Paleontology: Introduction: Measurement Abbreviations and Subscripts* for explanations.

AMNH loc.	(L)	(W)	(T)	Specimen type
3152	5.6	5.9	2.1	a.v.
3152	5.3	5.2	1.9	a.v.
3152	5.0	4.8	1.5	a.v.
3152	3.9	4.0	1.5	a.v.
3152	4.0	4.4	1.3	a.v.
3152	3.9	4.1	1.2	a.v.
3152	4.7	4.2	1.2	a.v.
3152	4.5	4.3	2.1	a.v.
3152	5.1	4.9	1.2	a.v.
3152	5.1	5.4	1.2	a.v.
3152	5.1	5.5	1.6	a.v.
3152	5.7	6.0	1.7	a.v.
3152	5.9	5.5	2.1	a.v.
3152	6.0	6.1	2.2	a.v.
3152	6.0	6.6	2.0	a.v.
3152	6.4	5.9	2.0	a.v.
3152	5.5	6.1	1.5	a.v.
3152	5.1	5.0	2.6	a.v.
3152	6.4	6.1	1.6	a.v.
3152	5.0	6.0	2.3	a.v.
3152	6.0	5.7	2.0	a.v.
3152	6.0	5.6	2.3	a.v.
3152	5.0	5.1	1.3	a.v.
3153	5.1	5.1	2.2	a.v.
3153	5.3	5.5	1.3	a.v.

Additional comparisons and references to the occurrence of *Coelospira camilla* in the United States can be found in Feldman (1985: 328) and Boucot and Johnson (1967: 1235).

Material: One hundred forty two articulated shells, tweny one pedicle valves, six brachial valves.

Occurrence: AMNH Locs. 3152, 3154.

Superfamily MERISTELLOIDEA Waagen, 1883
Family MERISTIDAE Hall and Clarke, 1895
Subfamily CAMAROPHORELLINAE Schuchert, 1929
Genus CAMAROSPIRA Hall and Clarke, 1893
Type species: Camarophoria eucharis Hall, 1867: 368.

Camarospira? sp., Plate 5, figures 1-5.

Description: Shell biconvex, elongate, with arcuate pedicle valve bearing shallow sulcus and terminating in tonguelike extension; pedicle beak suberect; brachial valve inflated posteriorly and bears correspondingly weak fold; brachial umbo incurved, partially concealing delthyrium; no deltidial plates preserved and no evidence of foramen; both valves show external evidence of median septum in the form of a thin line bisecting umbonal regions; shell smooth but for strong, irregularly spaced growth lines on anterior one-third of shell; specimen resembles *Camarophoria eucharis* (Hall, 1867: 368, pl. 57, figs. 40-45) collected from the Onondaga Limestone (Corniferous Limestone) of Canada West, Ontario, in its overall external morphology, but it is not as broadly ovate.

Material: One poorly preserved articulated shell.

Occurrence: AMNH Loc. 3153.

Order ATHYRIDIDA Boucot, Johnson and Staton, 1964
Suborder ATHYRIDIDINA Boucot, Johnson and Staton, 1964
Superfamily ATHYRIDOIDEA Davidson, 1881
Family ATHYRIDIDAE Davidson, 1881
Subfamily ATHYRIDINAE Davidson, 1881
Genus ATHYRIS M'Coy, 1844
Type species: Terebratula concentrica von Buch, 1834:
123 by subsequent designation of King, 1850: 136.

TABLE 14. Measurements (in mm) of *Athyris boucoti*. n. sp. See *Systematic Paleontology: Introduction: Measurement Abbreviations and Subscripts* for explanations.

AMNH loc.	(L)	(W)	(T)	Specimen type
3153 holotype AMNH 44227	15.3	14.8	11.3	a.v.
3153	12.9	12.2	7.2	a.v.
3152	11.4	12.9 dh	6.6 e	a.v.
3152	10.0 dh	9.6 dh	5.4	a.v.
3152	11.1	–	6.4	a.v.
3152	10.3	11.1	5.3	a.v.
3152	9.5 dh	9.3	3.0 c	a.v.
3152	9.3	11.6	5.3	a.v.
3152	9.0	10.3	2.9 c	a.v.
3152	9.0	9.3 dh	4.0	a.v.
3152	–	8.8	3.9	a.v.
3152	12.0	11.9	6.0	a.v.
3152	11.3	12.1	6.7	a.v.
3152	11.3	–	5.7	a.v.
3152	11.2	11.0	6.3	a.v.
3152	10.9	10.8 d	5.9	a.v.
3152	10.9	10.4	5.9	a.v.
3152	10.1	11.6	5.8 e	a.v.
3152	10.8	8.9	6.0	a.v.
3152	10.6	9.4	5.8	a.v.
3152	9.5	9.3 dh	5.3	a.v.
3152	9.2	10.1 dh	5.2	a.v.
3152	8.7	9.7	5.3	a.v.
3152	8.1	9.7	4.4	a.v.
3152	8.1	8.9	5.0	a.v.
3152	7.2	–	4.4	a.v.
3152	7.0	8.4	4.6	a.v.
3152	5.7	5.9	4.1	a.v.
3152	12.4 dh	12.2 dh	–	p.v.
3152	9.1	11.9	–	p.v.
3152	10.1	9.2	–	p.v.
3152	9.4	10.2	–	p.v.

Athyris boucoti n. sp., Plate 5, figures 6-15.

Athyris sp. A, Feldman, 1985: 331, figs. 38A-J

Diagnosis: Small athyrids with strong, concave, raised cardinal process bounded by truck lateral margins forming elevated ridges; narrow muscle field with striated adductor scars along two-thirds the valve length; small hinge teeth supported by weak, curved, divergent dental lamellae; shallow, narrow sulcus with corresponding fold extending to umbonal region; lamellose exterior.

Description.—Exterior: Shells small (Table 14), transversely suboval in outline and ventribiconvex; small, round pedicle foramen located at termination of suberect beak, which overhangs incurved brachial beak; shallow, narrow sulcus, beginning just anterior to umbo and widening slightly anteriorly, commonly present on pedicle valve, while brachial valve bears corresponding fold, resulting in weakly uniplicate anterior commissure; some specimens with an underdeveloped sulcus and fold have more rectimarginate anterior commissure; shell surface of well-preserved specimens covered by fine, concentric lamellose growth lines; occasional worn shells completely exfoliated.

The following descriptions of the pedicle and brachial valve interiors are based partly on a single silicified specimen, which has a small gap allowing an internal view. It was decided not to dissect the specimen and thus destroy the silicified shell, as the umbonal cavity and cardinalia were visible with a binocular microscope.

Pedicle valve interior: Small hinge teeth dorsally directed, pointed and supported by weak dental lamellae, which curve medially and diverge at about a 45 degree angle; umbonal cavity thickens laterally, forming slight ridge, but anteriorly merges with valve floor; no muscle scars preserved.

Brachial valve interior: Sockets short, U-shaped and diverge at about a 30 degree angle; cardinal process strong, typically concave and raised from valve floor; lateral margins thicken and form elevated ridges which border sockets; muscle field narrow, about one-fourth valve width, with striated adductor scars extending to about two-thirds valve length where they merge imperceptibly with valve floor; no jugum or spiralia preserved.

Discussion: Athyris boucoti also occurs in the mid-Hudson Valley and is illustrated by Feldman (figs. 38A-J) as *Athyris* sp. A. *Athyris boucoti* is homeomorphic with *Protathyris praecursor* from the Lower *Devonian Mitkov* beds of *Podolia,* in the Ukraine (Nikiforova et al., 1985: 55, pl. 15, figs. 1-4), but can be easily differentiated internally by the cardinal process, which in *Protathyris* is large and trilobate. *Athyris cora* from the Hamilton Group at Delphi, New York (Hall, 1867: 291, pl. 47, figs. 1-7) and the Middle Devonian Formosa Reef Limestone of southwestern Ontario (Fagerstrom, 1961: 34, pl. 11, figs. 37-41), a somewhat larger species, lacks the lamellose omamentation of *Athyris boucoti,* has fine concentric growth lines on the surface, and has a much narrower sulcus and fold. Specimens from the Skaneateles Formation in Hamilton, New York (USNM 447208-447211), are nonlamellose, much larger and more robust than the Ondaga shells, and range from weakly to strongly sulcate. *Athyris minuta* of the Formosa Reef Limestone (Fagerstrom, 1961: 34-35, pl. 11, figs. 42-44) is similar in size but differs in its equally biconvex valves, subpentagonal outline, and weak sulcus and fold. Internal comparisons must be deferred due to lack of *A. minuta* brachial interiors. *Athyris nuculoidea,* described by Cooper, from the St. Laurent Limestone, Missouri (1945: 485, pl. 64, figs. 12-19) closely resembles *Athyris boucoti* in its small size and ornamentation, but differs in that it has a wider and shorter pedicle sulcus (and corresponding fold), is pentagonal to heptagonal in outline, and has straight dental lamellae which rise from the pedicle valve floor at a ninety degree angle; in *Athyris boucoti* the dental lamellae are curved and divergent.

Etymology: After Professor Arthur J. Boucot, Oregon State University, Corvallis, Oregon.

Material: Forty articulated shells, five pedicle valves.

Occurrence: AMNH Locs. 3152, 3153.

Athyris leoni n. sp., Plate 5, figures 16-23.

Athyris sp. B, Feldman, 1985: 332, figs. 38k-l.

Diagnosis: Medium-sized athyrid with a triangular, almost flat cardinal process, without dorsal (visceral) foramen, narrowly elliptical brachial adductor impression, and pyriform diductor scar.

Description.—Exterior: Shells large (Table 15), transversely suboval in outline and ventribiconvex; pedicle beak suberect with poorly preserved foramen, while brachial beak is smaller, incurved; hinge line short, round; pedicle valve bears shallow sulcus, which originates just anterior to umbo and diverges widely as it nears anterior commissure, where it forms distinct but poorly preserved tongue; brachial valve bears corresponding fold that becomes more distinct anteriorly; ornamentation consists of concentric,

TABLE 15. Measurements (in mm) of *Athyris leoni*, n. sp. See *Systematic Paleontology: Introduction: Measurement Abbreviations and Subscripts* for explanations.

AMNH loc.	(L)	(W)	(T)	Specimen type
3152 holotype AMNH 44231	22.0 dh	25.8	16.5	a.v.
3152	–	29.3 d	–	b.v.
3152	22.9	30.0 d	–	p.v.

lamellose growth lines on both valves that become quite crowded anteriorly, especially at anterior one-third of shell.

Pedicle valve interior: Hinge teeth small, pointed, dorsally directed and supported by straight, stout dental lamellae; delthyrium relatively narrow, encompassing an angle of approximately 25 degrees; diductor(?) muscle field is an extension of umbonal cavity, clearly pyriform in outline, and merges with valve floor anteriorly; laterally it is bordered by a ridge that is continuous dorsally, with dental lamellae; bisecting the anterior part of the muscle field is a low myophragm; larger, transversely suboval adductor(?) muscle scar may occupy even greater area of the valve floor (two-thirds) but is not clearly preserved; both scars are striated.

Brachial valve interior. Sockets very deeply excavated, broadly U-shaped in cross section, slightly flattened at base, widen anterolaterally, and shallow out somewhat; cardinal process triangular, almost flat, with raised inner socket margins; no dorsal (visceral) foramen evident; anterior rim of cardinal process is also raised, although not as high, resulting in elevated border completely surrounding cardinal process; at base of cardinal process, below and posterior to raised rim, is an invagination, pointing posteriorly and ending in a small pit-like structure; adductor muscle impression narrowly elliptical and, although worn, appears to have been striated, with a low myophragm found only at posterior one-fifth of scar.

Discussion: Athyris leoni also occurs in the mid-Hudson Valley (AMNH Loc. 3138A) and is illustrated by Feldman (1985, figs. 38k-l) as *Athyris* sp. B. *Athyris leoni* from the Onondaga Limestone resembles *Athyris spiriferoides* from the Hamilton Group (AMNH 42351-42353) externally, but differs internally in the following ways: (1) the pedicle muscle field of *Athyris leoni* is pyriform in outline, whereas *Athyris spiriferoides* has an elliptical to suboval scar (also illustrated in Han and Clarke, 1894: pl. 35, fig. 5), and (2) The cardinal process of *Athyris leoni* is almost flat, whereas in *Athyris spiriferoides* it is concave to spoon-shaped, almost quadrilobed in appearance. Cooper (1944: 333, pl. 127, figs. 39-43) illustrated *Athyris spiriferoides* from the Upper Hamilton Group of New York; this form differs from the Onondaga species in its larger size (more than 38 mm in width), shallower sockets, and pedicle muscle field with straight lateral margins. Hall (1867: 288) noted that the lowest range of *Athyris spiriferoides* is in the Onondaga (Corniferous) Limestone, and it becomes abundant in the Hamilton. This is entirely possible, since I have observed specimens *in situ* (but not recoverable) in the field that more closely resemble *Athyris spiriferoides* but do not grade morphologically into *Athyris leoni.* These shells were seen in the mid-Hudson Valley but not in western New York, thus conforming with Hall's observation (1867: 288) that *Athyris spiriferoides* occurs in Albany and Schoharie counties but is rarely seen in equivalent rocks in the western part of the state.

Athyris vittata from the Hamilton Group (AMNH 37503) differs from *Athyris leoni* in its ovate to subquadrate outline, strong pedicle sulcus and brachial fold, and generally smaller size. Hall and Clarke (1894: 777, pl. 35, figs. 1-3) illustrated *Athyris vittata* from the Hamilton Group which distinctly differs from the Onondaga shells in that it has a trilobate cardinal process.

Etymology: After Dr. Leon A. Feldman, Professor Emeritus, Rutgers University, New Brunswick, New Jersey.

Material: Two pedicle valves, two brachial valves.

Occurrence: AMNH Loc. 3152.

Athyris sp. A, Plate 5, figure 24.

Description: This shell differs somewhat in outline and convexity from known athyrid species. The cardinal process is typically raised and concave, with bounding lateral ridges. The sockets are short, deeply excavated, and no muscle impression preserved.

Material: One brachial valve.

Occurrence: AMNH Loc. 3153.

Subfamily MERISTELLINAE Waagen, 1883
Genus MERISTINA Hall, 1867
Type species: Meristella maria Hall, 1863: 212.

Meristina cf. *nasuta* (Conrad, 1842) Plate 5, figures 25-31.

Atrypa nasuta Conrad, 1842: 265.

Meristella nasuta (Conrad, 1842), Hall, 1860: 93; Hall, 1867: 299, pl. 48, figs. 1-25; Fagerstrom, 1961: 33, pl. 11, figs. 1-4; Shimer and Shrock, 1944: 333, pl. 127, figs. 26-27.

Meristina nasuta Boucot and Johnson, 1968: B13-B14, pl. 4, figs. 26-43; Fagerstrom, 1971: 38, pl. 4, fig. 1; Feldman, 1985: 332, fig. 39.

Description.—Pedicle valve exterior: Pedicle valve medium-sized (Table 16), strongly convex, pyriform in outline; most specimens (17 of 19) incomplete to some extent, usually lacking anterior commissure; no brachial valves in collection; maximum width just anterior to midlength; open, triangular delthyrium; incomplete pedicle foramen due to lack of preservation of any delthyrial covering or plate; no articulated shells are available for study, so exact shape of the foramen is unknown, but appears to have been small and circular; beak erect to slightly incurved; no interarea and no indication of a sulcus in any shells studied, however, on one specimen growth lines near anterior commissure developed pointed tongue like projection that is anteriorly similar to Boucot and Johnson's (1968: B13) shells from the Bois Blanc Formation, on which anterior commissure extended into increasingly greater prolongation; irregularly spaced, numerous growth lines, increasing in number anteriorly; on most shells no growth lines evident due to lack of preservation and weak silicification.

Pedicle valve interior: Hinge teeth are small and range in shape from elongate, subpyriform to crescent shaped, with convex part of cresent projecting laterally; supported by thin dental lamellae that converge, often strongly, toward valve floor and then diverge again laterally at posterolateral boundary of muscle field; posterior region of shell, especially in adults, has been thickened by secondary shell material; diductor scars strongly impressed on valve floor in the form of a subtriangular outline with radial striae; center of muscle field in one specimen (AMNH 44236) has a medial depression; anterior boundary of muscle field is well-defined in most shells, but in some grades imperceptibly into valve floor; adult shells with secondary deposition show acute angle between muscle impressions and base of dental lamellae; although on some shells muscle impressions weak, there are none with no scars at all.

Discussion: Differentiation between *Meristina* and *Meristella* is difficult without preservation of the jugum. Boucot et al. (1964: 820) established the genus *Meristina* based on the nature of the dental lamellae and the muscle field configuration. They noted that there is some variation in large suites of the genus, particularly those containing large shells, in that

TABLE 16. Measurements (in mm) of *Meristina* cf. *nasuta* (Conrad, 1842). See *Systematic Paleontology: Introduction: Measurement Abbreviations and Subscripts* for explanations.

AMNH loc.	Distance between hinge teeth	Length of muscle field	Width of muscle field	Specimen type
3152	6.0	8.4	11.1	p.v.
3152	6.5	6.3	9.0	p.v.
3152	4.2	5.6	7.2	p.v.
3152	5.3	4.2	7.3	p.v.
3152	4.7	6.3	7.2	p.v.
3152	4.6	3.9	4.6	p.v.
3152	5.5	4.1	4.8	p.v.
3152	4.0	5.0	6.5	p.v.
3152	–	8.1	6.8	p.v.
3152	–	4.9	8.4	p.v.
3152	–	8.2	8.5	p.v.
3152	–	6.3	8.1	p.v.
3152	–	4.5	6.1	p.v.
3152	–	5.3	5.9	p.v.
3152	–	5.1	6.1	p.v.
3152	–	2.9	3.4	p.v.

some are found with the dental lamellae obsolete and others completely lack the bounding ridges adjacent to the muscle scar. These authors and Hall (1867) noted the resemblance of *"Meristella" nasuta* to *Meristina.* Bowen (1967: 35) noted that in *Meristella* the dental plates are characteristically short, whereas those in *Meristina* are long. The dental plates in the Onondaga specimens are shorter than those in various species of *Meristella* studied. The muscle field is not restricted to the region between the dental lamellae, as in *M. nasuta* from the Detroit River Group (Fagerstrom, 1971: 37). According to Amsden and Ventress (1963: 123), *Meristella* is characterized by deeply impressed muscle scars, with the dental lamellae becoming obscure in mature shells by the deposition of secondary shell material. This is not the case in the Onondaga shells where the dental lamellae are quite distinct even though there is secondary shell deposition. Based on overall morphology and the fact that the genus *Meristina* is found in the underlying Bois Blanc Formation (Boucot and Johnson, 1968: B13), the Onondaga shells are placed in the genus *Meristina* until more material, especially brachial interiors, becomes availble for study.

Boucot and Johnson (1968: B13) described *Meristina nasuta* from the Bois Blanc Formation of western New York and note that it has a faint sulcus on the pedicle valve, but is modified by the development of a low, rounded, medial plication that effectively extends the anterior commissure into increasingly greater prolongation on the larger specimens. The Onondaga shells lack this sulcus and medial plication. Also, *Meristina* cf. *nasuta* from the Onondaga has a slightly more elongate and less broad delthyrium and is somewhat smaller in size. *Meristina* cf. *nasuta* from the Onondaga of the mid-Hudson Valley (Feldman, 1985: 332, fig. 39) is similar to the shells described herein; both are convex, elongate, have no interareas, and one of the shells (AMNH 39900) has concentric growth lamellae along the internal margins. *Meristella nasuta*, described by Shimer and Shrock (1944: 333, pl. 127, figs. 26-27), is considerably larger, with a more rounded beak region than the Onondaga shells. Their articulated specimen precludes comparison of pedicle interiors. *Meristina haskinsi* from the Hamilton Group of Canandaigua Lake, New York (NYSM 1552), differs from the Onondaga shells in the following respects: (1) there is no secondary shell material, and consequently the shell is rather thin, (2) the muscle scars are barely impressed on the (pedicle) valve floor, (3) the outline in plan view is suboval rather than subpyriform, (4) the dental lamellae are longer, (5) the shell is faintly sulcate, and (6) the delthyrium is smaller and wider.

Meristella princeps from the Port Ewen Limestone, Gross Quarry, Rondout, New York (NYSM E2850) approximates *Meristina nasuta* in size and general morphology but differs in having considerably less prominent dental lamellae. *Meristella lentiformis* from the Glenerie Limestone, Glenerie, New York (NYSM E2900; USNM 163805-6 [Dutro, 1971: pl. 1, figs. 1-4]) has a much shallower pedicle valve and is transversely elliptical in outline, while *Meristina lata* from the same formation (NYSM E2902) has much more secondary shell material in the posterior region of the pedicle valve, which resorb, to some extent, the dental lamellae, has a more flaring muscle field and is more convex in lateral profile. (See also Amsden and Ventress, 1963: 120, pl. X, figs. 17-23.)

Material: Thirty pedicle valves.

Occurrence: AMNH Loc. 3153.

Meristina? sp., Plate 5, figures 32-34; Plate 6, figures 1-2.

Description: Valves assigned to genus based on size (AMNH 44237, L = 26.9 mm [est.], W = 29.4 mm; AMNH 44238, L = 30.5 mm, W = 29.0 mm [est.]) and morphology. Shells deeply concave, with small, round pedicle foramen and small triangular delthyrium; one specimen has faint sulcus; two specimens have numerous concentric growth lines; critical morphological structures such as muscle scars, dental lamellae and jugum not preserved; pedicle beaks not as prominent as in *Meristina nasuta.*

Material: Four pedicle valves.

Occurrence: AMNH Loc. 3154.

Genus CHARIONOIDES Boucot, Johnson, and Staton, 1964
Type species: Meristella doris Hall, 1860: 84.

Charionoides doris (Hall, 1860), Plate 6, figures 3-6.

Meristella doris Hall, 1860: 84.

Charionoides aff. *C. doris* Boucot, Johnson, and Staton, 1964: 817, pl. 127, figs. 14-20.

Charionoides doris Boucot, 1973: 64, pl. 20, figs. 14-22; Feldman, 1985: 335, fig. 40.

Description: Two specimens available for study: AMNH 44239 (L = 12.5 mm, W = 9.3 mm, T = 7.2 mm); AMNH 44240 (L = 10.5 mm, W = 8.7 mm, T = 5.0 mm). One shell biconvex and the other ventribiconvex; both pyriform to almost almond shaped in outline; maximum width attained approximately two-thirds distance from beak; beak region damaged and incomplete; palintrope poorly defined and slightly convex; neither fold nor sulcus present, although in one specimen (AMNH 44239) anterior commissure slightly plicate; no growth lines evident.

Discussion: These shells differ from *Charionoides doris* of the Onondaga Limestone in Williamsville, New York (NYSM 1546) only in their lack of growth lines, which may be due to exfoliation and weathering.

Material: Two articulated shells.

Occurrence: AMNH Loc. 3152.

Genus PENTAGONIA Cozzens, 1846
Type species: Atrypa unisulcata Conrad, 1841: 56.

Pentagonia unisulcata (Conrad, 1841), Plate 6, figures 7-13.

Atrypa unisulcata Conrad, 1841: 56.

Atrypa uniangulata Hall, 1861: 101.

Meristella? unisulcata (Conrad) Hall, 1862: 158, pl. 2, figs. 17, 20-23 (not figs. 19, 24, 25).

Meristella (Pentagonia) unisulcata (Conrad) Hall, 1867: 309, pl. 50, figs. 18-29 (not figs. 30-35).

Non *Meristella unisulcata* (Conrad) Nettleroth, 1889: 99, pl. 15, figs. 9-16.

Non *Pentagonia unisulcata* (Conrad) Savage, 1930: 47, 50, 53, 62; 1931: 242, pl. 30, figs. 17-18.

Pentagonia unisulcata (Conrad) Stauffer, 1915: 104, 245 (not pp. 160, 171, 175, 234); Goldring, 1935: 148, figs. 53B-D; Butts, 1941, pl. 115, figs. 17-21, 35; Cooper, 1944: 333, pl. 127, fig. 37; Dutro, 1971: 187-188, figs. 3, 5; Feldman, 1985: 335-337, fig. 41.

Description.—Exterior: Shells range from small to medium-sized (Table 17), nonstrophic, impunctate and pentagonal in outline, with suberect beak; shells dorsibiconvex with greatest width attained at about two-thirds to three-fourths shell length; raised, rounded fold bearing narrow median groove gives brachial valve cariniform appearance; groove originates at umbo and widens anteriorly, almost imperceptibly in one specimen, forming two subparallel ridges which end at anterior commissure; concave flanks drop steeply adaxially away from sulcate fold; pedicle valve bears broad sulcus that widens anteriorly; defining sulcus laterally are two ridges that extend from umbo across posterolateral margins of flanks to uniplicate anterolateral commissure; numerous, concentric growth lines evenly spaced on entire shell surface; on larger shells growth lines are coarser toward anterior third of the shell.

Pedicle valve interior: Hinge teeth short, blunt and supported by strong dental lamellae at base of which is deposited secondary shell material; muscle field broad, flabelliform.

Discussion: Due to lack of brachial interiors and relatively poor internal preservation, detailed comparisons must be deferred at this time. The Genesee Valley shells, however, are identical to *Pentagonia unisulcata* recovered from the mid-Hudson Valley (Feldman, 1985: fig. 41). *P. unisulcata* from the Genesee Valley is smaller than *Pentagonia peersi* (Dutro, 1971: pl. 1, figs. 9-12; pl. 2, figs. 1-3, 5-12) and larger than the subovate *P. lenta* (Dutro, 1971: pl. 1, figs. 5-8; pl. 2, fig. 4).

TABLE 17. Measurements (in mm) of *Pentagonia unisulcata* (Conrad, 1841). See *Systematic Paleontology: Introduction: Measurement Abbreviations and Subscripts* for explanations.

AMNH loc.	(L)	(W)	(T)	Specimen type
3152	8.6	11.0	5.7	a.v.
3152	14.1	17.3	9.0	a.v.
3152	–	20.7	11.3 dh	a.v.
3152	14.7	20.9	9.8	a.v.

Material: Five articulated shells, two pedicle valves.

Occurrence: AMNH Loc. 3152.

Superfamily NUCLEOSPIROIDEA Davidson, 1881
Family NUCLEOSPIRIDAE Davidson, 1881
Genus NUCLEOSPIRA Hall, 1859
Type species: Spirifer ventricosa Hall, 1857: 57.

Nucleospira ventricosa (Hall, 1857), Plate 6, figures 14-23.

Spirifer ventricosa Hall, 1857: 57, not figs. 1-2.

Nucleospira ventricosa Hall, 1859: 220-221, pl. 14, figs. 1a-h, pl. 28B, figs. 2-9; Hall and Clarke, 1894: pl. 48, figs. 2-6, 18; Weller, 1903: 209, pl. 30, figs. 19-22; Schuchert, 1913: 430, pl. 73, figs. 10-12; Bowen, 1967: 37-38, pl. 5, figs. 16-17.

Nucleospira sp. Boucot and Johnson, 1968: H14, pl. 5, figs. 1-11.

Nucleospira aff. *ventricosa* Feldman, 1985: 337-339, fig. 42.

Description.—Exterior: Biconvex shells small (Table 18) and transversely suboval in outline with curved hinge line; pedicle and brachial beaks erect with concave pseudodeltidium covering delthyrium in a few specimens; one large adult shell has short, extremely narrow and shallow pedicle sulcus with corresponding brachial fold; all other shells lack sulcus and fold; although radial ornamentation is lacking, there are concentric growth lines present, concentrated toward rectimarginate anterior commissure.

Pedicle valve interior: Hinge teeth small, pointed dorsally, unsupported by dental lamellae; delthyrium enclosed by concave pseudodeltidium; very low, thin median septum, most prominent in posterior half of valve, extends almost entire valve length beginning in umbonal cavity; muscle scars not preserved.

Brachial valve interior: Cardinal process relatively large with anterior margin corrugated such that it looks like an "M" in plan view; medial surface of cardinal process is scyphiform with lateral margins converging posteriorly, such that the apex of the cardinal process is deeper than in the anterior region; sockets shallow but appear to be deep due to raised cardinal process; muscle impressions, poorly preserved in only one specimen, consist of two slightly diverging striae (about 25 degrees) bisected by low myophragm; anterior border of muscle field extends to about 30 percent of valve length.

Discussion: These shells are identical to those collected from AMNH locs. 3137 and 3138A in the mid-Hudson Valley (Feldman, 1985: fig. 42). *Nucleospira ventricosa* from the Keyser Limestone (Bowen, 1967: 37, pl. 5, figs. 16-27) differs only in the shape of the cardinal process which may be due to intraspecific variation (see Feldman, 1985: 338 for further discussion).

Specimens of *Nucleospira ventricosa* from the New Scotland Formation and equivalents (Cooper, 1944: 331, pl. 127, figs. 8-9) have a somewhat different cardinal process in that they are subovate in outline. *Nuleospira* sp. from the Bois Blanc Formation (Boucot and Johnson, 1968: B14, pl. 5, figs. 1-11) is morphologically identical to the Genesee Valley shells. Specimens of *Nucleospira* aff. *ventricosa* from the Moose River Synclinorium, Maine (Boucot, 1973: 64, pl. 20, figs. 23-27), are too poorly preserved to be certain that they are indeed *ventricosa,* mainly because the cardinal process is not well preserved (see fig. 25); in all other respects they are identical to the Onondaga shells.

Material: Eighty articulated shells, eleven pedicle valves, three brachial valves.

Occurrence: AMNH Locs. 3152, 3153, 3154.

Order RETZIIDINA Alvarez and Jia-Yu
Suborder RETZIIDINA Boucot, Johnson, and Staton, 1964
Superfamily RHYNCHOSPIRINOIDEA Schuchert, 1929
Family RHYNCHOSPIRINIDAE Schuchert, 1929
Genus TREMATOSPIRA Hall, 1859
Type species: Trematospira gibbosa Hall, 1859: 272.

Trematospira gibbosa Hall, 1859, Plate 6, figures 24-33.

Trematospira gibbosa Hall, 1859: 272, pl. 45, figs. 7-15; Shimer and Shrock, 1944: 361, pl. 141, figs. 21-24.

TABLE 18. Measurements (in mm) of *Nucleospira ventricosa* (Hall, 1857). See *Systematic Paleontology: Introduction: Measurement Abbreviations and Subscripts* for explanations.

AMNH loc.	(L)	(W)	(T)	Specimen type
3152	8.7	9.4	5.3	a.v.
3152	8.3	8.7	4.7	a.v.
3152	7.6	7.9	4.5	a.v.
3152	8.0	7.5	4.5	a.v.
3152	4.8	5.3	2.8	a.v.
3152	8.8	8.7	4.6	a.v.
3152	10.1	10.3	6.1	a.v.
3152	8.7	8.4	5.3	a.v.
3152	7.4	7.0	4.9	a.v.
3152	8.5	8.7	5.5	a.v.
3152	7.9	8.3	5.6	a.v.
3152	5.7	5.9	2.9	a.v.
3152	6.7	7.0	3.7	a.v.
3152	5.8	6.8	3.2	a.v.
3152	8.0	7.3	4.1	a.v.
3152	6.4	7.0	3.3	a.v.
3152	8.6	8.9	5.1	a.v.
3152	6.9	7.7	4.5	a.v.
3152	9.5	10.0	5.0	a.v.
3152	6.9	7.9	4.1	a.v.
3152	8.9	9.6	5.4	a.v.
3152	8.5	8.5	5.1	a.v.
3152	8.2	7.6	4.4	a.v.
3152	9.9	10.1	5.0	a.v.
3152	7.4	7.1	4.4	a.v.
3152	7.0	7.5	3.7	a.v.
3152	7.7	7.8	4.2	a.v.
3152	7.8	8.1	4.3	a.v.
3152	8.4	9.5	5.2	a.v.
3152	7.1	8.3	4.8	a.v.
3152	9.4	8.6	5.5	a.v.
3152	6.6	6.9	3.7	a.v.
3152	7.3	9.3	4.8	a.v.
3153	10.4	10.7	6.7	a.v.
3153	9.1	9.4	5.6	a.v.
3153	9.5	9.3	5.9	a.v.
3153	10.8	12.5	7.7	a.v.
3153	9.6	10.6	5.3	a.v.

Description. Of four (articulated) specimens in the collection (three are silicified), one is biconvex, the second unequally biconvex with pedicle valve about twice as convex as brachial valve, the third crushed, and the fourth with only a trace of the pedicle valve; shell transversely subelliptical in outline, rostrate, with suberect beak; pedicle foramen mesothyridid; deltidium partially obscured by incurved brachial umbo, covers delthyrium; maximum width attained at approximately midlength; pedicle valve bears distinct sulcus, which becomes more clearly defined anteriorly, while brachial valve has corresponding, though somewhat less defined, fold; ornamentation consists of strong, angular, chevron-like costae, usually nine on brachial valve and 10 on pedicle valve; pedicle sulcus bears two costae while three are found on fold; two costae in sulcus distinctly smaller than others on pedicle valve exterior, especially the two adjacent to the sulcus; anterior commissure uniplicate and crenulated by costae.

Discussion: Trematospira camura from the Rochester Shale, Lockport, New York (NYSM E1890), is biconvex, less gibbous, and has less prominent costae (Table 19). *Trematospira multistriata* from the New Scotland Limestone, Schoharie County, New York (NYSM E2786) is considerably larger (L = 20.8 mm, W = 26.4 mm [pedicle valve]), more finely costate and has a broader, shallower sulcus. *Trematospira multistriata* from the Oriskany Sandstone of Becraft Mountain, Hudson, New York (NYSM 12122), is more

TABLE 19. Measurements (in mm) of *Tremtospira gibbosa* Hall, 1859 and *T. camura* (Hall, 1850). See *Systematic Paleontology: Introduction: Measurement Abbreviations and Subscripts* for explanations.

						Number of Plications			
AMNH loc.	Specimen type	(L) p.v.	(L) b.v	(W)	(T)	p.v.	b.v.	sulcus	fold
T. gibbosa									
3152	a.v.	9.4	8.2	11.0	5.9	10	9	2	3
3152	a.v.	10.4	8.9	10.5	7.12	10	9	2	3
3152	a.v.d.	8.5	7.3	9.0	–	9	9	2	3
3153	a.v.d.	–	10.1	11.9	–	–	9	–	3
T. camura									
NYSM E1890	a.v.	8.4	7.4	11.2	4.9	13	15	2	2
NYSM E1890	a.v.	7.4	6.9	8.7	5.6	11	10	2	2
NYSM E1890	a.v.	8.7	6.9	11.0	5.0	13	14	2	3
NYSM E1890	a.v.	6.1	5.9	8.2	4.2	10	10	2	3

gibbous and robust than *Trematospira gibbosa*. Specimens of *Trematospira* sp. from the Glenerie Formation just south of Glenerie, New York are medium-sized (average of two pedicle valves: L = 14.7 mm, W = 19.5 mm; average of two brachial valves: L = 13.2 mm, W = 19.0 mm), and possess more costae (14-15 per pedicle valve; 16-17 per brachial valve); this gives the appearance of being more finely costate than *Trematospira gibbosa*. These shells, however, lack a distinct sulcus and fold.

Johnson (1970: 179-181, pl. 52, figs. 8-25) described *Trematospira perforata* from the Great Basin of Nevada and noted that *Trematospira multistriata* is a related species from the Helderberg Group of New York. The specimens of *Trematospira perforata* from Nevada range in shape from elongate suboval to pyriform in juveniles, to transversely suboval in adults. The ornamentation differs from *Trematospira gibbosa* in that the costae are finer, more numerous (about 30 on USNM 157151 [brachial valve]) and display a greater degree of bifurcation (evident on *T. gibbosa* only at the lateral margins). One specimen of *Trematospira perforata* (NYSM E2862, L = 12.2 mm, W = 15.4 mm [brachial valve]) from the Port Ewen Limestone in Rondout, New York, has 20 costae on the brachial valve, which also increase by bifurcation at the lateral margins of larger ones. Another specimen of *Trematospira perforata* from the New Scotland Formation of Becraft Mountain, Hudson, New York (NYSM 12125) is not as alate as *Trematospira gibbosa* and has six costae in the sulcus.

Material: Twenty articulated shells.

Occurrence: AMNH Loc. 3152.

Order SPIRIFERIDA Waagen, 1883
Suborder DELTHYRIDINA Ivanova, 1972
Superfamily DELTHYRIDOIDEA Phillips, 1841
Family ACROSPIRIFERIDAE Termier and Termier, 1949
Subfamily ACROSPIRIFERINAE Termier and Termier, 1949
Genus ACROSPIRIFER Helmbrecht and Wedekind, 1923
Type species: Spirifer primaevus Steininger, 1853,
by subsequent designation of Wedekind, 1926: 202.

Acrospirifer duodenaria (Hall, 1843), Plate 7, figures 1-4.

Delthyris duodenaria Hall, 1843: 171, fig. 5.

Spirifer duodenaria Hall, 1867: 189, pls. 27-28; Landes, Ehlers, and Stanley, 1945: pl. 12, fig. 4.

Hysterolites (Acrospirifer) worthenanus? Amsden *in* Amsden and Ventress, 1963: 182, pl. 16, figs. 1-4, 6-8, 11-16, 5?, 9?, 10?

Acrospirifer duodenaria Boucot and Johnson, 1968: B14-15, pl. 5, figs. 12-39; Feldman, 1985: 341-342, fig. 46.

Description.—Exterior: Four fairly well preserved but incomplete silicified specimens are available for study and serve as the basis for the following description. Shells transversely subelliptical in outline with straight hinge line, at which point maximum width is attained; pedicle interarea low, narrow, Apsacline, and bears open delthyrium; brachial interarea apsacline and very narrow; triangular, smooth sulcus originates in umbonal area on pedicle valve; corresponding flattened fold found on brachial valve; four rounded plications with U-shaped interspaces found on each pedicle valve flank, while brachial valve bears five plications on each flank; one faint growth line evident on brachial valve exterior; no fine radial ornamentation preserved.

Pedicle valve interior: Hinge teeth short, blunt and unsupported by dental lamellae; area just below and posterior to teeth somewhat thickened; valve floor crenulated due to impress of plications.

Brachial valve interior: Notothyrial cavity chipped and incompletely preserved in all three brachial valves in the collection; no evidence of cardinal process, nor muscle scars; sockets short, deeply excavated, almost teardrop shaped and laterally directed; short, stubby crural bases which join inner margins of sockets evident in one specimen; valve floor crenulated due to impress of plications.

Discussion: The Genesee Valley shells are identical to those collected from the mid-Hudson Valley (Feldan, 1985: 341-342, fig. 46) but are not quite as well preserved. The occurrence of *Acrospirifer duodenaria* in western New York supports Hall's (1867: 189-190) claim that the species is known throughout "all the extent of the formation within the state."

Acrospirifer murchisoni, described by Boucot (1973: 41-46, pl. 16, figs. 19-25), from the Moose River Synclinorium, Maine, differs in its larger size and wider pedicle interarea. *Acrospirifer atlanticus,* also described by Boucot (1973: 46-47, pl. 17, figs. 1-9), is larger and more alate.

Boucot et al. (1970: 14, pl. 4, figs. 22-26) illustrated a form of *Acrospirifer?* sp. from the Green Pond Outlier that is considerably less transverse in outline. Johnson (1970: 189-190, pl. 56, figs. 5-13; pl. 57, figs. 1-6) described *Acrospirifer* aff. *murchisoni* from the Great Basin, Nevada, that is less alate, more transversely suboval, and commonly has six plications on the pedicle flank.

Material: One pedicle valve, three brachial valves.

Occurrence: AMNH Loc. 3152.

Family MUCROSPIRIFERIDAE Boucot, 1959
Subfamily MUCROSPIRIFERINAE Boucot, 1959
Genus MUCROSPIRIFER Grabau, 1931
Type species: Delthyris mucronatus Conrad, 1841: 54.

Mucrospirifer? sp., Plate 7, figures 5-6.

Description: A single silicified brachial valve is assigned to *Mucrospirifer* sp. based on the following description: Shell large (L = 22 mm [est.]; W = 57 mm [est.]), strongly mucronate (although missing lateral extremities), with prominent, low fold with slight concavity evident originating in umbonal region and widening and deepening anteriorly; 21-23 rounded costae crossed by two prominent growth lines; finer, numerous growth lines visible under low (10x) magnification, particularly on fold; sockets fairly long, U-shaped, widely divergent; cardinal process represented by low concavity to the sides of which extend narrow, concave (anacline) interarea; muscle field narrowly triangular, widening anteriorly and bisected by low, rounded myophragm, which merges imperceptibly with valve floor at about midlength; interior of valve floor crenulated due to impress of costae.

Material: One brachial valve.

Occurrence: AMNH Loc. 3152.

Superfamily CYRTOSPIRIFEROIDEA Termier and Termier, 1949
Family SPINOCYRTIIDAE Ivanova, 1959
Genus ALATIFORMIA Struve, 1963
Type species: Spirifer alatiformis Drevermann, 1907: 126.

Alatiformia? sp., Plate 7, figures 7-9.

Description: Shells small (Table 20), alate, transverse in outline, with flat striated interareas that were moderately high before compaction; delthyrium covered by chert, hence no deltidial plates evident; pedicle valve

TABLE 20. Measurements (in mm) of *Alatiformia?* sp. See *Systematic Paleontology: Introduction: Measurement Abbreviations and Subscripts* for explanations.

AMNH 1oc.	(L)	(W)	(T)	Specimen type
3152	8.8	9.1	6.1	a.v.
3152	10.7	18.6 d	5.5	a.v.
3152	8.1	15.2 dh	4.1	a.v.

bears well defined, deep sulcus, while brachial valve bears corresponding sharp, flat-topped fold; seven to eight costae cover each flank, crossed by numerous growth lamellae, clearly evident in sulcus and on fold; ornamentation identical to that of *Alatiformia varicosus* (Hall) (USNM 51202) from the Onondaga Limestone in Clarke County, Indiana. Morphology conforms to genus as described by Struve (1964: 326-328).

Material: Three articulated shells.

Occurrence: AMNH Loc. 3152.

Genus MEDIOSPIRIFER Bublichenko, 1956

Type species: Delthyris medialis Hall, 1843: 208.

Mediospirifer sp. A, Plate 7, figures 10-15.

Description.—Exterior: Shells incomplete anteriorly, but definitely transverse; largest pedicle valve very long making, outline almost semicircular; pedicle valve has characteristic high, very slightly concave interarea covered with lateral striae; delthyrium triangular with adjacent ridge *indicating* presence of delthyrial plates; well-developed shallow sulcus with about 20 fine costae, separated by narrow U-shaped interspaces, covering pedicle flank; corresponding low fold on brachial exterior; occasional growth lamellae cross costae.

Pedicle valve interior: No hinge teeth evident, but two distinct, strong dental lamellae present which diverge at an angle of about 55 degrees; muscle field, defined by base of dental plates, teardrop shaped; valve floor crenulated by impress of costae.

Brachial valve interior: Sockets shallow, widely divergent; no distinct cardinal process evident; valve floor crenulated by impress of costae.

Discussion: The shells more closely resemble *Mediospirifer fornaculus* (Hall) from Watson Station, Indiana (USNM 232452) than do typical *Mediospirifer audaculus* from the Hamilton Group of New York, studied in Hall's collection (AMNH), which have more widely divergent dental lamellae. The presence of dental lamellae supporting the hinge teeth rules out assignment to *Mucrospirifer*, which lacks dental plates.

Material: Three pedicle valves, three brachial valves.

Occurrence: AMNH 3152.

Mediospirifer? sp. B, Plate 7, figures 16-20.

Description: Fine external ornamentation not well preserved due to etching in acid; 13 rounded costae on each pedicle flank and 14 on each brachial flank; costae separated by shallow, almost V-shaped interspaces and crossed by numerous growth lines evident only adjacent to pedicle sulcus; brachial valve bears corresponding flat-topped fold; both sulcus and fold originate in beak region; shell transverse, almost semicircular in outline and ventribiconvex in lateral profile; pedicle interarea flat (with some indications of lateral striae) except near beak, where it becomes concave, and strongly apsacline; brachial interarea ribbon-like and appears to be anacline; delthyrium triangular and encloses angle of approximately 40 degrees; narrow ridge along one delthyrial margin indicating existence of plate.

Discussion: Specimens of *Mediospirifer audaculus* (Conrad) (USNM 275323) from the Wanakah Formation and *M. fornaculus* (Hall) (USNM 232542) from Watson Station, Indiana, have higher, flatter interareas, are more finely costate (17-25 costae per flank) and more robust than the Onondaga shell.

The Onondaga shell resembles *Spinocyrtia "euryteines"* from the Middle Devonian Silica Formation (UMMP 61090A-C) bit differs in its smaller size, more mucronate alae, and greater length.

Material: One articulated shell.

Occurrence: AMNH Loc. 3153.

Family CYRTINOPSIDAE Wwedekind, 1926
Subfamily CYRTINOPSINAE Wedekind, 1926
Genus MEGAKOZLOWSKIELLA Boucot, 1957
Type species: Spirifer perlamellosus Hall, 1857: 57.

Megakozlowskiella raricosta (Conrad, 1842), Plate 7, figures 21-26.

Spirifer perlamellosus Hall, 1857: 57.

Delthyris raricosta Conrad, 1842: 262, pl. 14, fig. 18.

Spirifer raricosta Hall, 1867: 192, pl. 27, figs. 30-34; pl. 30, figs. 1-9.

Kozlowskiella (Megakozlowskiella) raricosta Boucot, 1957: pl. 3, figs. 18-19.

Megakozlowskiella cf. *raricosta* Johnson, 1970: 204, pl. 70, figs. 26-28.

Megakozlowskiella raricosta Boucot and Johnson, 1968: B16, pl. 6, figs. 7-15; Feldman, 1985: 345-349, figs. 51-52.

Description: Shells medium-sized (Table 21), subtransverse in outline, strophic, and ventribiconvex; hinge line straight, with maximum width usually reached at or just anterior to hinge line, but in some shells maximum width occurs at midlength; pedicle interarea exposed on articulated shells very narrow and apsacline, but on free pedicle valves wide and concave; narrow brachial interarea concealed by overhanging pedicle umbo; pedicle valve bears strong U-shaped sulcus, while brachial valve bears corresponding, somewhat flattened fold; both sulcus and fold originate in umbonal area; commonly three plications on flanks of both valves that become narrower laterally; between plications are U-shaped interspaces; delthyrium, including an angle of about 60 degrees, open on all specimens, but on two shells there are ridges on sides of delthyrium, which may be indicative of plates; anterior commissure uniplicate; strong, coocentric growth lines possess anterior frills (numbering five to 13 per 5 mm) in well-preserved specimens; three juveniles have unusually large number of plications (eight on pedicle valve, nine on brachial valve), which leads me to suspect that as ontogeny progresses some lateral plications lost; adults, as a rule, have a maximum of six plications.

Pedicle valve interior: Small, pointed hinge teeth supported by thin dental lamellae that diverge medially from their point of attachment adjacent to median septum, in posterior portion of valve; thin median septum incomplete, but appears to have been fairly high, extending half the length of the valve; no muscle scars preserved; crenulations of plicae impressed on valve floor.

Brachial valve interior: Cardinal process bilobed and longitudinally striated; deeply excavated dental sockets are U-shaped, widening out anteriorly; outer socket ridges smooth and diverge at angle of about 30 degrees: crural plates small and somewhat concave, extending anteroventrally a short distance from notothyrial platform; no muscle scars preserved; interior of valve strongly corrugated, reflecting impress of plications.

Discussion: The shells are identical with those collected in the mid-Hudson Valley (Feldman, 1985, figs. 51-52) but not as well preserved.

Megakozlowskiella magnapluera from the Great Basin, Nevada (Johnson, 1970: 202, pl. 71, figs. 1-19) is more subquadrate in outline and has fewer plications. *Megakozlowskiella raricosta* was described by Hall (1867: 192, pl. 27, figs 30-34; pl. 30, figs. 1-9) from the Schoharie Grit (Helderberg Mountains and Schoharie, New York) and the Onondaga Limestone at Stafford, Caledonia, and Williamsville (western New York). He also reported occurrences at Columbus, Ohio, Falls of the Ohio, and Canada West.

Material: Twenty articulated shells, thirty four pedicle valves, sixteen brachial valves.

Occurrence: AMNH Locs. 3152, 3153.

Family HYSTEROLITIDAE Termier and Termier, 1949
Subfamily HYSTEROLITINAE Termier and Termier, 1949
Genus PARASPIRIFER Wedekind, 1926
Type species: Spirifer cultrijugatus Roemer, 1844:70.

Paraspirifer? sp., Plate 7, figure 27

Description: A large, robust spiriferid, suggestive of *Paraspirifer,* is embedded in a block of chert with one flank projecting out of the matrix.

TABLE 21. Measurements (in mm) of *Megakozlowskiella raricosta* (Conrad, 1842). See *Systematic Paleontology: Introduction: Measurement Abbreviations and Subscripts* for explanations.

AMNH loc.	(L)	(W)	(T)	Plications p.v.	Plications b.v.	Frills No./5 mm	Specimen type
3152	19.0 d	30.0	12.3	6	–	9	a.v.
3152	17.5	21.4	18.2	4	5	5	a.v.
3152	19.0	25.0 dh	11.9	4	5	9	a.v.
3152	7.2	14.0 dh	5.0	8	9	9	a.v.
3152	14.8 d	24.7	–	6	–	8	a.v.
3152	25.8	26.5 dh	16.2	6	5	13	a.v.
3152	–	28.2	10.9	–	–	12	a.v.
3152	15.3	17.3	8.0	5	4	8	a.v.
3152	21.2	21.5	13.8	–	6	–	a.v.
3152	8.7	16.2 dh	6.0	8	9	9 t	a.v.
3153	9.6	22.8 dh	7.2	20	21	13	a.v.

One-half of high fold is evident and there is assumed to be corresponding deep sulcus. Shell silicified with 10 coarse radial costae on one flank of brachial valve and three on fold; the four costae closest to fold bifurcate; bifurcations more pronounced anteriorly but faint due to shell erosion; costae with U-shaped interspaces; shell closely resembles *P. acuminatus* from the Hamilton Group collected near Fultonham, Schoharie County, New York (AMNH 5171), but is smaller. *P. acuminatus*, figured by Godefroid and Fagerstrom (1983), is almost identical to the shell illustrated here in terms of morphology and size. There are vague indications of concentric lamellae, but no granules are evident in this specimen.

Material: One articulated (?) shell.

Occurrence: AMNH Loc. 3153.

Family ELYTHIDAE Frederiks, 1924
Subfamily ELYTHINAE Frederiks, 1924
Genus ELITA Frederiks, 1918
Type species: Delthyris fimbriatus Conrad, 1842: 263.

Elita fimbriata (Conrad, 1842), Plate 7, figures 28-32; Plate 8, figures 1-3.

Delthyris fimbriatus Conrad, 1842: 263.

Spirifer fimbriata Hall, 1867: 214, pl. 33, figs. 1-21.

Elytha sp. Boucot and Johnson, 1968: B18, pl. 7, figs. 1-5.

Elytha fimbriata Goldring, 1943: 236, fig. 43J; Cooper, 1944: 327, pl. 126, figs. 1-3; Feldman, 1985: 349-350, fig. 54.

Description.—Exterior: Shells medium-sized (Table 22), ventribiconvex with short hinge line; beak erect but short; maximum width attained at about midlength; pedicle interarea small, low, and apsacline, while brachial interarea represented by thin strip, which appears to be apsacline; distinct shallow sulcus, originating in umbonal region, found on pedicle valve, while brachial valve bears corresponding low, rounded fold; low plications, U-shaped in cross section, cover lateral slopes, more clearly developed on ephebic forms; plications separated by U-shaped interspaces; concentric growth lamellae cross plications, becoming more numerous anteriorly; each lamella bears single row of medially grooved spines (11-12 per 5 mm), up to 2.7 mm in length, most of which are lost due to abrasion, in specimens still enclosed in limestone matrix, however, spines are visible (see Plate 8, Figure 3), anterior commisure uniplicate.

TABLE 22. Measurements (in mm) of *Elita fimbriata* (Conrad, 1842). See *Systematic Paleontology: Introduction: Measurement Abbreviations and Subscripts* for explanations.

AMNH loc.	(L)	(W)	(T)	No. of plications per flank	Specimen type
3152	19.4	28.5	13.3	5 t	a.v.
3152	21.5	30.6 dh	15.4	7	a.v.
3152	18.3	26.2	15.1	6	a.v.
3152	23.0	32.9	14.9	6 t	a.v.
3152	18.4	22.4	9.8	5 t	a.v.
3152	14.9	22.4 dh	–	6	a.v.
3152	15.6	18.6	7.3	6	a.v.
3152	17.8	27.0	11.5	–	a.v.
3152	18.4	24.3	11.2	–	a.v.
3152	14.0	20.8	–	–	a.v.
3152	15.6	18.6	7.3	–	a.v.
3152	17.0	25.2	10.2	–	a.v.
3152	14.8	18.1	6.9	–	a.v.
3152	12.0	15.9	–	–	a.v.

Pedicle valve interior: Hinge teeth short, pointed, supported by dental lamellae, which extend to valve floor and then anterolaterally; delthyrium open with no indication of modifying plates; low myophragm extends anteriorly from delthyrial cavity, just under one-third valve length, and passes through muscle field, which is very faintly impressed by anteriorly directed striations; valve floor crenulated due to impress of plications.

Brachial valve interior: Shallow sockets broaden and diverge anterolaterally; medially concave, broad crural plates partially adjoin valve floor but do not unite into aseptalium; neither muscle field nor myophragm preserved, although in specimens collected from mid-Hudson Valley, adult forms possessed short, low myophragm; valve floor, especially anteriorly, crenulated by impress of plications.

Discussion: Elita fimbriata described by Feldman (1985: fig. 54) is identical to the Genesee Valley shells. Boucot and Johnson (1968: B18, pl. 7, figs. 1-5) described specimens of *Elytha* sp. from the Bois Blanc Formation; these were assigned by Feldman (1985: 349-350) to *Elytha fimbriata* based on their oval shape, short hinge line, length of the myophragm, and presence of dental lamellae. Hall reported the occurrence of *Elita fimbriata* from the Oriskany Sandstone at Saugerties, New York, Knox in Albany County, in the

Schoharie Grit in Albany and Schoharie counties, and in the Onondaga Limestone in Cherry Valley, as well as at numerous other localities across the state and into Ohio.

Material: Thirteen articulated shells, thirty pedicle valves, eighteen brachial valves.

Occurrence: AMNH Locs. 3152, 3153.

Superfamily AMBOCOELIOIDEA George, 1931
Family AMBOCOELIIDAE George, 1931
Subfamily AMBOCOELIINAE George, 1931

Ambocoeliid indet., Plate 8, figure 4-6.

Description: Shell small (L = 6.4 mm; W = 7.3 mm; T = 4.3 mm), ventribiconvex, subcircular in outline; both valve exteriors smooth, nonspinose, with no growth lamellae evident; pedicle valve bears narrow, shallow sulcus that originates in umbonal region and extends to slightly uniplicate anterior commissure; greatest width attained at about midlength; beak and interarea poorly preserved.

Discussion. The Onondaga shell is similar to *Emanuella* described by Goldman and Mitchell (1990: 90-94) in its ventribiconvex shape and uniplicate anterior commissure, but differs in its lack of striations and large size. *Ambocoelia* (Goldman and Mitchell, 1990: 83) differs in that it is either planoconvex or concavoconvex and has a rectimarginate anterior commissure. *Crurispina,* also described by Goldman and Mitchell (1990: 95-96) is similar in size and shape (ventribiconvex) but is covered externally with numerous short spines. It is possible that the Onondaga specimen is a weathered *Crurispina,* but this seems unlikely, as Goldman and Mitchell (1990: 95) noted that the normally spinous shell is pitted when exfoliated. *Ambocoelia* sp. from the mid-Hudson Valley (Feldman, 1985: 350-351, fig. 56) differs in its much higher pedicle valve.

Material: One articulated shell.

Occurrence: AMNH Loc. 3152.

Suborder CYRTINIDINA Carter and Johnson, 1994
Superfamily CYRTINOIDEA Frederiks, 1911
Family CYRTINIDAE Frederiks, 1911
Genus CYRTINA Davidson, 1858
Type species: Cyrtina hamiltonensis Hall, 1857: 166.

Cyrtina hamiltonensis (Hall, 1857), Plate 8, figures 7-10.

Cyrtia hamiltonensis Hall, 1857: 166.

Cyrtina hamiltonensis Hall, 1867: 268, pl. 27, figs. 1-4; pl. 44, figs. 26-33, 38-52; Grabau and Shimer, 1909: figs. 393a-c; Prosser and Kindle, 1913: 185-187, pl. 17, fig. 109; Clarke and Swartz, 1913: 591-592, pl. 56. figs. 1-3; Branson and Williams, 1924: 149, pl. 19. figs. 5-8; Stewart, 1927: 43, pl. 3, figs. 27-28; Goldring, 1935: figs. 62c-d; Stumm, 1942: pl. 81, figs. 7-8; Ehlers, 1963: 198-199, pl. 1, figs. 1-12; Feldman, 1985: 351-353, figs. 57A-E.

Cyrtina hamiltonensis·var. *recta* Hall, 1867: 270, pl. 44, figs. 34-37; Hall and Clarke, 1895: pl. 28, figs. 21-22.

Cyrtina hamiltoniae var. *recta* Hall, Nettleroth, 1889: 97, pl. 13, figs. 13-16.

Description.—Exterior: Shells small (Table 23), ventribiconvex, hemipyramidal in outline; many slightly deformed due to compaction; maximum width attained just anterior to straight hinge line; pedicle interarea high, smooth, and apsacline; delthyrium obscured in shells studied, but appears to be covered by a convex pseudodeltidium; smooth, moderately deep sulcus originating deep in umbonal region found on pedicle valve, while brachial valve bears corresponding fold; in some specimens concentric growth lamellae are found concentrated near uniplicate anterior commissure.

Brachial valve interior: Sockets shallow, widely divergent; their outer boundary forms narrow interarea which overlaps socket posteriorly; cardinal process eroded posteriorly but appears to be bilobed, triangular in outline, and supported by thickened secondary shell material rising from valve floor, which is crenulated due to impress of plications; no muscle scars preserved.

Discussion: Cyrtina hamiltonensis from the Genesee Valley differs from *Cyrtina hamiltonensis* from the Traverse Group of Michigan (Keyes and Pitrat, 1978: pl. 1, figs. 11-15) only in that the Michigan specimens are essentially catacline. *Cyrtina alpenensis alpenensis* (Keyes and Pitrat, 1978: pl. 1, figs. 1-5) differs in that it has a smooth umbonal area, a character that separates it from all other species of *Cyrtina* in the Traverse Group. The specimens of *Cyrtina hamiltonensis* recovered from the mid-Hudson Valley (Feldman, 1985: figs. 57A-E) are identical to those found in western New York. *Cyrtina umbonata* (Cooper, 1944: pl. 140, figs. 40-42) is smaller and has a more incurved umbo. *Cyrtina* cf. *varia* Clarke, 1900 illustrated by Johnson (1970: pl. 73, figs. 1-14) has a shallower sulcus, more acute cardinal angles, and is less semi-circular in dorsal view.

TABLE 23. Measurements (in mm) of *Cyrtina hamiltonensis* (Hall, 1857). See *Systematic Paleontology: Introduction: Measurement Abbreviation and Subscripts* for explanations.

				Number of plications		
AMNH loc.	(L)	(W)	(T)	dorsal	ventral	Specimen type
3152	5.4	8.9	7.6	6	6	a.v.
3152	4.8	6.9	5.9	6	6	a.v.
3152	7.6	11.9	8.3	8	8	a.v.
3152	6.7	10.5	8.6	–	8	a.v.
3152	5.4	9.5	4.0	6	6	a.v.
3152	4.9	7.8	4.7	8	–	a.v.
3152	7.9	9.4	7.0	6	6	a.v.
3152	6.1	10.0	7.7	–	–	a.v.
3152	5.3	8.8 dh	–	–	6	a.v.

Material: Ten articulated shells, three pedicle valves, two brachial valves.

Occurrence: AMNH Loc. 3152.

Cyrtina sp. A, Plate 8, figure 8.

Description: A single highly compacted shell, much larger than *Cyrtina hamiltonensis* (L [brachial valve] = 10.3 mm, L [pedicle valve] = 18.6 mm, W = 16.4 mm) with 12 dorsal and 12 ventral plications, is assigned to *Cyrtina* sp. A until more material becomes available for study.

Material: One articulated specimen.

Occurrence: AMNH Loc. 3152.

Order TEREBRATULIDA Waagen, 1883
Suborder TEREBRATELLIDINA Waagen, 1883
Superfamily CRYPTONELLOIDEA Thomson, 1926
Family CRYPTONELLIDAE Thomson, 1926
Subfamily CRYPTONELLINAE Thomson, 1926
Genus CRYPTONELLA Hall, 1861
Type species: Terebratula rectirostra Hall, 1860: 88.

Cryptonella? sp., Plate 8, figures 12-17.

Description: Shells small (Table 24), smooth, elongate subpyriform to slightly subpentagonal in outline, biconvex to ventribiconvex and lenticular in profile; greatest width attained at about two-thirds valve length:

anterior commissure ranges from rectimarginate to sulcate; lateral margins rectimarginate, but in one specimen sinuate posteriorly; pedicle foramen permesothyridid; beak suberect to erect; delthyrium covered by conjunct deltidial plates, observable in only one specimen; no growth lamellae noticeable.

Discussion: The shells resemble *Cryptonella reimanni* (Cloud, 1942: 130-131, pl. 23, figs. 9-18) more than any other species, but differ in lack ofgrowth lamellae, which may be due to exposure and exfoliation. Specific designation must be deferred until better material becomes available for study.

Material: Six articulated shells.

Occurrence: AMNH Loc. 3152.

Family CRANAENIDAE Cloud, 1942
Subfamily CRANAENINAE Cloud, 1942
Genus CRANAENA Hall and Clarke, 1893
Type species: Terebratula romingeri Hall, 1863: 48.

Cranaena? sp., Plate 8, figures 18-19.

Description: Shell small (L = 13.4 mm; W = 9.0 mm; T = 5.6 mm), smooth, covered by occasional faint growth lamellae, ventribiconvex; pedicle valve bears shallow sulcus that originates at about midlength, while brachial valve bears corresponding fold that seems to be present throughout its entire length, although the right side of the brachial valve (when viewed dorsally) is compressed, making differentiation of fold somewhat questionable; anterior commissure sulcate and lateral commissures sinuate posteriorly; beak suberect and foramen mesothyridid; delthyrium covered by dorsal umbo; deltidial plates not visible.

TABLE 24. Measurements (in mm) of *Cryptonella?* sp. See *Systematic Paleontology: Introduction: Measurement Abbreviations and Subscripts* for explanations.

AMNH loc.	(L)	(W)	(T)	Specimen type
3152	15.9	11.5	4.6	a.v.
3152	15.8	12.3	6.6	a.v.
3152	14.4	10.4	6.0	a.v.
3152	10.8	6.8 dh	3.9	a.v.
3152	11.8	–	5.1	a.v.
3152	–	13.6	12.4	a.v.

Discussion: Cranaena romingeri (Cloud, 1942: 138-139, pl. 24, figs. 2-12) is smaller, more gibbous, subcircular to subovate in outline, and has a rectimarginate lateral commissure. The Genesee Valley shell more closely resembles *Cranaena schucherti* (Cloud, 1942: 139-140, pl. 23, figs. 24-31; pl. 24, fig. 1) in its lenticular appearance and size.

Material: One articulated shell.

Occurrence: AMNH Loc. 3152.

APPENDIX

Localities Cited in this Report

AMNH Locality 3152: Extensive bedding plane exposure in eastern portion of abandoned quarry, immediately west of Perry Road, 1.2 km (0.7 mi) north of B & O Railroad tracks, 3.2 km (2.0 mi) northeast of bridge in Le Roy, Genesee County, New York; USGS Le Roy quadrangle, 7.5 minute series [topographic], N4/4 Caledonia 15 ft quadrangle; Lower Moorehouse Member.

AMNH Locality 3153: Small bedding plane exposures along southeastern portion of active General Crushed Stone Co. Quarry, immediately south of Honeoye Falls Road, 5.5 km (3.4 mi) east of intersection with U. S. Route 15, 5.5 km (3.4 mi) WSW of Honeoye Falls, Monroe County, New York; USGS Rush quadrangle, 7.5 minute series [topographic], NW/4 Honeoye 15 ft quadrangle; Upper Moorehouse Member.

AMNH Locality 3154: Scattered exposures at southwest corner of active quarry operated by Penfield Dolomite Company, immediately south of Gulf Road, 2.1 km (1.3 mi) east of Perry Road, Le Roy, Genesee County, New York; USGS Le Roy quadrangle (see above); Lower Moorehouse Member.

REFERENCES

Amsden, T. W. 1964. Brachial plate structure in the brachiopod family Pentameridae. *Palaeontology* 7: 220-239.

Amsden, T. W., and Ventress, W. P. S. *1963. Early Devonian brachiopods of* Oklahoma. *Oklahoma Geological Survey Bulletin* 94: 1-216.

April, R., Selleck, B., and Aitaner, S. 1984. Clay mineralogy of the Onondaga Limestone, central and western New York State. *Northeastern Geology* 6: 83-87.

d'Archiac, E. J. A. D., and de Verneuil, M. E. 1842. On the fossils of the older deposits in the Rhenish Provinces, preceded by a general survey of the fauna of the Paleozoic rocks and followed by a tabular list of the organic remains of the Devonian System in Europe. *Transactions of the Geological Society of London* 6: 303-410.

Baird, G. C., and Brett, C. E. 1986. Submarine erosion on the dysaerobic seafloor: Middle Devonian corrasional disconformities in the Cayuga Valley region. *New York State Geological Association Field Trip Guidebook, Ithaca, New York* 58: 23-80.

Bassett, M. G. 1974. The articulate brachiopods from the Wenlock Series of the Welsh Borderland and South Wales. *Palaeontographical Society Monographs* 128: 79-122.

Bengtson, P. 1988. Open nomenclature. *Palaeontology* 31: 223-227.

Billings, E. 1860. Descriptions of some new species of fossils from the Lower and Middle Silurian rocks of Canada. *Canadian Naturalist* 5: 49-69.

Billings, E. 1874. Paleozoic fossils. *Geological Survey of Canada* 2: 1-144.

Boucot, A. J. 1957. Revision of some Silurian and Early Devonian spiriferid genera and erection of Kozlowskiellinae, new subfamily. *Senckenbergiana Lethaea* 38: 311-334.

Boucot, A. J. 1958. *Kozlowskiellina*, New name name for *Kozlowskiella* Boucot, 1957. *Journal of Paleontology* 32: 1030.

Boucot, A. J. 1959. Brachiopods of the Lower Devonian rocks at Highland Mills, New York. *Journal of Paleontology* 33: 727-769.

Boucot, A. J. 1973. Early Paleozoic brachiopods of the Moose River Synclinorium, Maine. *U. S. Geological Survey Professional Paper* 784: 1-81.

Boucot, A. J., and Harper. C. W. 1968. Silurian to Lower Middle Devonian Chonetacea. *Journal of Paleontology* 42: 143-176.

Boucot, A. J., and Johnson, J. G. 1967. Species and distribution of *Coelospira* (Brachiopoda). *Journal of Paleontology* 41: 1226-1241.

Boucot, A. J., and Johnson, J. G. 1968. Brachiopods of the Bois Blanc Formation in New York. *U. S. Geological Survey Professional Paper* 584B: 1-27.

Boucot, A. J., Gauri, K. L., and Southard, J. 1970. Silurian and Lower Devonian brachiopods, structure and stratigraphy of the Green Pond Outlier in southeastern New York. *Palaeontographica* 135: 1-59.

Boucot, A. J., Johnson, J. G., and Staton, R. D. 1964. On some atrypoid, retizoid and athyridoid brachiopods. *Journal of Paleontology* 38: 805-822.

Bowen, Z. P. 1967. Brachiopods of the Keyser Limestone (Silurian-Devonian) of Maryland and adjacent areas. *Geological Society of America Memoir* 102: 1-103.

Branson, E. B., and Williams, J. S. 1924. Fauna of the Devonian of southeastern Missouri, in The Devonian of Missouri. In Branson, E. B. [ed.], *Missouri Bureau of Geology and Mines*, ser. 2, vol. 17: 130-165.

Breger, C. L. 1906. On *Eodevonaria*, a new subgenus of *Chonetes*. *American Journal of Science* 22: 534-536.

Bronn, H. G. 1862. Die Klassen und Ordnungen der Welchthiere (Malacozoa) 3: 1-518.

Bublichenko, N. L. 1956. Nekotorye novyi predstaviteli brakhiopod Devona i Karbona Rudnogo Altaya i Sary-arka, Akademiia Nauk Kazakhskoy SSR. *Izvestiya. Seria Geologicheskaya* 23: 93-104.

Buch, L. von. 1834. Uber terebrateln. Akademie Wissenschaften zu Berlin, Abhandlungen, Jahrgang 1833, *Physikalische Klasse*, 21-144.

Butts, C. 1941. Geology of the Appalachian Valley in Virginia, part II: Fossils, plates and explanations. *Virginia Geological Survey Bulletin* 52: 1-271.

Caster, K. E. 1939. A Devonian fauna from Columbia. *Bulletins of American Paleontology* 24 (83): 1-128.

Clarke, J. M. 1908. Early Devonic history of New York and eastern North America. *New York State Museum Memoir* 9: 5-366.

Clarke, J. M., and Swartz, C. K. 1913. Systematic paleontology of the Upper Devonian deposits of Maryland. *Maryland Geological Survey, Middle and Upper Devonian*, 539-699.

Cloud, P. E., Jr. 1942. Terebratuloid Brachiopoda of the Silurian and Devonian. *Geological Society of America Special Paper* 38: 1-182.

Conrad, T. A. 1838. Report on the paleontological department of the survey. *New York State Geological Survey Annual Report* 2: 107-119.

Conrad, T. A. 1841. Fifth annual report on the paleontology of the State of New York. *New York State Geological Survey Annual Report* 5: 25-57.

Conrad, T. A. 1842. Observations on the Silurian and Devonian systems of the United States, with descriptions of new organic remains. *Academy of Natural Sciences Philadelphia Journal* 8: 228-280.

Cooper, G. A. 1942. New genera of North American brachiopods. *Journal of Washington Academy of Sciences* 32: 228-235.

Cooper, G. A. 1944. Phylum Brachiopoda. In H. W. Shimer and R. R. Schrock (eds.), *Index Fossils of North America.* Cambridge, MA: The MIT Press.

Cooper, G. A. 1945. New species of brachiopods from the Devonian of Illinois and Missouri. *Journal of Paleontology* 19: 479-489.

Cooper, G. A. 1955. New genera of Middle Paleozoic brachiopods. *Journal of Paleontology* 29: 45-63.

Cooper, G. A. 1956. Chazyan and related brachiopods. *Smithsonian Miscellaneous Collections* 127: 1-1245.

Cooper, G. A., and Dutro, J. T., Jr. 1982. Devonian brachiopods of New Mexico. *Bulletins of American Paleontology* 82-83 (315): 1-215.

Cozzens, I. 1846. Descriptions of three new fossils from the Falls of the Ohio. *New York Lyceum of Natural History Annals* 4: 157-159.

Dagis, A. S. *1974.* Triasovie Brakhiopody (Morfologia, Sistema, Filogenii, Stratigraficheskoe Znachenie i Biogeografiia). *Akademia Nauk SSR, Sibirskoe Otdelenie, Institut Geologii i Geofiziki, Trudy* 214: 1-386.

Dalman, J. W. 1828. Uppställning och Beskrfning af di i Sverige funne Terebratuliter. *Kongliga Svenska Vetenskaps-akademiens Handlinger, Upssala and Stockholm* (1827), 85-155.

Davidson, T. 1851-1886. *A Monograph of the British fossil Brachiopoda.* 6 vols. London: Palaeontographical Society Monographs.

Davidson, T. 1881. On genera and species of spiral-bearing Brachiopoda. *Geological Magazine* 8: 1-13.

Drevermann, F. 1907. Palaozoische Notizen. Senckenbergische Naturforschellde gesellschaft in Franfurt am Main. Bericht, 125-136.

Duméril, A. M. C. 1806. *Zoologie analytique ou méthode naturelle de classification des animaux.* Paris: Allais.

Dutro, J. T., Jr. 1971. The brachiopod *Pentagonia* in the Devonian of eastern United States. (Paleozoic Perspectives: a paleontologictribute to G. Arthur Cooper). *Smithsonian Contributions to Paleobiology* 3: 181-192.

Ehlers, G. M. 1963. *Cyrtina hamiltonensis* (Hall) and a new species of this brachiopod genus from New York. *Contributions from the Museum of Paleontology, University of Michigan* 18: 197-204.

Fagerstrom, J. A. 1961. The fauna of the Middle Devonian Formosa Reef Limestone of southwestern Ontario. *Journal of Paleonlology* 35: 1-48.

Fagerstrom, J. A. 1971. Brachiopods of the Detroit River Group (Devonian) from southwestern Ontario and adjacent areas of Michigan and Ohio. *Geological Survey of Canada Bulletin* 204: 1-113.

Feldman, H. R. 1980. Level-bottom brachiopod communities in the Middle Devonian of New York. *Lethaia* 13: 27-46.

Feldman, H. R. 1985. Brachiopods of the Onondaga Limestone in central and southeastern New York. *American Museum of Natural History Bulletin* 179: 289-377.

Feldman, H. R., and Lindemann, R. H. 1986. Fossils and facies of the Onondaga Limestone in central New York. *New York State Geological Association Field Trip Guidebook* 58: 145-156.

Frederiks, G. 1912. Bemerkungen über einige oberpalaeozoische Fossilien von Krasnoujimsk. *Kazani Protocoly Obshchestva estestvoisputatelei* 69: 1-9.

Frederiks, G. 1918. Diagnoses generum et specierum novum. *Annuaire de la Societe Paleontologique de Russie* 2: 87.

George, T. N. 1931. *Ambocoelia* Hall and certain similar British Spiriferidae. *Geological Society of London, Quarterly Journal* 87: 30-61.

Gill, G. H. 1871. Arrangement of the families of Molluscs prepared for the Smithsonian Institution. *Smithsonian Miscellaneous Collections* 10: 1-49.

Girty, G. H. 1904. New molluscan genera from the Carboniferous. *United States National Museum Proceedings* 27: 721-736.

Godefroid, J., and Fagerstrom, J. A. 1983. Le genre *Paraspirifer* Wedekind, R. 1926 dans le devonien moyen de la partie orientale de L'Amérique du Nord. *Institut royal des Sciences naturelles de Belgique Bulletin* 55: 1-61.

Goldman, D., and Mitchell, C. E. 1990. Morphology, systematics and evolution of Middle Devonian Ambocoeliidae (Brachiopoda), western New York. *Journal of Paleontology* 64: 79-98.

Goldring, W. 1935. Geology of the Berne Quadrangle. *New York State Museum Bulletin* 303: 1-238.

Goldring, W. 1943. Geology of the Coxsackie Quadrangle. *New York State Museum Bulletin* 332: 1-374.

Grabau, A. W. 1906. Geology and paleontology of the Schoharie Valley. *New York State Museum Bulletin* 92: 1-386.

Grabau, A. W. 1931, 1933. Devonian brachiopods of China; 1. Devonian brachiopods from Yunnan and other districts in South China. *Palaeontologica Sinica*, ser. 8, vol. 3: 1-752.

Grabau, A. W., and Shimer, H. W. 1909 (1910). *North American Index Fossils, Invertebrates.* 2 vols. New York: A. G. Seiler.

Grant, R. E. 1965. The brachiopod superfamily Stenoscismatacea. *Smithsonian Miscellaneous Collections* 148: 1-192.

Guensburg, T. E. 1984. Echinodermata of the Middle Ordovician Lebanon Limestone. central Tennessee. *American Paleontology Bulletin* 86: 1-100.

Hall, J. 1843. Geology of New York, Part 4, comprising of the survey of the fourth geological district. *Natural History of New York, Albany.*

Hall, J. 1850. On the Brachiopoda of the Silurian: particularly the Leptaenida. *American Association for the Advancement of Science* 2: 347-351.

Hall, J. 1852. Descriptions of the organic remains of the lower middle division of the New York System. *New York State Geological Survey, Paleontology* 2.

Hall, J. 1857. Appendix, Descriptions of Paleozoic fossils. *New York State Cabinet 10th Annual Report, Part C,* 41-186.

Hall, J. 1859, 1861. Paleontology of New York. *New York Geological Survey* 3: 1-532 (1859), pls. 1-120 (1861).

Hall, J. 1860. Contributions to palaeontology. *New York State Cabinet of Natural History Annual Report* 13: 55-125.

Hall, J. 1861. Descriptions of new species of fossils from the Upper Helderberg, Hamilton and Chemung groups; with observations upon previously described species. *New York State Cabinet of Natural History, 14th Annual Report,* 99-109.

Hall, J. 1862. Contributions to palaeontology. *New York State Cabinet of Natural History, 15th Annual Report,* 1-158.

Hall, J. 1863. Notice of some new species of fossils from a locality of the Niagara Group in Indiana, with a list of identified species from the same place. *Transactions of the Albany Institute* 4: 195-228.

Hall, J. 1867. Descriptions and figures of the fossil Brachiopoda of the Upper Helderberg, Hamilton, Portage and Chemung groups. *New York Geological Survey, Palaeontology of New York* 4: 1-428.

Hall, J., and Clarke, J. M. 1892. An introduction to the study of the genera of Paleozoic Brachiopoda, part I. *New York Geological Survey, Palaeontology of New York* 8: 1-367.

Hall, J., and Clarke, J. M. 1893, 1895 (1894). An introduction to the study of the genera of Paleozoic Brachiopoda, part II. *New York Geological Survey, Palaeontology of New York* 8: 1-317, pl. 2 (1893); 319-394, pls. 21-84 (1895).

Harper, C. W., Jr., and Boucot, A. J. 1978. The Stropheodontacea. *Palaeontographica* 161: 55-175; 162: 1-80.

Havlicek, V. 1953. O Nékolika nových rameononožcích českého a moravského stredniho devonu. *Ustředniho, Ustavu Geologickeho, Věstník* 28: 4-9.

Havlicek, V. 1960. Bericht über die Ergebnisse der Revision der böhmischen altpaläozioschen Rhynchonelloidea. *Ustředniho, Ustavu Geologickeho, Věstník* 35: 241-244.

Helmbrecht, W., and Wedekind, R. 1923. Versuch einer biostratigraphischen Gliederung der Siegener Schichten auf Grund von Rensselaerien und Spiriferen. *Gluckauf, Berg. -und Hütlenmännische Zeitschr., Jahrgang* 59: 949-953.

Hoover, P. R. 1981. *Paleontology, taphonomy, and paleoecology of the Palmarito Formation (Permian of Venezuela).* American Paleontology Bulletin 80 (313): 1-138.

Imbrie, J. 1959. Brachiopods of the Traverse Group (Devonian) of Michigan. American Museum of Natural History Bulletin 116: 349-409.

Johnson, J. G. 1966. Middle Devonian brachiopods from the Roberts Mountains, central Nevada. *Palaeontology* 9: 152-181.

Johnson, J. G. 1970. Great Basin Lower Devonian Brachiopoda. *Geological Society of America, Memoir* 121.

Kelly, F. B. 1967. Silurian leptaenids (Brachiopoda). *Palaeontology* 10: 590-602.

Keyes, S. W., and Pitrat, C. W. 1978. Spiriferid brachiopods from the Traverse Group of Michigan: Cyrtinacea. *Journal of Paleontology* 52: 221-233.

Kindle, E. M. 1901. The Devonian fossils and stratigraphy of Indiana. *Indiana Department of Geology and Natural Resources, Annual Report* 25: 529-758, 773-775.

King, W. 1846. Remarks on certain genera belonging to the class Palliobranchiata. *Annals and Magazine of Natural History, London* 18: 26-42, 83-94.

King, W. 1850. A monograph of the Permian fossils of England. *Palaeontographical Society Monographs* 3: 1-258.

Kissling, D. L., and Moshier, S. O. The subsurface Onondaga Limestone: Stratigraphy, facies, and paleogeography. *New York State Geological Association Guidebook for Fieldtrips in South-Central New York* 53: 279-280.

Kornicker, L. S. 1979. The question mark in taxonomic literature. *Journal of Paleontology* 53: 761.

Kuhn, O. 1949. *Lehrbuch der Paläozoologie.* Stuttgart: E. Schweizerbart'sche Verlagsbuchhandlung.

Landes, K. K., Ehlers, G. M., and Stanley, G. M. 1945. Geology of the Mackinac Straits region and the subsurface geology of the northern southern peninsula. *Michigan Geological Survey Division*, Publication 44, Geology Series 37: 1-204.

Lenz, A. C. Upper Silurian and Lower Devonian brachiopods of Royal Creek, Yukon, Canada. Part I: Orthoidea, Strophomenida, Pentamerida, Rhynchonellida. *Palaeontographica* 159: 37-109.

Lindemann, R. H., and Feldman, H. R. 1981. Paleocommunities of the Onondaga Limestone (Middle Devonian) in central New York State. *New York State Geological Association Guidebook for Fieldtrips in South-Central New York* 53: 79-96.

Lindemann, R. H., and Feldman, H. R. 1987. Paleogeography and brachiopod paleoecology of the Onondaga Limestone in eastern New York. *New York State Geological Association Field Trip Guidebook* 59: D1-D30.

Linnaeus, C. 1758. *Systema Naturae.* 10th ed. Stockholm: Laurentius Salvius.

Linnaeus, C. 1767. *Systema Naturae.* 12th ed. Stockholm: Laurentius Salvius.

Lucas, S. C. 1986. Proper syntax when using aff. and cf. in taxonomic statements. *Journal of Vertebrate Paleontology* 6: 202.

Martin, W. 1809. *Petrificata derbiensia; or figures and descriptions of petrifactions collected in Derbyshire.* London: Wigan.

Matthews, S. C. 1973. Notes on open nomenclature and on synonomy lists. *Palaeontology* 16: 713-719.

Mayr, E. 1976. *Evolution and the Diversity of Life.* Cambridge, MA: The Belknap Press of Harvard University Press.

McLaren, D. J. 1965. Paleozoic Rhynchonellacea. In R. C. Moore (ed.), *Treatise on Invertebrate Paleontology, Part H, Brachiopoda*, H552-H597. Lawrence, KS: University of Kansas Press and Geological Society of America.

M'Coy, F. 1844. *A Synopsis of the Characters of the Carboniferous Limestone Fossils of Ireland.* London: Williams & Norgate.

Moore, R. C. 1952. Brachiopods. In R. C. Moore, C. G. Lalicker, and A. G. Fischer (eds.), *Invertebrate Fossils*, 197-267. New York: McGraw-Hill.

Muir-Wood, H. M. *1955. A History of the Classification of the Phylum Brachiopoda.* London: British Museum [Natural History].

Muir-Wood, H. M. 1962. *On the Morphology and Classification of the Brachiopod Suborder Chonetoidea.* London: British Museum [Natural History].

Nettleroth, H. 1889. Kentucky fossil shells, a monograph of the fossil shells of the Silurian and Devonian rocks of Kentucky. *Kentucky Geological Survey.*

Nikilorova, O. I., Modzalevskaya, T. L., and Bassett, M. G. 1985. Review of the Upper Silurian and Lower Devonian articulate brachiopods of Podolia. *Special Papers in Palaeontology*, no. 34. London: The Palaeontological Association..

Oehlert, D. P. 1887. Brachiopodes. In P. H. Fischer, *Manuel de Conchyliologue et du Paleontologie Conchyliologieque au histoire naturelle des Mollusques vivants et f ossiles*, 1189-1334. Paris: F. Savy.

Oliver, W. A., Jr. 1954. Stratigraphy of the Onondaga Limestone (Devonian) in central New York. *Geological Society of America Bulletin* 65: 621-652.

Oliver, W. A., Jr. 1956. Stratigraphy of the Onondaga Limestone in eastern New York. *Geological Society of America Bulletin* 67: 1441-1474.

Öpik, A. A. 1934. Über Klitamboniten. *Acta et Commeniationes Universitatis Tartuensis (Dorpatensis)*, ser. A, 26: 1-239.

Ozol, M. A. 1963. Alkali reactivity of cherts and stratigraphy and petrology of cherts and association limestones of the Onondaga Formation of central and western New York. Unpublished PhD dissertation, Rensselaer Polytechnic Institute, Troy, New York.

Phillips, J. 1841. *Figures and Descriptions of the Paleozoic Fossils of Cornwall, Devon, and West Somerset.* London: Geological Survey of Great Britain.

Pitrat, C. W. 1965. Spiriferidina. InR. C. Moore (ed.), *Treatise on Invertebrate Paleontology. Part H, Brachiopoda*, H667-H728. Lawrence, KS: University of Kansas Press and Geological Society of America.

Prosser, C. S., and Kindle, E. M. 1913. Systematic paleontology of the Middle Devonian deposits of Maryland: Pelecypoda. In *Middle and Upper Devonian. Maryland Geological Survey.*

Racheboeuf, P. R., and Feldman, H. R. 1990. Chonetacean brachiopods of the "Pink *Chonetes*" Onondaga Limestone, (Devonian, Eifelian), central New York. *American Museum Novitates* 2974: 1-86.

Raup, D. M., and Stanley, S. M. 1978. *Principles of Paleontology.* New York: W. H. Freeman and Co.

Richter, R. *1943. Einfuhrung in die Zoologische Nomenclature.* Frankfurt: Senckenbergische Naturforschende gesellschaft.

Richter, R. 1948. *Einfuhrung in die Zoologische Nomenclature.* 2nd edition. Frankfurt: Kramer.

Roemer, C. F. 1844. *Das Rheinische Uebergangsgebirge. Eine paläontologisch-geognostiche Darstellung.* Hanover: Hahn'sche Hofbuchhandlung.

Rzhonsnitskaya, M. A. 1964. On Devonian atrypids of the Kuznetsk Basin. Paleontologiya i Stratigraphiya, Nauchno-isseldovatel'shii Institut Lesnogo Khozyaistva. *Trudy*, vol. 93: 91-112.

Sartenaer, P. 1961. Etude nouvelle, en deux parties, du genre *Camarotoechia* Hall et Clarke, 1893. Deuxieme partie: Cupularostrum recticostatum n. gen., n. sp. *Institut royal des Sciences naturelles de Belgique Bulletin* 37: 1-15.

Savage, T. E. 1930. The Devonian rocks of Kentucky. *Kentucky Geological Survey* 33: 1-161.

Savage, T. E. 1931. The Devonian fauna of Kentucky. *Kentucky Geological Survey, Paleontology of Kentucky* 36: 217-246.

Schuchert, C. 1894. A revised classification of the spire-bearing Brachiopoda. *American Geologist* 13: 102-107.

Schuchert, C. 1913. Systematic paleontology. Lower Devonian, Brachiopoda. In Lower Devonian Volume (with T. P. Maynard). *Maryland Geological Survey*, 290-449.

Schuchert, C., and Cooper, G. A. 1931. Synopsis of the brachiopod genera of the suborders Orthoidea and Pentameroidea with notes on the Teleotremata. *American Journal of Science* 22: 241-251.

Schuchert, C., and Cooper, G. A. 1932. Brachiopod genera of the suborders Orthoidea and Pentameroidea. *Peabody Museum of Natural History, Memoir* 4: 1-270.

Schuchert, C., and LeVene, C. M. 1929. Brachiopoda (Generum et Genotyporum Index et Bibliographica). *Fossilium Catalogus I, Animalia* 42: 1-140.

Selleck, B. 1985. Chert and dolomite in the Onondaga Limestone (Devonian) of New York State. *Northeastern Geology* 7: 136-143.

Shaler, N. S. 1865. List of Brachiopoda from the Island of Anticosti sent by the Museum of Comparative Zoology to different institutions in exchange for other specimens, with annotations. *Harvard University Museum of Comparative Zoology Bulletin* 1: 61-70.

Sheehan, P. M. 1971. Silurian Brachiopoda, community ecology and stratigraphic geology in western Utah and eastern Nevada, with a section on Late Ordovician stratigraphy. PhD Thesis, University of California, Berkeley.

Shimer, H. W., and Shrock, R. R. 1944. *Index Fossils of North America.* Cambridge, MA: MIT Press.

Sohn, I. G. 1983. Ostracodes of the "Winifrede" Limestone (Middle Pennsylvanian) in the region of the proposed Pennsylvanian System Stratotype, West Virginia. *American Paleontology Bulletin* 84 (316): 1-53.

Sokolskaya, A. N. 1960. Otriad Strophomendia. In T. G. Sarycheva, [asst. ed.], Mshanki Brakhiopody [Bryozoa, Brachiopoda], Y. A. Orlav, ed., *Osnovy Paleontologii*, 7: 206-220. Moskva: Akademia Nauk SSSR.

Stauffer, C. R. 1915. Stratigraphy of southwestern Ontario. *Geological Survey of Canada Memoir* 34: 1-341.

Stehli, F. G. 1965. Paleozoic Terebratulida. In R. C. Moore (ed.), *Treatise on Invertebrate Paleontology, Part H, Brachiopoda*, H730-H762. Lawrence, KS: University of Kansas Press and Geological Society of America.

Steininger, J. 1853. *Geognostische Beschreibung der Eifel.* Trier: F. Lintz'schen.

Stewart, G. A. 1927. Fauna of the Silica Shale of Lucas County, Ohio. *Ohio Journal of Science Bulletin* ser. 4., vol. 32: 1-76.

Stoll, N., Dollfus, R. Ph., Forest, J., Riley, N. D., Sabrosky, C. W., Wright, C. W., and Melville, R. Y. (eds.). 1961. *International Code of Zoological Nomenclature.* London: International Trust for Zoological Nomenclature.

Struve, W. 1963. Bcitragc zur Kenntnis devonischer brachiopoden, 3: *Alatiformia* n.g. (Spiriferacea). *Senckenbergiana Lethaea* 44: 499-500.

Struve, W. *1964.* Über *Alatiformia*—Arten und andere, ausserlich ähnlich Spiriferacea. *Senckenbergiana Lethaea* 45: 325-346.

Stumm, E. C. 1942. Fauna and stratigraphic relations of the Prout Limestone and Plum Brook Shale of northern Ohio. *Journal of Paleontology* 16: 549-563.

Thomson, J. A. 1926. A revision of the subfamilies of the Terebratulidae (Brachiopoda). *Annals and Magazine of Natural History, London* 18: 523-530.

Waagen, W. H. 1882-1885. Salt Range fossils, Part 4 (2), Brachiopoda. *Palaeontologica Indica, Memoir*, ser. 13, vol. 1: 329-770.

Wedekind, R. 1926. Die devonische Formation. In W. Salomon (ed.), Grundzüge der Geologie 2: 194-226. Stuttgart: E. Schweizerbastische, .

Weller, S. 1903. The Paleozoic faunas. *New Jersey Geological Survey Report on Paleontology* 3: 1-462.

White, C. A. 1862. Description of new species of fossils from the Devonian and Carboniferous rocks of the Mississippi Valley. *Boston Society of Natural History, Proceedings* 9: 8-33.

Williams, A. 1953. North American and European stropheodontids: Their morphology and systematics. Geological Society of America Memoir 56: 1-67.

Williams, A., Rowell, A. J., Brunton, C. H. C., and Carlson, S. J. 1997. Brachiopoda. In R. L. Kaesler (ed.), *Treatise on Invertebrate Paleontology, Part H.* 6 vols. Boulder, CO and Lawrence, KS: The Geological Society of America and University of Kansas Press.

Williams, H. S., and Breger, C. L. 1916. The fauna of the Chapman Sandstone of Maine, including descriptions of some related species from the Moose River Sandstone. *U. S. Geological Survey Professional Paper* 89: 1-347.

Winchell, A. 1866. *The Grand Traverse Region.* A report on the geological industrial resources in the Lower Peninsula of Michigan. Ann Arbor: Dr. Chase's Steam Printing House.

Woodrow, D. L., Brett, C. E., and Selleck, B. 1989. Sedimentary sequences in a foreland basin: The New York System. *28th International Geological Congress, American Geophysical Union, Field Trip Guidebook* T156: 1-43.

EXPLANATION OF PLATE 1

Figure .... Page

1-4. *Levenea* cf. *subcarinata* (Hall, 1857). Pedicle exterior, interior, posterior, anterior, AMNH 44150, x1.5, AMNH Loc. 3152..

5-12. Dalejina cf. alsa (Hall, 1863).

5-6. Brachial exterior, interior, AMNH 44151, x3.5.

7. Pedicle exterior, AMNH 44152, x2.

8-11. Pedicle, brachial exterior, anterior, posterior, AMNH 44153, x1.5.

12. Detail of ornamentation AMNH 44154, x2 (all specimens from AMNH Loc. 3152).

13-15. Discomyorthis? sp..

13-14. Brachial valve exterior, pedicle valve exterior, AMNH 44155, x1.5.

15. Pedicle interior with large flabellate(?) muscle field, AMNH 44156, x2 (both from AMNH Loc. *3152).*

16-25. Schizophoria cf. *multistriata* (Hall, 1859-1861)..

16-18. Pedicle, brachial exterior, posterior, AMNH 44157, x1.75, AMNH Loc. 3152.

19-22. Pedicle, brachial exterior, anterior, posterior, AMNH 44158, x3, AMNH Loc. 3152.

23. Pedicle interior, AMNH 44159, x2.5, AMNH Loc. 3152.

24. Pedicle interior, AMNH 44160, x2.5, AMNH Loc. 3153.

25. Brachial interior, AMNH 44161, x4.5, AMNH Loc. 3152.

26-29. Pentamerella arata (Conrad, 1841). Pedicle, brachial exterior, posterior, anterior, AMNH 44162, x2, AMNH Loc. 3152..

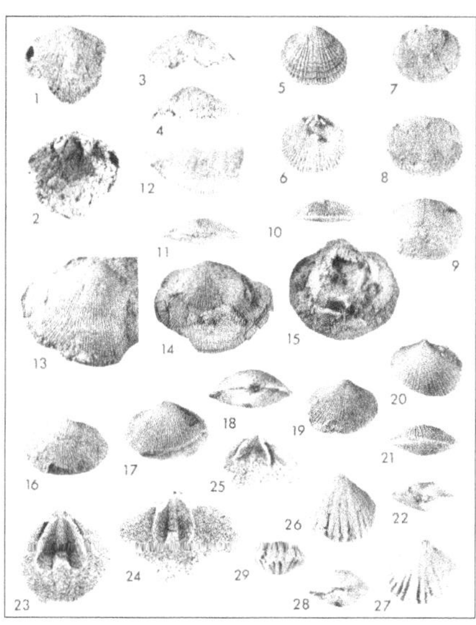

PLATE 1

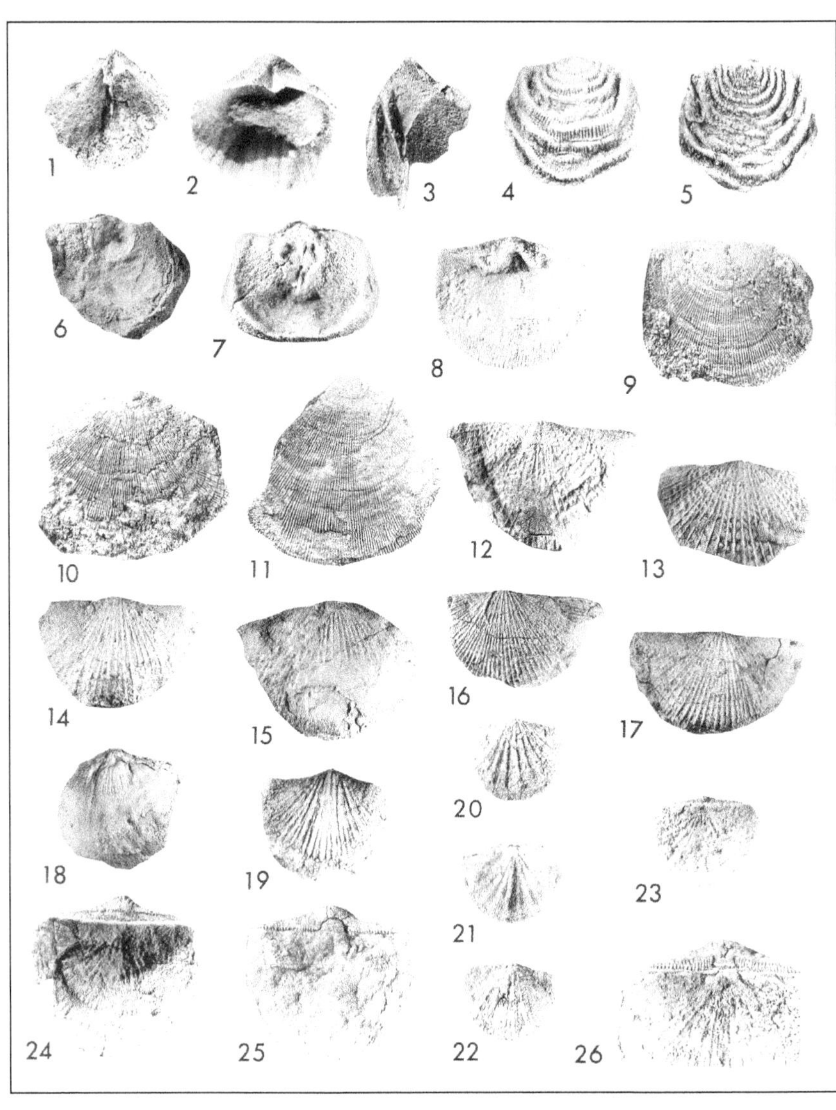

PLATE 2

EXPLANATION OF PLATE 2

Figure .... Page

1-3. *Pentamerella arata* (Conrad, 1841). ....

1. Pedicle interior, AMNH 44163, x2.5.
2. Pedicle interior, AMNH 44164, x2.
3. Brachial interior, AMNH 44265, x3 (all from AMNH Loc. 3152).

4-7. *Leptaena* sp. ....

4-5. Brachial, pedicle exterior, AMNH 44166, x 2.
6. Pedicle interior, AMNH 44167, x1.
7. Brachial interior, AMNH 44168, x1 (all from AMNH Loc. 3152).

8. *"Schuchertella"* sp. Pedicle interior, AMNH 44169, x1.2, AMNH Loc. 3152.

9-11. *Protoleptostrophia?* sp. (Conrad, 1842). ....

9. Pedicle exterior, AMNH 44170, x2.
10. Pedicle exterior, AMNH 44171, x2.
11. Pedicle exterior, AMNH 44172, x2 (all from AMNH Loc. 3152).

12-13. *"Brachyprion"* cf. *mirabilis* (Johnson, 1970). ....

12. Pedicle exterior, AMNH 44173, x1.
13. Pedicle exterior, AMNH 44174, x2.5 (both from AMNH Loc. 3152).

14-17. *Brachyprion*? sp. ....

14. Pedicle exterior, AMNH 44175, x2.
15. Pedicle exterior, AMNH 44176, x1.3.
16. Pedicle exterior, AMNH 44177, x1.2.
17. Pedicle exterior, AMNH 44178, x1.3 (all from AMNH Loc. 3152).

18. *Megastrophia*? sp. Pedicle interior (note impression of muscle field) AMNH 44179, x1.3, AMNH Loc. 3152. ....

19. *Plicostropheodonta*? sp. Pedicle exterior, AMNH 44180, x2, AMNH Loc. 3152.

20-26. *Strophodonta demissa* (Conrad, 1842). ....

20-21. Pedicle, brachial exterior, AMNH 44181, x2.5. ....

22-23. Pedicle, brachial exterior, AMNH 44182, x2.5. ....

24. Brachial exterior, AMNH 44183, x2.
25. Pedicle interior filled with chert, AMNH 44184, x2.
26. Brachial exterior (note psuedodeltidium and pseudoteeth[?]), AMNH 44185, x2 (all from AMNH Loc. 3152).

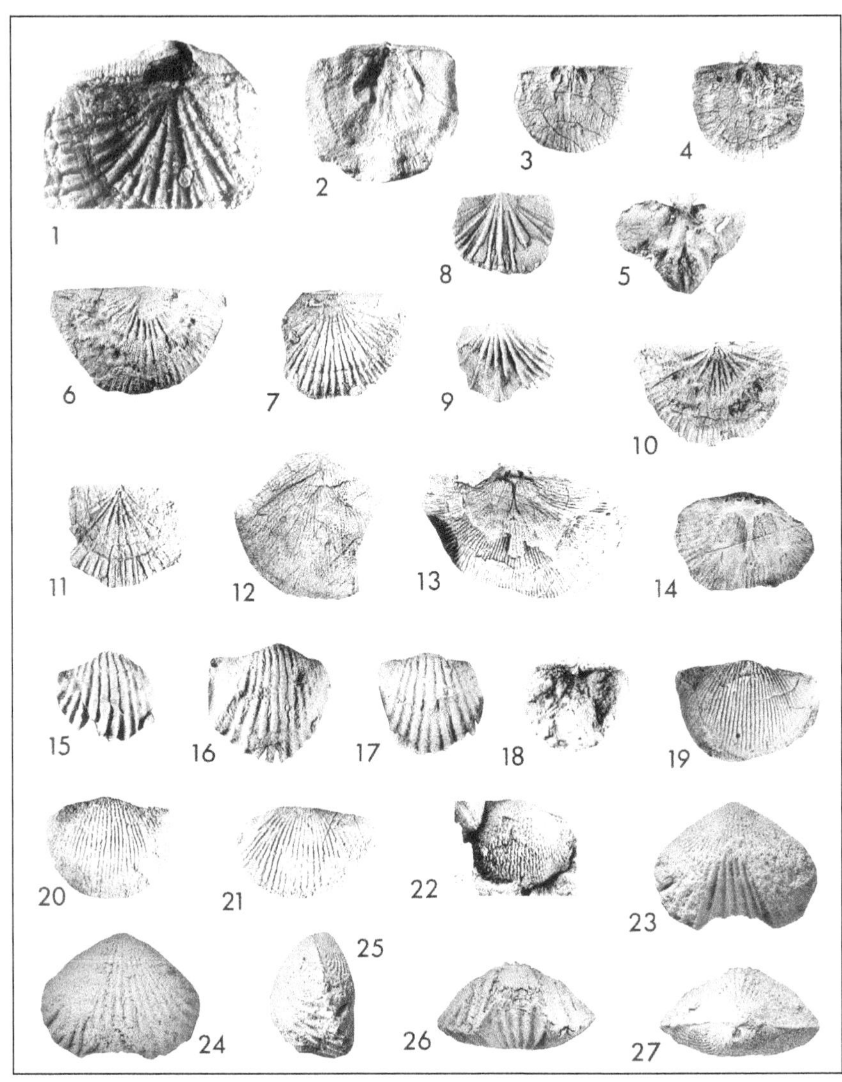

PLATE 3

EXPLANATION OF PLATE 3

Figure .... Page

1-5. *Strophodonta demissa* (Conrad, 1842). ....

1. Brachial exterior, AMNH 44186, x3.
2. Pedicle interior (compacted and deformed), AMNH 44187, 1.5.
3. Brachial interior (cardinal process broken off), AMNH 44188, x1.5.
4. Brachial interior, AMNH 44189, x1.5.
5. Brachial interior, AMNH 44190, x2 (all from AMNH Loc. 3152).

6-11. *Strophonella punctulifera* (Conrad, 1838)..................................

6. Pedicle exterior, AMNH 44191, x1.3.
7. Pedicle exterior, AMNH 44192, x1.3.
8. Pedicle exterior, AMNH 44193, x2.
9. Pedicle exterior, AMNH 44194, x2.
10. Brachial exterior, AMNH 44195, x1.3.
11. Brachial exterior, AMNH 44196, x1.5 (all from AMNH Loc. 3152).

12-14. *Strophonella ampla* (Hall, 1857). ....

12. Pedicle exterior, AMNH 44197, x1.
13. Brachial interior (mold), AMNH 44198, x9.
14. Rubber impression of brachial muscle field, AMNH 44199, x1.3 (all from AMNH Loc. 3152).

15-18. *Longispina mucronata* (Hall, 1843). ....

15. Pedicle exterior, AMNH 44200, x6.
16. Pedicle exterior, AMNH 44201, x4.
17. Pedicle exterior, AMNH 44202, x4.
18. Pedicle interior, AMNH 44203, x5 (all from AMNH Loc. 3152).

19-21. *"Eodevonaria"* cf. *hemispherica* (Hall, 1857).

19. Pedicle exterior, AMNH 44204, x2.
20. Pedicle exterior, AMNH 44205, x2.
21. Pedicle exterior, AMNH 44206, x2.5 (all from AMNH Loc. 3152).

22. *Eodevonaria* cf. *arcuata* (Hall, 1857). Pedicle exterior (in chert matrix), AMNH 44207, x2, AMNH Loc. 3153.

23-27. *Machaeraria* sp. Pedicle, brachial, lateral, anterior, posterior views, AMNH 44208, x2, AMNH Loc. 3152.

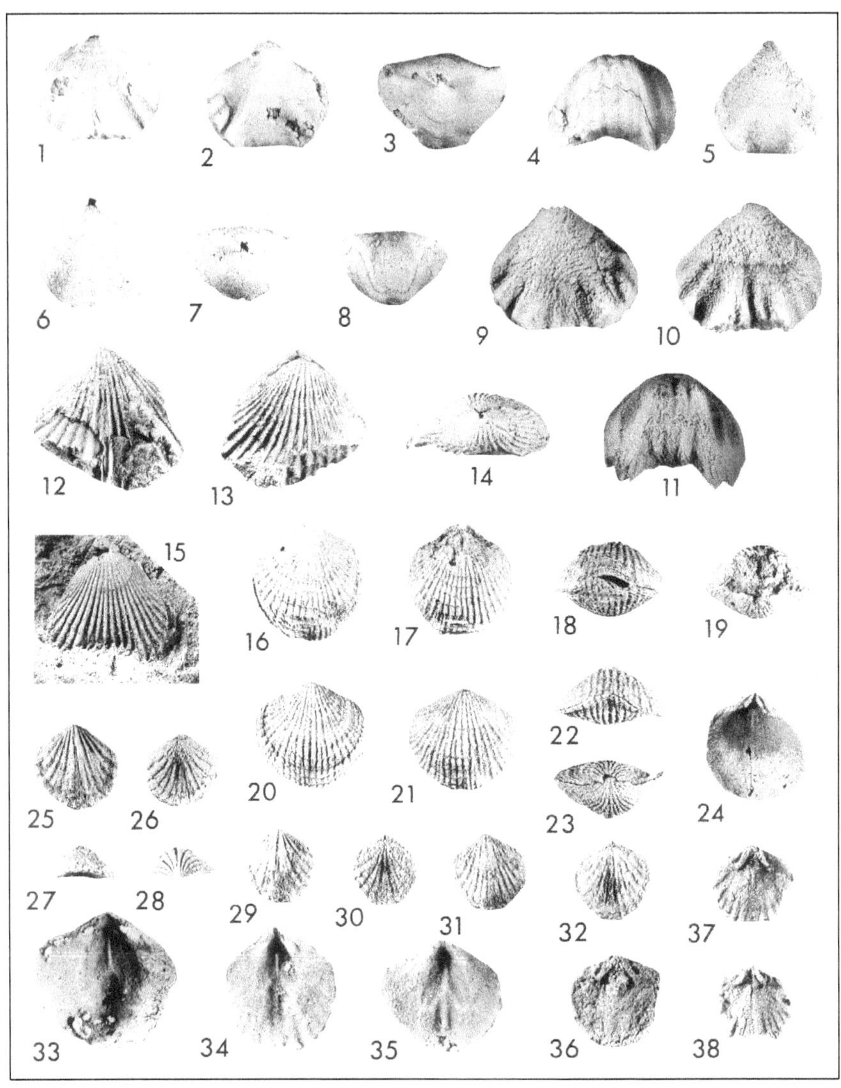

1
2
3
4
5
6
7
8
9
10
12
13
14
11
15
16
17
18
19
22
24
25
26
20
21
23
27
28
29
30
31
32
37
33
34
35
36
38

PLATE 4

EXPLANATION OF PLATE 4

Figure .... Page

1-11. *Atribonium halli* (Fagerstrom, 1961). ....

1-4 Brachial, pedicle, posterior, anterior views, AMNH 44209, x3, AMNH Loc. 3152.

5-8. Pedicle, brachial, posterior, anterior views, AMNH 44210, x4, AMNH Loc. 3152.

9-11. Brachial, pedicle, anterior views, AMNH 44211, x3, AMNH Loc. 3153.

12-15. *Cupularostrum?* sp. ....

12-14. Pedicle, brachial, posterior views, AMNH 44212, x2, AMNH Loc. 3152.

15. Pedicle exterior, AMNH 44213, x2.3, AMNH Loc. 3153.

16-24. *Atrypa "reticularis"* (Linnaeus, 1767). ....

16-19. Pedicle, brachial, anterior, posterior views, AMNH 44214, x1.5, AMNH Loc. 3153.

20-23. Pedicle, brachial, anterior, posterior views, AMNH 44215, x1.5, AMNH Loc. 3153.

24. Brachial interior, AMNH 44216, x1.5, AMNH Loc. 3152.

25-38. *Coelospira camilla* Hall, 1867 ....

25-28. Pedicle, brachial, anterior, posterior views, AMNH 44217, x4.

29-30. Pedicle, brachial exterior, AMNH 44218, x3.

31-32. Pedicle, brachial exterior, AMNH 44219, x3.5.

33. Pedicle interior, AMNH 44220, x5.

34. Pedicle interior, AMNH 44221, x5.

35. Pedicle interior, AMNH 44222, x6.

36. Brachial interior, AMNH 44223, x4.

37. Brachial interior, AMNH 44224, x4.

38. Brachial interior, AMNH 44225, x3 (all from AMNH Loc. 3152).

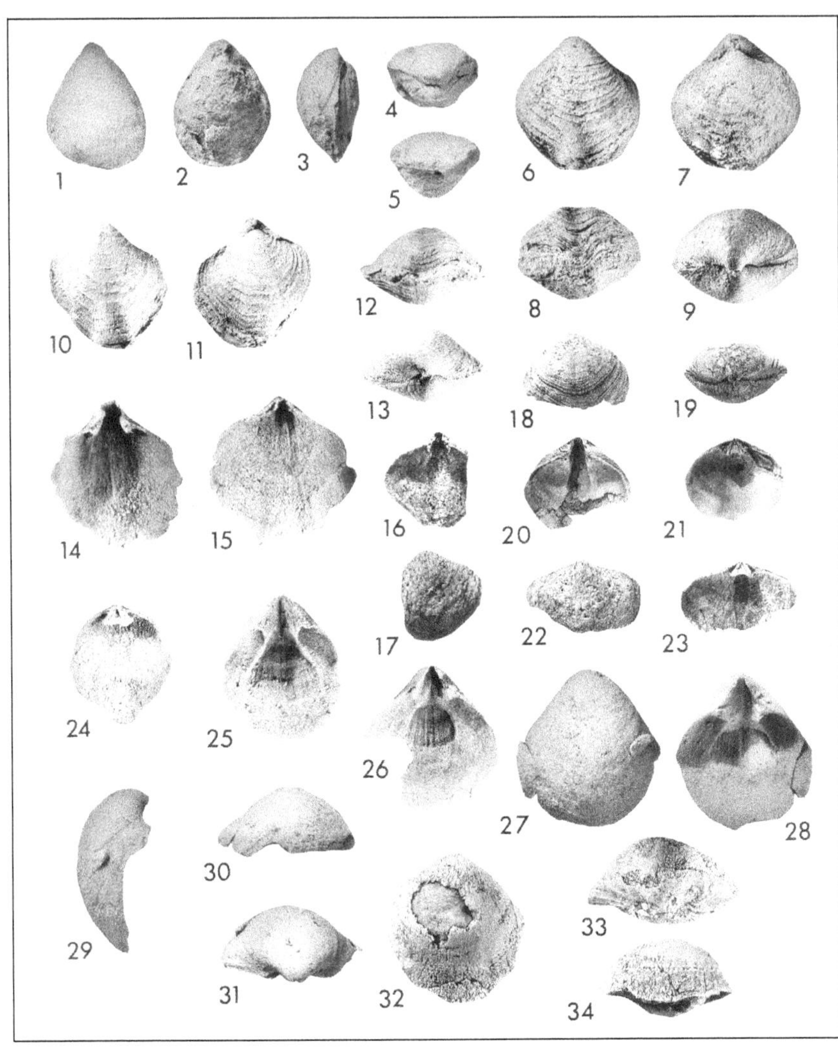

1
2
3
4
5
6
7
10
11
12
8
9
13
18
19
14
15
16
20
21
17
22
23
24
25
26
27
28
29
30
31
32
33
34

PLATE 5

EXPLANATION OF PLATE 5

Figure .... Page

1-5. *Camarospira?* sp. Pedicle, brachial, lateral, posterior, anterior views, AMNH 44226, x2, AMNH Loc. 3153........................

6-15. *Athyris boucoti* new species........................

6-9. Pedicle, brachial, anterior, posterior views, AMNH 44227 (holotype), x2, AMNH Loc. 3153.

10-13. Pedicle, brachial, anterior, posterior views, AMNH 44228 (paratype), x2.5, AMNH Loc. 3153.

14-15. Pedicle, brachial interior, AMNH 44229 (paratype), x4, AMNH Loc. 3152.

16-23. *Athyris leoni* new species........................

16-17. Pedicle interior, exterior, AMNH 44230, x1.

18-21. Pedicle exterior, posterior, pedicle, brachial interior, AMNH 44231, x1.

22-23. Brachial exterior, interior, AMNH 44232, x1 (all from AMNH Loc. 3152).

24. *Athyris* sp. A. Brachial interior, AMNH 44233, x 3, AMNH Loc. 3153........................

25-31. *Meristina* cf. *nasuta* (Conrad, 1842)........................

25. Pedicle interior, AMNH 44234, x2.

26. Pedicle interior, AMNH 44235 , x2.

27-31. Pedicle exterior, interior, lateral, anterior, posterior, AMNH 44236, x1.5 (all from AMNH Loc. 3153).

32-34. *Meristina?* sp. Pedicle exterior, posterior, anterior, AMNH 44237, x1.2, AMNH Loc. 3152. ....

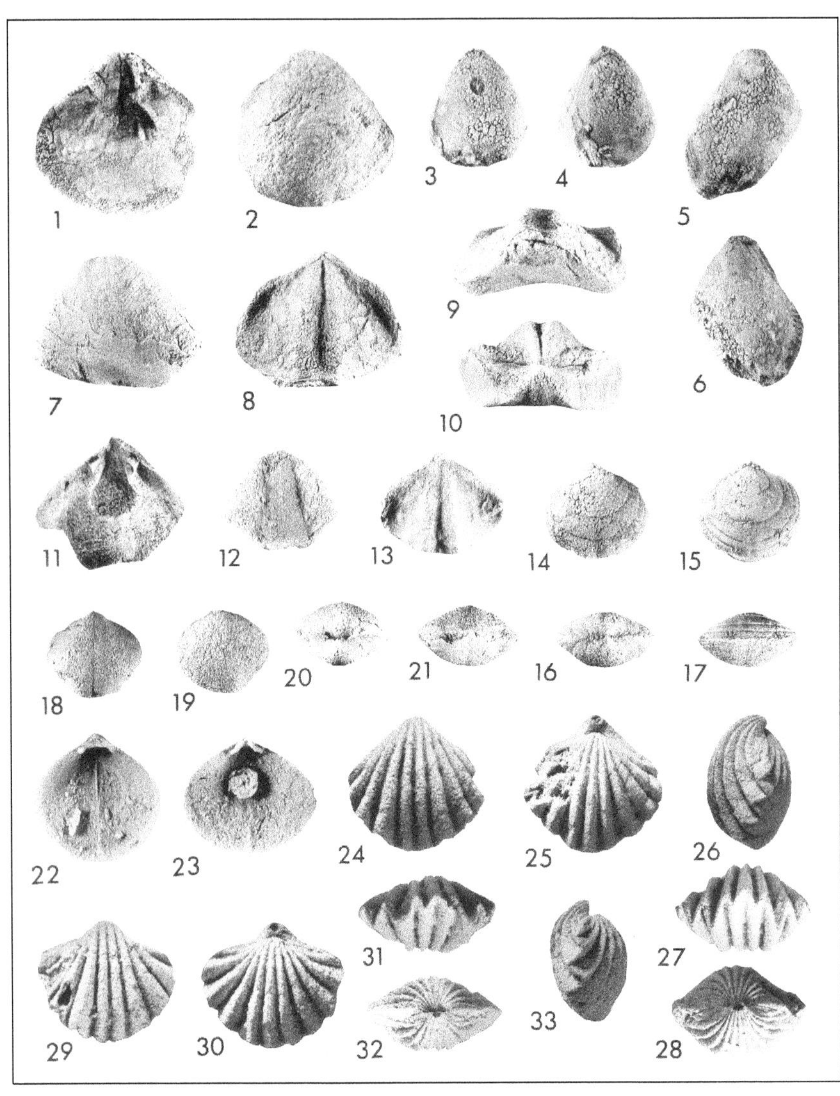

1
2
3
4
5
6
7
8
9
10
11
12
13
14
15
16
17
18
19
20
21
22
23
24
25
26
27
28
29
30
31
32
33

PLATE 6

EXPLANATION OF PLATE 6

Figure .... Page

1-2. *Meristina?* sp. Pedicle interior, exterior, AMNH 44238, x1.5, AMNH Loc. 3152. ....

3-6. *Charionoides doris* (Hall, 1860). ....
- 3-4. Pedicle, brachial exterior, AMNH 44239, x2.5.
- 5-6. Pedicle, brachial exterior, AMNH 44240, x3 (both from AMNH Loc. 3152).

7-13. *Pentagonia unisulcata* (Conrad, 1841). ....
- 7-10. Pedicle brachial, anterior, posterior views, AMNH 44241, x2.
- 11. Pedicle interior, AMNH 44242, x2.
- 12-13. Brachial exterior, interior, AMNH 44243, x3 (all from AMNH Loc. 3152).

14-23. *Nucleospira ventricosa* (Hall, 1857). ....
- 14-17. Pedicle, brachial, posterior, anterior views, AMNH 44244, x2.5, AMNH Loc. 3152.
- 18-21. Pedicle, brachial, posterior, anterior views, AMNH 44245, x1.8, AMNH Loc. 3153.
- 22. Pedicle interior, AMNH 44246, x4.
- 23. Brachial interior, AMNH 44247, x5 (both from AMNH Loc. 3152).

24-33. *Trematospira gibbosa* Hall, 1859. ....
- 24-28. Pedicle, brachial, lateral, anterior, posterior views, AMNH 44248, x3.5.
- 29-33. Pedicle, brachial, anterior, posterior, lateral views, AMNH 44249, x3 (both from AMNH Loc. 3152).

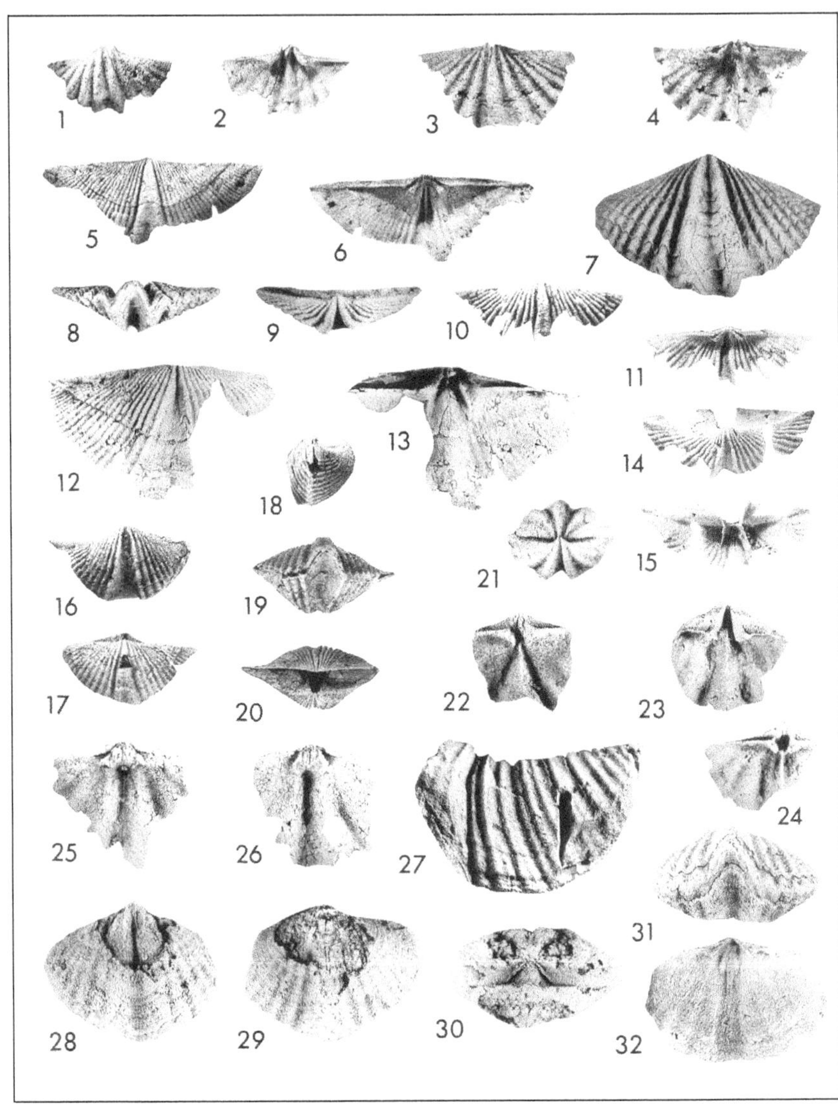

1
2
3
4
5
6
7
8
9
10
11
12
13
14
15
16
17
18
19
20
21
22
23
24
25
26
27
28
29
30
31
32

PLATE 7

EXPLANATION OF PLATE 7

Figure ....Page

1-4. *Acrospirifer duodenaria* (Hall, 1843)...

1-2. Pedicle exterior, interior, AMNH 44250, x2.

3-4. Brachial exterior, interior, AMNH 44251, x2 (both from AMNH Loc. 3154).

5-6. *Mucrospirifer?* sp. Brachial exterior, interior, AMNH 44252, x1, AMNH Loc. 3154. ....

7-9. *Alatiformia?* sp. Pedicle exterior (x3), anterior (x2), posterior (x2), AMNH 44253, AMNH Loc. 3154...

10-15. *Mediospirifer* sp. A. ....

10-11. Brachial exterior, interior, AMNH 24454, x1.5.

12-13. Pedicle exterior, interior, AMNH 44255, x1.5.

14-15. Pedicle exterior, interior, AMNH 44256, x1.5 (all from AMNH Loc. 3154).

16-20. *Mediospirifer?* sp. B. Pedicle, brachial, lateral, anterior, posterior views, AMNH 44257, x1, AMNH Loc. 3154...

21-36. *Megakozlowskiella raricosta* (Conrad, 1842)...

21. Posterior, AMNH 44258, x1.

22. Pedicle interior, AMNH 44259, x1.

23. Pedicle interior, AMNH 44260, x2.

24. Pedicle interior, AMNH 44261, x2.

25. Brachial interior, AMNH 44262, x1.5.

26. Brachial interior, AMNH 44263, x1.5 (all from AMNH Loc. 3154).

27. *Paraspirifer?* sp. Brachial exterior, AMNH 44262, x2, AMNH Loc. 3153. ....

28-32. *Elita fimbriata* (Conrad, 1842)...

28-31. Pedicle, brachial, posterior, anterior views, AMNH 44265, x1.5, AMNH Loc. 3152.

32. Brachial interior, AMNH 44266, x2, AMNH Loc. 3153.

EXPLANATION OF PLATE 8

Figure .... Page

1-3. *Elita fimbriata* (Conrad, 1842).
- 1. Pedicle interior, AMNH 44267, x3, AMNH Loc. 3153.
- 2. Brachial interior, AMNH 44268, x6.5, AMNH 3154.
- 3. Detail of ornament (note spines) on pedicle exterior, AMNH 24469, x2, AMNH Loc. 3152.

4-6. Ambocoeliid indet. Pedicle, brachial, anterior views, AMNH 44270, x5, AMNH Loc. 3152. ....

7-10. *Cyrtina hamiltonensis* (Hall, 1857). ....
- 7-8. Pedicle, brachial exterior, AMNH 24471, x2.5.
- 9. Pedicle exterior, AMNH 24472, x3.5 (both from AMNH Loc. 3152).
- 10. Brachial interior, AMNH 24473, x5, AMNH Loc. 3153.

11. *Cyrtina* sp. A. Pedicle exterior, AMNH 24474, x2, AMNH Loc. 3152.....................

12-17. *Cryptonella?* sp. ....
- 12-14. Pedicle, posterior, brachial views, AMNH 44275, x2.3.
- 15-17. Brachial, posterior, pedicle views, AMNH 44276, x3.7 (both from AMNH Loc. 3152).

18-19. *Cranaena?* sp. Brachial, pedicle views, AMNH 44277, x3, AMNH Loc. 3152....

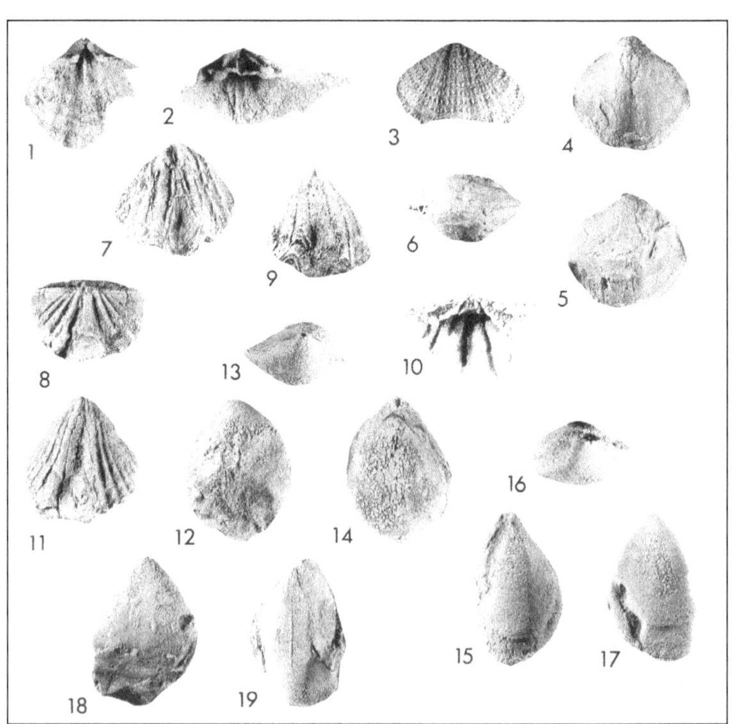

PLATE 8

Paleoecology and Morphologic Variation of a Paleocene Terebratulid Brachiopod (*Oleneothyris harlani*) from the Hornerstown Formation of New Jersey

ABSTRACT

Study of topoclines and chronoclines reveals the presence of three morphotypic end members of the brachiopod *Oleneothyris harlani* (Morton) in the Upper Hornerstown biostrome of the New Jersey Coastal Plain near New Egypt. Most of the shells can be categorized into two distinct morphotypes, whereas the third represents a relatively insignificant intrapopulation variant. Striking internal variation occurs within the umbonal cavity of the individuals studied, which can be statistically correlated with the relative size of the pedicle foramen. Variation in the cardinal process is also noted. Paleoecological conclusions reached through studying intrapopulation variation within the *Oleneothyris* community are as follows: the upper part of the Hornerstown Formation was deposited in shallow water at an extremely slow rate; at times the depositional environment was one of strong current and wave action particularly affecting the biostrome at the top of the formation; a strong tolerance to current and wave action on the part of *Oleneothyris harlani* is suggested along with the probable evolution of a sturdier pedicle and more robust shell; and the trophic structure of the *Oleneothyris* community was found to conform in some ways to that of Recent Arctic and Boreal communities. In addition, the biostrome represents a life assemblage (biocoenose) with minimal post-mortem transport. *Oleneothyris harlani* appears to be an explosive opportunist, limited both geographically and stratigraphically.

INTRODUCTION

Preserved in the Tertiary sediments of the Atlantic Coastal Plain in New Jersey, Delaware, and Maryland are specimens of one of the largest brachiopods of the family Terebratulinae Gray, 1840, *Oleneothyris harlani* (Morton). A biostrome of *Oleneothyris harlani* shells is present as a lens within the

Hornerstown Formation (Paleocene) of New Jersey (Text-figure 1), where the fossils are exceptionally well-preserved. The biostrome consistently occurs in the uppermost beds of the formation and may be regarded as a marker for the top of the Hornerstown. Reworked greensand and fragments of *Oleneothyris harlani* have been reported by Dorf and Fox (1957) from the basal 91 cm of the overlying Vincentown sand.

The Hornerstown Formation strikes in an approximately N55E direction in a belt along the Coastal Plain of New Jersey. Its upper contact is gradational with the light buff sands of the Vincentown Formation, while the stratigraphic position of the beds underlying the lower contact varies geographically. In the northern sections of Monmouth County, the Hornerstown Formation overlies the Tinton Formation, whereas in the New Egypt

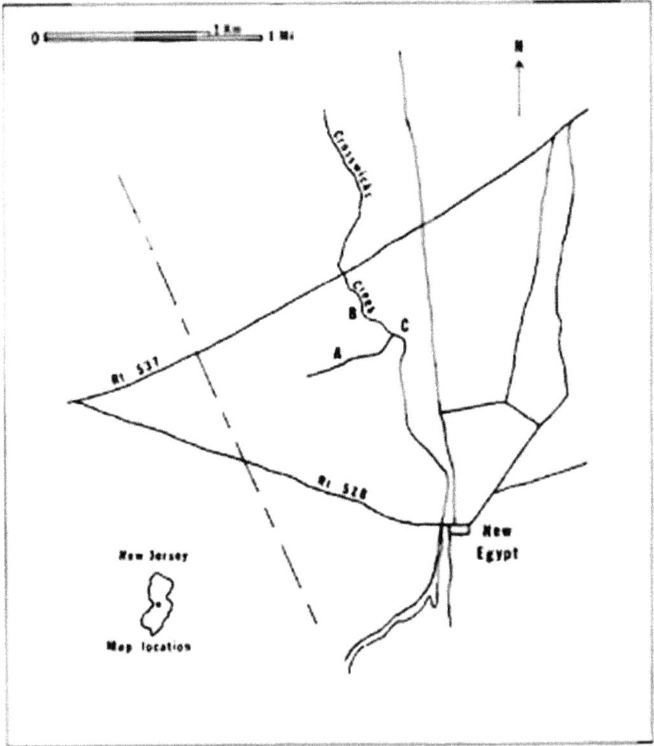

TEXT-FIGURE 1. Index map to collecting localities of *Oleneothyris harlani* in the Hornerstown Formation (Paleocene) of New Jersey. A, B, and C represent outcrops in the vicinity of Crosswicks Creek at which the brachiopod is well represented along with other faunal constituents of the formation.

area it overlies the New Egypt Formation, and south of New Egypt it overlies the Navesink Formation. The formation extends from the Atlantic Highlands through Freehold, Hornerstown, Birmingham, Mullica Hill, Sewell, and Woodstown to the Delaware River. The average dip is 5.7 m per km (30 ft per mi) (Richards, Groot, and Germeroth, 1957).

This study deals with the problem of recognizing intrapopulation variation within the *Oleneothyris* community. A reconstruction of the trophic structure of the community is presented along with an autecological and synecological analysis. Variation within the population is tied to environmental stress caused by a regressing sea during the Early Tertiary.

Specimens are deposited in the U. S. National Museum, Washington, D.C. (USNM), the Rutgers University Geological Museum, New Brunswick, New Jersey (RU), and the American Museum of Natural History, New York City (AMNH).

AGE OF HORNERSTOWN FORMATION

Until recently, workers have not agreed upon the correct age of the Hornerstown Formation of the Atlantic Coastal Plain. For the last one hundred years the Hornerstown has been, at various times, designated as a Cretaceous, Paleocene, and Eocene formation.

Clark, Bagg, and Shattuck (1897) listed the Sewell Marls (Hornerstown marls) as part of the Rancocas Group of the Eocene, directly overlying the Cretaceous Monmouth Formation. Based on molluscan and brachiopod faunal associations, Cooke and Stephenson (1928) assigned an Eocene age to the Hornerstown, Vincentown, and Manasquan formations and concluded that an overlapping unconformity existed between them and the underlying Cretaceous sediments. Colbert (1948) considered the Hornerstown to be Eocene, whereas Richards (1948) presented further evidence for an Eocene age: "Only three genera were recognized which are of sufficiently short geologic range to corroborate the Eocene age of the Hornerstown, Vincentown and Manasquan formations." (Data from a well in Neptune Township, Monmouth County, New Jersey, elev. 80 ft, depth 475 ft, NJGS: ANSP.—Wardell Dairy: 3 mi west of Bradley Beach.)

In 1950, Spangler and Peterson assigned a Paleocene age to the Hornerstown Formation without citing any paleontological evidence.

Johnson, Meredith, and Richards (1952) considered the Hornerstown Formation to be Eocene, with the upper part Wilcox and the lower part

Midway in age. They also noted an unconformity between the Hornerstown and the underlying Cretaceous strata.

McLean (1953) observed that both the Hornerstown and Vincentown formations contained benthonic foraminifera that he identified as Paleocene.

The unconformity mentioned by Johnson, Meredith, and Richards is again noted by Fox and Olsson (1955) in a paper dealing with the stratigraphy of Late Cretaceous and Early Tertiary formations in New Jersey. They state that the Hornerstown Formation is clearly Eocene in age; it contains no diagnostic Paleocene forms.

A description of the macrofauna of the uppermost Hornerstown Formation is provided (Fox and Olsson, 1955) in which the presence of 31 species restricted to the Midway of the Gulf Coast is noted, thus implying strong Paleocene affinities. Fox and Olsson (1955) also describe a macrofauna of 70 species from the basal part of the Vincentown Formation containing a reworked mixture of typical Paleocene forms in the lower 1.5 m of the unit. They occur with fragments of *Oleneothyris harlani* and with considerable reworked typical Hornerstown greensand.

Foraminifera of the Vincentown Formation were analyzed in 1955 by Hofker and found to be of Paleocene age, thus eliminating the possibility of an Eocene age for the Hornerstown.

Miller (1956) concluded that the Hornerstown Formation is Paleocene (Midway) on the basis of paleontological evidence (Table 1) and that the Vincentown Formation is Lower Eocene. Richards et al. (1957) also concluded that the Hornerstown Formation is Paleocene, but instead of assigning an Eocene age to the Vincentown Formation they include it in the Paleocene.

Loeblich and Tappan (1957) studied the planktonic foraminifera from the uppermost Hornerstown and assigned the formation a Paleocene age. Olsson (1959, 1963) recognized a zone of *Globorotalia compressa-Globigerinoides daubjergensis* in the basal part of the Hornerstown, which correlates with the Lower Midway Group (Early Paleocene) of the Gulf Coast. A subzone of *Globorotalia pseudobulloides* of the *Globorotalia angulata* zone, characterizing the middle parts of the Hornerstown Formation, was correlated with the Late Paleocene Upper Midway Group of Alabama. Nevertheless, according to Olsson (1963), the Paleocene-Eocene boundary is subject to a degree of uncertainty until future studies can demonstrate more positive evidence. This evidence was forthcoming in studies made in 1967 and 1968. Richards (1967) reviewed the stratigraphy of the Atlantic Coastal Plain between Long Island and Georgia and concluded that the Hornerstown and Vincentown formations are Paleocene in age, whereas the Manasquan Formation is Eocene. Workers in the field subsequent to 1968 have generally agreed with Richards' findings and have considered the Hornerstown Formation to be Paleocene (Minard et al., 1969).

SYSTEMATIC PALEONTOLOGY

Order TEREBRATULIDA Waagen, 1883
Suborder TEREBRATULIDINA Waagen, 1883
Superfamily TEREBRATULACEA Gray, 1840
Family TEREBRATULIDAE Gray, 1840
Genus OLENEOTHYRIS Cooper, 1942
OLENEOTHYRIS HARLANI (Morton), Plates 1-3

Terebratula harlani Morton, 1828a: 73-76, pl. 3, figs. 1-4, 7-8; 1830 250, pl. 3, fig. 16-17; 1834: 70, pl. 3, fig. 1, pl. 9, fig. 8-9; Marcou, 1853: 47, pl. 7, fig. 8; Cook, 1868: 375, two fig.; Whitfield, 1886: 6, pl. 1, fig, 15-23; Clarke and Martin, 1901: 204, pl. 58, fig. 2-3; Weller and Knapp, 1907: 357-360, pl. 28, fig. l-8.

Terebratula fragilis Morton, 1828a: 75, pl. 3, fig. 3-4; 1830: 250, pl. 3, fig. 17; 1834, (*non* Sowerby 1825): 70, pl. 3, fig. 2.

Terebratula perovalis Morton, 1828a (*non* Sowerby 1825): 77, pl. 3, fig.7-8.

Terebratula atlantica Morton, 1828b: 214, in text; Whitfield, 1886, pl. 1, fig. 10-13.

Terebratula camella Morton, 1834: 70, in text.

Terebratula subfragilis D'Orbigny, 1850: 258, in text.

Oleneothyris harlani Cooper, 1942: 233, in text; Cooper, 1942, 365, pl. 143, fig. 44-49.

Exterior: Shell large, rostrate, astrophic, mesothyrid to permesothyrid; symphitium present; elongately subovate in outline and elongately ovate in lateral profile; pedicle valve of adult specimens strongly convex; brachial valve planoconvex to convex; in anterior view both valves equally biconvex; when viewed posteriorly pedicle valve considerably more excurvated.

Anterior commissure recurrently folded, varying from uniplicate to sulciplicate, beak slightly incurved to erect; ventral umbo swollen with faint median ridge in many adults; ventral beak ridges subangular to rounded; cardinal margin highly variable; palintrope weakly defined on brachial valve bu more developed on pedicle valve (the terms palintrope and cardinal margin have been adopted by Rudwick, 1959, to represent the "interareas" and "hinge lines" of astrophic brachiopods. "Palintrope" is now used to define the curved surface of a shell, bounded by the beak ridges and cardinal margin of astrophic shells [see Williams et al., 2002: 434]). Shell extremely thick in umbonal region but tapering and thinning anteriorly. Growth lines evident on both valves; variably distributed with a greater concentration noticeable toward anterior commissure of mature specimens; shell finely punctate; radial ornamentation absent.

Interior of pedicle valve: Umbonal cavity well developed, lying between two dental plates originating anterior to pedicle collar; plates terminating with constriction of umbonal cavity at varying distances from pedicle foramen. Two strong, well developed hinge teeth located at dorsal part of dental plates; teeth cyrtomatodont with hook-like process on posterior margin; denticular cavity well defined and progressively deeper posteriorly. Muscle field narrow and elongate, fanning out triangularly with broad base of field anterior. Point of attachment of adjustor muscles represented by shallow, broad subovate scar at base of umbonal cavity; impressions of adductor muscles located immediately anterior to constriction of umbonal cavity.

Interior of brachial valve: Exterior hinge plates short, slightly concave, and joined with crural base posteriorly, almost to cardinal process. Cardinalia distinct with massive inner socket ridges fluted on surface facing sockets; cardinal process slightly cup-shaped with distinct median ridge on well preserved specimens; crural processes relatively long, blunt and thick. Dental sockets deep, diverging strongly from cardinal process. Muscle field large with strongly impressed scars; about one-half as long and one-fourth as wide as the valve; elongate, deeply impressed adductor muscle scars extending from crural base throughout three-fourths of muscle field.

Remarks: Oleneothyris harlani was first described by Morton in 1828 as *Terebratula harlani*, at which time he referred to two distinct species—*T. harlani* and *T. fragilis. T. harlani* was described by Morton as being a large shell, about twice as long as wide, with straight but imperfectly parallel sides. The dorsal valve was said to be planoconvex (a term no longer used in the same context; it now refers to the entire brachiopod rather than just one valve), obscurely biplicated except near the margin; the ventral valve was very convex, with a longitudinal ridge and slight lateral depressions; the beak was incurved and the umbo prominent. *T. fragilis* was described as a more cylindrical shell with very strongly marked ridges and sinuses, the ridges often forming strong lobes extending almost to the beak on the flat dorsal valve, and almost as far on the ventral valve. The shell was described as very thin and quite fragile.

In his synopsis (1834), Morton clarified his original descriptions of the variant forms of *T. harlani;* upon further examination of a few hundred specimens he decided that two different species did not exist, rather, that there were two "varieties" which graded into each other.

Terebratula fragilis was renamed *subfragilis* by d'Orbigny (1850), as "fragilis" had been previously used by Schlotheim (1813) (see d'Orbigny, 1850: 258) for a species of *Terebratula.*

Very few complete specimens fitting the description of *T. subfragilis* (*T. fragilis* Morton) have been found by the author. Of the specimens that have been studied, none show significant morphologic variance from the brachiopods of the *Oleneothyris harlani* population to be considered a subspecies. The only important difference observed between the two brachiopods is the deeper sulcus and more prominent ridges on the dorsal valves of specimens fitting the description of *T. subfragilis.* The shells of *T. subfragilis* have not been found to be significantly more fragile than those of the majority of the *Oleneothyris* population. Thus, I would consider d'Orbigny's *T. subfragilis* specimens to be one of the two, possibly three, morphotypic end members of the *Oleneothyris harlani* population.

According to Whitfield (1886), *T. perovalis M*orton (1828) was originally thought to be a variety of *T. perovalis S*owerby, 1825, a European shell. However, it was simply a two-thirds grown form of *T. harlani.*

If *T. perovalis* Morton was a valid species the name, *T. camella* would have applied, since *T. perovalis* was only a provisional name. Morton (1834) had, after erecting the invalid homonym *T. perovalis,* proposed *T. camella* for it if found to be distinct from *T. harlani.*

In 1861 Gabb cited *T. atlantica* Morton as a junior synonym of *T. harlani* as he believed it to be a young individual of that species, very small and clearly marked with radiating striae, which are never found in *Oleneothyris harlani.*

TABLE 1. Macrofauna of the Hornerstown Formation (Modified from Miller, 1956).

PORIFERA	*Cucullaea compressirostra* Whitfield
Eudea dichotoma	*Cucullaea* cf. *neglecta* Gabb
Entobia	*Cucullaea vulgaris* Morton
	Cucullaea sp.
COELENTERATA	*Ostrea tecticosta* Gabb
Paracyathus vaughani Weller	*Ostrea bryani* Gabb
Trochocyathus conoides Gabb and Horn	*Gryphaeostrea vomer* Morton
Flabellum mortoni Vaughan	*Gryphaea vesicularis* Lamarck, var. *dissimilaris*
BRACHIOPODA	*Anomia* cf. *argentaria* Morton
Terebratulina cf. *manasquanensis* Stenzel	*Crassatella* cf. *obliquata* Whitfield
Oleneothyris harlani	*Crassatella rhombia* Whitfield
	Caryatis veta Whitfield
ANNELIDA	*Veniella decisa* Morton
Hamulus sp.	*Veniella rhomboidea* Conrad
Longitubus lineatus Weller	*Cardium knappi* Weller
	Cardium nucleolus Whitfield
MOLLUSCA	*Polorthis tibialis* Morton
CEPHALOPODA:	*Teredo irregularis* Gabb
Aturoidea paucifex Cope	
	CHORDATA
GASTROPODA:	*Lamna cuspidata* Agassiz
Pleurolomaria crotaloides Morton	*Myliobatis* sp.
Anchura abrupta Conrad	*Edaphodon* sp.
Anchura sp.	*Euchodus* sp.
Turritella vertebroides Morton	*Chelonia* fragments
Turritella sp.	*Osteopygis emarginatus* Cope
Lunatia halli Gabb	*Thoracosaurus* sp.
Rostellaria cf. *fusiformis* Whitfield	Crocodile bone fragments
Pyropsis cf. *trochiformis* Tourney	Shark teeth and vertebra
PELECYPODA:	MISCELLANEOUS
Modiola subinflata Whitfield	Bryozoa—unidentified
Venericardia smithii Aldrich	Decapod crustacean fragments
Arca quindecemradiata Gabb	Echinoid spines

PALEOECOLOGY

SYNECOLOGY

There is a vast difference in the quality of the fossils found in the upper and lower parts of the Hornerstown Formation. In addition to a wide range of microfossils found in the formation, many groups of macrofossils are represented, including: gastropods, pelecypods, coelenterates, sponges, brachiopods, annelid worms, and vertebrate remains (see Table 1). Although the lower part of the formation contains a wider variety of fossils they are poorly preserved and few in number (Olsson, 1963). A large number of the fossils found throughout the formation were capable of existing on muddy bottoms, but the genus *Cardium,* because of its short siphon, is limited to a sandy substrate very much like the adult

M. mercenaria (Stanley, 1972). A sandy bottom is thus implied for the Upper Hornerstown Formation, since (1) *Cardium* is present throughout the upper part, and (2) although there is a large quantity of clay-sized grains in the lower part of the formation (Adams, 1963), the upper part is generally free of clay-sized particles.

Upon applying Walker's (1972) study (based on Turpaeva's work, 1957) of trophic analysis of Ordovician-Mississippian fossil communities to the Paleocene *Oleneothyris* biostrome, several important paleoecological conclusions may be reached.

According to Walker, ancient communities may be subdivided into several trophic levels: high level filter feeders, low level filter feeders, collectors, and swallowers. Brachiopods fall into the category of either high level or low level filter feeders, depending upon the feeding position occupied in life. Generally, those with a functional pedicle, such as *Oleneothyris,* feed several centimeters above the sediment-water interface and are classified as high level filter feeders. Brachiopods without functional pedicles, such as the strophomenids of the Paleozoic (e. g. *Pholidostrophia, Plectodonta*), rested on the surface of the sediment and thus fed several centimeters lower, as low level filter feeders.

The megafossil assemblage of the biostrome of the Upper Hornerstown Formation is comprised of more than 99 percent *Oleneothyris harlani*, with occasional specimens of *Gryphaea dissimilaris* (Olsson, 1963; Adams, 1963), *Polorthis tibialis* (Olsson, 1957), *Cucullaea vulgaris* (Miller, 1956) and *Ostrea* (Miller, 1956). That the living community was dominated by *Oleneothyris* is certain; to exactly what extent, however, is not as certain. Assuming that the *Oleneothyris* brachiopod community had a significant percentage of other organisms that fed at the same trophic level (e.g. other brachiopods or molluscs)—even if we allow for the possibility that the community was transported and/or reworked—why do we not find any significant trace of these other organisms *in situ?* It is possible that the smaller shells were winnowed away and that the larger *Oleneothyris* shells are residual. However, the author believes that given the tremendous variability in size, shape, and weight of the *Oleneothyris* shells found together in the biostrome, and taking into consideration the fact that the hydrodynamic properties of invertebrate shells are so variable, it would be logical to assume that if any other shelled invertebrates existed with *Oleneothyris* as part of the *Oleneothyris* community, there would remain some significant evidence of their position within the community as reflected by the number of individuals present.

As in Walker's *Clorinda* community, all of the organisms of the *Oleneothyris* community derived nourishment from a single source. However, the number of different species in the biostrome is far less than that found in the *Clorinda* community. The *Oleneothyris* community was preserved at a stage past the competitive situation commonly associated with many organisms of the same type feeding at only one trophic level, thus resulting in selection pressures favoring *Oleneothyris.*

Upon closer analysis of the trophic structure of the *Oleneothyris* community, several factors appear to conform to Turpaeva's (1957) feeding structure in Recent Arctic and Boreal communities: (1) the community was dominated by one trophic group—high level filter feeders; (2) the dominant species, *Oleneothyris harlani,* belonged to one trophic group, while the next most dominant species, *Gryphaea,* in this writer's opinion, belonged to a different trophic group—low level filter feeders; and (3) one species, *Oleneothyris harlani,* dominates in terms of biovolume (biomass).

The similarity in trophic structure between the *Oleneothyris* and Arctic and Boreal communites may lead one to assume that the waters of the Atlantic in Paleocene times were cool, whereas, according to current thought (Menzies et al., 1973), they were tropical. This apparent paradox may be explained as follows: (1) the *Oleneothyris* community existed in deeper water than mid-neritic; accordingly, the water temperature was lower; (2) a tongue of cooler water locally invaded the Atlantic coast in a limited geographic region (comparison of New Jersey foraminiferal faunas with those of other sections in the north Atlantic indicates a cool climatic shift in the Late Paleocene-Olsson, 1970); and (3) a condition of environmental stress existed that produced a trophic structure similar to that of cold-water faunas. (This possibility is probably correct, as it conforms to the idea that *Oleneothyris* was an explosive opportunist, appearing at only one point stratigraphically, and limited geographically to the Upper Hornerstown Formation.)

In conclusion, the macrofauna of the Upper Hornerstown Formation is characterized by great population density and low diversity. In rapidly shifting sand deposits, the invertebrate fauna generally is one of low population density (Purdy, 1964). Here, however, the substratum was, for the most part, stable, resulting in a high degree of density. The diversity of the macrofauna can be broken down into two components: (1) variety and (2) equitability. The variety, or species-richness component, in the author's estimation, is moderate. The equitability component, based on the relative numbers of individuals within each species, is extremely low. For example, if we were to hypothesize an equitability ratio for *Oleneothyris harlani* and the five most populous species within the formation, the result would most likely be 95:1:1:1:1:1. Any diversity index estimated for the Hornerstown Formation would be very low.

Autecology: Rates of sedimentation and water activity probably had as significant effect on the distribution of Lower Tertiary brachiopods as they do on the distribution of modern brachiopods. Rudwick (1962) determined that five terebratellaceans and one rhynchonellid on the coast of New Zealand can tolerate turbidity and temperature fluctuations but are intolerant of sedimentation and wave action. Since very slow rates of sedimentation (Adams, 1963) characterized the paleoenvironment of *Oleneothyris* the question of tolerance to rapid sedimentation does not exist. However, the robustness of *Oleneothyris*, the large pedicle foramen, and the probable evolutionary trend with regard to morphology (Text-figure 2) suggested by statistical analysis tend to indicate a strong tolerance to wave or current action. Thus, at some point during the evolution of the *Oleneothyris* population, the brachiopods were exposed to a form of selection pressure (current or wave action) that favored those types which were able to withstand and adapt to changing environmental conditions. Accordingly, it is noted that a large pedicle foramen is generally associated with a more robust specimen, implying that the robust type possessed a sturdier pedicle, which afforded it stronger attachment to its substratum.

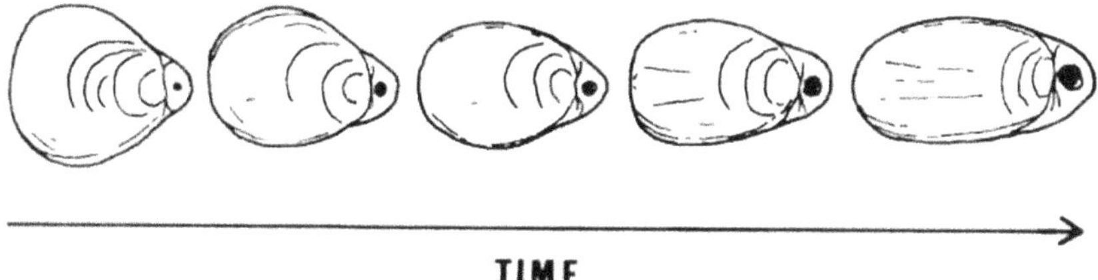

TEXT-FIGURE 2. A suggested evolution (chronocline) of *Oleneothyris harlani*. Note the relative enlargement of the pedicle foramen and proportionate increase in length.

Oleneothyris harlani was a ciliary suspension feeder, as are all living terebratulid brachiopods. Accordingly, food particles had to be sorted out from suspended inedible material in the surrounding water. The intestine, as in all articulate brachiopods, ended blindly with fecal pellets egested from the mouth.

Was it possible for the *Oleneothyris* community to have remained stationary while the sea regressed? In other words, are not these organisms affected and limited by depth such that in an environment that is changing they would have to change also? There are two possible answers to this problem. Firstly, the brachiopods might have adapted to the new depth. Evidence for adaptation is observed in the morphologic differences between Type A and Type B brachiopods (Text-figures 3 and 4) discussed below. Secondly, regression of the sea at that time (Schlanger, 1954) was perhaps

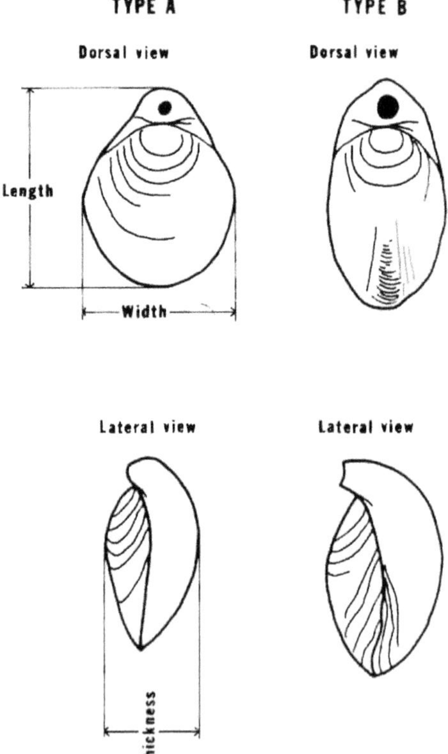

TEXT-FIGURE 3. Diagram illustrating morphotypic variants of *Oleneothyris harlani* population: Type A, flat and anteriorly lobate with a relatively small pedicle foramen and Type B. cylindrical, with a relatively large pedicle foramen.

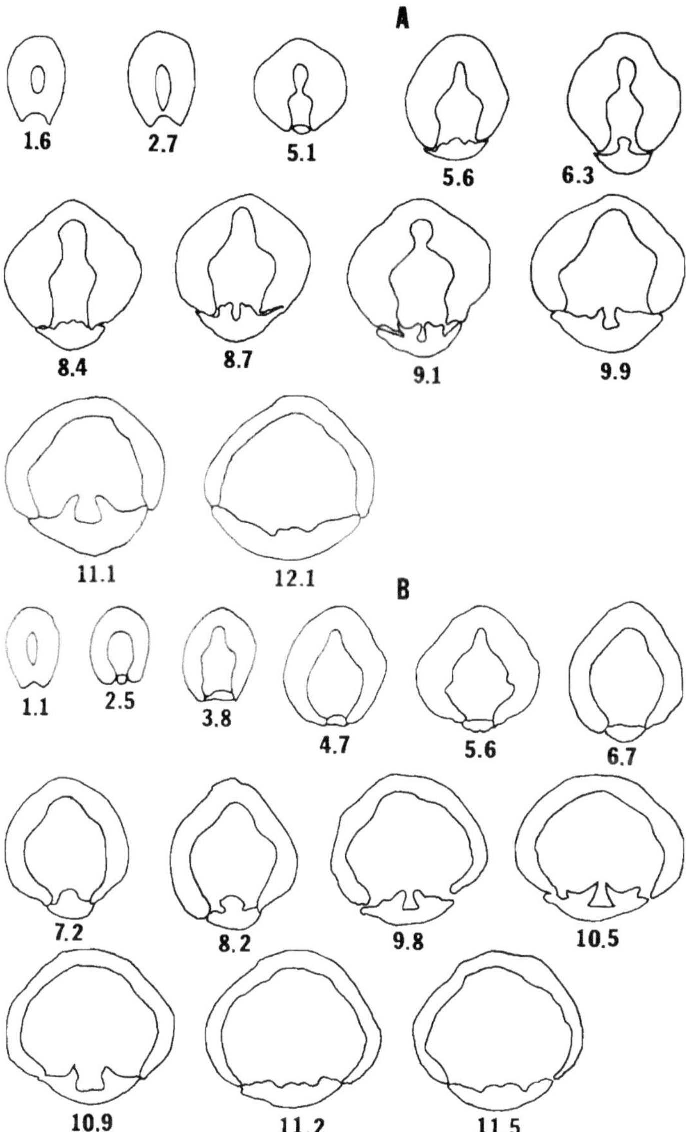

TEXT-FIGURE 4. Serial sections of two specimens of *Oleneothyris harlani* illustrating variation within the umbonal cavity and relative thickness of shell in Types A and B. Note the constriction of the umbonal cavity and the posterior thickening of the shell in specimen A. There is a distinct lack of constriction in the umbonal cavity of specimen B, associated with a relatively thinner shell in the posterior region. Specimen A had a smaller pedicle foramen than did specimen B. Numbers indicate the distance in millimeters of the section anterior to the beak of the pedicle valve. The loop is not preserved (x1).

not rapid enough to affect the brachiopods inhabiting the sea floor below. The strand line did not retreat so rapidly that the organisms inhabiting the sea floor were unable to migrate with the receding shoreline. A change of depth of a few meters one way or the other may have a negligible consequences.

Assuming the first of the above-mentioned theories explaining the morphologic variation in *Oleneothyris* to be correct, one would expect to find some sort of gradation in morphology of the brachiopods relevant to the ideas stated. Either specimens with a narrower pedicle foramen should be found below those with a wider foramen, or vice-versa. Since the Hornerstown Formation consists almost exclusively of unconsolidated greensands (Colbert, 1948; Fox and Olsson, 1955; Owens and Minard, 1960; Minard and Owens, 1960), the slight reworking of which would have destroyed any biostratigraphic gradation within the biostrome, one would have to find a gradation in larger layers. In other words, the bottom 15 cm of the shell bed may consist of valves with mostly one size pedicle foramen. The next few centimeters may consist of fewer valves of that size and more of the other type. This trend could continue until the top of the shell bed is reached, whereupon the reverse of the bottom 15 cm is found.

Upon trying to establish whether or not any gradation existed, a recently exposed section of the Hornerstown Formation was studied in detail. A 98 cm stratigraphic section, exposed in outcrop immediately after a rainstorm, was divided into four subunits, each measuring 25.5 cm (Table 2). Fifteen brachiopods were selected, at random, from each of the four subunits I, II, III, and IV. The mean diameter (D) of the pedicle foramen was calculated along with the mean distance representing the narrowest space at the anterior end of the umbonal cavity (C). Upon analyzing the data, a relatively greater number of specimens with a larger pedicle foramen were found as one progressed up the section from IV to I.

Based on the preceding paragraph, an evolutionary trend may have existed in which the type of *Oleneothyris* that possessed a relatively small pedicle foramen in conjunction with a narrow umbonal cavity (Type A) was selected against.

There is a complete morphological gradation between the two types of *Oleneothyris* described above, thus ruling out the possibility of the existence of two subspecies. The existence of two distinct species is negated on two counts: (1) there is not sufficient morphological evidence to distinguish two species within the population, and (2) according to the competitive exclusion principle (Gause's axiom), two different species cannot for long simultaneously occupy the same niche in the same place. The latter would be valid only if the assemblage was not transported.

Although some reworking of shells probably occurred (evidenced by occasional valves and fragments in the overlying basal Vincentown Formation), there is convincing evidence that the shell bed represents a life

TABLE 2. Data from a 98 cm stratigraphic section in the Hornerstown Formation at Shingle Run, which was divided into four subunits, each measuring 24.5 cm.

Subunit I				Subunit II				Subunit III				Subunit IV			
Speci-men	D	C	D/C	Spec-imen	D	C	D/C	Speci-men	D	C	D/C	Spec-imen	D	C	D/C
1	4.8	7.3	.65	1	3.6	6.5	.55	1	3.0	3.0	1.00	1	4.6	8.6	.53
2	4.2	8.7	.48	2	1.7	1.4	1.21	2	2.1	1.6	1.31	2	1.9	1.8	1.06
3	4.9	8.4	.58	3	3.5	6.8	.51	3	1.7	1.6	1.06	3	4.7	6.4	.73
4	4.0	9.0	.44	4	3.7	7.2	.51	4	3.7	7.4	.50	4	2.6	3.0	1.30
5	4.5	7.5	.60	5	3.6	7.1	.50	5	4.6	5.4	.85	5	2.5	1.5	1.66
6	3.9	8.0	.49	6	4.0	6.5	.62	6	2.3	1.9	1.21	6	2.0	1.5	1.33
7	2.0	1.4	1.43	7	3.6	9.3	.39	7	4.3	5.4	.80	7	2.5	1.8	1.39
8	2.6	2.4	1.08	8	3.8	8.0	.48	8	2.7	2.0	1.35	8	2.2	1.5	1.47
9	4.4	5.9	.74	9	2.0	1.9	1.05	9	2.3	2.5	.92	9	2.5	1.7	1.47
10	3.5	6.2	.56	10	4.0	7.3	.55	10	4.0	6.0	.67	10	2.1	1.5	1.40
11	3.5	7.5	.47	11	2.5	2.5	1.00	11	2.6	1.5	1.73	11	2.4	1.4	1.71
12	3.7	8.0	.46	12	2.7	2.3	1.17	12	4.5	8.8	.51	12	2.7	2.0	1.35
13	5.5	7.6	.72	13	2.3	2.0	1.15	13	2.7	1.5	1.80	13	4.4	8.5	.52
14	1.9	1.2	1.58	14	2.8	1.6	1.75	14	2.4	1.5	1.60	14	2.0	1.4	1.43
15	2.2	1.9	1.16	15	2.1	1.7	1.24	15	3.8	6.7	.57	15	2.0	1.0	2.00
Σ X	55.6	91.3	11.44		45.9	72.1	12.68		46.7	56.8	15.88		41.1	42.6	19.35
$\bar{x}$	3.7	6.1	.76		3.1	4.8	85		3.1	3.8	1.06		2.7	2.8	1.29

Summary of Data

Subunit	$\bar{X}$D	$\bar{X}$C	$\bar{X}$D/C
I	3.7	6.1	.76
II	3.1	4.8	.85
III	3.1	3.8	1.06
IV	2.7	2.8	1.29

D = the diameter of the pedicle foramen; C = the width of the constriction of the muscle field at the anterior end of the umbonal cavity of the pedicle valve; Σ X = the sum of the variable x; $\bar{x}$ = the mean. All measurements in mm.

PLATE 1

assemblage (biocoenose). Most specimens are not found in exact life positions (probably due to the nature of the substrate) but post-mortem transport was minimal for the following reasons: (1) many shells exhibit deformation due to crowding, thus negating the idea that the biostrome represents an accumulation of transported shells or a "bank" fauna, and supporting the notion that the shells grew in place; (2) although there are signs of fragmentation due to predation most valves are not abraded; (3) many valves remained articulated; and (4) a complete size gradation exists, thus precluding transport of juveniles (see Text-figure 5; Plate 3, figure 11). Although Text-figure 6 reveals an apparent deficiency (not due to collecting bias) of juveniles of 0-35 mm in length, this can be explained by post-burial selective solution, which acted upon smaller shells whose periostracum disintegrated more rapidly than that of larger ones (Frederick C. Shaw, personal communication). This idea is supported further by the fact that very few juvenile pelecypod valves are found in the shell bed. Thus, the two types of *Oleneothyris* represent end members of one population adapting to changing environmental conditions.

PALEOECOLOGICAL SUMMARY

The lower and middle parts of the Hornerstown Formation were deposited on a stable continental shelf at a very slow rate of deposition. Lithologic evidence points to the existence of a sandy bottom at the time of deposition of most of the lower and middle parts and fossils characteristic of sandy bottoms are found throughout. Also, the presence of large amounts of clay-size grains indicates an unagitated depositional environment for the lower part of the formation.

EXPLANATION OF PLATE 1

Figures 1-10*: Oleneothyris harlani* (Morton). 1-3, 7, Dorsal, ventral, anterior, and lateral views of Type B specimen. Note relatively large pedicle foramen, cylindrical shell, and general robust appearance, x1, USNM 222599. 4-6, 10, Anterior, dorsal, ventral, and lateral views of Type A specimen. Generally lobate in appearance (compare Figure 5 with Figure 1), x1. USNM 222600.8, Internal structure of ventral valve of Type A specimen showing constriction at anterior end of umbonal cavity, narrow pedicle foramen, and closely-spaced hinge teeth, x1.5. USNM 222601. 9, Internal structure or articulated specimen intermediate between Type A and Type B. Note massive crura, x1.5, USNM 222602.

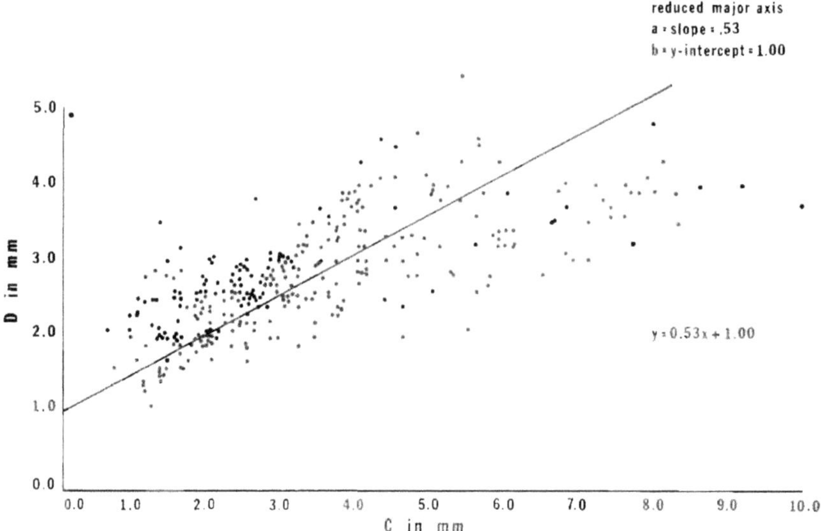

TEXT-FIGURE 5. Scattergram showing relationship of diameter of the pedicle foramen (D) to the constriction at the anterior end of the umbonal cavity (C) in *Oleneothyris harlani.*

The upper part of the Hornerstown Formation was deposited in shallower water at an extremely slow rate. At times the depositional environment was one of strong current or wave action particularly affecting the biostrome at the top of the formation.

A strong tolerance to current or wave action on the part of *Oleneothyris harlani* is suggested, along with the probable evolution of a sturdier pedicle and more robust shell.

The *Oleneothyris* community was found to conform in some ways to Recent Arctic and Boreal communities analyzed by Turpaeva (1957): (1) The community was dominated by one trophic group, (2) The dominant species belonged to one trophic group, while the next most dominant species belonged to a different trophic group, (3) One species domintated in terms of biovolume.

In conclusion, increased turbulence during the time of deposition of the Hornerstown Formation can be correlated with the morphologic changes noticeable in the population of the *Oleneothyris* community. The lower part of the biostrome represents deposition in somewhat calmer waters, while the upper part represents deposition in more turbulent waters. In exact agreement with this is the gradational change observed among the members

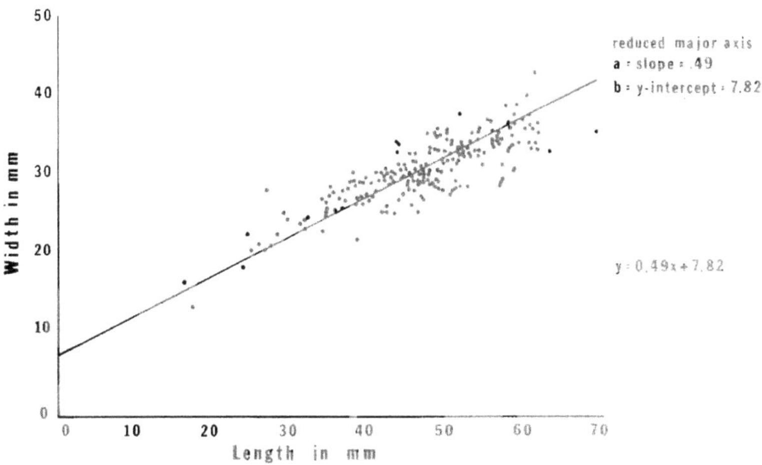

TEXT-FIGURE 6. Scattergram showing width-length dimensions of *Oleneothyris harlani.*

of the brachiopod population; more flat shells with a narrower pedicle foramen are found at the lower part of the shell bed than cylindrical shells with relatively wider pedicle openings.

VARIATION

VARIATION IN OLENEOTHYRIS HARLANI

The major part of the Upper Hornerstown Formation biostrome is composed of three major morphologic variants (morphotypes) of the brachiopod *Oleneothyris harlani,* two of which constitute the major portion of the shell bed (Plate 1, figures 1-3, 7; 4-6, 10) while the third (Plate 2, figures 1-4) represents a rather insignificant intrapopulation variant referred to by Morton (1834) as the species *subfragilis.* The vast majority of the shells can be categorized into two distinct morphotypic groups; a long, narrow cylindrical variant (Type B) and a shorter flat form (Type A) (see Text-figure 3). Several basic differences between the two are readily apparent: (1) the pedicle foramen is larger in the narrow valves (Type B) and smaller in the flatter, wider valves (Type A), (2) Type A valves are smaller and thinner shelled anteriorly than Type B valves, (3) in Type A valves there is more of a constriction in the umbonal area than in Type B valves (Text-figure 4), and (4) a flatter cardinal process seems to be associated with brachial valves of Type B (Plate 2, Figure 8), whereas a deeper,

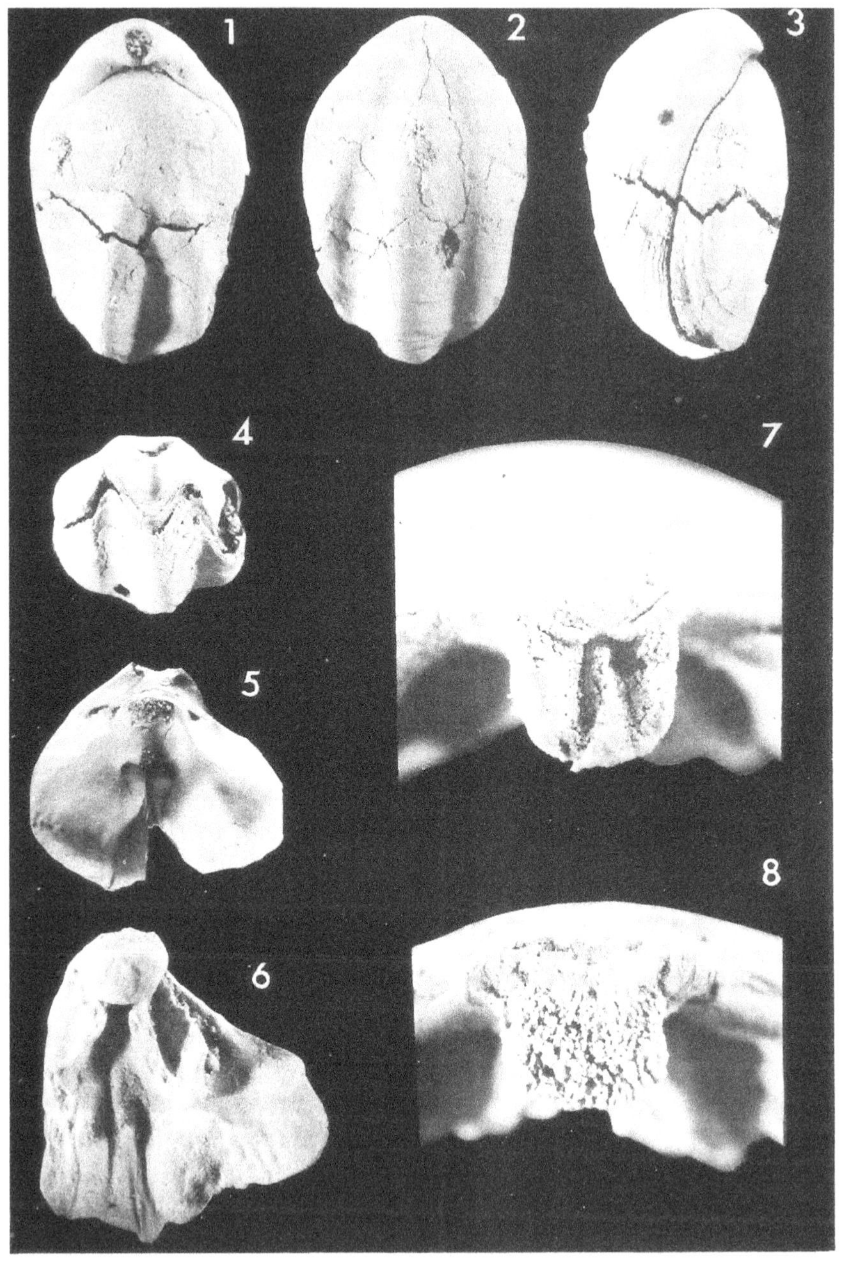

PLATE 2

cup-shaped cardinal process with a median ridge is associated with the brachial valves of Type A (Plate 2, Figure 7).

The writer believes that there are two possible explanations for the aforementioned differences in the two predominant types of *Oleneothyris.* One has already been outlined above. As these differences are related, they can be tied together by explaining the reasons for the difference in the relative size of the pedicle foramen between Type A and Type B.

During deposition of the Hornerstown Formation the Coastal Plain was undergoing several cycles of transgression and regression (Schlanger, 1954). The water level was constantly changing, becoming deeper and shallower at various intervals. The *Oleneothyris* community might have reacted to these changes in sea level in various ways, one of which could have been an increase or decrease in the size of the pedicle foramen. If, for example, the water level dropped along the shoreline, as it would in a regressing sea, the *Oleneothyris* population would have been subjected to a change in current strength and wave action. Due to this gradual shift in location of turbulence, and thus greater force exerted against the attachment potential (e .g., pedicle) of the brachiopods, a compensatory enlargement of the pedicle, and accordingly the pedicle foramen, developed due to natural selection; selection pressure favored those brachiopods (Type B) with the larger, stronger pedicles and those with better ability to withstand the attempts of displacement by waves and currents. Conversely, if the water level along the shoreline rose due to transgression by the sea, (causing a shift in the zone of turbulence) less force would have been exerted against the pedicle. This decrease in stress might have resulted in the favoring of a reduced pedicle, which in turn would result in a narrower pedicle foramen as encountered in Type A.

EXPLANATION OF PLATE 2

Figures 1-8. *Oleneothyris harlani* (Morton) *1-4,* Dorsal, ventral, lateral, and anterior views of Morton's *T. fragilis* from Shingle Run, New Egypt, New Jersey. Note cylindrical shell with deep sulcus and fold and longitudinal ridge extending almost to the beak on the ventral valve, x1. RU 5700. 5, Ventral interior of Type B valve showing lack of constriction at anterior end of umbonal cavity, relatively large pedicle foramen. and widely spaced hinge teeth, x1.5. USNM 222603. (Compare with Plate 1, Figure 8.) 6, Dorsal interior with bulbous cardinal process, x3.5, USNM 222604. 7, Posterior view of cup-shaped cardinal process with median ridge associated with Type A specimens, x7, USNM 222605. 8, Posterior view of flat cardinal process associated with Type B specimens, x7, USNM 222606.

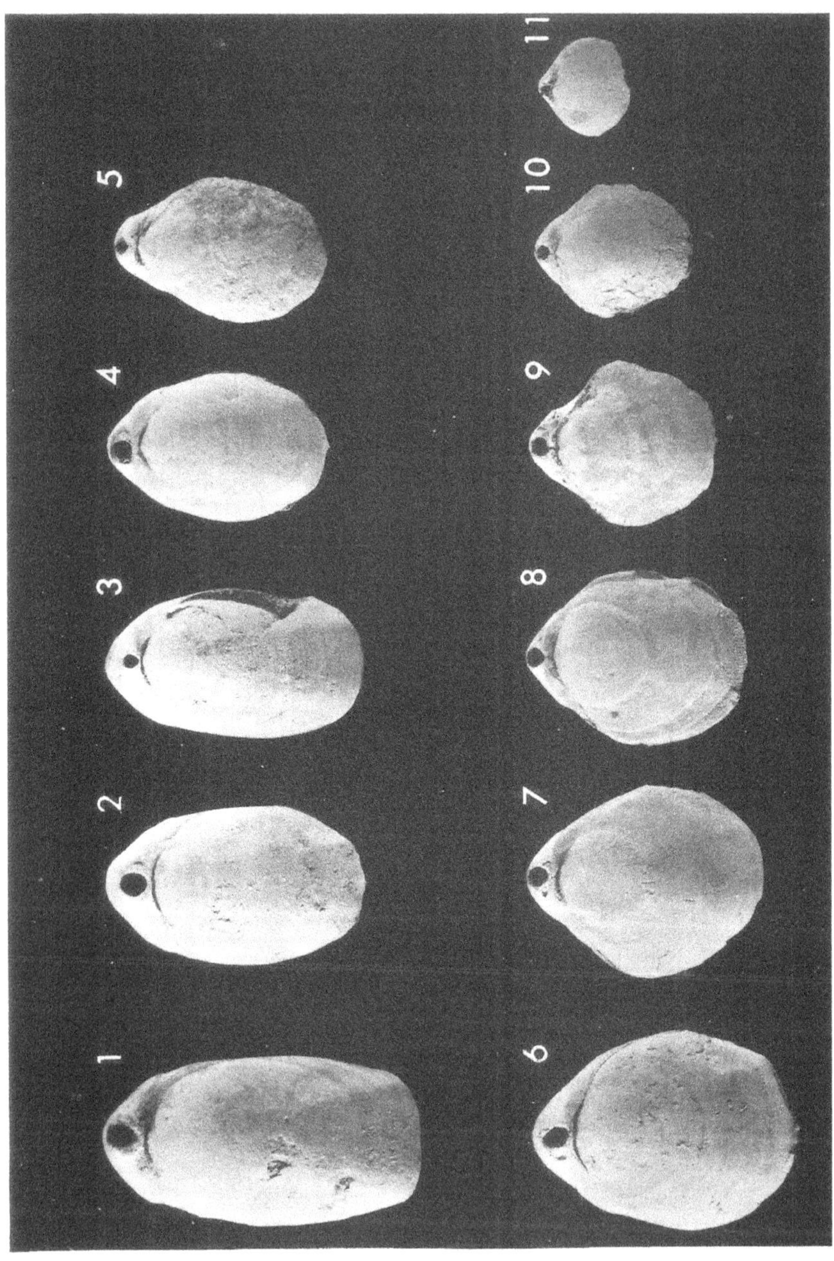

PLATE 3

TABLE 3. Summary of data used in determining the correlation between (D) and (C), where (D) = the diameter of the pedicle foramen and (C) = the width of the constriction of the muscle field at the anterior end of the umbonal cavity. A positive linear correlation was found with r = 0.73. All measurements in millimeters.

	(D)	(C)
N	320	320
OR	1.1–4.9	0.1–10.0
$\bar{x}$	2.820	3.373
r_{cd}	0.73	0.73
Z	16.60	16.60

N = total number of measurements; OR = observed range; $\bar{x}$ = the mean; r = correlation coefficient; and Z = test for null hypothesis. Measurements of all specimens can be obtained from the author on request. Sample (N = 320)—AMNH 32697.

In conjunction with this theory, the degree of correlation between the width of the constriction (C) of the muscle field at the anterior end of the umbonal cavity of the pedicle valve and the diameter (D) of the pedicle foramen was investigated (Table 3). For calculation of the correlation coefficient (R_{cd}) the formula used was:

$$r_{xy} = \frac{\sum_{i=1,2}^{K} (x_i - \bar{x})(y_i - \bar{y})}{\sqrt{\sum_{i=1,2}^{K} (x_i - \bar{x})^2 \sum_{i=1,2}^{K} (y_i - \bar{y})^2}}$$

Where K represents the total number of specimens investigated: x_i represents the width of the constriction of the muscle field in each specimen: $\bar{x}$ represents the mean of the constriction of the muscle field for all specimens investigated; y_i represents the diameter of the pedicle foramen for each specimen; $\bar{y}$ represents the mean of the diameters of the pedicle foramen for all specimens investigated.

EXPLANATION OF PLATE 3

Figures 1-11. *Oleneothyris harlani* (Morton), 1-5, Dorsal view of ontogenetic growth series of Type B specimens, x0.9. USNM numbers 222607-222611 inclusively. 6, dorsal view of largest adult specimen in ontogenetic growth series of Type A specimens, x0.9, USNM 222600. Note external variation in comparison with figure 1, which represents the largest adult specimen of the Type B series. 7-11, Dorsal view of remaining specimens in growth series (Type A), x0.9, USNM numbers 222612-222616 inclusively.

$$\bar{x} = 3.4$$
$$\bar{y} = 2.8$$
$$K = 320$$

r_{xy} was found to equal 0.73. Since r is an estimate of the true but unknown correlation coefficient, the fact that r = 0.73 suggests a strong correlation between (C) and (D) (Text-figure 5). In order to determine the significance of the positive correlation between (C) and (D) the null hypothesis (ρ) = 0 was tested against the alternative hypothesis $\rho \neq 0$. The formula used results in a value assumed by a random variable having approximately the standard normal distribution—the z test (Freund, 1962). Thus, upon substituting r = 0.73, N = 320 and $\rho = 0$ into

$$z = \frac{N-3}{2} \log \frac{1+r}{1-r} \frac{1-\rho}{1+\rho}$$

where z = distribution within a bivariate normal population; N = the total number of specimens investigated; r = the correlation coefficient; and ρ = the correlation between two variables upon application of the null hypothesis to (C) and (D), z was found to equal 16.60. Since this is greater than z = 1.64 (obtained from the table), where z is equal to the critical value (which equals the cutoff point such that when a certain probability of the z statistic exceeds that value the null hypothesis is not true) we must reject the null hypothesis (e.g., if $\rho = 0$ there is no correlation; if $\rho \neq 0$ there is a positive correlation). Thus, there is significant linear relationship between (C) and (D).

Variation in the cardinal process The vast majority of specimens of *Oleneothyris harlani* studied possessed a cup-shaped cardinal process divided anteroposteriorly by a well-defined median ridge. The only significant variant from this type of cardinal process is one in which the median ridge is nonexistent, thus giving the cardinal process a flat appearance. It is entirely possible, however, that the brachial valves of this type underwent such a great amount of wear that the median ridge originally present was removed. Often well preserved articulated specimens, the flat cardinal process is associated with the brachial valves of Type B whereas the deeper, cup-shaped cardinal process with the median ridge is associated with the brachial valves of Type A.

The significance of statistical variation studies in brachiopods A recurring problem in paleontology has been the limited number of characters preserved as hard parts that can be used to define a species. Thus, the morphological characters of the shells of fossil brachiopods that are used to determine taxonomic status may be interpreted differently by different workers, quite often in an arbitrary manner. As of now, no standard approach to the definition of the taxonomic distance between morphological species of brachiopods has yet been agreed upon; the tendency is to rely more on the genus as the basic workable unit, both for evolutionary studies and for the purposes of stratigraphic correlation (Wright, 1972). It would prove useful in paleontological studies to standardize the characterizations of taxa studied. Whenever characters selected for study can he expressed quantitatively, the use of statistical techniques best meets the desired objectives. In order to adequately characterize any pair of variates, only seven quantities require calculation: N, $\bar{x}$, $\bar{y}$ S_x, S_y, r, and OR_x (Imbrie, 1956). The true value of biometrical techniques will not be fully realized until the majority of species are both qualitatively and quantitatively diagnosed.

In statistically characterizing a population (Table 4, Text-figure 6), the mean and standard deviation (the measure of dispersal of measured values about the mean), are extremely useful for analyzing frequency distribution and for comparing distributions of the same type. For example, in comparing length, width, and thickness in the *Oleneothyris harlani* population, I find the standard deviations to be 10.91, 5.35, and 6.35 respectively. It is evident that the various lengths of the *Oleneothyris* brachiopods within the population are less homogeneous than are the widths and thicknesses. If the frequency distribution of the values mentioned above has a bell-shaped form of a normal curve, approximately 68 percent of the measurements will lie in the range from $\bar{x} - 2s$ to $\bar{x} + 2s$, where ($\bar{x}$) is the mean and (s) is the standard deviation (Freund, 1962). Thus, if any measurement is expressed as $\bar{x} = s$, the position of the individual relative to the population would be clearly defined.

The correlation coefficient measures the intensity of the relationship between two variables expressed as the ratio between calculated and observed dispersion from the regression line:

TABLE 4. Statistical Characterization of *Oleneothyris harlani*

	X = length Y = width	X = length Y = thickness	Formula used
N	206	206	
$\bar{x}$	46.98	46.98	$\bar{x} = \frac{\Sigma(x)}{N}$
$\bar{y}$	30.84	25.312	
S_x	10.91	10.91	$s = \sqrt{\frac{\Sigma(d^2)}{N-1}}$ where $d = x - \bar{x}$
S_y	5.35	6.35	
r	0.82	0.93	$r = \frac{\Sigma(x-\bar{x})(y-\bar{y})}{\sqrt{\Sigma(x-\bar{x})^2\Sigma(y-\bar{y})^2}}$
OR_x	16.0–69.9	16.0–69.9	$a = \frac{S_y}{S_x}$
OR_y	13.9 -43.2	8.5–388.9	
a	0.49	0.58	$\sigma a = a\sqrt{\frac{1-r^2}{N}}$
σa	0.02	0.01	
b	7.82	1.94	$b = \bar{y} - \bar{x}a$
D_d	13.10	8.80	$D_d = 100\sqrt{\frac{2(1-r)(S_{x2}S_{y2})}{\bar{x}^2\bar{y}^2}}$
V_x	23.22	23.22	$V = \frac{100S}{\bar{x}}$
V_y	17.35	25.09	
$\sigma\bar{x}_{(x)}$	0.77	0.77	$\sigma\bar{x} = \frac{S}{\sqrt{N}}$
$\sigma\bar{y}_{(y)}$	0.38	0.45	

N = total number of measurements; $\bar{x}$ = mean of x; $\bar{y}$ = mean of y; S_x = standard deviation of x; S_y = standard deviation of y; r = coefficient of correlation; OR = observed range; a = slope of the growth line (growth ratio); σa = standard error of the slope; b = initial growth index; D_d = coefficient of relative dispersion from the reduced major axis; V_x = coefficient of variation for x; V_y coefficient of variation for y; $\sigma\bar{x}_{(x)}$ = standard error of the mean for x; $\sigma\bar{y}_{(y)}$ = standard error of the mean for y. All measurements in millimeters. Sample (N = 206)—AMNH 32698.

$$r_{xy} = \frac{S_{yc}}{S_{yo}}$$

where ($S_{yc)}$ is the variance of the calculated values of the variable (y) corresponding with the values of the variable (x) along the line Y = ax – b (Beerbower, 1968). As (r) approaches – 1 from zero the degree of negative linear correlation increases, while as (r) approaches + 1 from zero the degree of positive linear correlation increases. The significance of (r) for the *Oleneothyris* population is discussed in more detail above.

Upon analysis of the variants within the *Oleneothyris harlani* population, it was necessary to define the line of growth that best fits the observed trends. In order to define this line the growth ratio (a) and initial growth index (b) were required. In the scattergrams, an algebraic solution for the line of "best fit" through the middle of the distribution was obtained by calculation of the reduced major axis. This line minimizes the sum of the areas of the triangles formed by lines drawn from each point to the desired line and parallel with the x and y axes. The advantages to using this method of analysis is fourfold (Imbrie, 1965): (1) it makes no assumptions of independence; (2) it is invariant under change of scale; (3) it is simple to compute; and (4) results obtained from its use are intuitively more reasonable than corresponding results obtained from regression analysis.

According to Imbrie (1956), calculation of the reduced major axis in paleontology avoids the major drawback of regression analysis of (y) on (x) and of (x) on (y); namely, that in regression analysis the assumption exists that all the dispersion involved is due to deviations in one variate. From the paleontological and neontological point of view this assumption is never justified, for biological variability and observational errors are always involved in both variates (Kermak and Haldane, 1950, cited in Imbrie, 1956: 230). Shortcomings of the reduced major axis method are discussed by Simpson, Roe, and Lewontin (1960: 402).

After plotting the scatter of points around the reduced major axis, one can measure the total dispersion in the bivariate morphological field by computing the vector sum of (d_x) and (d_y), where (d_x) is the distance from a single point along the x axis to the reduced major axis and (d_y) is the distance from a single point along the y axis to the reduced major axis. Once this vector sum is calculated, the total dispersion about the reduced major axis can be expressed as the standard deviation (S_d) of all the vector sums (Imbrie, 1956: 239):

$$S_d = \sqrt{2(1-r)(S_{x^2} - S_{y^2})}$$

Thus, (S_d) can be used as a measure of absolute variation. It is also a measure of shape variability, since any deviation in (x) or (y) affects (S_d). By using (S_d) one is able to calculate the coefficient of relative dispersion around the reduced major axis (D_d), which is the ratio, expressed in percent, between the standard deviation of the vector sums of the deviations from the reduced major axis in the x and y directions and the distance from the origin to the joint mean of the sample (Imbrie, 1956). In paleontology this statistic is useful in interpreting the degree of shape variation as a proportion of the average shape of the sample.

The coefficient of variation (V) is significant in invertebrate paleontology because most organisms tend to show an absolute variation that is proportional to absolute size. Thus, when comparing two fossil populations of different average size the calculation of (V) can help in comparing inherent morphological variability, provided that each sample is homogeneous with respect to growth stage (Imbrie, 1956). If *Oleneothyris subfragilis* is ever erected as a subspecies or new species, the coefficient of variation for the *Oleneothyris harlani* population will be invaluable in characterizing the new population. However, in addition to the calculation of (V), one must compute (S_d) and (D_d), since (V) is a measure of shape variation in only one dimension.

CONCLUSIONS

1) The macrofauna of the Upper Hornerstown Formation is characterized by high density and low diversity; the diversity is such that the variety is moderate while the equitability is extremely low.
2) Deposition of the upper part of the formation occurred in shallow water at a slow rate; the depositional environment was one of strong wave and current action.
3) The *Oleneothyris* community represents a biocoenose; *post-mortem* transport was minimal.
4) The *Oleneothyris* community conforms in some ways to Recent Arctic and Boreal communities analyzed by Turpaeva (1957). Namely, the community was dominated by one trophic group; the dominant species belonged to one trophic group while the next most dominant species

belonged to a different trophic group; one species dominated in terms of biovolume.

5) The similarity in trophic structure between the *Oleneothyris* community and cold water faunas is the result of environmental stress, probably in the form of current and wave action, which coincided with a regression of the sea. *Oleneothyris harlani* was an explosive opportunist limited geographically to the Upper Hornerstown Formation and appearing at only one point stratigraphically.
6) As deposition of the Hornerstown Formation progressed turbulence increased. Morphologic changes in the *Oleneothyris* community can be correlated with this increasing turbulence such that more flat shells with a narrower pedicle foramen and greater umbonal constriction (Type A) are found at the lower part of the biostrome than the cylindrical variants with relatively wider pedicle openings (Type B). As selection pressures (e.g. turbulence) increased, Type A shells were selected against, resulting in a greater proportion of Type B shells at the top of the biostrome.

ACKNOWLEDGMENTS

Special thanks are due to Dr. Joaquin Rodriguez of Hunter College, CUNY, who originally suggested this study and without whose encouragement and assistance this paper would not have been possible. I should also like to express appreciation to Dr. G. A. Cooper of the U. S. National Museum, Washington, D.C., for invaluable suggestions and criticism. Dr. Serge Wind assisted in checking the statistical calculations. Drs. R. K. Olsson and S. K. Fox of Rutgers University and Dr. Frederick C. Shaw of Lehman College, CUNY, provided additional useful criticism and advice. Mrs. Yvette Vogel deserves thanks for drafting the text-figures.

REFERENCES

Adams, J. K. 1963. Petrology and origin of the Lower Tertiary formations of New Jersey. *Journal of Sedimentary Petrology* 33: 587-603.

Ager, D. V. 1965. Serial grinding techniques. In B. Kummel and D. Raup (eds.), *Handbook of Paleontological Techniques*, 212-224. San Francisco: Freeman.

Ager, D. V. 1967. Brachiopod paleocology. *Earth Science Reviews* 3: 157-179.

Ager, D. V., and Riggs, E. A. 1964. The internal anatomy, shell growth and asymmetry of a Devonian spiriferid. *Journal of Paleontology* 3: 749-760.

Baker, P. G. 1972. The development of the loop in the Jurassic brachiopod *Zeilleria leckenbyi. Palaeontology* 15: 450-472.

Beerbower, J. R. 1968. *Search for the Past.* Englewood Cliffs: Prentice-Hall.

Bergquist, H. R., and Coggan, W. A. 1957. Mollusks of the Cretaceous. In H. Ladd (ed.), Treatise on marine ecology and paleontology. *Geological Society of America Memoir* 67 (2): 871-884.

Blackwelder, R. E. 1967. *Taxonomy: A text and reference*. New York: Wiley.

Boucot, A. J. 1975. *Evolution in Extinction Rate Controls.* Amsterdam: Elsevier.

Bowen. Z. P. 1966. Intraspecific variation in the brachial cardinalia of *Atrypa reticularis. Journal of Paleontology* 40: 1017-1022.

Burma, B. J. 1948. Studies in quantitative paleontology 1: Some aspects of the theory and practice of quantitative invertebrate paleontology. *Journal of Paleontology* 22: 725-761.

Burt, W. H. 1954. The subspecies categories in mammals. *Systematic Zoology* 3: 99-104.

Chilingar, G. V. 1956. Joint occurrence of glauconite and chlorite in sedimentary rocks. *American Association of Petroleum Geologists Bulletin* 40: 394-398.

Clark, B. L. 1945. Problems of speciation and correlation as applied to mollusks of the marine Cenozoic. *Journal of Paleontology* 19: 158-172.

Clark, R. B. 1956. Species and systematics. *Systematic Zoology* 5: 110-120.

Clark, W. B. 1907. The classification adopted by the U. S. Geological Survey for the Cretaceous deposits of New Jersey, Delaware, Maryland, and Virginia. *Johns Hopkins University Circular, New Series* 7, Whole No. 199.

Clark, W. B., Bagg, R. M., and Shattuck, G. B. 1897. Upper Cretaceous formations of New·Jersey, Delaware, and Maryland. *Geological Society of America Bulletin* 8: 315-358, pl. 40-50.

Clark, W. B., and Martin, G. C. 1901. *Maryland Geological Survey, Eocene.* Baltimore: Johns Hopkins.

Cloud, P. E. 1942. Terebratuloid brachiopoda of the Silurian and Devonian. *Geological Society of America Special Paper* 38.

Colbert, E. H. 1948. A hadrosaurian dinosaur from New Jersey. *Academy of Natural Science, Philadelphia, Proceedings* 100: 23-38.

Cook, G. H. 1868. *Geology of New Jersey.* Newark: Geological Survey of New Jersey.

Cooke, C. W., and Stephenson, L. W. 1928. The Eocene age of the supposed Upper Cretaceous greensand marls of New Jersey. *Journal of Geology* 36: 139-148.

Cooper, G. A. 1942. New genera of North American brachiopods. *Journal of Washington Academy of Science* 32: 228-235.

Cooper, G. A. 1944. Phylum Brachiopoda. In H. W. Shimer and R. R. Shrock (eds.), *Index Fossils of North America*. New York: Wiley.

Cooper, G. A. 1957a. Brachiopods (annotated bibliography). In J. W. Hedgepeth (ed.), Treatise on marine ecology and paleoecology. *Geological Society of America, Memoir* 67(1).

Cooper, G. A. 1957b. Loop development of the Pennsylvanian Terebratulid *Cryptacanthia. Smithsonian Miscellaneous Collection* 134: 18, pl. 1-2.

Cope, T. F. 1965. Statistical analysis. Queens College, New York.

Dorf, E., and Fox, S. K. 1957. Cretaceous and Cenozoic of the New Jersey Coastal Plain. *Geological Society of America Guidebook for Field trips, Atlantic City meeting*, 3-15.

Doutt, J. K. 1955. Terminology of microgeographic races in mammals. Systematic Zoology 4: 179-185.

Durrant, S. D. 1955. In defense of the subspecies. Systematic Zoology 4: 186-190.

Easton, W. H. 1960. *Invertebrate Paleontology*. New York: Harper Bros.

Edward, J. G. 1954. A new approach to infraspecific categories. *Systematic Zoology* 3: 1-20.

Enright, L. C. E., Jr. 1969. The stratigraphy, micropaleontology and paleoenvironmental analysis of the Eocene sediments of the New Jersey Coastal Plain. PhD thesis. *Dissertation abstracts* 30: 3706B.

Enright, R. 1969. The stratigraphy and clay mineralogy of the Eocene sediments of the northern New Jersey Coastal Plain. In S. Subitsky (ed.), *Geology of Selected Areas in New Jersey and Eastern Pennsylvania*. New Brunswick: Rutgers University Press.

Feldman, H. R. 1974. Morphologic variation in a Paleocene Terebratulid brachiopod from the Hornerstown Formation of New Jersey. *Geological Society of America Northeastern Section Meeting, Baltimore, Maryland.*

Fox, R. M. 1955. On subspecies. *Systematic Zoology* 4: 93-95.

Fox, S. K., Jr., and Olsson, R. K. 1955. Stratigraphy of Late Cretaceous and Early Tertiary formations in New Jersey. Journal of Paleontology 29: 736-757

Freund, J. E. 1962. *Mathematical Statistics.* Englewood Cliffs: Prentice-Hall.

Gabb, W. M. 1861. Synopsis of American Cretaceous brachiopoda. Academy of Natural Science, Philadelphia Proceedings 13: 18-19.

Galliher, E. W. 1935. Geology of glauconite. *American Association of Petroleum Geologists Bulletin* 19: 1569-1601.

Gosline, W. A. 1954. Further thoughts on subspecies and trinomials. *Systematic Zoology* 3: 92-94

Graecen, K. 1941. The stratigraphy and correlation of the Vincentown Formation. *New Jersey Geological Survey Bulletin* 52.

Hagmeier, E. M. 1958. Inapplicability of the subspecies concept to North American Marten. *Systematic Zoology* 7: 1-7.

Herrick, S. M. 1962. Marginal sea of Middle Eocene age in New Jersey. Geological survey research. *U. S. Geological Survey Professional* Paper 450: B56-58.

Hofker, J. 1955. The foraminifera of the Vincentown Formation. *McLean Foraminifera Lab Report* 2: 1-21, pl. 1-6.

Imbrie, J. 1956. Biometrical methods in the study of invertebrate fossils. *American Museum of Natural History Bulletin* 108: 217-252

Jennings, P. H. 1936. A microfauna from the Monmouth and basal Rancocas groups of New Jersey. *American Paleontology Bulletin* 23: 161-234, pl. 28-34.

Johnson, C. W. 1905. Annotated list of the types of invertebrate CretaceouS fossils in the collection of the Academy of Natural Science. *Academy of Natural Science, Philadelphia Proceedings* 7: 4-28.

Johnson, M. E., Meredith, E., and Richards, H. G. 1952. Stratigraphy of the Atlantic Coastal Plain. American *Association of Petroleum Geologists Bulletin* 36: 2150-2160.

Kermack, K. A., and Haldane, J. B. S. 1950. Organic correlation and allometry. *Biometrika* 37: 30-41.

Le Grande, H. E. 1961. Summary of geology of Atlantic Coastal Plains. *American Association of Petroleum Geologists Bulletin* 45: 1557-1571.

Loeblich, A. R., Jr., and Tappan, H. 1957a. Correlation of the Gulf and Atlantic Coastal Plains Paleocene and Lower Eocene formations by means of Planktonic foraminifera. *Journal of Paleontology* 31: 1109-1137.

Loeblich, A. R., Jr., and Tappan, H. 1957b. Planktonic foraminifera of Paleocene and Eocene from the Gulf and Atlantic Coastal Plains. *U. S. Natural Museum Bulletin* 125: 173-198, pl. 40-64.

Mansfield, G. R. 1922. Potash in the greensands of New Jersey. *U. S. Geological Survey Bulletin* 727.

Marcou, J. 1853. (A) *Geological Map of the United States, and the British Provinces of North America; with an Explanatory Text, Geological Sections, and Plates of the Fossils which Characterize the Formations*. Boston: Gould and Lincoln, Boston.

Mayr, E. 1942. *Systematics and the Origin of Species*. New York: Columbia University Press.

Mayr, E. 1948. Speciation and systematics. In G. L. Jepsen and G. A. Cooper (eds.), *Princeton University bicentennial conference*, series 2, conference 3: 281-298.

Mayr, E. 1954. Notes on nomenclature and classification. *Systematic Zoology* 3: 86-89.

Mayr, E. 1963. *Animal Species and Evolution*. Cambridge, MA: Belknap Press of Harvard University Press.

Mayr, E. 1969. *Principles of Systematic Zoology*. New York: McGraw-Hill.

Linsley, E. G., and Usinger, R. L. 1953. *Methods and Principles of Systematic Zoology*. New York: McGraw-Hill.

McKerrow, W. S. 1952. Notes on the species and subspecies in paleontology. *Geology Magazine* 89: 148-151.

McKerrow, W. S. 1953. Variation in the Terebratulacea of the Fuller's Earth rock. *Geological Society Quarterly Journal* 109: 97-124.

McLean, J. D. 1953. A summary of the guide fossil foraminifera of the Atlantic Coastal Plain between New Jersey and Georgia: A revision. *McLean Foraminifera Lab Report* 1: 1-6.

Menzies, R. J., George, R. Y., and Rowe, G. T. 1973. *Abyssal Environment and Ecology of the World Oceans*. New York: Wiley.

Miller, H. 1956. Correlation of Paleocene and Eocene formations and Cretaceous-Paleocene boundary in New Jersey. *American Association of Petroleum Geologists Bulletin* 40: 722-736.

Minard, J. P., and Owens, J. P. 1960. Differential subsidence of the southern part of the New Jersey Coastal Plain since Early Late Cretaceous time. *U. S. Geology Survey Professional Paper* 400-B: BI84.

Minard, J. P., Owens, J. P., Sohl, N. F., Gill, H. E., and Mello, J. F. 1969. Cretaceous-Tertiary boundary in New Jersey, Delaware, and Maryland. *U. S. Geology Survey Bulletin* 1274-H.

Morton, S. G. 1828a. Description of the fossil shells which characterize the Atlantic Secondary Formation of New Jersey and Delaware; including four new species. *Academy of Natural Science Philadelphia, Journal* 6: 73-76, pl. 3. fig. 1-4, 7-8.

Morton, S. G. 1828b. Description of some new species of organic remains of the Cretaceous group of the U. S.: with a tabular view of the fossils hitherto discovered in this formation. *Academy of Natural Science Philadelphia, Journal* 8: 207-227.

Morton, S. G. 1830. Synopsis of the organic remains of the ferruginous sand formation of the U. S.: with geological remarks. *American Journal of Science* 17: 274-295, continued in 18: 243–250, pl. 3, fig. 16-17.

Morton, S. G. 1834. *Synopsis of the Cretaceous group of U. S.* Philadelphia: Key and Biddle.

Muir-Wood, H. M. 1965. Strophomenida. In R. C. Moore (ed.), *Treatise on Invertebrate Paleontology*, Part H, Brachiopoda 1: 361-521. Lawrence, KS: University of Kansas Press and Geological Society of America.

Newell, N. D. 1947. Infraspecific categories in invertebrate paleontology. *Evolution* 1: 163-171.

Newell, N. D. 1956. Fossil populations. In P. C. Sylvester-Bradley (ed.), *The Species Concept in Paleontology*, 63-82. London: The Systematic Associaion pub. no. 2.

Olsson, R. K. 1957. Late Cretaceous and Early Tertiary stratigraphy of New Jersey (abstract). *Geological Society of America Bulletin* 68 (12) pt.2: 1776.

Olsson, R. K. 1959. Late Cretaceous-Early Tertiary stratigraphy of New Jersey. *Dissertation Abstracts* 19: 2063-2064.

Olsson, R. K. 1963. Latest Cretaceous and Earliest Tertiary stratigraphy of the New Jersey Coastal Plain. *American Association of Petroleum Geologists Bulletin* 47: 643-665.

Olsson, R. K. 1970. Paleocene planktonic foraminiferal biostratigraphy and paleozoogeography of New Jersey. *Journal of Paleontology* 44: 589-597.

Olsson, R. K, and Gaffney, E. S. 1970. The Cretaceous-Tertiary datum in New Jersey. *Geological Society of America Northeastern Section Meeting*, Pittsburgh, Pennsylvania.

d'Orbigny, A. A. 1850. Prodrome de paleontologie stratigraphique universelle des animaux mollusques et rayonnes. *Victor Masson* 2: 1-427.

Owens, J. P., and Minard, J. P. 1960. Some characteristics of glauconite from the Coastal Plain formations of New Jersey. *U. S. Geological Survey Professional Paper* 400-B: 430-432.

Owens, J. P., Minard, J. P., and Sohl, N. F. 1969. Shelf and deltaic paleoenvironments in the Cretaceous-Tertiary formations of the New Jersey

Coastal Plain. In S. Subitsky (ed.), *Geology of Selected Areas in New Jersey and Eastern Pennsylvania*, 235-378. New Brunswick, NJ: Rutgers University Press.

Parkes, K. C. 1955. Sympatry, allopatry and the subspecies in birds. *Systematic Zoology* 4: 35-39.

Pratt, D. M., and Campbell, D. A. 1956. Environmental factors affecting growth in *Venus mercenaria. Limnology and Oceanography* 1: 2-17.

Purdy, E. G. 1964. Sediments as substrates. In J. Imbrie (ed.), *Approaches to Paleoecology*, 238-271. New York: Wiley.

Raup, D. M., and Stanley, S. M. 1971. *Principles of Paleontology.* San Francisco: W. H. Freeman.

Richards, H. G. 1947. Invertebrate fossils from deep wells along the Atlantic Coastal Plain. *Journal of Paleontology* 21: 23-37

Richards, H. G. 1948. Studies on the geology and paleontology of the Atlantic Coastal Plain. *Academy of Natural Science, Philadelphia Proceedings* 100: 39-76.

Richards, H. G. 1967. Stratigraphy of the Atlantic Coastal Plain between Long Island and Georgia: Review. American Association of Petroleum Geologists Bulletin 51: 2400-2429.

Richards, H. G., Groot, J. J., and Germeroth, R. M. 1957. Cretaceous and Tertiary geology of New Jersey, Delaware, and Maryland. *Geological Society of America Guidebook for Fieldtrips, Atlantic City Meeting* 1957—Fieldtrip no. 6: 18-230.

Rudwick, M. J. S. 1959. The growth and form of brachiopod shells. *Geological Magazine* 94: l-24.

Rudwick, M. J. S. 1962. Notes on the ecology of brachiopods in New Zealand. *Royal Society of New Zealand Transactions and Proceedings* 1: 327-335

Schlanger, S. O. 1954. The petrology of the Vincentown formation. *Journal of Sedimentary Petrology* 24:·212-217.

Schuchert, C. 1897. A synopsis of American fossil brachiopods including biblioliography and synonymy. *U. S. Geological Survey Bulletin* 87: 1-464.

Schuchert, C. and Levene, C. M. 1929. Brachiopoa (Generum et Genotyporum Index et Bibliographia). *Fossilium Catalogus 1, Animalia*, 42: 1-140.

Simpson, C. G. 1943. Criteria for genera, species, and subspecies in zoology and paleontology. *New York Academy of Science Annals* 44: 170-177.

Simpson, C. G., Roe, A., and Lewontin, J. 1960. *Quantitative Zoology.* New York: McGraw-Hill.

Spangler, W. B., and Peterson, J. J. 1950. Geology of the Atlantic Coastal Plain in New Jersey, Delaware, Maryland and Virginia. *American Association of Petroleum Geologists Bulletin* 34: 1-99.

Stanley, M. 1972. Functional morphology and evolution of Byssally attached bivalve mollusks. *Journal of Paleontology* 46: 165-212.

Starrett, A. 1958. What is the subspecies problem? *Systematic Zoology* 7: 111-115.

Stehli, F. G. 1956. Evolution of the loop and lophophore in terebratuloid brachiopods. *Evolution* 10: 187-200.

Sowerby, J. 1825. *Mineral Conchology of Great Britain*. 5: 49-54

Swinnerton, H. H. 1940. The study of variation in fossils. *Geological Society of London Quarterly Journal* 96: 77-118.

Sylvester-Bradley, P. C. 1951. The subspecies in paleontology. *Geological Magazine* 88: 88-102.

Tilden, J. W. 1961. Certain comments on the subspecies problem. *Systematic Zoology* 10: 17-23.

Trueman, A. E. 1924. The species concept in paleontology. *Geological Magazine* 61: 355-360.

Trueman, A. E., and Weir, J. 1946. A monograph of the British Carboniferous non-marine Lamellibranchia pt. 1. *Palaeontological Society* 99: 1-18.

Turpaeva, E. P. 1957. Food interrelationships of dominant species in marine benthic biocoenoses. In B. N. Nikitkin (ed.), *Transa*, Institute of Oceanography, Marine Biology USSR Academy Sciences Press 20: 137-148. (Published in the United States by the American Institute of Biological Science, Washington, D.C.)

Walker, K. R. 1972. Trophic analysis: A method for studying the function of ancient communites. *Journal of Paleontology* 46: 82-93.

Weller, S., and Knapp, G. N. 1907. A report on the Cretaceous paleontology of New Jersey, based upon stratigraphic studies of George N. Knapp. *New Jersey Geological Survey, Paleontology* Ser. 4.

Wermund, E. G. 1961. Glauconite in Early Tertiary sediments of Gulf Coastal Province. *American Society of Petroleum Geologists Bulletin* 45: 1667-1696.

Whitfield, R. P. 1886. Brachiopoda and lamellibranchiata of the Raritan clays and greensand marls of New Jersey. *U. S. Geological Survey Monograph* 9: 6-9.

Williams, A. 1957. Evolutionary rates of brachiopods. *Geological Magazine* 94: 201-211.

Williams, A., and Wright, A. D. 1961. The origin of the loop in articulate brachiopods. *Palaeontology* 4: 149-176.

Williams, A., Rowell, A. J., Brunton, C. H. C., and Carlson, S. J. 1997. Brachiopoda. In R. L. Kaesler (ed.), *Treatise on Invertebrate Paleontology, Part H.* 6 vols. Boulder, CO and Lawrence, KS: The Geological Society of America and University of Kansas Press.

Wilmarth, M. G. 1938. Lexicon of geologic names of the United State (including Alaska). *U. S. Geological Survey Bulletin* 896 Part 1.

Wilson, E. O., and Brown, W. L., Jr. 1953. The subspecies concept and its taxonomic application. *Systematic Zoology* 2: 92-111.

Wood, A., and T. Barnard. 1946. Ophthalmidium: A study of nomenclature, variation and evolution in the foraminifera. *Geological Society of London Quarterly Journal* 102: 77-113.

Wright, A. D. 1972. The relevance of zoological variation studies to the generic identification of fossil brachiopods. *Lethaia* 5: 1-13.

CHAPTER SEVEN

NOTES ON AND DESCRIPTION OF *OLENEOTHYRIS FRAGILIS* (MORTON) 1828 (BRACHIOPODA, TEREBRATULIDAE)

ABSTRACT

The *Oleneothyris* biostrome, a well-known stratigraphic marker at the top of the Hornerstown Formation in the Atlantic Coastal Plain, contains two distinct species of a large terebratulid brachiopod. These two species, *O. harlani* (Morton) and *O. fragilis* (Morton), may be differentiated by the prominent sulcus and fold, along with two pronounced dorsal ridges on the less robust *O. fragilis.* The latter species is confined vertically to a lens within the biostrome and geographically to the area around New Egypt, New Jersey. A narrowly triangular, strongly arched loop, characteristically terebratulid, is found to differ slightly from that of *O. harlani.* Since the holotype of *O. fragilis,* first described in 1828 by S. G. Morton, is either no longer in existence or lost, a neotype (RU 5700) is herein designated. The neotype is housed in the Rutgers University Geological Museum, New Brunswick, New Jersey.

INTRODUCTION

The present study was undertaken in the spring of 1973 when I discovered a small lens of brachiopods in the *Oleneothyris* biostrome (Figure 1), an important stratigraphic marker in the Hornerstown Formation of the Atlantic Coastal Plain. The brachiopods in the lens differed significantly from the well-known terebratulid *O. harlani.* Subsequently, I found additional specimens of this "variant" in collections at the Rutgers University Geological Museum and the American Museum of Natural History. In both cases the brachiopods were labeled *O. harlani* and designated as having been collected from the biostrome at the top of the Hornerstown Formation in the vicinity of New Egypt, New Jersey (Figure 2). Upon re-examination of several hundred specimens of *O. harlani* and comparison with 93 specimens of *O. fragilis,* no significant morphological intergradation between the two was found. The existence of a prominent sulcus and fold along with two

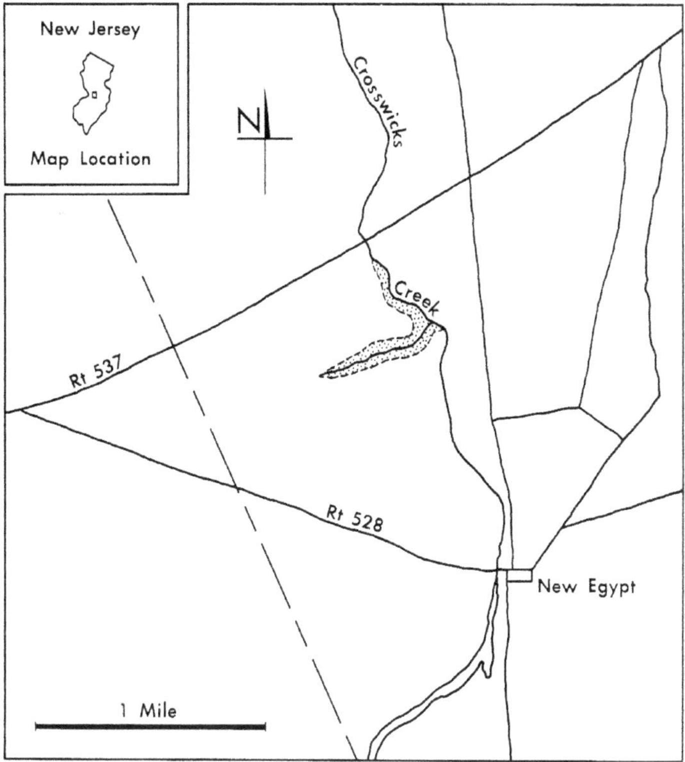

FIGURE 1. Type locality of *Oleneothyris fragilis* in the Hornerstown Formation (Paleocene) of the New Jersey Atlantic Coastal Plain. Stippled area approximately in the center of map, along Crosswicks Creek and northeasterly flowing tributary, represents outcrop of Hornerstown glauconite at the top of which is the *Oleneothyris* biostrome. Immediately above the biostrome is the gradational contact between the Hornerstown and overlying Vincentown formation.

distinct ridges on the brachial valve of *O. fragilis* possibly imply the existence of two distinct genera, but the internal morphology of all specimens in the sample was identical to that of *Oleneothyris harlani.*

Oleneothyris harlani was first described by Morton (1828) as *Terebratula harlani*, at which time he referred to two distinct species—*T. harlani* and *T. fragilis.* Morton (1834) clarified his original descriptions of the variant forms of *T. harlani;* upon further examination of a few hundred specimens he decided that two different species did not exist: rather there were two "varieties" that graded into each other. *T. fragilis* was renamed *subfragilis* by d'Orbigny (1850), as *fragilis* was preoccupied by Schlotheim (1813) for a species of *Terebratula.* The significance of the name *fragilis* was related to the supposedly fragile nature of the shell. However, I have found the shells of *fragilis* no more or less fragile than those of *Oleneothyris harlani.* Perhaps some valves were affected by dissolution to a larger extent than others, thus making them appear to be more delicate, hence fragile. In 1942 Cooper erected the genus *Oleneothyris,* which at the time was monotypic. Accordingly, after reinvestigation of available material, Morton's (1828) *Terebratula fragilis* is now placed in the genus *Oleneothyris* and the specific name *subfragilis* is no longer applicable. Morton (1828: 75) based his description of *T. fragilis* on one specimen, which is no longer extant. Correspondence with various museums and search of collections, such as those in the American Museum of Natural History, the Academy of Natural Sciences, and the Paleontological Research Institution, has provided no evidence for the existence of Morton's holotype. However, Morton had access to another specimen of the same species that belonged to Conrad, which I located in Conrad's collection in the Academy of Natural Sciences, Philadelphia. Upon careful analysis of the brachiopod, which is slightly deformed but in a good to excellent state

STAGE		AGE	FORMATION	THICKNESS	ENVIRONMENT	LITHOLOGY
Ypresian	Wilcox	Eocene	Manasquan Fm. Farmingdale Mbr.	40 ft.	Mid-outer shelf	Slightly clayey, medium coarse quartzose glauconitic sand.
Thanetian	Wilcox / Midway	Paleocene	Vincentown Fm.	15-100 ft.	Mid-inner shelf	Quartz sands with calcarenite facies. Silt facies in shallow subsurface.
Montian / Danian	Midway	Paleocene	biostrome Hornerstown Fm.	20-30 ft.	Mid-inner shelf	Almost pure glauconitic sand. Little fine-grained matrix.
Maestrichtian	Navarro	Cretaceous	Tinton Fm.	22 ft.	Inner shelf	Argillaceous, quartz, glauconite sand.

FIGURE 2. Upper Cretaceous and Lower Tertiary formations of northeastern New Jersey (modified from Olsson, 1975), illustrating the stratigraphic setting of the Hornerstown Formation and the *Oleneothyris* biostrome.

of preservation, the following may be concluded: the shell definitely possesses the morphologic characters of *Oleneothyris fragilis* and is not intermediate between *O.fragilis* and *O. harlani.* Morton was correct in originally erecting *fragilis* as a new species, even though his decision was based upon a single specimen. The clarification of the morphology of the species based upon a population study is presented herein, along with an amplification of Morton's description and designation of a neotype.

SYSTEMATICS

TEREBRATULIDA Waagen, 1883
TEREBRATULIDINA Waagen, 1883
TEREBRATULACEA Gray, 1840
TEREBRATULIDAE Gray, 1840
TEREBRATULINAE Gray, 1840
OLENEOTHYRIS Cooper, 1942

Oleneothyris fragilis, Figures 3-7 A-D; 8.

Terebratula fragilis Morton, 1828: 75, pl. 3, figs. 3-4; Morton, 1830: 250, pl. 3, fig. 17; Morton, 1834 *(non* Sowerby 1825): 70, pl. 3, fig. 2.
Terebratula camella Morton, 1834: 70, in text.
Terebratula subfragilis d'Orbigny, 1850: 258, in text.
Terebratula harlani Whitfield, 1886: 6-9, pl. 1, figs. 15-18.
Oleneothyris harlani Feldman *(pars)* (1977): pl. 2, figs. 1-4.

Type species: Terebratula fragilis Morton, 1828.

Type specimens: The following figured type specimens are deposited in the American Museum of Natural History, the Rutgers University Geological Museum, and the Academy of Natural Sciences. Not illustrated are four additional specimens labeled by F. B. Meek in the collection at the National Museum of Natural History (G. A. Cooper, personal communication):

Neotype. RU 5700 (figs. 3D-F; 8F). Collected from the biostrome at the top of the Hornerstown Formation near New Egypt, New Jersey. Collector: H. R. Feldman, 1973.

Topotype. ANSP 19476 (fig. 3A-C). Same formation and locality. Collector. S. W. Conrad, 1828 (?).

Topotype. AMNH 32680 (figs. 4A-C; 8D). Same formation and locality. Collectors: H. R. Feldman and S. Feldman, 1974.

Topotype. AMNH 32681 (figs. 4D-F; 8A). Same formation and locality. Collectors: H. R. Feldman and M. Perlman, 1974.

Topotype. AMNH 32682 (figs. 5A-C; 8E). Same formation and locality. Collector: unknown.

Topotype. AMNH 32683 (figs. 5D-F; 8C). Same formation and locality. Collector: H. R. Feldman, 1975.

Topotype. AMNH 32684 (figs. 6A-C; 8B). Same formation and locality. Collector: H. R. Feldman, 1975.

Whitfield's (1886) collection, housed at the American Museum of Natural History, was studied in order to determine whether the designation of a neotype was necessary. Since Morton's (1828) holotype was either lost or destroyed, Whitfield's types might have helped establish the identities of the closely similar species, as required by the Code (Art. 75ai) (Stoll et al., 1961). However, Whitfield's specimen AMNH 32697 is incorrectly described as f*ragilis* var.; it is actually a form of *O. harlani.* Also confusing is AMNH 32698, which is described as *fragilis* var. but appears to be intergradational between *O. fragilis* and *O. harlani.* It possesses a dorsal sulcus and weakly defined median ridge on the ventral valve, but is extremely lobate, closely resembling Feldman's (1977) Type A. In addition, it is a juvenile and its morphology may change significantly during ontogeny. Whitfield's remaining types labeled *fragilis* var. are:

AMNH 32699: a neanic specimen of *O. fragilis.* AMNH 32700, 32701: elongately deformed specimens of *O. fragilis.* AMNH 32702, 32703, 32704, 32705: Specimens of *O. fragilis* that closely resemble the types of the present paper.

Specimens designated *O. harlani* by Whitfield (1886) include:

AMNH 32706: labeled var. *perovalis* (Morton) but is actually *O. harlani,* Type A of Feldman (1977). AMNH 32707: resembles AMNH 32698 but has a less distinct sulcus and fold and lacks a ventral median ridge. It is also intergradational between *O. fragilis* and *O. harlani.*

Subsequent to the discovery of Whitfield's collection in the American Museum of Natural History, I located the specimen from S. W. Conrad's collection, which was referred to by Morton (1828: 76) in his description of the new species *T. fragilis:* "I possess but a single specimen, from which the annexed drawing was taken: another in S. W. Conrad's collection has its sides much more rounded, but appears to be specifically the same." Although slightly deformed and a bit damaged, Conrad's shell (ANSP 19476) is definitely *O. fragilis* and is not at all intergradational with *O. harlani.*

Although Morton did not use the word "holotype," his specimen served the same function, which can be inferred from his description of a new species based on a single shell. As stated in Article 73(b) of the Code (1961), if an author says in the description of a new nominal species that one specimen and only one is "the type" or uses some equivalent expression, that specimen is the holotype (italics mine). Thus, Morton's single specimen was, in actuality, the holotype. Since the holotype is either lost or no longer in existence, and since Morton had no type series (hypodigm)—only the single specimen—no paratypes, lectotype, or syntypes could possibly exist. Conrad's specimen, although seen and mentioned by Morton, was not part of the original hypodigm. Therefore, in accordance with the Code (ibid., Art. 75), I herein designate RU 5700 the neotype for *Oleneothyris fragilis.* The neotype is deposited in the Rutgers University Geological Museum, New Brunswick, New Jersey.

Type Locality: Shingle Run, an easterly flowing tributary of Crosswicks Creek near New Egypt, New Jersey (See Figure 1).

Type Horizon: A lens approximately in the middle of a biostrome of *Oleneothyris harlani* shells situated at the top of the Hornerstown Formation. The matrix is almost pure glauconitic sand.

Material: Ninety-three specimens: fifty-seven articulated (twenty-seven well preserved, thirty damaged), thirty pedicle valves, six brachial valves, most of which are in an excellent state of preservation.

Geographic Distribution: The species is known only from the biostrome at the top of the Hornerstown Formation in the area of New Egypt, New Jersey. Although *O. harlani* has been found in Delaware, Maryland, and Alabama, there are no reports or references in the literature to the existence of *O. fragilis* in areas other than the type locality.

DESCRIPTION

Exterior: The shells are medium-sized and range from slightly lobate to cylindrical. All forms are rostrate and astrophic. The pedicle valve of adult specimens is strongly convex; the convexity is accentuated when viewed laterally. The brachial valve is planoconvex to convex. The maximum depth of the shell is reached at or posterior to midlength. The anterior commissure, strongly paraplicate, is alternately folded such that the brachial valve has a deep sulcus bounded laterally by two distinct ridges which extend posteriorly at least halfway to the end of the dorsal umbo.

A large fold is evident in the pedicle valve, which often extends the entire length of the valve as a median ridge. In some gibbous adult specimens, a slight incurvedness is present at the lateral and anterior commissures. The beak is slightly incurved to erect and the ventral umbo is swollen in many adults. The ventral beak ridges range from angular to rounded but are most often rounded. A symphitium is present due to the fusion of the deltidial plates on the posterior edge of the pedicle valve. The pedicle foramen is divided by the beak ridges such that a mesothyrid to permesothyrid condition may be present. The cardinal margin is highly variable. The hinge axis intersects the cardinal margin at only two points, each posterolateral of the concealed dorsal umbo. The palintrope is weakly defined on the brachial valve but is much more developed on the pedicle valve. Secondary shell material deposited in the umbonal regions of many specimens thins anteriorly. Growth lines, strongly defined in most instances, are evident on both valves and are variably distributed with a greater concentration noticeable toward the anterior commissure of mature specimens. The shell is finely punctate with a distinct lack of radial ornamentation.

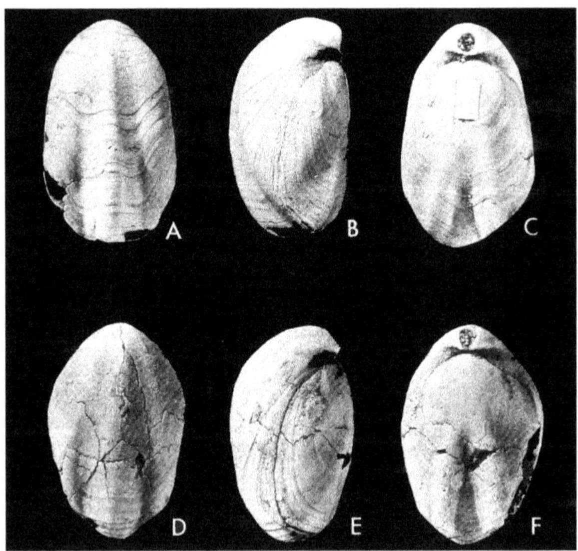

FIGURE 3. *Oleneothyris fragilis* (Morton). A, B, C. Ventral, lateral, and dorsal views, ANSP 19476. x1. D, E, F. Ventral, lateral, and dorsal views, RU 5700. x1.

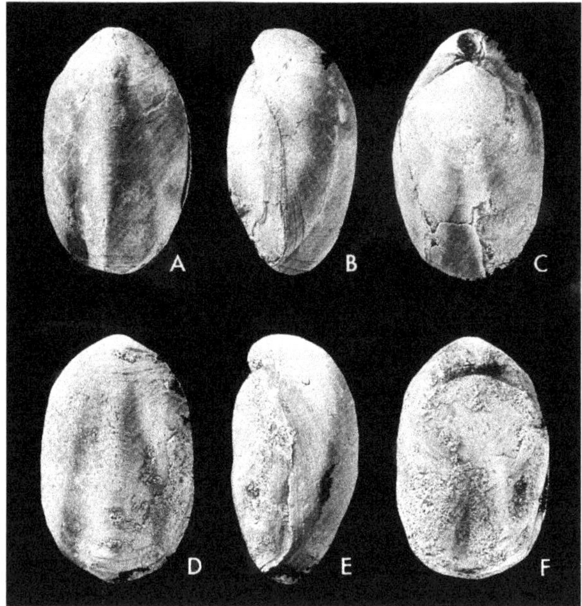

FIGURE 4. *Oleneothyris fragilis* (Morton) A, B, C. Ventral, lateral, and dorsal views, AMNH 32680. x1. D, E, F. Ventral, lateral, and dorsal views, AMNH 32681. x1.

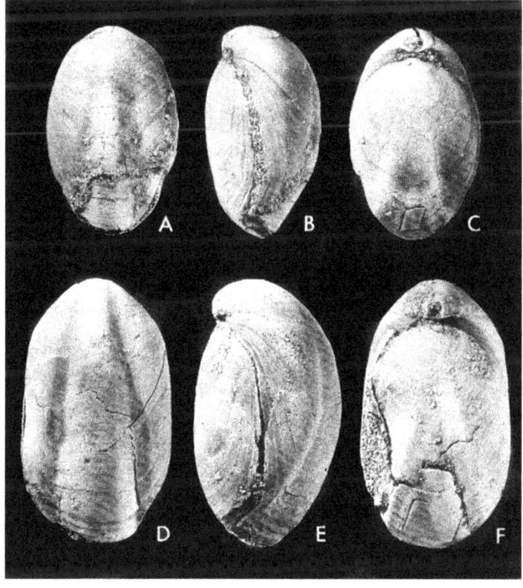

FIGURE 5. *Oleneothyris fragilis* (Morton). A, B, C. Ventral, lateral, and dorsal views, AMNH 32682. x1. D, E, F. Ventral, lateral, and dorsal views, AMNH 32683. x1.

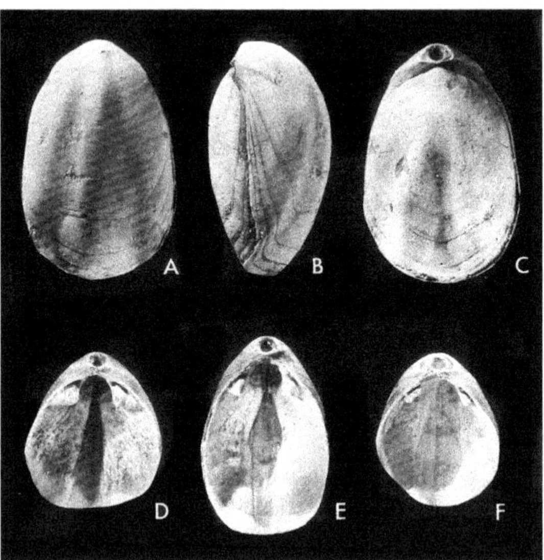

FIGURE 6. *Oleneothyris fragilis* (Morton). A, B, C. Ventral, lateral, and dorsal views, AMNH 32684. x1. D. Pedicle interior. Note deposit of secondary shell material and constriction at anterior end of umbonal cavity, AMNH 32773. x1. E. Pedicle interior, AMNH 32769. x1. F. Pedicle interior, AMNH 32768. x1.

Pedicle interior: Two well-developed dental plates originate anterior to the pedicle collar. The plates terminate with a constriction of the umbonal cavity at varying distances from the pedicle foramen. Two strong hinge teeth are dorsal to the dental plates. The teeth are cyrtomatodont, resulting in strong articulation with the brachial valve. The denticular cavity is well defined and progressively deeper posteriorly.

The muscle field is narrow and elongate, fanning out triangularly with the broad base of the field anterior. Immediately anterior to the constriction of the umbonal cavity are found the impressions of the adductor muscles, which shallow out as the anterior commissure is approached.

Brachial interior: The exterior hinge plates are short, slightly concave, and joined with the crural base posteriorly, almost to the cardinal process. The cardinalia are distinct with massive inner socket ridges. The cardinal process is variable, the most common form being cup-shaped with a distinct median ridge on well-preserved specimens. The crural processes are long, blunt, and thick. Deep dental sockets diverge strongly from the cardinal process.

The muscle field is about one-half as long and one-fourth as wide as the valve, with strongly impressed scars. Elongate adductor muscle scars extend from the crural base throughout three-fourths of the muscle field.

The Loop: The loop (Figure 7C) is about four-tenths of the total valve length. It is narrowly triangular with a transverse band, strongly arched, forming an inverted U, the base of which points posteroventrally. The crura are large, ending in sharp, ventrally-directed, hooklike apophyses. The loop is characteristically terebratulid, and more angular than that of *O. harlani.*

Comparison: Oleneothyris fragilis differs from *O. harlani* primarily in the deep sulcus and associated fold in the brachial and pedicle valves, respectively. Shells of *O. fragilis* are less robust and generally smaller in size than those of *O. harlani* (Table 1). Internally, the two species are identical. There is, however, a characteristic deposit of secondary shell material in the posterior areas of the pedicle interiors of most specimens, which may possibly be related to life orientation on the sea floor (Paul Boyer, personal communication). This feature is discussed in greater detail in conjunction with the morphologic variation of *O. harlani* (Feldman, 1977).

Variation in Oleneothyris fragilis: Less than one percent of the hypodigm displayed any lobate characteristics. Aside from AMNH 32686, which is a distinctly lobate neanic specimen, all adult forms were characteristically cylindrical. Growth lines are evident on most shells.

When viewed dorsally, many specimens appear to be pyriform at the posterior portion of the shell (AMNH 32680). Some shells are oval, almost egg-shaped (AMNH 32682). In lateral view, truncation of the pedicle foramen varies largely due to weathering and erosive action. No consistent conclusions can be drawn regarding this feature, other than to note that in the majority of specimens the pedicle foramen is truncated such that the angle of truncation parallels the lateral commissure. The incurvature of the dorsal umbo ranges from suberect (AMNH 32680) to erect (AMNH 32682, 32683) to slightly incurved (AMNH 32681). The size of the pedicle foramen is unusually small in some specimens (AMNH 32683). This feature probably correlates with the degree of turbulence or some other type of stress applied to the attachment potential (e.g., the pedicle) as described by Feldman (1974). However, a comparison and statistical analysis of the internal morphology of several hundred specimens of *Oleneothyris fragilis* would be necessary before any valid conclusions can be drawn and the above hypothesis substantiated.

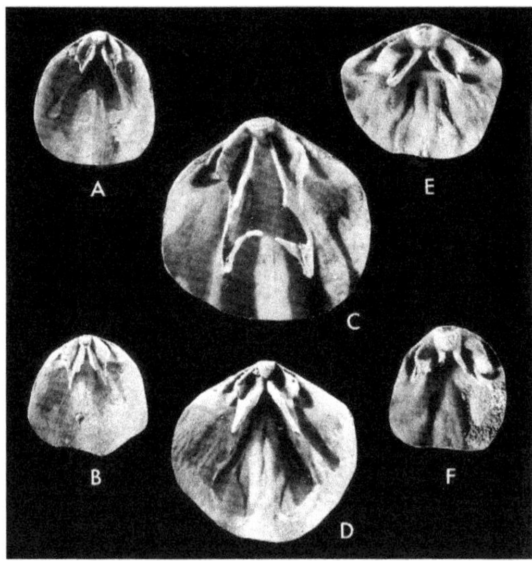

FIGURE 7. A-D. *Oleneothyris fragilis* (Morton). A. Brachial interior, AMNH 32772. x1. B. Brachial interior, AMNH 32771. x1. C. Brachial interior. Note the characteristically terebratulid loop, significantly more angular than that of *O. harlani,* AMNH 32708. x1. D. Brachial interior, AMNH 32770. x2. E, F. *Oleneothyris harlani* (Morton). E. Brachial interior, AMNH 32774. x1. F. Brachial interior, AMNH 32775. x1.

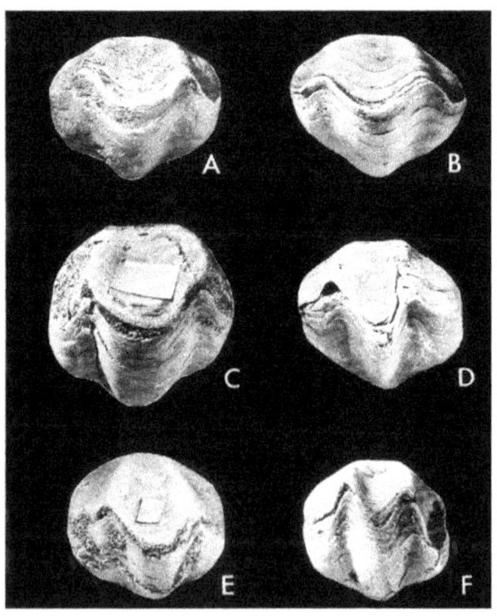

FIGURE 8. *Oleneothyris fragilis* (Morton). A. Anterior view, AMNH 32681. x1. B. Anterior view, AMNH 32684. x1. C. Anterior view of gerontic specimen, AMNH 32683. x1. D. Anterior view of slightly deformed specimen, AMNH 32680. x1. E. Anterior view, AMNH 32681. x1. F. Anterior view of neotype, RU 5700. x1.

TABLE 1. Statistical Characterization of *Oleneothyris fragilis* and *O. harlani*. Measurements in millimeters where applicable.

	Oleneothyris fragilis		*Oleneothyris harlani*	
Character	X = length Y = width	X = length Y = thickness	X = length Y = width	X = length Y = thickness
N	26	26	206	206
$\bar{X}$	43.7	43.7	46.98	46.98
$\bar{y}$	28.1	24.1	30.84	25.31
S_x	7.16	7.16	10.91	10.91
S_y	3.55	4.19	5.35	6.35
r	0.68	0.74	0.82	0.93
OR_x	29.6-56.9	29.6-56.9	16.0-69.9	16.0-69.9
OR_y	21.6-35.7	15.2-34.1	13.9-43.2	8.5-38.9
a	0.50	0.59	0.49	0.58
σ_a	0.07	0.08	0.02	0.01
b	6.25	-1.68	7.82	1.94
D_d	1.66	1.80	13.10	8.80
V_x	16.38	16.38	23.22	23.22
V_y	12.63	17.39	17.35	25.09
$\sigma X_{(x)}$	1.40	1.40	0.77	0.77
$\sigma Y_{(y)}$	0.70	0.82	0.38	0.45

Because they were incomplete, three specimens of *O. fragilis* (AMNH 32689, 32690, 32691) were not included in length measurements; therefore, for calculations of $\bar{X}$, S_x, and OR_x, N = 23. For all other calculations, N = 26. For calculation of rx,y the mean ($\bar{X}$) was substituted for the missing data (note: $r_{w,t}$ = 0.81). ANSP 19476 was not included in the above analysis, as it was obtained on loan after the completion of all calculations. Data for characterization of *O. harlani* was taken from Feldman (1977). *Abbreviations:* N = total number of measurements; $\bar{X}$ = mean of X; $\bar{Y}$ = mean of Y; S_x = standard deviation of X; S_y = standard deviation of Y; r = coefficient of correlation; OR = observed range; a = slope of the growth line (growth ratio); σ_a = standard error of the slope; b = initial growth index; D_d = coefficient of relative dispersal from the reduced major axis; V_x = coefficient of variation for X; V_y = coefficient of variation for Y; $\sigma\bar{X}_{(x)}$ = standard error of the mean for X; $\sigma\bar{Y}_{(y)}$ = standard error of the mean for Y.

Most specimens, in anterior view, display a characteristic deep sulcus and prominent sharp fold (AMNH 32680; ANSP 19476). In gerontic individuals, the shell becomes robust and the sulcus broad and deep; prominent growth lines are usually visible, especially near the anterior commissure (AMNH 32683). Variation in the sulcus and fold in adults ranges from sharp and deep to broad and shallow (AMNH 32681). The dorsal median ridges in all specimens are distinct and in most instances extend more than 50 percent of the valve length (Table 2).

A characteristic by which *Oleneothyris fragilis* may be distinguished from *O. harlani* is the nature of the alternate folding (plication) along the

TABLE 2. Morphologic Characters of Articulated Type Specimens of *Oleneothyris fragilis*. Measurements in millimeters.

Type Specimen		(L)	(W)	(T)	DVMR	SP	MP	WP	R>50%	R = 50%	R<50%
Neotype	RU 5700	38.5	24.7	23.3	X	X	--	--	X	--	--
Topotype	AMNH 32680	42.9	26.5	23.2	X	X	--	--	X	--	--
Topotype	AMNH 32681	42.2	27.4	22.9	X	X	--	--	X	--	--
Topotype	AMNH 32682	40.1	26.2	22.4	X	X	--	--	X	--	--
Topotype	AMNH 32683	51.4	29.4	29.4	X	X	--	--	X	--	--
Topotype	AMNH 32684	44.2	28.9	23.4	X	X	--	--	X	--	--
Topotype	AMNH 32685	37.2	21.6	19.5	X	X	--	--	X	--	--
Topotype	AMNH 32686	29.6	23.6	15.2	X	--	X	--	--	--	X
Topotype	RU 5701	29.1	24.6	21.0	X	X	--	--	--	X	--
Topotype	AMNH 32687 (*)	44.9	28.4	24.9	X	X	--	--	X	--	--
Topotype	AMNH 32688	38.5	26.6	21.6	X	X	--	--	X	--	--
Topotype	AMNH 32689	--	25.5	25.0	X	X	--	--	X	--	--
Topotype	AMNH 32690	--	29.9	25.9	X	X	--	--	X	--	--
Topotype	RU 5702	47.3	31.7	24.6	X	X	--	--	X	--	--
Topotype	RU 5703	48.8	33.0	26.6	X	X	--	--	X	--	--
Topotype	AMNH 32691	--	28.9	25.9	X	X	--	--	X	--	--
Topotype	AMNH 32692	47.6	32.2	26.9	X	X	--	--	X	--	--
Topotype	AMNH 32693	54.4	35.7	29.6	X	X	--	--	X	--	--
Topotype	AMNH 32694 (*)	56.9	31.4	30.4	X	X	--	--	X	--	--
Topotype	AMNH 32695	55.2	30.0	30.7	X	X	--	--	--	X	--
Topotype	RU 5704	53.6	33.9	34.1	X	X	--	--	X	--	--
Topotype	RU 5705	41.3	24.0	20.1	X	--	X	--	X	--	--
Topotype	RU 5706	42.4	30.7	20.22	X	--	X	--	X	--	--
Topotype	RU 5707	36.8	25.1	21.4	X	X	--	--	--	X	--
Topotype	RU 5708	36.4	25.9	19.3	X	X	--	--	X	--	--
Topotype	AMNH 32696 (*)	35.7	25.0	18.8	X	--	X	--	X	--	--
Topotype	ANSP 19476	41.4	25.1	23.3	X	X	--	--	X	--	--

Abbreviations: (L), length; (W), width; (T), thickness; DVMR, distinct ventral median ridge; SP, strongly paraplicate; MP, moderately paraplicate; WP, weakly paraplicate; R>50%, the two dorsal ridges extend more than 50 percent of the valve length; R = 50%, the two dorsal ridges extend approximately one-half of the valve length; R<50%, the two dorsal ridges extend less than 50 percent of the valve length; (*), internal molds.

anterior commissure. Whereas *O. harlani* ranges from rectimarginate to uniplicate to sulciplicate, *O. fragilis* is distinctly paraplicate. The distribution pattern that emerges when the two species are compared (Figure 9) provides strong support for the differentiation into separate species.

No significant variation was noted in the cardinal processes of studied specimens. Some evidenced a greater amount of weathering, resulting in a concave depression, but the vast majority were essentially flat.

Ontogeny: Few juveniles were collected in the field, and only two were present in any of the museum collections (AMNH 32686, 32698). Ontogenetic changes, from the juvenile state, are therefore difficult to analyze. The paucity of younger forms may be due to anyone or a combination of several factors: (1) winnowing, (2) dissolution of smaller shells, and (3) collecting bias.

Adult forms, which comprise most of the sample, seem to have a greater range in length than in either width or thickness (Figure 10a). Whereas in the juvenile forms studied the sulcus and fold were not well developed (but still recognizable), most of the adult forms possessed a sharp, well defined sulcus and fold. Gerontic forms were relatively few (AMNH 32683, 32693, 32694, 32719, and RU 5704), and all individuals had a smoothed-out, shallow sulcus and associated fold. Growth lines were very strongly defined in gerontic specimens, present in adults, and barely noticeable in juveniles.

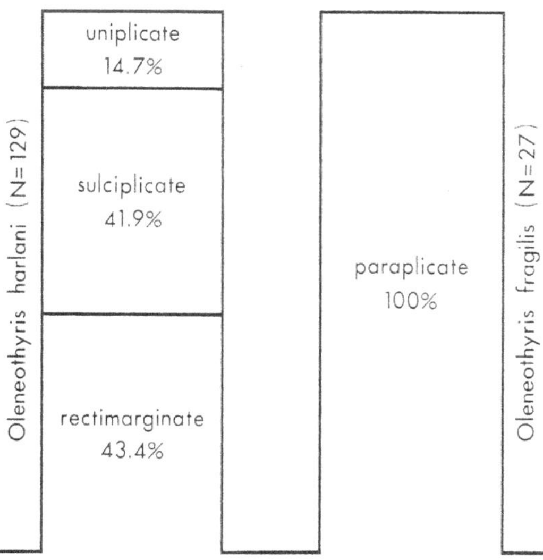

FIGURE 9. Distribution of alternate folding in the genus *Oleneothyris*.

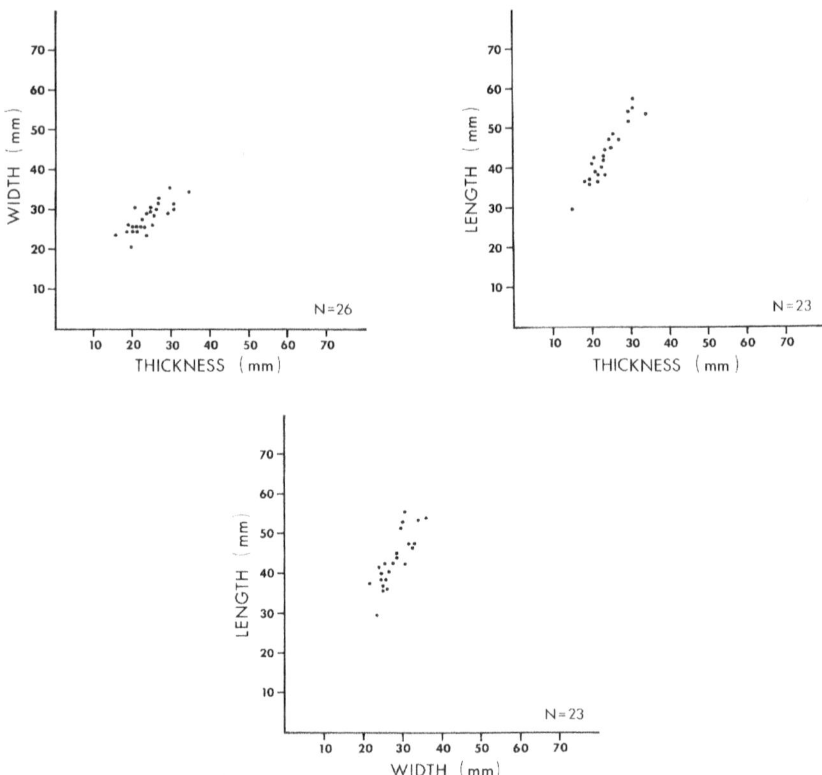

FIGURE 10. Scatter diagrams showing measurements of *Oleneothyris fragilis*. Compare with measurements of *O. harlani* (Feldman, 1977) for evaluation of the two populations.

The deltidial plates, fused into a symphitium in the juvenile stage, remained a constant morphologic characteristic throughout all ontogenetic stages. Shells of older individuals are generally more rounded laterally than in younger specimens and the ventral median ridge becomes less prominent.

ACKNOWLEDGMENTS

I thank the late Dr. G. A. Cooper, National Museum of Natural History, Smithsonian Institution, Washington, D.C., for discussion and critical review of the manuscript. Dr. Cooper made available for study two rare silicified specimens of the loop of *Oleneothyris harlani*. I am grateful to Dr. E. A. Shapiro, Academy of Natural Sciences (ANSP), Philadelphia, Pennsylvania, for the loan of material from S. W. Conrad's collection. Drs. N. Eldredge, W. K. Emerson, E. Eisenmann, and Mr. S. Horenstein, all of the American Museum of Natural History (AMNH), rendered invaluable suggestions and advice. Drs. R. K. Olsson and the late S. K. Fox of Rutgers University (RU), New Brunswick, New Jersey, kindly provided material from the Rutgers University Geological Museum and allowed me to search through their large collection of Tertiary faunas. I am indebted to Mr. G. R. Adlington for the excellent photographs and to Mr. F. Lombardi for his technical skill in reconstructing damaged specimens. Mr. S. Shank deserves thanks for his draftsmanship.

I am especially grateful to Dr. N. Eldredge and the American Museum of Natural History for providing space and facilities in the Department of Fossil and Living Invertebrates where I was able to do the research for this investigation.

REFERENCES

Cooper, G. A. 1942. New genera of North American brachiopods. *Journal of Washington Academy of Science* 32: 228-235.

Feldman, H. R. 1974. Morphologic variation in a Paleocene terebratulid brachiopod from the Hornerstown Formation of New Jersey. *Geological Society of America Northeastern Section Meeting, Baltimore, Maryland.*

Feldman, H. R. 1977. Paleoecology and morphologic variation of a Paleocene terebratulid brachiopod *(Oleneothyris harlani)* from the Hornerstown Formation of New Jersey. *Journal of Paleontology* 51: 86-107.

Morton, S. G. 1828. Description of the fossil shells which characterize the Atlantic Secondary Formation of New Jersey and Delaware; including four new species. *Academy of Natural Science Philadelphia, Journal* 6: 73-76, figs. 1-4, 7-8.

Morton, S. G. 1830. Synopsis of the organic remains of the ferruginous sand formation of the U.S.; with geological remarks. *American Journal of Science* 17: 274-295, continued in 18: 243-250.

Morton, S. G. 1834. *Synopsis of the Organic Remains of the Cretaceous Group of the United States*. Philadelphia: Key and Biddle.

Olsson, R. K. 1975. Upper Cretaceous and Lower Tertiary stratigraphy New Jersey Coastal Plain. *Petroleum Exploration Society of New York, Second Annual Field Trip Guidebook.*

d'Orbigny, A. A. 1850. Prodrome de paleontologie stratigraphique universelle des animaux et rayonnes. *Victor Masson* 2: 1-427.

Schlotheim, E. F. 1813. *Taschenbuch fur die gesammte Mineralogie mit Hinicht auf die neuesten Entdeckungen* 7: 1-611. C. C. Leonard (ed.). Frankfurt Am Main.

Stoll, N. R., Dollfus, R. P., Forest, J., Riley, N. D., Sabrosky, C. W., Wright, C. W., Melville, R. V. (eds.). 1961. *International Code of Zoological Nomenclature.* London: International Trust for Zoological Nomenclature.

Whitfield, R. P. 1886. Brachiopoda and lamellibranchiata of the Raritan clays and greensand marls of New Jersey. *Geological Survey of New Jersey* 1: 6-9.

Chapter Eight

Oleneothyris subfragilis (d'Orbigny, 1850), a Replacement Name for the Brachiopod *Oleneothyris fragilis* (Morton, 1828)

Upon investigation of the *Oleneothyris* biostrome (Feldman, 1977), an important stratigraphic marker in the Hornerstown Formation of the Atlantic Coastal Plain, I found two distinct species of the previously monotypic genus *Oleneothyris* erected by Cooper (1942). Morton (1828) first described the two species, *Terebratula harlani* and *T. fragilis,* but subsequently (Morton, 1834) decided that only one species, which displayed extreme variation and intergradation, existed. After morphologic analysis of representative specimens of *Oleneothyris,* I was able to substantiate Morton's (1828) recognition of two species. *Terebratula fragilis* was re-named *T. subfragilis* by d'Orbigny (1850: 258) as *T. fragilis* was preoccupied by Sclotheim (1813) for a species of *Terebratula.* Consequently, my use of the specific epithet (Feldman, 1977) is a resurrected junior primary homonym and must therefore be rejected as a valid name (Stoll et al., 1964: Article 59a). The proper name for the species should be *Oleneothyris subfragilis* (d'Orbigny, 1850).

REFERENCES

Cooper, G. A. 1942. New genera of North American brachiopods. *Journal of Washington Academy of Sciences* 32: 228-235.

Feldman, H. R. 1977. Notes on and description of *Oleneothyris fragilis* (Morton) 1828 (Brachiopoda, Terebratulidae). *American Museum Novitates* 2621.

Morton, S. G. 1828. Description of the fossil shells which characterize the Atlantic Secondary Fonnation of New Jersey and Delaware; including four new species. *Academy of Natural Sciences of Philadelpha, Journal,* 6: 73-76.

Morton, S. G. 1834. *Synopsis of the Organic Remains of the Cretaceous Group of the United States.* Philadelphia: Key and Biddle.

D'Orbigny, A. A. 1850. *Prodrome de Paleontologie Stratigraphique Universelle des Animaux et Rayonnes*, vol. 2. Paris: Victor Masson.

Schlotheim, E. F. 1813. Beiträge zur Naturgeschichte der Versteinerungen in geognostischer Hinsicht. In K. C. R. von Leonhard (ed.), Tascbenbuch für die gesarnmte Mineralogie mit Hinsicht auf die neuesten Entdeckungen, vol. 1, 1-134. Frankfurt am Main: Hermann.

Stoll, N., et al. (eds.). 1964. *International Code of Zoological Nomenclature*. London: International Trust for Zoological Nomenclature.

CHAPTER NINE

The Shawangunk and Martinsburg Formations Revisited: Sedimentology, Stratigraphy, Mineralogy, Geochemistry, Structure, and Paleontology

INTRODUCTION

In southeastern New York the Middle Silurian Shawangunk Formation (Figure 1), containing gray conglomerate, sandstone and shale, lies unconformably above the Ordovician Martinsburg Formation, consisting of shales and graywackes. In southwestern New York, near the Port Jervis area, the Shawangunk Formation is overlain by the Bloomsburg Red Beds, the same stratigraphic sequence that occurs in Pennsylvania and New Jersey to the southwest. The Shawangunk Formation thins gradually from Port Jervis to its pinchout near Hidden Valley and Binnewater, New York. Two tongues of the upper part of the Shawangunk are: the Ellenville Tongue that extends from the Ellenville-Accord area to its feather edge just southwest of the New York-New Jersey border, and the High View Tongue that is restricted to the Wurtsboro area (Epstein and Lyttle, 1987; Epstein, 1993).

Early in the Paleozoic, carbonate banks lay along the east coast of the ancient North American continent. During the Ordovician, plate convergence commenced in the closing of the Iapetus Ocean and a deep basin developed into which thick muds and dirty sands were deposited. These were later lithified into the Martinsburg Formation. Eventually, with continued compression, these sediments were folded and faulted during the complex deformation of the Taconic Orogeny. The trend of these folds in southeastern New York is approximately N20E. As one proceeds westward across the Wallkill Valley these structures become less intense. Subsequent to the Taconic Orogeny, mountains rose to the east and coarse sediments were transported westward and deposited as the conglomerates and sandstones of the Shawangunk Formation across the beveled folds of the Martinsburg. Deposition occurred on a plain of alluviation and in a marine basin to the northwest. Erosion of the source area was intense, and the climate, based on the mineralogy of the rocks, was warm and at least semiarid. The source was composed predominately of sedimentary and low-grade metamorphic rocks with exceptionally abundant quartz veins and small local areas of gneiss and granite. As the source highlands were eroded, the steep braided streams of the Shawangunk gave way to more gentle-gradient streams of the Bloomsburg Red Beds.

THE TACONIC OROGENY

During the Ordovician Period there existed an ocean, known as the Iapetus or proto-Atlantic, to the east of the Shawangunk Mountains. Bisecting this body of water was a narrow, mountainous landmass called the Taconic Island Arc, similar in shape to today's Caribbean or Aleutian islands. Landmasses that bordered the Iapetus Ocean began to move toward one another. On the east the landmass was present-day Western Europe composed largely of granitic rocks, whereas on the west it was present-day North America. The movement of these landmasses, or tectonic plates, resulted in the shrinking of the Iapetus Ocean and eventual collision of the Taconic Island Arc with eastern proto-North America, resulting in the formation of large mountains as high as the present-day Himalayas. The collision resulted in the (new) Taconic Mountains riding up over the edge of the proto-North American continent and pushing it down to form a basin. The sediments shed by these mountains were deposited in this basin that was more than 500 m (1,500 ft) deep. This process, known as the Taconic Orogeny, or mountain-building episode, occurred during the Middle Ordovician Period about 450 million years ago. The Taconian event was the result of shelf, slope, and island arc accretion. It was during this time period that the Martinsburg Formation was deposited.

THE ACADIAN OROGENY

During the Devonian Period (about 418-362 million years ago), the Iapetus Ocean became much narrower and shallower, finally becoming a shallow basin. Landmasses representing North America and Western Europe began to move toward one another. This collision, known as the Acadian Orogeny, took tens of millions of years to occur and resulted in the formation of large mountains, the roots of which are actually what we now call the

Berkshires in Massachusetts. The Avolonian Terrane was accreted during the Acadian Orogeny. Eventually, these mountains were weathered and eroded. The sediment that they shed accumulated to the west, forming the Catskill Delta during the middle of the Devonian Period. Some workers estimate that the volume of sediment that was dumped into this basin approached 70,000 cubic miles. The collision of two large landmasses known as Laurasia and Gondwana began after the Devonian and lasted through the Permian Period. This collision, locally called the Alleghany Orogeny, overprinted much of the deformation produced during the Acadian and Taconic orogenies and produced the major structures of the central and southern Appalachians. With Laurasia and Gondwana now sutured together, the supercontinent Pangaea had been formed. This was the last major orogenic event to affect the present-day east coast of North America. The relationship between the overlying Shawangunk and the underlying Martinsburg formations is an angular unconformity. That is, there is a significant gap in the rock record that resulted from a change that caused deposition of the Martinsburg to cease for a considerable amount of time during which there was uplift and erosion with loss of the previously formed record. In other words, although the Shawangunk overlies the Martinsburg, it is not in stratigraphic succession; there is a hiatus of 10 to 30 million years between the formation of the Martinsburg and the deposition of the Shawangunk.

STRATIGRAPHY

The following is a list of formations that occur in the field trip area from the mid-Hudson Valley to Port Jervis, listed from youngest to oldest. The last carbonate unit is the Onondaga Limestone, above which occur the clastics of the Middle Devonian Hamilton Group.

Plattekill Formation of Fletcher (1962) (Middle Devonian): red and gray shale, siltstone and sandstone; 500+ ft thick.

Ashokan Formation (Middle Devonian): thin- to thick-bedded olive gray sandstone with minor siltstone and shale; 500-700 ft thick.

Mount Marion Formation (Middle Devonian): olive gray to dark gray, platy, very fine- to medium-grained sandstone, siltstone and shale; >1,000 ft thick.

Bakoven Shale (Middle Devonian): dark gray shale; 200-300 ft thick.

Onondaga Limestone (Middle Devonian): fossiliferous limestone with the following members: Edgecliff at base (light-weathering chert), Nedrow above (dark-weathering chert; shaly), and Moorehouse at top (dark weathering chert). The Seneca Member, missing in the Hudson Valley, can be observed at Cherry Valley; 100 ft thick.

Schoharie Formation (Lower Devonian): thin- to medium-bedded, calcareous mudstone and limestone, more calcareous toward top. From top to bottom: Saugerties, Aquetuck, and Carlisle Center members; 180-215 ft thick.

Esopus Formation (Lower Devonian): dark, laminated, and massive noncalcareous, siliceous, argillaceous siltstone and silty shale; 200 ft thick (thickens to southwest).

Glenerie Formation (Lower Devonian): thin- to medium-bedded siliceous limestone, chert and shale; 50-80 ft thick.

Connelly Conglomerate (Lower Devonian): dark, thin- to thick-bedded pebble conglomerate, quartz arenite, shale, and chert; 0-20 ft thick.

Port Ewen Formation (Lower Devonian): dark, fine- to medium-grained, sparsely fossiliferous, calcareous, partly cherty, irregularly bedded mudstone and limestone; 0-175(?) ft thick (180 ft thick near Port Jervis).

Alsen Formation (Lower Devonian): fine- to coarse-grained, irregularly bedded, thin- to medium-bedded, argillaceous to partly cherty limestone; 20 ft thick.

Becraft Limestone (Lower Devonian): massive, very light to dark gray and pink, coarse-grained crinoidal limestone with thin-bedded limestone and shaly partings near the bottom in places; 30-50 ft thick (thins toward High Falls; 3 ft thick near Port Jervis).

New Scotland Formation (Lower Devonian): calcareous mudstone and silty, fine- to medium grained, thin- to medium-bedded limestone; may contain some chert; 100 ft thick.

Kalkberg Limestone (Lower Devonian): thin- to medium-bedded, moderately irregularly bedded limestone, finer grained than the Coeymans Formation below, with abundant beds and nodules of chert and interbedded calcareous and argillaceous shales; 70 ft thick.

Ravena Limestone Member of the Coeymans Formation (Lower Devonian): wavy bedded, fine- to medium-grained and occasionally coarse-grained limestone with abundant thin shaly partings; 15-20 ft thick.

Thacher Member of the Manlius Limestone (Lower Devonian): laminated to thin-bedded, fine-grained, cross-laminated, graded, microchanneled, mudcracked, locally biostromal limestone with shale partings; 40-55 ft thick.

Rondout Formation (Lower Devonian and Upper Silurian): Fossiliferous, fine- to coarse-grained, thin- to thick-bedded limestone and barren laminated, argillaceous dolomite. Limestone lentils come and go but the more persistent ones have been named (from top to bottom): Whiteport Dolomite, Glasco Limestone, and Rosendale members; 10-50 ft thick.

Binnewater Sandstone (Upper Silurian): fine-grained, thin- to thick-bedded, cross-bedded and planar cross-bedded, rippled quartz arenite, with gray shale and shaly carbonate. Probably grades southwestward into the Poxono Island Formation; 0-35 ft thick.

Poxono Island Formation (Upper Silurian): poorly exposed gray and greenish dolomite and shale, possibly with red shales in the lower part; 0-500 ft thick.

High Falls Shale (Upper Silurian): red and green, laminated to massive, calcareous shale and siltstone, occasional thin argillaceous limestone and dolostone; ripple marks, dessication cracks; 0-80 ft thick.

Bloomsburg Red Beds (Upper Silurian): grayish-red and gray shale, siltstone and sandstone; 0-700 ft thick.

Tongue of the Bloomsburg Red Beds: grayish-red siltstone and shale and slightly conglomeratic, partly cross-bedded sandstone with pebbles of milky quartz, jasper, and rock fragments, and gray sandstone; 0-300 ft thick.

Shawangunk Formation (Middle Silurian): cross-bedded and planar-bedded, channeled, quartz-pebble conglomerate (rose quartz conspicuous in upper part), quartzite, minor gray shale and siltstone and lesser red to green shale. Lower contact unconformable; 0-1,400 ft thick.

Tongue of the Shawangunk Formation: cross-bedded, cross-laminated (distinctive very light and medium-dark gray laminae), planar-bedded, thin- to thick-bedded, medium-grained quartzite and conglomerate with quartz pebbles as much as 2 inches long and greenish-gray silty shale and siltstone; 0-350 ft thick.

Diamictite (Lower Silurian or Upper Ordovician): diamictite (colluviums and shale-chip gravel with exotic pebbles and fault gauge of sheared clay and quartz veins; lower contact unconformable; less than 1 ft thick.

Martinsburg Formation (Upper and Middle Ordovician): greater than 10,000 ft thick.

Shale and Graywacke at Mamakating: dominantly thick sequences of thin- to medium-bedded, medium-dark gray shale interbedded with very thin- to thick-bedded graywacke (as much as 6 ft thick) alternating with thinner sequences of medium-bedded graywacke interbedded with less thin- to medium-bedded shale. Grades downward and laterally into the sandstone at Pine Bush.

Sandstone at Pine Bush: medium-grained, medium- to thick-bedded, medium gray, speckled light olive gray- and light olive brown-weathering quartzitic sandstone interbedded with and containing rip-up clasts of thin- to medium-bedded, medium dark gray, greenish-gray-weathering shale and fine-grained siltstone. Lower contact with Bushkill Member is

FIGURE 1. Mohonk Lake lies in a faulted, glacially scoured basin bounded by the light, quartz pebble Shawangunk Conglomerate. Beyond the Shawangunk lies the Port Jervis trough underlain by glacial sediments and occasional outcrops of Onondaga Limestone (see Feldman, 1985). Note the Catskill Mountains in the distance. The valleys in the Catskills are aligned along linears that are presumably controlled by structural weakness.

FIGURE 2. The Shawangunk Conglomerate, a quartz pebble conglomerate that is quartzitic in places, cross-bedded and planar-bedded, with minor gray shale and siltstone.

interpreted to be conformable, but in many places it is marked by a thrust fault; grades upward and laterally in the shale and graywacke Mamakating.

Bushkill Member: laminated to thin-bedded shale and slate containing fine-grained graywacke siltstone; bed thickness of shales does not exceed 2 inches and bed thickness of graywackes rarely exceeds 12 inches; lower contact conformable with underlying Balmville Limestone of Holzwasser (1926), but often disrupted by thrust faulting.

ROAD LOG AND STOP DESCRIPTIONS

We will be traveling through tick-infested areas; take the proper precautions and be sure to use insect repellent.

Cumulative Miles	Mileage from Last Point	Route Description
0.0	0.0	Leave parking lot at SUNY New Paltz. This locality is on blacktop of Holocene age, conformably overlying fill of unknown age brought in by contractors. Make a left turn on Plattekill Avenue and continue to stop sign; bear right.
0.4	0.4	Turn left at stop sign onto Main Street (Starbucks on left) and proceed through town; cross bridge over Wallkill River.
0.8	0.4	Make first right turn onto Springtown Road.

1.3	0.5	Make first left onto Mountain Rest Road. Continue until stop sign at four corners.
2.5	1.2	Continue straight up steep hill. Several sharp curves will signal approach to the top of hill. Note outcrops of Martinsburg Formation on right, consisting of thin-bedded shale, minor siltstone, and very rusty weathering with fine-grained graywacke. Cleavage is poorly developed.
4.7	2.2	Turn left at Gatehouse onto Mohonk Mountain House grounds. Stop at gate house and check in. Proceed to guest parking about 2 mi ahead.
6.9	2.2	Park cars in last guest parking lot.

Stop 1. We will be spending about two hours here examining various outcrops on the Mohonk Mountain House grounds. Please, no hammers, only cameras!

The Shawangunk (pronounced *Shongum)* Formation (Figure 1) was deposited during the Silurian Period, approximately 438-418 million years ago. At that time a shallow sea covered the southeastern part of New York State into which drained rivers and streams. The bottoms of these streams were layered with pieces of abraded quartz. Constant motion of the water eroded the chunks of quartz into sub-rounded "pebbles" that were eventually "glued" together by silica-rich cement carried by percolating ground water. The name of the sedimentary rock formed by this process is conglomerate (Figure 2). The exact source of the quartz that makes up the conglomerate is not clear. It may have formed by braided streams or alluvial fan deposition. A braided stream divides into an interlacing network of several small branching and reuniting shallow channels. An alluvial fan is a low, outspread, flat to gently sloping mass of rock material shaped like an open fan, deposited by a stream at the place where it issues from a narrow mountain valley upon a plain (Jackson, 1997). The Shawangunk Mountains (Figure 3) extend to the southwest into New Jersey, where they are called the Kittatinny Mountains. The Shawangunk Formation ranges in thickness from 733 m (2,200 ft) near Ellenville, to the southwest, to less than 1 m (about 1 ft) near Hidden Valley, to the northeast, finally "pinching out" and disappearing. Stratigraphically below the Shawangunk lies the Middle to Late Ordovician age (489-439 million years old) Martinsburg Formation, consisting mostly of shale, up to a thickness of almost 3,333 m (10,000 ft).

FIGURE 3. The Shawangunk Ridge that continues southwestward into New Jersey and Pennsylvania where it is overlain by the Bloomsburg Red Beds. In the foreground note the famous Trapps (Gunks) that, according to Van Diver (1985), as all mountain climbers know, constitute the best rock-climbing region of the eastern United States.

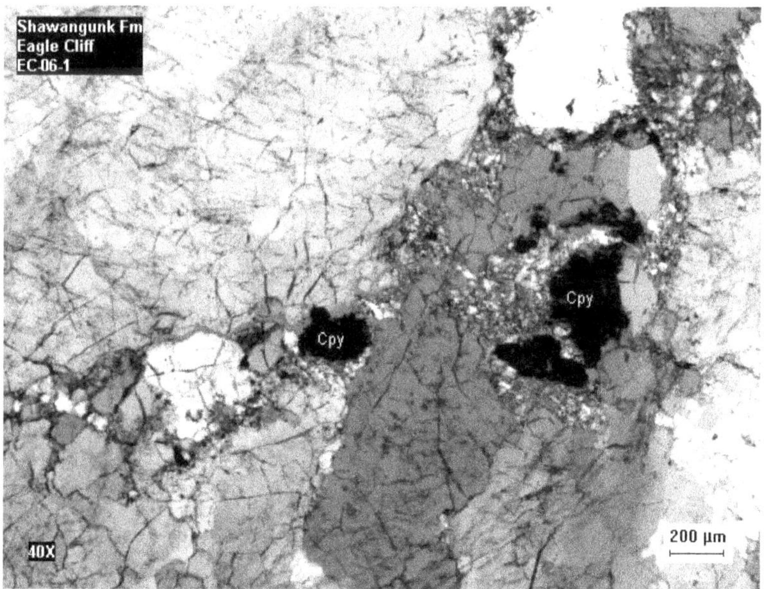

FIGURE 4. Chalcopyrite (copper-iron sulfide, Cpy) in the Shawangunk Conglomerate along Eagle Cliff Road, lake side.

Glacial Geology

Note the beautiful glacially paved surface with striations and chattermarks. Chattermarks are small, closely spaced curved scars or cracks made by chipping a brittle bedrock surface, in this case the conglomerate, by rock fragments carried in the base of a glacier. Each chattermark is roughly transverse to the direction of ice movement and its "horns" point in the direction the glacier moved. Compare the direction of the striations here (about 10 degrees to the northeast) with the direction of the striations at the top of the crevice (about 20 degrees to the northwest). As the glacier moved down through the valley, a small lobe moved around the ridge, changing direction slightly, resulting in a change in strike.

Mineralization within the Shawangunk Formation

The sulfide mineralization observed within the Shawangunk Formation along Eagle Cliff consists of predominately marcasite and pyrite with small amounts of chalcopyrite (Figure 4). The sulfides are disseminated within the conglomerate. Emplacement appears to be associated with fault structures. SEM/EDS analysis of the sulfides also shows trace amounts of lead and zinc. Gossan in the form of goethite, hematite, and jarosite has formed in places as a result of sulfide weathering. In some areas the water soluble sulfate minerals melanterite ($Fe[SO_4]\ 7H_2O$) and rozenite ($Fe[SO_4]\ 4H_2O$) are observed as a white to yellowish encrustation on the surface of sulfide containing exposures, when weather conditions permit (Figure 5). One sample from Eagle Cliff (valley side) contained two hydrated Fe-sulfate phases: melanterite ($Fe[SO_4]\ 7H_2O$) and rozenite ($Fe[SO_4]\ 4H_2O$). Rozenite is a lower hydrate containing 4 moles of water, as compared to melanterite which contains 7 moles of water. Rozenite can be formed in one of two ways: (1) dehydration of melanterite, or (2) during lower relative humidity conditions; at room temperature rozenite is the stable phase below 70-80 percent RH. Examination by polarized light microscopy of the sample shows euhedral (well formed) crystals of rozenite encased in melanterite. This relationship indicates that the rozenite crystallized first, followed by the melanterite. Additionally, no alteration/dehydration textures were observed in the melanterite. From this we conclude that the- rozenite crystallized during lower RH conditions, followed by melanterite, which crystallized during increased RH conditions (i.e., above 70-80 percent RH depending on ambient temperature).

FIGURE 5. Melanterite, a white precipitate, can be found in the Shawangunk Conglomerate along Eagle Cliff Road, valley side.

FIGURE 6. Mohonk Lake with a pH of 7 near the surface. This is one of five sky lakes on the ridge; the others are: Minnewaska, Awosting, Mud Pond, and Maratanza.

The Sky Lakes

Looking toward the southwest, observe the ridge forming the "backbone" of the Shawangunks. Continuing along the ridge, located at its highest point about 30 km (10 mi) away, is Sam's Point and the Ice Caves, so called because of the cold air trapped at the bottom of open vertical joints. Within the deep recesses of the caves one can find ice that lasts all summer long. Along the top of the Shawangunk Ridge are located four "Sky Lakes" in addition to Mohonk Lake (Figure 6) that are so called because they receive their water supply solely from rainwater. These lakes are Minnewaska (closest to Mohonk), Awosting, Mud Pond, and Maratanza. The lakes most likely overlie faults that weakened the bedrock, which was then scooped up by the passing glaciers. Interestingly, Mohonk Lake (Figure 6) is the only one of the five sky lakes that has greenish water; the others have clear blue water. The reason for this is that the acidic pH of the water of the four sky lakes (4.0) precludes algae and fish from living in them. However, Mohonk Lake's water is buffered by the alkaline rocks of the underlying Martinsburg Formation that floors the southeastern portion of the lake and has a pH of 7-7.5. This raises the pH level and creates conditions more conducive to life. Between the Shawangunks and the Catskill Front is a broad valley known as the Port Jervis trough. The trough was formed by erosion and glacial scouring of the relatively weak sediments beneath the Catskill Formation. The floor of the valley is fairly flat due to the deposition of glacial debris (till) and lake sediments. Looking toward the southeast one can see the Hudson Valley (the Hudson River is about 18 km [6 mi] to the east).

Continue to the Huntington Lookout summer cottage. The view here is spectacular. Toward the southwest we see the Shawangunk Ridge and the Trapps (famous among technical rock climbers). To the northwest one can see the Catskill Mountains, also known as the Catskill Front, rising in the distance to a height of 700 m (2,100 ft) above the Hudson River. The highest peak in the Catskill Mountains is Slide Mountain, which rises 1,282 m (3,846 ft.) above sea level. These mountains began to rise in New England and the Canadian Maritime Provinces during the Devonian Period (due to a collision of tectonic plates known as the Acadian Orogeny). Throughout the Middle and Late Devonian Period, the mountains were eroded by streams and rivers that flowed toward the west, carrying vast quantities of mud and sand that were deposited on the floor of a great inland sea as the rivers lost velocity. Closer to the ancient shoreline, the heavier particles settled out first, leaving mostly sands and some shale, whereas farther from the shoreline the deposits consisted of mainly shales with some sand, and farthest from the shoreline we find only shale. Thus, as the rivers and streams lost velocity, the heaviest particles (larger, sand-sized grains) settled out first and the smallest, finer particles settled out last, forming the fine-grained shales.

8.2	1.3	Make a right turn out of parking lot and proceed toward Gatehouse. On the right observe the (covered) contact between the Shawangunk and Martinsburg at Woodland Bridge. Note the relatively shallow dip of the shales typical of the proximity to the contact with the Shawangunk
8.3	0.1	Make right turn into shale pit.

Stop 2. Shale Pit.

In this shale pit (Figure 7), the Martinsburg contains fault-related structures interpreted to be of two different ages. A mélange, or broken

FIGURE 7. Shale pit in the Martinsburg Formation dominated by medium-dark gray shale with fine-grained graywacke. The graywackes are fossiliferous and contain a fair amount of pyrite. The Martinsburg here contains fault-related structures interpreted to be of two different ages. A melange zone that is Taconic in age has been carried westward over the Silurian Shawangunk in the hanging wall of the younger Kleine Kill thrust (see Epstein and Lyttle, 1987) that is Taconic in age with a strike of N5E-NI0E. Most of the small faults (see eastern wall of the pit) and slickensided surfaces that are common at this location are probably related to this younger thrust fault. The trace of the Kleine Kill thrust has a strike of N15E and is interpreted to be Alleghanian in age.

formation, is part of a Taconic-age mélange zone that has been carried westward over the Silurian Shawangunk in the hanging wall of the younger Kleine Kill thrust. Most of the small faults and slickensided surfaces that are common in this locality are probably related to this younger thrust fault. The trace of the Kleine Kill thrust nearby has a strike of N15E and is interpreted to be Alleghanian in age. The fault zones in the Martinsburg which clearly do not cut Silurian and younger rocks generally have: (1) a trend more northerly than the northeast-trending folds and faults in the Silurian rock, (2) a diagnostic scaly cleavage, (3) tightly folded "floating" knockers of graywacke whose axes plunge predominantly to the northeast with variable azimuth, and (4) very little, if any, vein quartz. The faults that definitely cut both the Ordovician Martinsburg and the Silurian Shawangunk, and which we interpret to be Alleghanian in age, generally have: (1) a slightly more easterly strike, (2) well-developed "pencils" in the shale formed by the intersection of bedding and cleavage, and (3) vein quartz parallel to bedding and/or cleavage that commonly contains shale fragments.

The exposure here consists of predominantly dark gray shales and siltstones interbedded with fine-grained graywacke beds, occasional prominent pyrite layers and disseminated sphalerite, chalcopyrite, and galena. Oscillation ripples (Figure 8) occur on some bedding surfaces.

FIGURE 8. Oscillation ripples in Martinsburg Formation, shale pit.

Carbonaceous material occurs mostly as fine-grained patches throughout the matrix. The studied section is tectonically stressed with shiny quartz slickensided surfaces, parallel cross-laminated strata and ripple marks. Crinoid stems, some disarticulated, and free columnals occur on different bedding surfaces, indicating a possible change in current regime. Scattered linear to sinusoidal horizontal burrow structures ranging in diameter from 0.5-3 cm are found on the silty beds. Some of the burrows are infilled with coarse quartz grains. The faunal constituents include brachiopods (93%), crinoids (*Ectenocrinus;* 3%), bivalves (3%), ostracodes (<1%), corals (<1%), trilobites (*Cryptolithus, Isotelus* <1%), conulariids (<1%) and unidentified burrowers (<1%). The brachiopods are represented by a low diversity assemblage of dalmanellids and what may be a new species of *Sowerbyella.* The fauna can be classified into distinct trophic groups: (1) high-level suspensions feeders (crinoids, corals); (2) low-level suspension feeders (brachiopods, bivalves); (3) animals that collect food from the sediment surface (ostracodes, trilobites); and (4) animals that feed within the sediment (burrowers). This partition of feeding niches leads to a

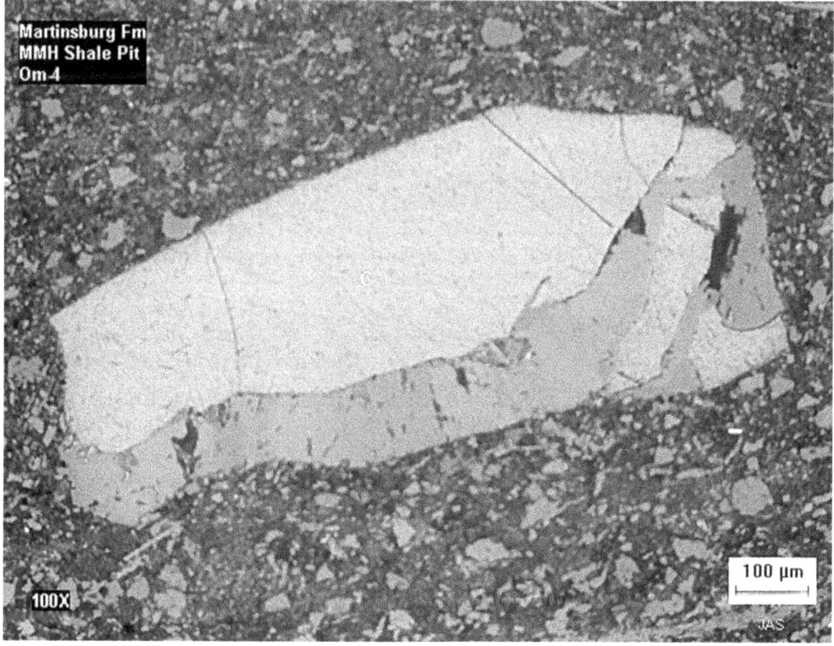

FIGURE 9. Carbonaceous material (thin section) in Martinsburg Formation, shale pit.

reduced competitive trophic structure and therefore increased community stability.

Mineralization within the Martinsburg Formation

Within the Martinsburg Formation at the Mohonk shale pit, sulfide mineralization has also been observed. The sulfides include pyrite, chalcopyrite, galena, and sphalerite. The sulfides are found associated with calcite veins and as disseminations throughout the shale. Carbonaceous material has also been observed in thin section (Figure 9).

		Make right turn out of shale pit and proceed to gatehouse. At stop sign make right turn onto Mountain Rest Road. RESET ODOMETER TO ZERO!
0.0	0.0	Turn right at stop sign just past Gatehouse (Mountain Rest Road).
2.1	2.1	Make right turn at the bottom of the hill onto Butterville Road. As we progress in a southerly direction on Butterville Road, observe cliffs of the extremely resistant Shawangunk Formation on the right. The lowland and hills in the foreground are underlain by the shales and graywackes of the Martinsburg Formation, striking toward the northeast. Note Sky Top (tower) from which one can see six states (New York, New Jersey, Pennsylvania, Connecticut, Vermont, and Massachusetts). Sky Top sits atop a rock scramble and steep climb through the "Crevice" and ending with an egress called the "Lemon Squeeze." This is the signature hike of Mohonk Mountain House that is adjacent to the 6,400 acre Mohonk Preserve.
3.7	1.6	At the stop sign make a right turn onto NY 299 heading west. On the right note the overturned shales and thin graywackes of the Martinsburg Formation. Here there are several narrow fault zones with vein quartz.
4.8	1.1	On the right note the steeply dipping and overturned Martinsburg shales and thin graywackes. Ahead, view the "Trapps," a world famous site for technical rock climbers located on the Mohonk Preserve. The Mohonk Preserve, founded in 1963, was established to protect the northern Shawangunk Ridge. Its mission is to protect the ecology of the area and to provide for public environmental education and recreation.
5.0	0.2	Jenkins-Lueken orchards.
5.5	0.5	Steeply dipping and overturned Martinsburg shales and thin graywackes on right. View of the Shawangunk cliffs at the Trapps straight ahead.
6.0	0.5	Note Martinsburg with graywackes on right. The bedding appears to be right-side-up, however, the poorly developed cleavage in the shales dips steeply to the west, suggesting that both bedding and cleavage have been rotated by faulting.
7.3	1.3	At stop sign make right turn onto US 44/55. We will be climbing to the top of the Shawangunk Mountains for the next mile or so.

7.8	0.5	Visitor's Center, Mohonk Preserve on right. Shortly we will be making a sharp left turn with the Martinsburg (on the right) and shale interbedded with thin-bedded (up to 5 in), cross-bedded to planar laminated siltstone, and minor fine-grained sandstone. Soft-sediment slump folds are common in the siltstones. Cleavage is absent or very poorly developed, except near tight folds and narrow fault zones. Most of these faults and folds do not affect the overlying Shawangunk Formation and must be Taconic in age. They trend from N5E to N20E, whereas structures in the overlying Shawangunk trend in a more easterly direction.
8.6	0.8	On the left is a scenic overview of the Wallkill Valley.
8.7	0.1	The contact between the Shawangunk and Martinsburg formations is exposed at two spots approximately 400 ft southwest of the road at the base of the cliffs. Fairly regular and linear mullions can be observed in the basal Shawangunk at the contact.
8.8	0.1	As we pass under the Trapps Bridge, note the conglomerates and quartzites of the Shawangunk dipping to the northwest (31 NW).
10.2	1.4	Cross Coxing Kill. As we cross Coxing Kill note that the Coxing Kill Valley coincides with the trough of a broad, open syncline that plunges gently to the northeast. This is also a favorite site for skinny dipping. For the next 0.5 mi we will begin to cross a broad open anticline that exposes a window of Martinsburg that is about 2 mi long. To the right note an outcrop of basal Shawangunk with unconformably underlying shales of the Martinsburg exposed about 40 ft away. Dips in both units are very gentle and the angle of unconformity is as little as 29 degrees. The divergence in strike, however, is as much as 38 degrees, with the strike in the Martinsburg being more northerly. The Martinsburg here is in the broad open-fold Taconic tectonic zone (zone 3 of Epstein and Lyttle, 1987).
10.7	0.5	Peters Kill parking area on right.
11.8	1.1	Entrance to Minnewaska State Park on the left.
12.0	0.2	Trailhead leading to Lake Awosting on left.
12.5	0.5	Cross Sanders Kill.
13.8	1.3	Turn into small parking area with scenic overview of the Rondout Valley underlain by Upper Silurian through Middle Devonian rocks that are buried by a variety of glacial sediments. In the distance, note the Middle and Upper Devonian clastics forming the Catskill Mountains. There is a thrust fault with nice slickensides across the road from the parking area. BE CAREFUL OF ONCOMING CARS!

Stop 3. Mysterious "Pods."

Walk back along NY 44/55 (north side of road) and observe the mysterious "pods" of Feldman et al. (2009) that follow the beds (N20E; 38NW) in the upper part of the Shawangunk Formation. The pods (Figures 10-11) are often conical with flat bottoms and associated with quartz pebbles.

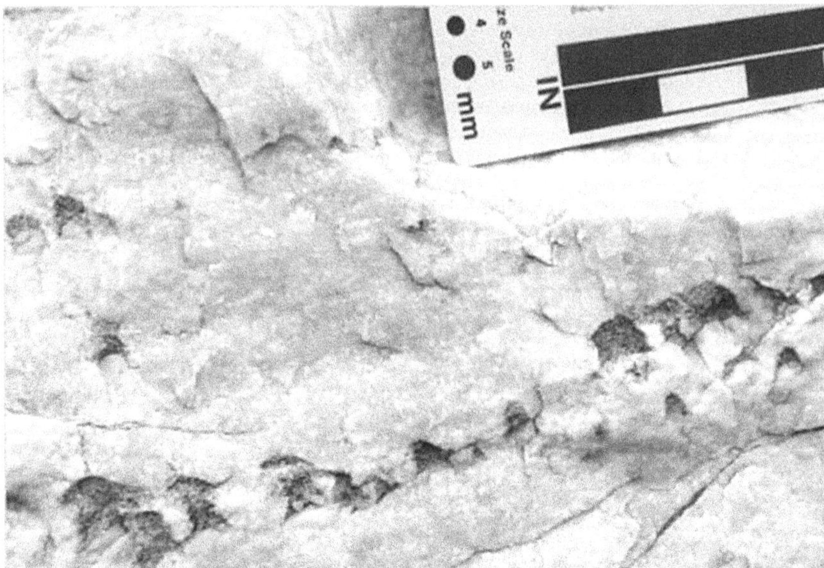

FIGURE 10. "Pods" in Shawangunk Conglomerate are often conical with flat bottoms and associated with quartz pebbles. Note that the quartzite is recrystallized. The matrix filling the pods is dark gray and seems to follow the bedding and/or crossbedding and cleavage.

FIGURE 11. Large "pod" in Shawangunk Conglomerate. Note quartz pebbles at bottom right.

Note that the quartzite is recrystallized. The matrix filling the pods is often dark gray and seems to follow the bedding and/or crossbedding. WHAT ARE THEY? The pods range from 0.2 to 7 cm across (at the base) and 0.2 to 6 cm in height. Internally, the "pods" are mostly non-recrystallized quartz grains with interstitial mica and possibly clays (Figure 12). In contrast, the areas outside the "pods" consist of pressure solution welded quartz grains consistent with the quartzitic nature of the formation. The pods are restricted to a relatively narrow stratigraphic interval and show no evidence of internal bedding. The pods may be: (1) microbial mounds/algal mats, (2) sponges, (3) mud balls, or SOMETHING ELSE! They are similar in general outline to some thrombolites in that the internal texture is non-laminated but there is no indication of microbial activity such as clotting. The possibility that they are sponges or algal mats is somewhat doubtful because the braided stream depositional environment was one of relatively fast-moving fresh water draining mountains to the southeast; in addition, most sponges live in a marine environment. The conical shape does not favor a mud ball origin.

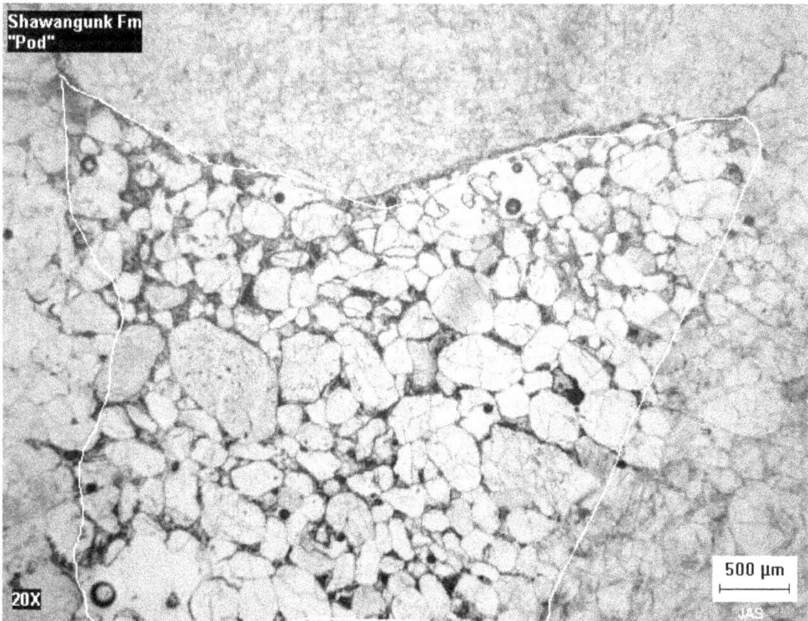

FIGURE 12. Thin section of "pod" containing non-recrystallized quartz grains with interstitial mica and possibly clays. In contrast, the areas outside the pod consist of pressure solution welded quartz grains.

Hint: note that the gray matrix follows the cleavage. (PLEASE DO NOT COLLECT SAMPLES; ACTIVE RESEARCH SITE!)

END OF TRIP

REFERENCES

Epstein, J. B. 1989. Regional stratigraphic relations of Silurian rocks and an enigmatic Ordovician diamictite, southeastern New York. Appalachian Basin Symposium-program and extended abstracts, *U. S. Geological Survey Circular* 208: 3-5.

Epstein, I. B. 1993. Stratigraphy of Silurian rocks in Shawangunk Mountain, southeastern New York, including a historical review of nomenclature. *U. S. Geological Survey Bulletin* 1839-L: L1-L40.

Epstein, J. B and Lyttle, P. T. 1987. Structure and stratigraphy above, below and within the Taconic unconformity, southeastern New York. In R. H. Waines (ed.), *New York State Geological Association, 59th Annual Meeting*, Kingston, New York, November 6-8, 1987, Field Trip Guidebook: New Paltz, New York, State University of New York, College at New Paltz, C1-C78.

Feldman, H. R. 1985. Brachiopods of the Onondaga Limestone in central and southeastern New York. *American Museum of Natural History Bulletin* 179: 289-377.

Feldman, H. R. and Thompson, J. 2008. Top of the Gunks. *Natural History Magazine* 117: 36-38.

Feldman, H. R., Smoliga, J., Wilson, M. A., Schemm-Gregory, M., and Starr, J. 2009. Mysterious "pods" in the Middle Silurian Shawangunk Formation, mid-Hudson Valley, New York. *Geological Society of America Northeastern Section Meeting, Portland, Maine* 41: 88.

Fletcher, F. W. 1962. Stratigraphy and structure of the "Catskill group" in southeastern New York. *New York State Geological Association Guidebook, 34th Annual Meeting*, D-4.

Gray, C. 1961. Zinc and lead deposits of Shawangunk Mountains. *New York Academy of Sciences* 23: 315-331.

Heroy, W. B. 1974. History of Lake Wawarsing. In D. R. Coates (ed.), *Glacial geomorphology*, 277-292. State University of New York at Binghamton, Special Publications in Geomorphology.

Holzwasser, F. 1926. Geology of Newburgh and vicinity. *New York State Museum Bulletin* 270.

Ingham. A. I. 1940. The zinc and lead deposits of Shawangunk Mountain, New York. *Economic Geology*, 35: 751-760.

Jackson, J. A. 1997. *Glossary of Geology* (4th ed.). Alexandria: American Geological Institute.

Lindemann, R. H. and Feldman, H. R. 1987. Paleogeography and brachiopod paleoecology of the Onondaga Limestone in eastern New York. *New York State Geological Association Guidebook for Fieldtrips, 59th Annual Meeting* 59: 1-30.

Rich, J. L. 1934. Glacial geology of the Catskills. *New York State Museum Bulletin* 299: 1-180.

Van Diver, B. B. 1985. *Roadside Geology of New York*. Missoula: Mountain Press Publishing Company.

Wolff, M. P. 1977. Tectonic origin and redefinition for the type section of a Middle Devonian conglomerate within the Marcellus Fm. (Hamilton Group) of southern New York: The Alcove Conglomerate; a sandy debris flow (abstract). *Geological Society of America, Northeaster Section Meeting, Binghamton, New York*, 331.

Chapter Ten

Paleocommunities of the Onondaga Limestone (Middle Devonian) in Central New York State

INTRODUCTION

The Onondaga Formation of New York State is a 21 to over 50 m unit of Middle Devonian marine limestones. The formation's outcrop belt runs east to west from Buffalo to the Helderbergs and southward from the latter to Port Jervis, a distance in excess of 550 km. Over much of this area the limestone strata have been eroded into extensive escarpments and flats.

Onondaga stratigraphy and paleogeography have received a great deal of attention during the past three decades. Formational lithofacies have been identified and interpreted. The lithofacies and paleocommunities of at least ten Onondaga bioherms have been investigated. However, until recently non-reef paleocommunities have been neglected. Feldman (1978, 1980) and Lindemann (1980) filled this gap through separate studies of Onondaga paleosynecology, using similar yet divergent approaches. It is the purpose of this trip to familiarize participants with Onondaga stratigraphy and with the results of these recent paleoecologic investigations. To these ends, this article describes the formation and its faunas. In the field, we will consider the formation's paleoenvironmental setting and the degree to which temporal changes in environmental conditions influenced and are revealed by vertical successions of Onondaga paleocommunities.

STRATIGRAPHY

The limestone strata now known as the Onondaga Formation were first recognized to be a discrete stratigraphic unit by Eaton (1828). Hall (1839) applied the name Onondaga Limestone to this unit. Oliver (1954, 1956a) formally divided the formation into four members. As described by Oliver, the members extend statewide and are characterized by a combination of lithologic and faunal criteria. This use of dual criteria facilitates recognition of the members in areas where lithologies differ significantly from type descriptions. A fifth member, the Clarence (Ozol, 1963; Oliver, 1966), is characterized by an anomalous abundance of chert. This member is not recognized east of Avon, New York and will receive no further attention in this chapter.

Unless otherwise noted, the descriptions contained in this section are derived from Oliver's (1954, 1956a) work on the type localities of Onondaga members in the trip area. Representative thicknesses and spatial relationships of the members as seen in outcrop between Buffalo and the Helderbergs are schematically depicted in Figure 1.

Edgecliff Member

At its type locality, Split Rock (Stop 2), the Edgecliff Member is a light-gray, very coarse-grained limestone. Beds range in thickness from 15 cm to over 1.5 m. A basal section, 1 cm to 1.8 m thick, ranges from a quartz arenite to a sandy limestone. Quartz abundance rapidly decreases upward. Light-colored chert nodules are found throughout the member. The Edgecliff is characterized by an abundant fauna of rugose corals, tabulates, and large crinoid columnals. This fauna forms a coral biostrome throughout much of the state (Oliver, 1956b). Coral bioherms of the formation are restricted to this member.

Nedrow Member

In central New York, the Edgecliff Member is overlain by a 3-4.3 m unit of thin-bedded, very fine-grained argillaceous limestone, referred to as the Nedrow Member. The type section of this unit is Indian Reservation Quarry (Stop 1), just south of Nedrow, New York.

Though chert is uncommon in the Nedrow, the upper beds locally contain some scattered medium to dark gray nodules.

The Nedrow may be recognized by its typical recessed weathering caused by numerous shaly beds within the matrix. In outcrop this feature makes the member easy to recognize; however, in a fresh cut (such as found in the Jamesville Quarry, Stop 3) it becomes difficult to differentiate between the Nedrow and Edgecliff strictly on lithologic criteria. For this reason, recognition of a Nedrow fauna is critical. The Lower Nedrow bears a brachiopod-dominated fauna with several species of platyceratid gastropods and very few corals. Two characteristic coral species are *Amplexiphyllum hamiltonae* and a turbinate growth form of *Heliophyllum halli*.

The Upper Nedrow has a less diverse brachiopod fauna with only a few platyceratids.

Moorehouse Member

At its type locality in the Jamesville Quarry (Stop 3), the Moorehouse Member is a medium gray, very fine-grained limestone with numerous shaly partings. This unit gradationally overlies the Nedrow Member, making their contact difficult to place. However, in the central New York region the top of the Nedrow is considered to be the uppermost shaly bed, usually separated by about 15 to 20 cm from the underlying shaly bed. Dark gray chert is common throughout the Moorehouse. Chert increases in abundance in the upper half of the member, where it commonly forms beds or anastomosing networks. The upper half of the Moorehouse is also less shaly and more fossiliferous than the lower half. The entire member in central New York is strongly dominated by brachiopods; nowhere are corals abundant.

Seneca Member

The type section of the Seneca Member is at Union Springs, New York. There the basal part of the member is a fine to medium-grained limestone, which overlies a greenish-gray to ochre colored clay layer 15 cm thick. This clay layer, the Tioga Bentonite, separates the base of the Seneca from the lithologically similar Upper Moorehouse Member throughout much of the state. Approximately 3 m above the bentonite layer occurs a zone of "*Hallinetes*"aff. *lineatus* (Zone J of Oliver, 1954). The Seneca is a "muddy" limestone, highly argillaceous, and poorly fossiliferous except for the "*Hallinetes*" Zone. The Seneca grades upward into the Marcellus Shale. The gradation is represented by a 2+ m section of increasing shale content within the limestone and by alternating beds of shale and lime.

Formational Contacts

Over its exposure area, the Onondaga overlies several older formations which generally increase in age westward. In eastern New York, the Onondaga conformably and gradationally overlies the Schoharie Formation. Between Cobleskill and Richfield Springs, the Onondaga overlies the Carlisle Center Formation. There the contact is marked by phosphorite nodules and glauconite grains, and represents a minor unconformity. Within the field trip area, the Onondaga unconformably overlies one or the other of the Lower Devonian Oriskany, Coeymans, or Manlius Formations. Further west, erosional remnants of the Lower Devonian Bois Blanc and the Silurian Akron Formations underlie the Onondaga.

The Onondaga Limestone is overlain by the Marcellus Shale. West of Cherry Valley the Marcellus Formation rests on the Seneca Member of the Onondaga Formation. In central New York, the contact is both interbedded and gradational. The limestone-shale contact is not exposed in western New York. However, it appears to be more abrupt than in the type area. East of Cherry Valley and north of Catskill, the Marcellus Formation rests on the Moorehouse Member of the Onondaga Formation. Their contact is abrupt, marking a minor unconformity which is either erosional (Chadwick, 1927) or non-depositional (Cooper, 1930; Flower, 1936). Oliver (1956a) tended to support the latter conclusion. Subsurface data indicates that the Seneca is the uppermost Onondaga member south of Catskill (L. V. Rickard, personal communication, 1980).

Age and Correlation

The Onondaga Limestone was deposited during early Middle Devonian time. Age determinations based on coral (Oliver, 1954, 1956a), cephalopod (Oliver, 1956c), and conodont (Klapper et al., 1971) faunas indicate that the

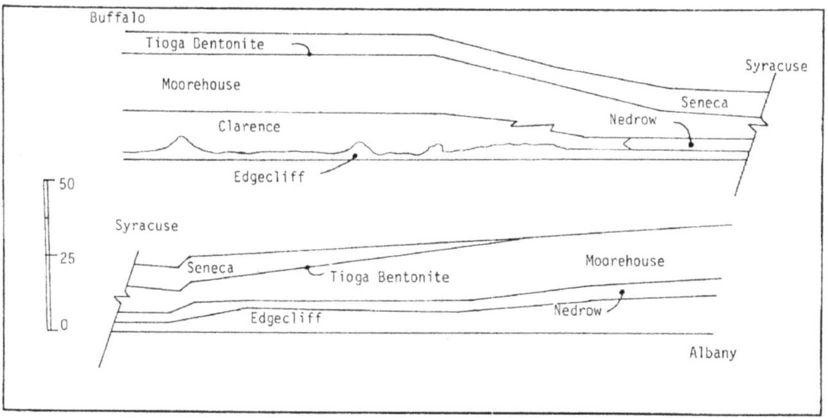

FIGURE 1. Generalized cross section of the Onondaga Formation members as seen in outcrop between Buffalo and the Helderbergs, showing thicknesses and physical relationships. Adapted from Oliver (1954, 1956a). The scale is in meters.

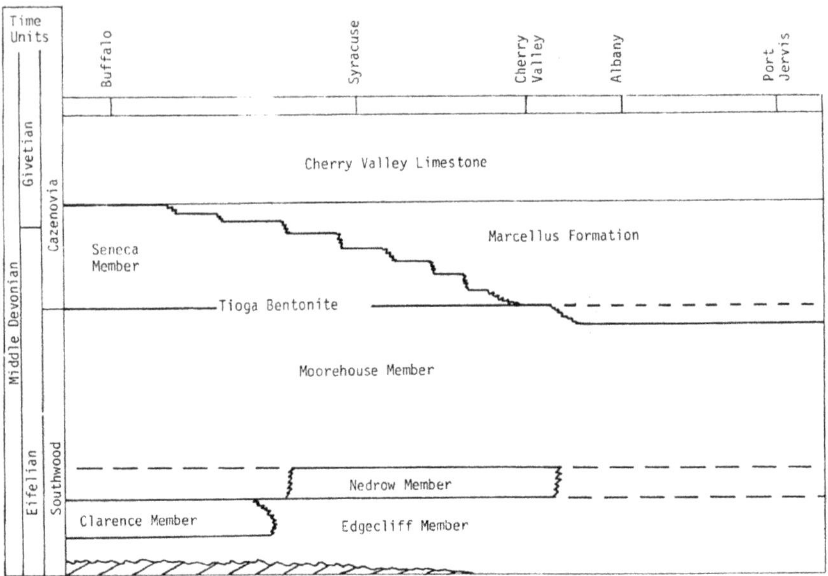

FIGURE 2. Chronostratigraphic correlation of the Onondaga Formation throughout New York State. Horizontal dimension not to scale. Adapted from Oliver and others (1969) and Rickard (1975).

Edgecliff, Nedrow, and Moorehouse members are of Southwood age (Rickard, 1975). The Seneca Member was deposited during Cazenovia time (Rickard, 1975). The formation is entirely correlative to the Eifelian of Europe. Intraformational chronostratigraphic relations are presented in Figure 2.

The basal Edgecliff Member marks the earliest record of the Middle Devonian throughout the state (Oliver, 1954, 1967). The Edgecliff is conformable with the Schoharie Formation in eastern New York and unconformable with the Bois Blanc Formation in the western area. Oliver (1967) concluded that the Bois Blanc and Schoharie Formations are age equivalent, and that the base of the Onondaga is nearly, but not perfectly, isochronous.

The Tioga Bentonite, which is found at the Moorehouse-Seneca contact, delineates the end of Southwood (Onesquethaw) time (Oliver et al., 1967), placing the Seneca Member in the Cazenovian stage. Proceeding from west to east in outcrop, the Tioga and the top of the Seneca converge and become coincident just east of Cherry Valley. In eastern New York, outcrops north of

Catskill, the Tioga presumably lies within the Marcellus Shale. In the subsurface of southeastern New York, the Tioga lies between the Moorehouse and Seneca members as it does in central New York outcrops (L. V. Rickard, personal communication, 1980). Rickard has also found that in western New York the Tioga occupies three separate horizons, one of which is found at the Moorehouse-Seneca contact. Thus, the exact chronostratigraphic character of the uppermost Onondaga is not as straightforward as was previously thought. However, as seen in outcrop in the trip area, the top of the Onondaga is considered to be of early Cazenovian age (Rickard, 1975).

METHODS OF COMMUNITY ANALYSIS AND DESCRIPTIONS OF ONONDAGA PALEOCOMMUNITIES

In independent studies of Onondaga paleosynecology, the authors generally applied Fager's (1963: 45) definition of paleocommunities: "recurrent organized systems of organisms with similar structures in terms of species presence and abundances." However, Feldman (1978, 1980) concentrated on the distribution of brachiopod genera, whereas Lindemann (1980) stressed coral distribution and dealt with the remaining fauna bionomically or at high taxonomic levels. While these divergent approaches partly resulted from differences in research objectives and observational scales, both attempted to recognize and reconstruct the fossil equivalents of once-living communities of organisms. For this reason we feel that it is instructive to separately describe our analytic methods and the resultant communities.

Communities Recognized by Feldman

More than 50 localities within the outcrop belt of the Onondaga Limestone were studied, but only 30 were sampled in detail. Silicified outcrops were encountered mainly in southeastern New York, specifically in the mid-Hudson valley. Silicification was poor to nonexistent in the central region. The most productive outcrops were those that consisted of a large expanse of limestone in which single bedding planes were exposed laterally. In many cases, it was possible to identify the specimens in the field. However, if preparation of a specimen was necessary or if a particular specimen was needed for comparison and further study, it was often possible to crack out small slabs of limestone. This is especially true of the Nedrow Member.

Collecting in the Jamesville Quarry (stop 3) was relatively easy in this respect due to blasting and subsequent jointing. However, the fresh unweathered surfaces provided little in the way of good fossil material.

Nine brachiopod communities have been recognized in the Onondaga Limestone from Syracuse to southeastern New York (see Feldman, 1980: 31). Four of these, briefly discussed below, are found in the Syracuse area. The vertical distribution of these communities in the trip area is presented in Figure 3.

1) *Leptaena-Megakozlowskiella* community: This community is found in Nedrow-Moorehouse age rocks at the Onondaga Indian Reservation Quarry (stop 1) in Nedrow, New York and the Jamesville Quarry (Stop 3). A mid-neritic environment of deposition is probable, along with a moderately to highly argillaceous substratum. *Leptaena* and *Megakozlowskiella* are the dominant brachiopod genera (29.9 and 28.4%, respectively) of a total of 17 genera. Other taxa present include tabulate and rugose corals (common), gastropods (very rare), a cephalopod (*Foordites*, very rare), trilobites (rare), camerate crinoid columnals (rare), and bryozoan fragments (rare). The trophic nucleus of this community is one of low-level suspension feeders (*Leptaena-Megakozlowskiella*). Noticeably absent are the numerous platyceratid gastropods of the mid-Hudson valley.

Locality description: Onondaga Indian Reservation, Nedrow, New York (Stop 1), southwest of the junction of Route 11 and highway I-81. Almost the entire Onondaga Limestone is exposed at this locality. This is the type section of the Nedrow Member (Oliver, 1954). The Edgecliff Member is best observed on the quarry floor at the southwest end of the quarry. Large crinoid columnals and stems are visible. The Nedrow-Edgecliff contact is best seen at the south face of the quarry wall, where interbedded shaly beds mark the Nedrow. The Moorehouse-Nedrow contact is at the topmost shaly bed. The Seneca-Moorehouse contact is represented by the Tioga Bentonite, which forms a re-entrant on the face of the quarry wall. The top of the Seneca is eroded.

2) Pacificocoelia community: we have found evidence of this community at only one outcrop within the Nedrow Member. The limestone is dominant. Trilobites are represented by *Odontocephalus* and *Phacops*. Rugose and tabulate corals are rare, of low diversity, and primarily consist of juvenile and/or stunted individuals. Gastropods, primarily *Platystoma*, are also rare. The orthoconic nautiloid, *Michelinoceras*, is present, as is the goniatite, *Foordites*.

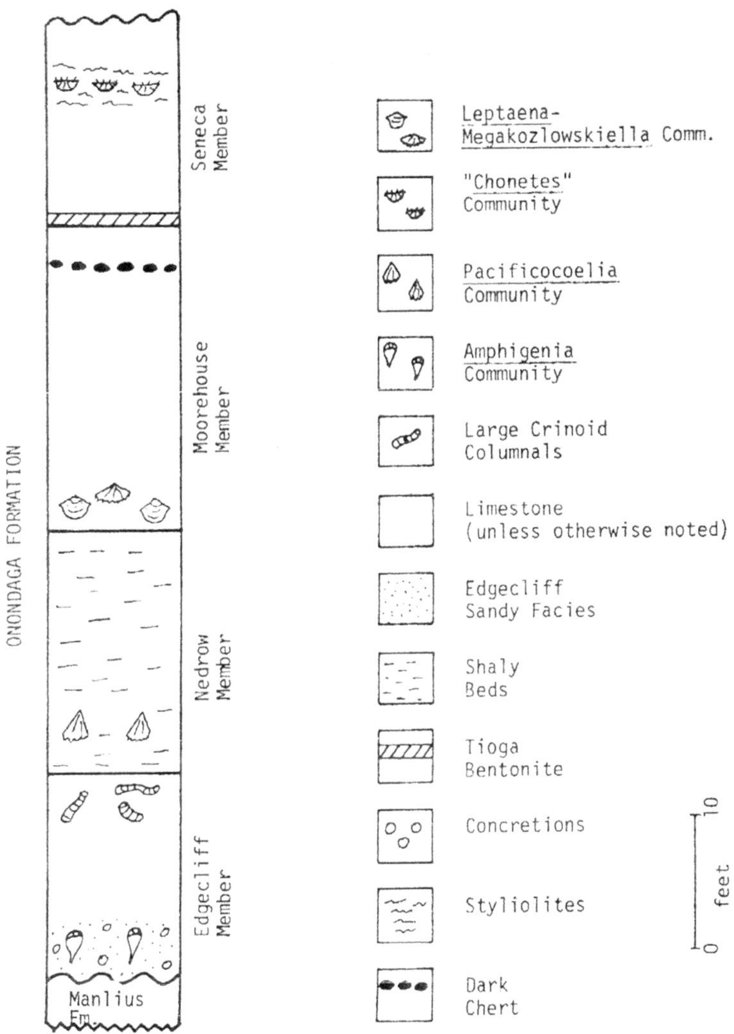

FIGURE 3. Biostratigraphy of the Onandaga Limestone in central New York showing distribution of Feldman's brachiopod communities.

Bioturbation is prevalent. However, recognizable ichnofossils are uncommon. A *Chondrites* measuring 2-3 mm in diameter is present. A vertical trace measuring 5 mm in diameter is also present, but uncommon. The most common tract fossil is an epichnial groove 4 cm in depth and 2 cm in width. Each epichnion occurs at the top of the host bed, extending downward as a flat-bottomed trough of fossiliferous sediment.

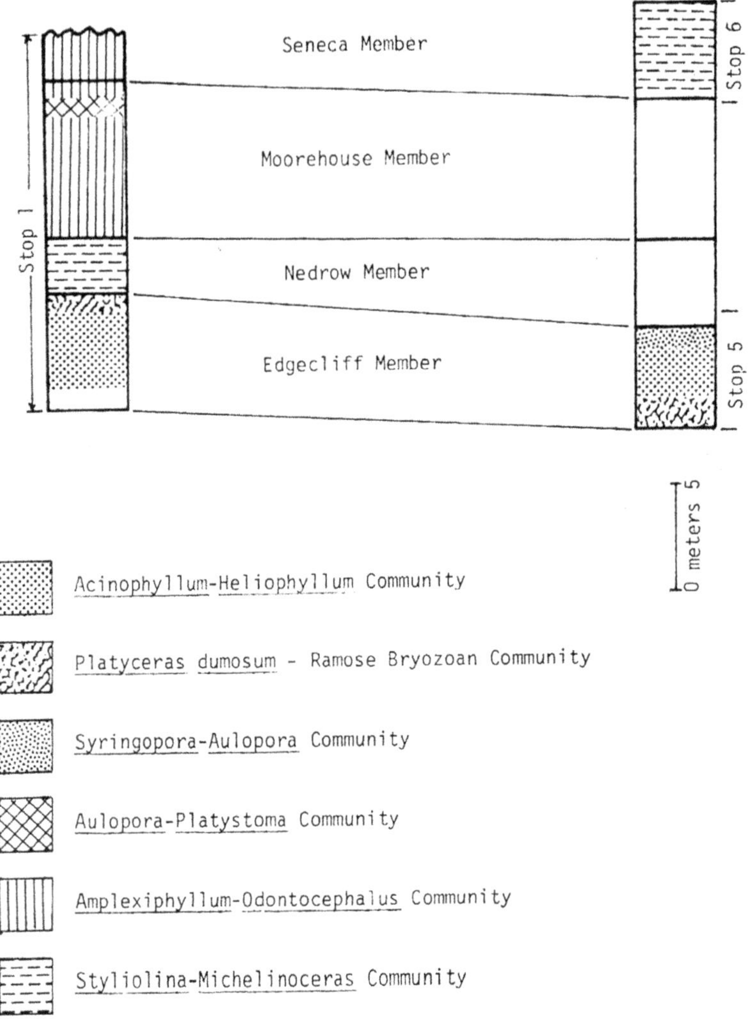

FIGURE 4. Biostratigraphy of the Onondaga Limestone in central New York showing the distribution of Lindemann's communities.

ROAD LOG FOR PALEOCOMMUNITIES OF THE ONONDAGA LIMESTONE

Cumulative Miles	Mileage from Last Point	Route Description
0.0	0.0	Leave SUNY Binghamton campus and head east on Route 434 to Interstate Route I-81. Take I-81 north toward Syracuse. Log mileage begins at this point.

66	66	Leave Route I-81 at Exit 16 and turn left onto U. S. Route 11 north.
66.7	0.7	Very shortly after passing beneath I-81 turn left from Route 11 onto Quarry Road.
67	0.3	STOP 1. Park on the road shoulder and walk into the quarry which lies between Quarry Road and the interstate.

Stop 1. This is the Indian Reservation Quarry described in connection with Feldman's *Leptaena-Megakozlowskiella* community.

67.3	0.3	Return to Route 11 and turn left (north) toward Nedrow.
70.1	2.8	Leave Route 11 turning left (west) onto Route 173.
75.2	5.1	Turn left onto Split Rock Road. The road sign may not be visible so watch for a black-on-white D.O.T. sign and a yellow-and-blue state historical marker.
75.9	0.7	STOP 2. Continue to the end of Split Rock Road and into the quarry entrance.

Stop 2. This is the type section of the Edgecliff Member, which is named for Edgecliff Park, located just to the west. The member's full thickness of 2.3 m is exposed in the upper areas of the main quarry. Its erosional base is marked by the presence of phosphate nodules. In the southern quarry wall, 3 m of the Nedrow Member are exposed. The top of the Nedrow has been removed by erosion.

76.6	0.7	Upon leaving the quarry, return to Route 173 and turn right.
86.7	10.1	In Jamesville, turn right (south) onto NY Route 91.
87.5	0.8	STOP 3. Turn left into the broad flat quarried area adjacent to Route 91 and park. The Onondaga is exposed in the upper quarry. Walk up into it by way of the unpaved road which comes down to Route 91 just south of where you've parked.

Stop 3. A nearly complete section of the Onondaga is exposed in this quarry. Only the top of the Seneca is gone. Thicknesses of the members are given below. An excellent exposure of the "Springvale" can be found to the west of the large storage building seen from Route 91. Here the "Springvale" contains a sparse brachiopod-bryozoan fauna and phosphate nodules. The main quarry floor is in the Edgecliff Member and is best seen along the south wall; it is at your right as you enter. Most of the quarry wall consists of the Nedrow and Moorehouse Members. This is the type locality of the latter. Fracturing and jointing of the rock have made hand and foot holds

insecure. Please avoid falling or dropping rocks (or yourself) on other trip participants. The Seneca Member can be closely observed above and to the east of the main quarry. You can get to it by walking around the quarry to the north.

Seneca Member 1.5 m
Moorehouse Member 7.0 m
Nedrow Member 4.4 m
Edgecliff Member 5.2 m
"Springvale" 1.5 m

88.3	0.8	Take Route 91 back to Route 173 and turn right (east) uphill toward the prison.
88.8	0.5	Turn right (south) onto Taylor Road.
89.6	0.8	STOP 4. Park along road shoulder.

Stop 4. East of Taylor Road in the clump of trees across from the first house on the left is exposed a small section of a reef in the Edgecliff Member. While exposure does not permit a three-dimensional study of the reef, the fauna, which includes *Cylindrophyllum*, *Acinophyllum*, *Emmonsia*, and *Favosites*, indicates that what we are seeing is part of the core facies.

90.4	0.8	Return to Route 173 and turn right.
95.8	5.4	STOP 5. Drive through Manlius on Route 173. Going uphill and out of town there will be what appears to be a wooded and scrubby area on the left (north) side of the road. This is sharply broken by the school athletic field. Turn left into the western corner of the field and park near the storage shed. Frog Pond Quarry is the scrubby area west of the athletic field. You can get into the quarry by way of any of the foot paths which lead in that direction.

Stop 5. The full 5.8 m of the Edgecliff Member are exposed here as is the lower 1 m of the Nedrow. The base of the Edgecliff is exposed on the middle of the quarry face, which overlooks the frog pond in the southwest corner of the quarry. Rock weathering and extensive exposure of bedding surfaces make sampling and fossil collection here excellent.

114.1	18.3	Continue east on Route 173. At Chittenango, get onto Route 5 headed east.
126.7	12.6	At the intersection of Routes 5 and 46 turn right (south) onto Route 46 toward Munnsville.
135.2	8.5	Proceed through Stockbridge and Munnsville and bear right on Phillips Road.
135.4	0.2	Turn right onto Phillips Drive.
135.8	0.4	Turn right onto Stockbridge Falls Road.
137.3	1.5	STOP 6. Proceed uphill until you see black shale to the right and Oneida Creek to the left. Look for and park in the wide road shoulder to your left as the road leans right. Watch out for eastbound traffic!

Stop 6. While virtually all of the Onondaga can be pieced together in the creek bed downstream from here, we will observe only the Seneca Member, 3.7 m of which are present in the stream. If water levels and recent sediment transport have been kind, we will be able to see a rare exposure of the contact between the Onondaga and the Union Springs Shale. For those who are interested, the full thickness of the Cherry Valley Limestone, about 1 m, is exposed on the north side of the road. There is no need to scramble up the shale for it, just trace the bed along until you meet it near road level.

138.8	1.5	Take Stockbridge Falls Road downstream to Pratts Road and turn right.
143.8	5	Turn right from Pratts Road onto Route 20.
174.8	31	At Lafayette get onto I-81 headed south and return to Binghamton.
237	~62	(totals)

ACKNOWLEDGMENTS

We thank Drs. Richard A. Park of Rensselaer Polytechnic Institute, Lawrence V. Rickard of the New York State Museum and Science Service, and John J. Thomas of Skidmore College for their consultations and for reviewing various sections of that which became this chapter. Special thanks go to Faith Grey Lindemann for proofing the final draft, driving during preparation of the trip log, and patiently coping with RHL during the production of this article.

REFERENCES

Chadwick, G. H. 1927. New points in New York stratigraphy (abstract). *Geological Society of America Bulletin* 38: 160.

Cooper, G. A. 1930. Stratigraphy of the Hamilton Group of New York. *American Journal of Science* 19: 116-135; 214-236.

Eaton, A. 1824. A geological and agricultural survey of the district adjoining the Erie Canal, in the State of New York, taken under the direction of the Hon. Stephen Van Rensselaer, Part I, containing a description of rock formations, together with a geological profile, extending from the Atlantic to Lake Erie. Albany, New York: Packard and Van Benthuysen.

Fager, E. W. 1963. Communities of organisms. In M. H. Hill (ed.), *The Sea* 2: 415-437. New York: Wiley.

Feldman, H. R. 1978. Brachiopods and community ecology of the Onondaga Limestone. Unpublished PhD thesis, Rutgers University, New Brunswick, New Jersey.

Feldman, H. R. 1980. Level-bottom brachiopod communities in the Middle Devonian of New York. *Lethaia* 13: 27-46.

Flower, R. H. 1936. Cherry Valley cephalopods. *American Paleontology Bulletin* 76.

Hall, J. 1839. Third annual report of the survey of the fourth geological district. *New York Geological Survey Annual Report* 2: 286-339.

Johnson, R. G. 1964. The community approach to paleoecology. In J. Imbrie and N. Newell (eds.), *Approaches to Paleoecology*, 107-134. New York: Wiley.

Klapper, G., Sandberg, C. A., Collinson, C., Huddle, J. W., Orr, R. E., Rickard, L. V., Schumacher, D., Seddon, G., and Uyeno, T. T. 1971. North American Devonian conodont biostratigraphy. In W. C. Sweet and S. M. Bergstrom (eds.), *Symposium on Conodont Biostratigraphy: Geological Society of America Memoir* 127: 285-316.

Lindemann, R. H. 1980. Paleosynecology and paleoenvironments of the Onondaga Limestone in New York State. Unpublished PhD thesis, Rensselaer Polytechnic Institute, Troy, New York.

Lindholm, R. C. 1967. Petrology of the Onondaga Limestone (Middle Devonian), New York. Unpublished PhD thesis, Johns Hopkins University, Baltimore, MD.

Oliver, W. A. Jr. 1954. Stratigraphy of the Onondaga Limestone (Devonian) in central New York. *Geological Society of America Bulletin* 65: 621-652.

Oliver, W. A. Jr. 1956a. Stratigraphy of the Onondaga Limestone in eastern New York. *Geological Society of America Bulletin* 61: 1441-1474.

Oliver, W. A. Jr. 1956b. Biostromes and bioherms of the Onondaga Limestone in eastern New York. *New York State Museum and Science Service Circular* 45.

Oliver, W. A. Jr. 1956c. *Tornoceras* from the Devonian Onondaga Limestone of New York. *Journal of Paleontology* 30: 402-405.

Oliver, W. A. Jr. 1966. Clarence Member of Onondaga Limestone. In *Changes in Stratigraphic Nomenclature by the U. S. Geological Survey 1965: U. S. Geological Survey Bulletin* 1244-A: A48-A49.

Oliver, W. A. Jr. 1967. Stratigraphy of the Bois Blanc Formation in New York. *U. S. Geological Survey Professional Paper* 584-A.

Oliver, W. A. Jr., de Witt, W. Jr., Dennison, J. M., Hoskins, D. M., and Huddle, J. W. 1960. Devonian of the Appalachian basin, United States. In D. H. Oswald (ed), *International Symposium on the Devonian System: Alberta Society of Petroleum Geologists* 1: 1001-1040.

Ozol, M. A. 1963. Alkali reactivity of cherts and stratigraphy of the Onondaga Formation of central and western New York. Unpublished PhD thesis, Rensselaer Polytechnic Institute, Troy, New York.

Rickard, L. V. 1975. Correlation of the Silurian and Devonian rocks in New York State. *New York State Museum and Science Service Map and Chart Series* 24.

Chapter Eleven

Fossils and Facies of the Onondaga Limestone in Central New York

INTRODUCTION

Four members of the Onondaga limestone are recognized in the central part of New York State: Edgecliff, Nedrow, Moorehouse, and Seneca. Detailed descriptions of these members may be found in Oliver (1954, 1956) and Feldman (1985). Paleocommunity analyses of the non-reefal aspects of the Onondaga Limestone have been made by Lindemann (1980) and Feldman (1980) while Lindemann and Feldman (1981) have summarized the stratigraphy, lithofacies, and paleocommunities of the Onondaga in the Syracuse area. We believe that the formation in this area has been studied thoroughly; consequently, the purpose of this trip is to sample the fauna in order to: (1) become familiar with the fossils that are characteristic of the Onondaga, and (2) attempt to correlate the various taxa with their paleoenvironments. This trip, therefore, will concentrate on collecting the fossils of the Onondaga Limestone in the Syracuse area.

The most abundant invertebrate fossils of the Onondaga that lend themselves to collecting in this extremely dense limestone are the brachiopods and corals. Brachiopod collecting is usually best accomplished in the shaly Nedrow Member, at the base of shale lenses, followed by well weathered Moorehouse and Seneca strata. Corals are most abundant in the Edgecliff Member. Since the Onondaga weathers very slowly, it is best to locate bedding planes that have not been touched for several years, if possible.

EARLY HISTORY

Students of New York geology probably require little introduction to the Onondaga Limestone. Firmly established in the geologic literature as the State's lowermost Middle Devonian formation, the Onondaga is one of its most prominent units, due in part to the fact that it is laterally continuous from the central Hudson River Valley north to the Helderbergs and west to the Niagara River and the Ontario Peninsula. Between Buffalo and the Helderbergs, the formation's resistance to erosion, slight southerly dip, and position in the section place it atop an escarpment defining the southern margins of the Ontario lowlands and the Mohawk Valley; a situation that provided early explorers and geologists with a wealth of natural exposures.

Following on the heels of the American Revolution, the westward expanding settlement of New York prompted exploration for transportation routes, mineral resources, and building materials. In the bad old days long before central heating, finished lumber, and fiberglass insulation, quarried-stone and kiln-produced mortar and plaster were, where available, valuable building materials. Historic records document the former workings of hundreds of quarries and kilns by which the Onondaga supplied construction industries of the early 1800s. This gives the formation a role in human history. As 1986 marks the sesquicentennial of the New York Geological Survey, it seems appropriate to note some of the highlights in the history of geologic activities leading to our current understanding of the Onondaga Limestone.

Geologic observations made during the pre-Survey years are summarized by Wells (1963). Even a casual reading of Wells' book leaves little doubt that fossils figured significantly among descriptions recorded during the early 1800s. In 1803, Compte de Volney collected fossil specimens from the strata now referred to as the Onondaga and sent them to Lamarck in France. The first published illustration of a fossil from the New York Devonian, which appeared in 1807, is that of a "sheep horn" collected from the Onondaga. Wells (1963: 14), reaching a slightly different conclusion, considered the specimen to be the gyroconic nautiloid *Goldringia cyclops*. The year 1810 found future New York governor De Witt Clinton in the field checking on potential routes for a proposed canal linking Lake Erie and the Hudson River. Noting details and fossils of the limestone escarpments which traverse the State, Clinton made some of the earliest interpretations of Devonian water depths (1810: 18). Furthermore, Wells (1963: 38) reported that in 1822 Clinton anonymously published a series of geologic observations, among which can be found the statement ". . . a great limestone ridge runs through the whole of this country, east to west . . . north of it a ledge of gypsum commences. . . ." Obviously this is not the first observation of the Onondaga Formation, but it may be the earliest to demonstrate knowledge of a discrete limestone body extending fully across the State. During this time Amos Eaton, Senior Professor of the Rensselaer Polytechnic Institute (RPI), was hard at work trying to document and make sense of New York's geologic column. The conceptual necessity to reproduce the strata of Europe severely hampered Eaton's work. However, in 1824 (Wells, 1963: 42) he recognized a limestone unit with abundant "hornstone" (chert) which he named the Corniferous limerock. Four years later, Eaton (1828: 153) recognized "lower or compact" and "upper or shaly" divisions within the unit. This was the status of what would later become the Onondaga Limestone in 1836, when Governor William Marcy established the Survey of New York. Within the next decade, the work of the Survey's principal geologists, Emmonds, Hall, Mather, and Vanuxem, would alter the course of geologic investigations in North America. Eaton did not live to see his work completed; he died on Tuesday, May 10, 1842.

The completed reports of the original four district surveys brought the Onondaga Limestone into a fairly modern aspect. Since Emmond's Second District lies beyond the formation's limits and Mather contributed little original information in his First District report, we will concentrate on the reports of Vanuxem (1842) and Hall (1843). In examining the writings of these two pioneer geologists, we should bear in mind that Vanuxem considered the term "formation" to be ambiguous and eschewed its use, whereas Hall, holding to the principle that similar products result from similar processes, saw no such ambiguity and promoted the term. This dichotomy is clearly shown by the ways in which the two geologists treated similar observations. Relative to the Onondaga Limestone, Vanuxem was a "splitter" and Hall a "lumper."

Following Hall's (1841) nomenclatural lead, Vanuxem (1842: 132) recognized the Onondaga Limestone and described its "light grey color, crystalline structure, toughness, and its organic remains which are very numerous." He noted that the Onondaga contained numerous "smooth encrinal stems" exceeding a half inch in diameter and that the unit maintained a thickness of 10-14 ft. Succeeding the Onondaga, Vanuxem (1842: 139) recognized, as a discrete unit, the cherty Corniferous Limestone, measuring 60-80 ft in thickness at Cherry Valley. Above the Corniferous, Vanuxem (1839: 275) identified a set of chert-free limestone beds with abundant *Strophomena* ("*Chonetes*") *lineata*, (now *Hallinetes lineatus*) which he referred to as the Seneca Limestone. It is evident from his annual report (1839) and final report (1842) that while in print he treated it as a subdivision of the Corniferous, Vanuxem considered the Seneca to be a separate unit. He only refrained from completing the divorce in the final report due to uncertainties resulting from the absence of a Seneca lithology and fauna in the Helderberg area.

Working in the Fourth Geological District, which covers much of New York's western half, Hall (1843) expanded on the detail of Vanuxem's report. Beneath the Onondaga Limestone Hall (1843: 151) reported on "a few inches of sandstone," which he referred to the Schoharie Grit. He (1843: 152) described the Onondaga as 1-40 ft of lime stone with "light grey color often approaching to white, more or less crystalline in structure, and containing numerous fossils." Noting extreme variations in chert abundance as well as alternating chert-rich and chert-free strata, Hall included the "cherty mass" with the Onondaga rather than with the Corniferous Limestone above. Attending to details, Hall described the now famous Onondaga reefs, complete with various facies, and devoted considerable attention to interpreting depositional environments. Working at a time when many people, geologists included, still believed the earth to be very young, he noted (156) that "the simple fact of the successive growth of coral upon deposits covering other corals, of itself proves a great lapse of time; for the growth of all these forms is exceedingly slow."

Above the Onondaga, Hall found that the Corniferous Limestone thickened westward from less than 30 ft in Seneca County to more than 71 ft at Leroy, and that its "hornstone content" showed considerable variation between localities. Recognizing, as had Vanuxem, that the Upper Corniferous strata lacked chert, it is interesting to note that, in a quarry near Waterloo, Hall (1843: 163) found a "separation of the higher and lower strata by a 'wayboard,' or a seam of (yellowish) clay about four inches thick." This is an early description of the Tioga Bentonite, which, in modern stratigraphic terminology, separates the Moorehouse and Seneca members of the Onondaga Formation. As we have seen, the Onondaga of today was not recognized at the time that Hall wrote the report of the Fourth District. However, noting vertical and lateral lithologic variations within both the Onondaga and Corniferous (including Seneca) limestones which approximate the differences between them, Hall was adamant that they be united as "one formation" (1843: 152) with divisions based on fossil content. This brought the earlier observations of De Witt Clinton to full circle in establishing a single unit and set the stage for today's concept of the Onondaga Limestone.

LITHOSTRATIGRAPHY AND BIOSTRATIGRAPHY

In time, the formational name Onondaga Limestone came to encompass the 1836 Geological Survey's Onondaga, Corniferous, and Seneca limestones. Based on observations made in Hall and Vanuxem's type areas of Onondaga and Seneca counties, Oliver (1954) formally divided the Onondaga into four members. In vertical succession, these are the Edgecliff (formerly Onondaga), Nedrow, Moorehouse (formerly Corniferous), and Seneca limestones.

The Edgecliff is a thick-bedded to massive, light gray to pink, poorly washed to unsorted biosparite. The basal bed(s) of the member contains quartz sand which, in some places, is sufficiently abundant to constitute a quartz arenite. The Edgecliff fauna abounds with rugose corals, tabulate corals, and crinoids. At Split Rock, the type locality, the Edgecliff is 8 ft (2.3 m) thick. Its thickness is highly variable to the west and generally increases eastward to about 48 ft (15 m) in the mid-Hudson Valley.

Succeeding the Edgecliff in central and eastern New York, the Nedrow Member is a 10-14 ft (3-4.3 m) sequence of thinly-bedded, dark gray, argillaceous, fossiliferous to sparse biocalcisiltites. While brachiopods dominate the megafauna and the unit is typified by platyceratid gastropods and the small rugosan *Amplexiphyllum hamiltoniae*, thin sections reveal that in many places the Nedrow is numerically and volumetrically dominated by the microfossil *Styliolina fissurella*. The Nedrow undergoes an eastward facies change and thickens to approximately 43 ft (13 m) near Catskill. In western New York it overlies the Clarence Member and is of uncertain thickness.

Gradational with and succeeding the Nedrow, the Moorehouse Member is a 24 ft (7.3 m) unit of laminated to thick-bedded, dark gray, argillaceous, cherty, sparse biocalcisiltite. Shale partings typically separate the limestone strata. The fauna is dominated by a diversity of brachiopods with lesser numbers of trilobites and gastropods. Both east and west of its central New York type section at Jamesville, the Moorehouse thickens and undergoes a facies change to cleaner and coarser grained textures with crinoid-coral faunas.

The Moorehouse is overlain by 4-6 ft (10-15 cm) of yellowish clay, the Tioga Bentonite, which is the lowermost bed of the Seneca Member. A 26 ft (7.8 m) unit of dark-gray, fossiliferous and sparse biocalcisiltites, the Seneca becomes increasingly argillaceous upward. In central New York, the fauna is dominated by the diminutive strophomenid brachiopod "*Hallinetes*" *lineatus*, which is found in a zone approximately 10 ft (3 m) above the Tioga. The Seneca is overlain by black shales of the Marcellus Formation.

The base of the Onondaga Limestone has long been considered to coincide with the base of the Middle Devonian in New York State (Rickard, 1975). The biozones of conventional megafossils such as brachiopods, corals, and cephalopods are summarized by several authors in Oliver and Klapper (1981); all support this age assignment relative to other North American faunas. Unfortunately, these megafossils are restricted to North America and direct correlation with European biozones is not possible. This difficulty in correlation was compounded when the International Union of Geological Sciences ratified the decision of the Subcommission on Devonian Stratigraphy that the base of the Middle Devonian Series and of the Eifelian Stage coincide with the first occurrence of the conodont *Polygnathus costatus partitus* (Ziegler and Klapper, 1985). The subspecies *partitus* is the second in a lineage of three that are sequentially related. Thus, the bottom of the *patulus* Zone is high in the Emsian Stage and the bottom of the *costatus* Zone is well within the Eifelian. Klapper (in Oliver and Klapper, 1981) reported that the Upper Nedrow beds at Cherry Valley yield both *Polygnathus costatus costatus* and *P. c. patulus*, placing the member's top well within the *partitus* Zone. Noting this along with the fact that *P. c. partitus* is unknown from the Onondaga, Ziegler and Klapper (1985) suggested, with question marks, that the Edgecliff Member is within the *patulus* Zone and correlative to the Emsian Stage of the Lower Devonian Series. At this time, the absence of definitive taxa from the Edgecliff, Clarence, and Lower Nedrow members leaves the precise age of the Lower Onondaga in a very gray (N3?) area.

ONONDAGA LITHOFACIES

Lindemann (1980) identified six carbonate lithofacies of the Onondaga Formation on the basis of relative abundances of calcisiltite, bioclasts, cement, argillaceous mud, and pyrite as point-counted in thin section. Four of these lithofacies are well represented in central New York and figure in this discussion. Mean abundances of their constituents are shown in Table 1 and descriptions follow. As the lithofacies do not exactly correspond to formally named carbonate lithologies, they are referred to by Roman numerals.

Lithofacies VI: Lithofacies VI consists of thick-bedded to massive, light gray, poorly washed to sorted biosparites. Varying abundances of quartz sand are present in samples from the lower beds of the Edgecliff Member. Comminuted crinoids and bryozoans dominate the fossils seen in thin-sectioned samples from central New York. While evidence of bioturbation is rarely observed, vertical burrows are common as are shell lag concentrates and cross-laminae. Rare cryptalgal laminae, oncolites, and the calcareous alga *Asphaltina* are present. This lithofacies is characteristic of the Edgecliff throughout the state and also occurs higher in the formation in eastern and western New York.

Lithofacies VI is interpreted as having been deposited under shallow shelf conditions in wave-agitated waters of very low turbidity.

Lithofacies II: Lithofacies II consists of medium-bedded to massive, medium gray packed biocalcisiltites. Crinoids and bryozoans dominate the fossils in thin section and in the field. Intimately associated with Lithofacies VI and occurring in the Edgecliff throughout the state and the Upper Onondaga to the east and west, Lithofacies II differs primarily in its paucity of interparticulate cement. There are additional faunal differences, but these do not figure in lithologic descriptions

Lithofacies II is interpreted as having been deposited under nonturbid carbonate shelf conditions, quieter than but similar to those of Lithofacies VI. Stratigraphic distribution and association with other moderate energy lithofacies indicate that II was deposited just offshore from or in slightly deeper water than VI. Lagoonal conditions are not indicated in central New York.

Lithofacies III: Lithofacies III consists of laminated to medium-bedded, dark gray, argillaceous calcisiltite, fossiliferous calcisiltite, and sparse biocalcisiltite. The pyrite content of this lithofacies does not exceed 1 percent in central New York. Comminuted crinoids and trilobites dominate the megafossils in thin section and the microfossil *Styliolina fissurella* reaches its maximum abundance. Fossils are occasionally concentrated in thin stringers associated with argillaceous laminae. However, most fragments were scattered by intense bioturbation. This lithofacies predominates in the Nedrow of central New York and in the Moorehouse elsewhere in the state.

Lithofacies III is interpreted as having been deposited in quiet, moderately turbid water offshore from Lithofacies VI and II. Restricted circulation and low oxygen levels are not indicated. The sediment's fine-grained nature suggests a flocculent or soupy sediment-water interface, a condition not

TABLE 1. Mean percent abundances of lithofacies constituents in the Onondaga Limestone, central New York.

Facies	Calcisiltite	Bioclasts	Cement	Detrital Mud	Pyrite
VI	15	61	21	4	1
II	38	53	3	4	1
III	84	2	0	10	2
V	67	10	0	23	2+

particularly conducive to colonization by the larvae of sessile organisms. This accounts for the relative abundance of calcisiltites and planktonic styliolines.

Lithofacies V: Lithofacies V consists of laminated to medium-bedded, dark-gray, highly argillaceous fossiliferous calcisiltites and sparse biocalcisiltites. Trilobites dominate the fossils in thin section and share dominance with brachiopods in field observations. *Chondrites* and general bioturbation are abundant. This facies is virtually restricted to the Moorehouse and Seneca members of central New York, where it is intimately associated with Lithofacies III. It differs from III in that it contains about twice as much argillaceous mud and slightly more pyrite (Table 1).

Lithofacies V is interpreted as having been deposited in quiet, relatively deep and turbid water in and near the subsiding axis of the Appalachian Basin. Because this facies occurs in the area of the formation's lowest rate of sedimentation, probably less than half that of some sites to the east for example, the magnitude of real day-to-day turbidity required to attain its approximately 20 percent argillaceous content is uncertain. While fluctuations in argillaceous influx are evident as shale laminae, it appears that the depositional conditions of lithofacies V differ from those of III primarily in geographic proximity to the relatively carbonate-starved and more restricted axis of the Appalachian Basin.

DEPOSITIONAL HISTORY

The succession of Onondaga members is closely associated with specific lithofacies, as represented by outcrops in the general Syracuse vicinity. The interpretation below pertains only to this area and does not apply elsewhere.

Onondaga deposition began in clean, shallow, wave-agitated waters on a relatively flat erosional surface during a generally westward transgression of the sea. Quartz sand eroded from Early Devonian units was rounded, sorted, and redeposited as a quartz arenite at the formation's base. As the shoreline progressed westward, the shallow carbonate shelf conditions of Lithofacies VI were established and the supply of quartz sand diminished, eventually ceasing all together. Northward (today) migration of the Appalachian Basin into the area beginning in late Edgecliff time produced a gentle down-warp in the sea floor, in which progressively increasing water depths are indicated by the succession of Lithofacies VI, II and III. Flocculent bottom conditions reduced the sessile benthonic fauna, thus minimizing carbonate production and paving the way for the Lithofacies III to V succession. In the face of argillaceous influx coupled with increasingly deleterious benthonic conditions, shale deposition gradually became dominant over carbonate deposition until the latter virtually ceased. The Marcellus Shale gradually prograded westward over the Onondaga Limestone.

MEGAFOSSILS OF THE ONONDAGA LIMESTONE IN CENTRAL NEW YORK

The predominant megafossils of the Onondaga Limestone in the Syracuse area that will be collected on this trip are brachiopods (Table 2, Figures 1-4) and corals (Table 3). Therefore, these groups will be treated in more detail than other, less common fossils, such as gastropods, trilobites,

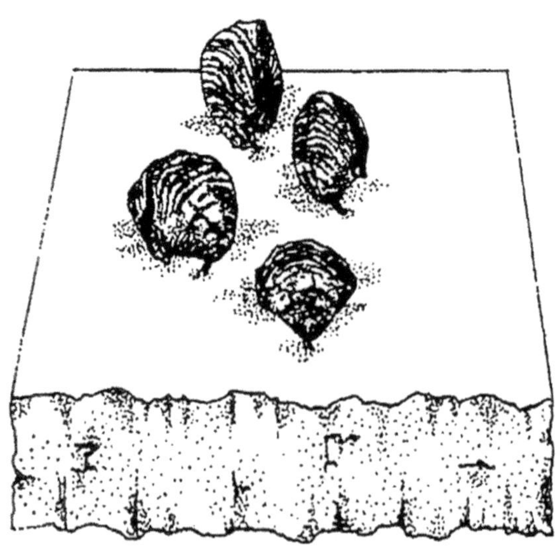

FIGURE 1. *Athyris* sp. A in plenipedunculate life position on the sea floor. This life strategy refers to those brachiopods in which the pedicle is a single, unbranched muscular structure apart from its distal tip.

TABLE 2. Brachiopods of the Onondaga Limestone in central New York from Cherry Valley to Syracuse.

Taxon	common	rare	Very rare
Acrospirifer duodenaria			X
Athyris sp. A		X	
Atrypa "reticularis"	X		
Amphigenia sp.			X
Atlanticocoelia acutiplicata		X	
"Hallinetes" aff. *lineatus*	X		
Charionoides aff. *doris*			X
Costistrophonella cf. *punctulifera*			X
Coelospira camilla		X	
Dalejina aff. *alsa*		X	
Gypidula sp.			X
Leptaena aff. *"rhomboidalis"*	X		
Levenea aff. *subcarinata*		X	
Meristina cf. *nasuta*			X
Megakozlowskiella raricosta	X		
Megastrophia sp.		X	
"Mucrospirifer" cf. *macra*			X
Nucleospira aff. *ventricosa*			X
Orthotetacid indet.	X		
Pentagonia unisulcata			X
Pentamerella arata		X	
Schizophoria cf. *multistriata*			X
Strophodonta cf. *demissa*			X
Stropheodontid indet.			X
Trematospira sp.			X

Note: Although this table denotes relative abundance of brachiopod taxa in the central part of the state in terms of common, rare, and very rare, it should be noted that some species are more abundant in specific horizons or beds and are relatively rare throughout the remainder of the formation. For example, *"Hallinetes"* aff. *lineatus* occurs abundantly 10 ft above the Tioga Bentonite, but rarely in the rest of the Onondaga, and *Amphigenia* sp. occurs only in the basal Edgecliff Member, where it is locally abundant.

cephalopods, crinoids, bryozoans, and sponges (Table 4). Some common corals and a lithistid sponge, illustrated in Shimer and Shrock (1944), are shown in Figures 5 and 6.

Brachiopods: Since most of the brachiopods that will be collected loose consist of articulated specimens, the descriptions below do not include interior morphologies. For more detailed descriptions, including internal features, refer to Feldman (1985). Collecting brachiopods from the Onondaga Formation in central New York is best accomplished by carefully

TABLE 3. Corals of the Onondaga Limestone in central New York from Cherry Valley to Syracuse.

Taxa	common	rare
Tabulates		
Aulopora	X	
Syringopora	X	
Favosites	X	
Lecfedites	X	
Rugosans		
Amplexiphyllum	X	
"Heterophrentis"	X	
Acinophyllum	X	
Breviphrentis		X
cf. *Syringaxon*		X
Cystophylloides	X	
Siphonophrentis	X	
Heliophyllum	X	

searching weathered outcrops in the Nedrow Member, especially re-entrants, which form due to the more rapid weathering of the shale as compared to the dense limestone. Collecting in the other members varies, but is generally difficult. As a rule, the longer the outcrop has weathered, the better the chances of finding suitable specimens. Occasionally, due to local diagenetic factors, it is possible to locate silicified material ranging in quality from poor to fair. As one approaches the western and especially eastern areas of the outcrop belt, silicification tends to increase in amount and quality.

Acrospirifer duodenaria: Biconvex shells transversely subelliptical in outline; hinge line long and straight, medial open delthyrium with no preserved deltidial plates; pedicle valve bears narrow, triangular, moderately deep, noncostate sulcus; brachial valve bears corresponding fold; five to six rounded plications on each pedicle flank with U-shaped interspaces; anterior commissure uniplicate.

Athyris sp. A: Shells transversely suboval in outline and subequally biconvex, with the pedicle valve slightly deeper than the brachial valve; ventral beak suberect, terminating in a small round foramen; brachial beak smaller and less noticeable; pedicle valve bears a shallow sulcus with corresponding low fold on brachial valve; anterior commissure weakly uniplicate; some forms nonsulcate and rectimarginate; fine, concentric growth lines present on both valves.

Atrypa "reticularis": Dorsibiconvex shells with well-rounded radial costellae which increase in size and number anteriorly; costellae separated by U-shaped interspaces; concentric growth lamellae cross the costellae, becoming more distinct and frilly anteriorly; anterior commissure rectimarginate or slightly deflected toward brachial valve.

Amphigenia sp.: Large shells, costellate, elongate in outline; anterior commissure rectimarginate to broadly sulcate; delthyrium wide, triangular, and unmodified by deltidial plates; round pedicle foramen present.

Atlanticocoelia acutiplicata: Subcircular in outline with length almost equal to width; brachial valve gently convex, pedicle valve slightly more so; weak pedicle sulcus sometimes noticeable on larger specimens; no corresponding dorsal fold; hinge line very short and becomes rounded anteriorly; no interareas present, anterior and lateral commissures crenulate; 10 to 12 plications with U-shaped interspaces; concentric growth lines, two or three per shell, common on ephebic forms.

"Hallinetes" aff. *lineatus*: Shells small, subsemicircular in outline and concavoconvex in lateral profile; interareas very narrow; no delthyrial structures preserved; greatest width at hinge line or anterior to midlength; valves covered with fine capillae which increase anteriorly by bifurcation.

Charionoides aff. *doris*: Biconvex, elongate shells; fold and sulcus often poorly defined; pedicle beak slightly incurved with round foramen; delthyrium triangular; valve exteriors smooth with fine, concentric growth lines present at lateral margins of both valves.

Costistrophonella cf. *punctulifera*: Shells wider than long, subsemicircular in outline; hinge line long, straight; interarea evident; numerous, subangular, radiating costae, which increase anteriorly by intercalation and bifurcation; interspaces broad and shallow with 11 to 12 costae per 5 mm near anterior commissure at midline; costae crossed by about 7 evenly spaced growth lines.

Coelospira camilla: Small, concavoconvex to planoconvex, subcirclar to suboval in outline: small, distinct pedicle foramen on incurved pedicle beak; no interarea evident; maximum width about one-third of the valve length in adults; pedicle valve bears two medial plications usually at least as large as remaining radial plications on flanks; interspaces U-shaped; brachia valve bears medial plication which generally bifurcates at one-third the valve length; median interspace usually flat but sometimes bears small ridge; plications broader on flanks and thinner toward lateral commissure; several well-defined, concentric growth lines evident near anterior commissure in adult forms.

Dalejina aff. *alsa*: Shells ventribiconvex, transversely suboval to subcircular in outline; hinge line very short and straight in apical area but becomes rounded as lateral margins are approached; maximum width at or just anterior to midlength; pedicle valve bears slight median depression; brachial valve often bears corresponding median ridge; anterior commissure most often rectimarginate to slightly sulcate; ventral interarea short, narrow; numerous radial costellae which increase anteriorly both by intercalation and bifurcation; at anterior commissure, there are 18 to 20 costellae per 5 mm, near midline; costellae occasionally crossed by concentric growth lines near anterior margins.

Gypidula sp.: Elongate oval to subcircular in outline; pedicle valve swollen; costate to multicostate; almost identical to *Pentamerella arata* (see description below) but can be differentiated by a pedicle fold and brachial sulcus, whereas *Pentamerella* has a pedicle sulcus and brachial fold.

Leptaena aff. "*rhomboidalis*": Transversely subquadrate in outline, concavo-convex to slightly biconvex with pedicle valve strongly geniculate at anterior and lateral commissures; brachial valve correspondingly geniculate within pedicle trail; hingline straight, pedicle interarea flat; ornamentation consists of radial costellae which extend past the point of geniculation and continue on trail of valves; concentric rugae cross costellae, becoming larger anteriorly.

Levenea aff. *subcarinata*: Shells small to medium sized, transversely suboval in outline ventribiconvex in lateral profile; brachial valve bears shallow, rounded sulcus, which broadens anteriorly; length slightly greater than width; maximum width at or just anterior to midlength; ventral interarea short, slightly incurved; triangular delthyrium encloses angle of approximately 60 degrees; delthyrium often widens apically into small, circular foramen; ornamentation consists of rounded, radial costellae which increase in number anteriorly by bifurcation.

Megakozlowskiella raricosta: Shells subtransverse in outline, strophic, medium to large, ventribiconvex; hinge line straight; pedicle interarea moderately narrow with striae which parallel hinge line; brachial interarea extremely narrow; distinct, slightly flattened fold on brachial valve and corresponding deep, U-shaped sulcus on pedicle valve; commonly three plications on flanks; delthyrium includes angle of approximately 60 degrees; no deltidial plates preserved; anterior commissure uniplicate; strong, concentric growth lamellae with anterior frills; radial ornamentation consists of very fine striae or capillae.

Megastrophia sp.: Medium to large, subsemicircular to transversely suboval in outline; somewhat alate, concavoconvex in lateral profile; maximum width attained at hinge line; unequally parvicostellate to subuniformly costellate; psuedodeltidium flat, complete with narrow median ridge; chilidium flat, complete, with median ridge; hinge entirely denticulate.

Meristina cf. *nasuta*: Convex, elongate, and suboval in outline with no noticeable interarea; unequally biconvex, with pedicle valve much deeper than brachial valve; maximum width commonly anterior to midlength; delthyrium broad, triangular, and opening apically into semicircular foramen; faint pedicle sulcus modified by development of low, rounded medial plication that extends anterior commissure in tongue-like projection; concentric growth lamellae evident at anterior portion of valves, but remainder of shell smooth.

"*Mucrospirifer*" cf. *macra*: Small to large alate shells transversely subtrigonal to subsemicircular in outline; biconvex in lateral profile with brachial valve slightly flatter than pedicle valve, ventral interarea moderately high, long, somewhat curved; ventral beak posterior to interarea, short and stubby; open, triangular delthyrium present that divides interarea medially; dorsal interarea long, thin, ribbon-like; brachial valve bears high, medial fold flattened at top; pedicle valve bears corresponding U-shaped sulcus; surface of shells covered by sharply defined plications ranging from U-shaped to subangular in cross section; numerous, concentric, frilly growth lines present; no fine radial ornamentation.

Nucleospira aff. *ventricosa*: Small, transversely suboval in outline, biconvex in lateral profile with pedicle valve slightly deeper than brachial valve; hinge line curved; brachial beak fits into anterior end of delthyrium, which is partially covered by concave pseudodeltidium in some specimens; both beaks erect, no interarea evident; shell surface lacks radial ornamentation; no fold or sulcus present; pedicle valve shows faint median depression in some specimens; concentric growth lamellae present, more concentrated toward rectimarginate anterior commissure.

Orthotetacid indet.: Small to medium sized shells, generally poorly preserved as internal impressions; hinge line straight; ornamentation finely costellate.

Pentagonia unisulcata: Medium sized, nonstrophic, pentagonal in outline when viewed posteriorly; beak suberect, dorsibiconvex with greatest width attained between midlength and anterior commissure, brachial valve cariniform due to presence of raised, rounded fold bearing narrow, median groove; in some forms groove widens slightly anteriorly, forming two parallel to subparallel ridges extending almost half the valve length; flanks concave, dropping steeply away from sulcate fold; sulcus broad, shallow, with two distinct ridges which define sulcus laterally and extend from umbo across posterolateral margins of flanks to uniplicate anterolateral commissure; vague, concentric growth lines on anterior portion of shell.

Pentamerella arata: Subglobose and broadly pyriform in outline with strongly convex pedicle valve and weakly convex brachial valve; hinge line short, curved, narrow; no interarea evident; pedicle valve beak short, strong, incurved, not closely pressed against brachial beak; brachial beak small, less erect and less incurved; maximum width attained at or about midlength; weak sulcus present on anterior half of pedicle valve, with corresponding fold on brachial valve (note: this morphological feature is the key to differentiating *Gypidula* from *Pentamerella*; see *Gypidula* above); both valves ornamented with numerous, rounded, bifurcating plications, which become narrower on lateral slopes than near midline; interspaces between plications U-shaped and wider than plications which tend to become slightly V-shaped in cross section on some specimens: about 5 plications in sulcus and 6 on fold; concentric growth lines more numerous anteriorly.

Schizophoria cf. *multistriata*: Shells medium sized, suboval to subquadrate in outline, unequally biconvex, brachial valve deeper and more uniformly convex; in juveniles both valves become almost equally biconvex; pedicle valve develops broad, shallow sulcus on adult forms; brachial valve bears indistinct fold; hinge line short, slightly rounded; maximum width attained at or just past midlength; ventral interarea triangular, fairly high in larger shells, relatively narrow in younger ones; dorsal interarea narrower; interareas of both valves equal to about one-half width: ornamentation consists of rounded to subangular radial costellae with broad, flat inter spaces; about 11 costellae in a 5 mm space near anterior commissure at midline.

Strophodonta cf. *demissa*: Subcircular to shield shaped shells, concavoconvex in lateral profile; shells wider than they are long; point of maximum width at hinge line; lateral margins almost straight posteriorly; anterior

TABLE 4. Other faunal constituents of the Onondaga Limestone in central New York.

Taxa	Common	Rare	Very rare
Gastropods			
Platyceras		X	
Tropidodiscus			X
Straparollus			X
Liospira			X
Ecculiomphalus			X
Loxonema			X
Platystoma			X
Euomphalacean fragments			X
Cephalopods			
Foordites			X
Halloceras			X
cf. *Goldringia*			X
Trilobites			
Phacops cristata	X		
Odontocephalus	X		
Dechenella			X
cf . *Otarion*			X
cf. *Proetus*			X
Dalmanitid fragments			X
Crinoids			
Camerate columnals	X		
Calyces			X
Bryozoans			
Dyoidophragma			X
Fistulipora			X
Sponges			
Hindia	X		

margins evenly rounded: all margins crenulate; anterior commissure rectimarginate; costellae coarse, bifurcating with angular interspaces in cross section.

Stropheodontid indet.: Small, subcircular in outline, alate; smooth exterior with irregularly spacea growth lines.

Trematospira sp.: Elongate, suboval, moderately biconvex; dorsal fold and ventral sulcus; hinge line curved narrow interarea; conjunct deltidial plates; pedicle foramen mesothyrid; approximately 45 rounded bifurcating costae on dorsal exterior and 19 on ventral exterior; beak pointed and apically twisted.

DISTRIBUTION OF MEGAFOSSILS

Corals dominate the shallow water habitat of the Edgecliff and Lithofacies VI. Common rugosans include *Acinophyllum*, *Cystiphyloides*, *Heliophyllum*, *Heterophrentis*, and *Siphonophrentis*, while the tabulates are represented by *Emmonsia*, *Favosites*, and *Lecfedites*. Brachiopods include *Amphigenia*, *Atrypa*, *Elita*, *Leptaena*, and *Levenea*. The trilobite

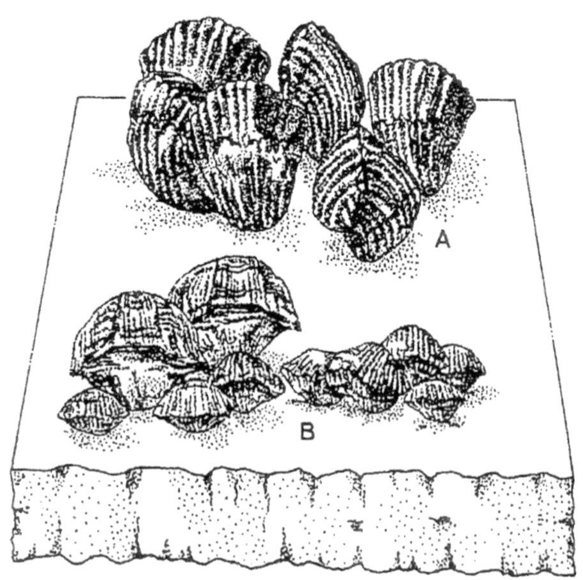

FIGURE 2. A. *Pentameralla arata* and B. *Atrypa "reticularis"* shown in cosupportive life habit. This refers to those ambitopic brachiopods which maintain an umbo-down posture and are packed tightly together, often growing on one another.

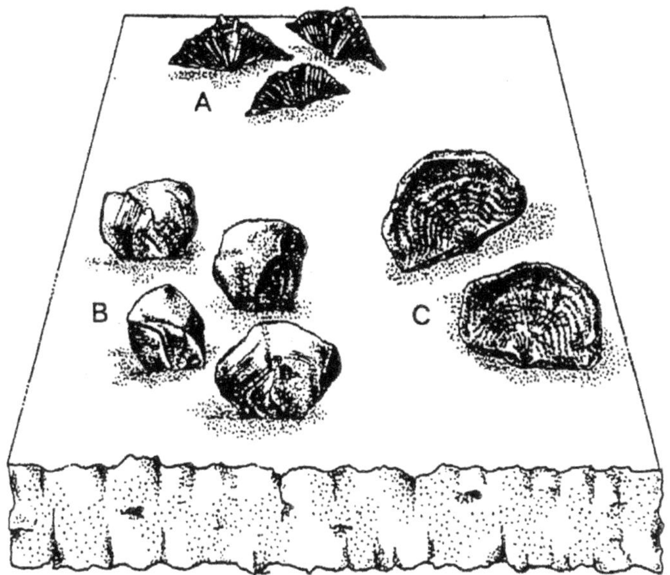

FIGURE 3. A. *Mucrospirifer* cf. *macra* (rhizopedunculate); B. *Pentagonia unisulcata* (ambitopic) C. *Leptaena* aff. "*rhomboidalis*" (quasi-infaunal).

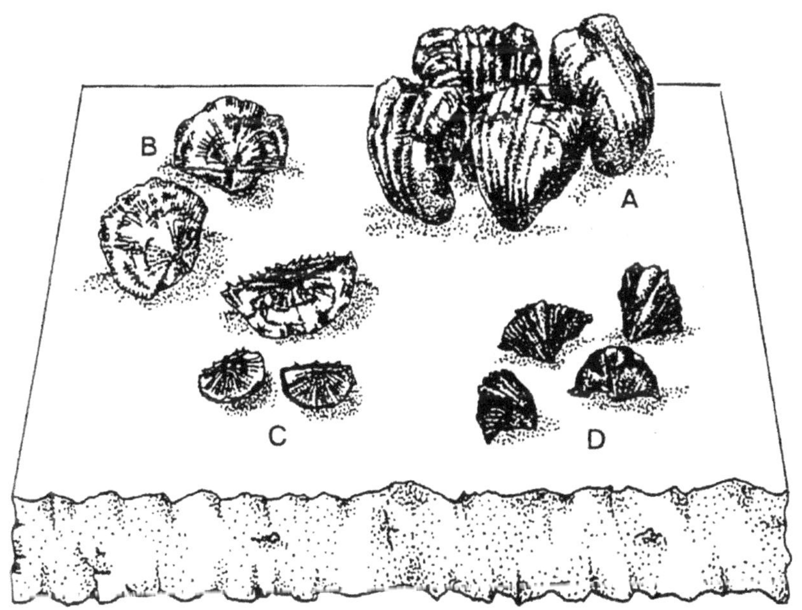

FIGURE 4. A. *Gypidula* sp. (cosupportive); B. Orthotetacid indet.; C. "*Chonetes*" sp.; D. *Cyrtina hamiltonensis* (extremely rare in central New York) (all ambitopic).

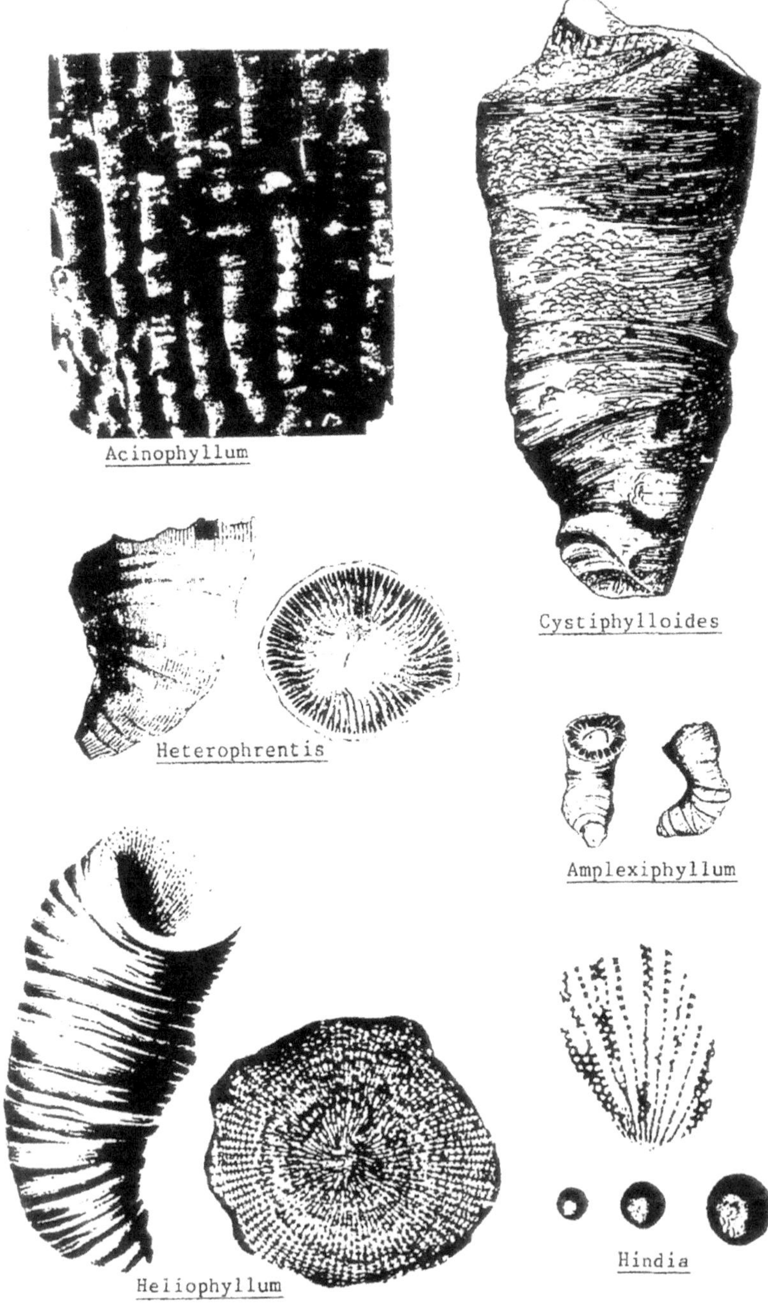

FIGURE 5. Corals and sponges of the Onondaga Limestone.

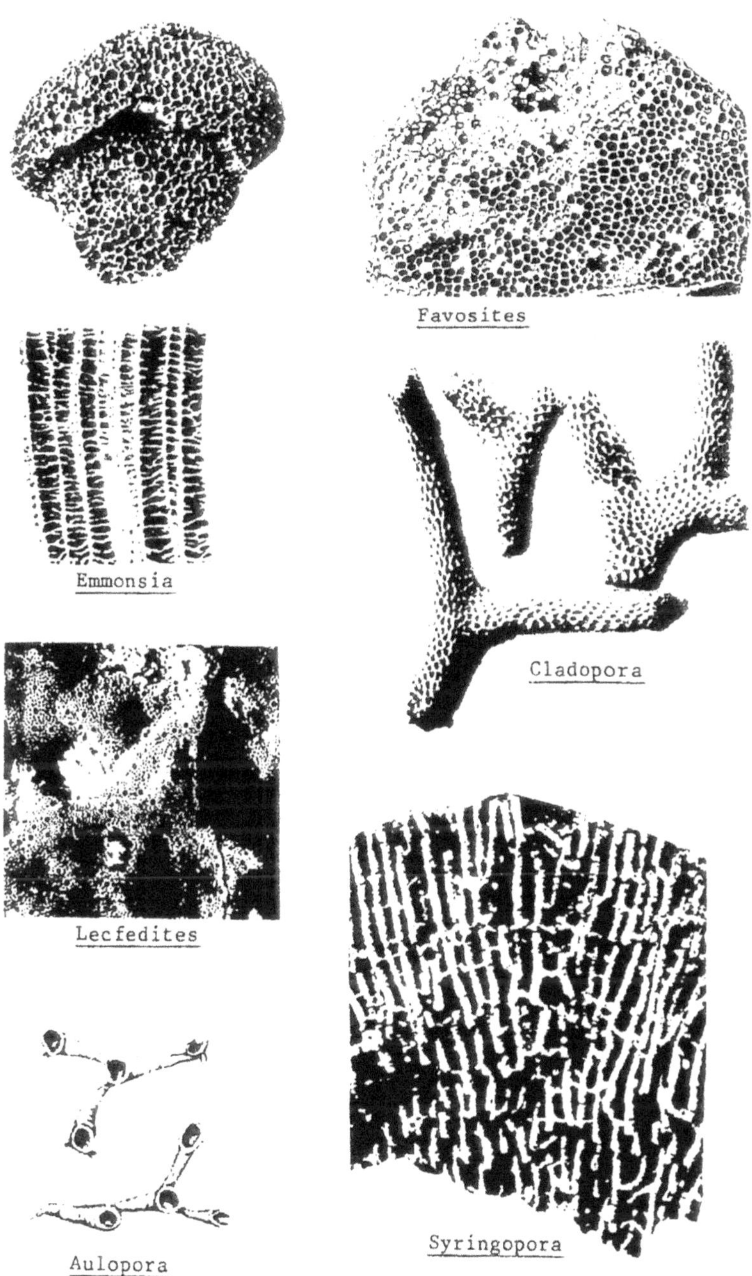

FIGURE 6. Corals of the Onondaga Limestone.

Phacops is present in small numbers; platyceratid gastropods are locally common and crinoid columnals abound. Bryozoans are rare and almost never intact.

In the field trip area, the fauna of the Upper Edgecliff and Lithofacies II differs little from that of Lithofacies VI. While most of the fauna remains unchanged, though somewhat less abundant, *Lecfedites* and fenestrate bryozoans become more common. This appears to be in response to diminishing water agitation.

Brachiopods, including *Atrypa*, *Megakozlowskiella*, and *Megastrophia*, dominate the fauna of the Nedrow Member and Lithofacies III. The lithistid sponge *Hindia* is quite common low in the section. The rugose corals *Amplexiphyllum*, *Heliophyllum*, and *Heterophrentis* are present, though not particularly common in central New York. *Phacops* and *Odontocephalus* are common trilobites while the gastropods are represented by *Platyceras* and *Platystoma* and the cephalopods by *Goldringia*? and *Foordites*.

In the field trip area, the offshore habitat of the Moorehouse and Seneca members and Lithofacies V is dominated by trilobites (*Phacops*, *Odontocephalus*) and a diversity of brachiopods including *Atrypa*, *Athyris*, *Coelospira*, *Leptaena*, *Mucrospirifer* and *Strophodonta*. "*Hallinetes*" is relatively common in the Moorehouse and extremely abundant high in the Seneca. The rugose coral *Heterophrentis* as well as the tabulates *Aulopora* and *Syringopora* are often present, though not at all common.

ROAD LOG FOR THE ONONDAGA LIMESTONE IN CENTRAL NEW YORK

Cumulative Miles	Mileage from Last Point	Route Description
		Leaving Ithaca take NY Route 13 east to I-81 north, approximately 27 mi.
0.0	0.0	Leave I-81 at Exit 16 and turn left onto U.S. Route 11 north, approximately 24 mi. Log mileage begins at this point.
0.7	0.7	Immediately after passing beneath I-81, turn left from Route 11 onto Quarry Road
1.0	0.3	STOP 1: Park on the road shoulder and walk into the south end of the quarry which lies between Quarry Road and the Interstate.

Note: This quarry is on the Onondaga Indian Reservation. Do not, under any circumstances, enter without prior permission from the Tribe!

The quarry's south end is floored by the Edgecliff Member. Above this about 13 ft of Nedrow is exposed (type section). The Moorehouse is 19 ft thick and overlain by an incomplete section of the Seneca Member. The Tioga Bentonite horizon is evident as the glacially scoured surface above the quarry's northeast wall where it forms a re-entrant high in the southeast wall (it can be located between two chert horizons).

1.3	0.3	Return to Route 11 and park in the lot on the road's west side. Walk across the road to the outcrcp between Route 11 and the I-81 exit ramp.

The entire Edgecliff and Lower Nedrow are exposed here in a weathered condition which may be more suitable for collcting than the quarry exposures.

		Leave the parking lot and head left (north) on Route 11.
4.2	2.9	Turn right (east) onto Route 173.
8.3	4.1	In Jamesville, turn left (north) onto Solvay Road.
9.0	0.7	Turn right (east) onto the entrance road to the Jamesville Quarry.
		STOP 3: Continue along the entrance road and stop at the quarry office.

This is reputed to be the largest quarry in New York State. Hard hat and prior permission are required. The entire thickness of the Onondaga, from the phosphatic "Springvale Sandstone" to the Seneca and Marcells Shale, is exposed near the quarry's southeast corner. However, quarry operations have left only the "Springvale," Edgecliff, and Nedrow sections readily accessible. The Tioga Bentonite and Seneca/Marcellus at the quarry top can be safely seen only from a distance. Exercise extreme caution in proximity to the quarry walls; hard hats offer little protection from large falling rocks.

9.7	0.7	Leave the quarry via Solvay Road, return to Route 173 and turn right (west) back toward Syracuse.
18.9	9.2	Turn left onto Split Rock Road. The road sign may not be visible, so watch for a yellow-and-blue state historical marker.
19.6	0.7	STOP 4: Continue to the end of Split Rock Road and into the quarry entrance.

This is the type section of the Edgecliff Member which is named for Edgecliff Park, located just to the west. The member's full thickness of 2.3 m is exposed in the upper areas of the main quarry and its erosional base is marked by the presence of phosphate nodules. It is interesting to note the pronounced difference between the "Springvale" horizon here and in the Jamesville Quarry. In the southern quarry wall, 3 m of Nedrow are exposed, but the top of the Nedrow has been removed by glacial scour.

END OF TRIP

REFERENCES

Eaton, A. 1828. Geological nomenclature, exhibited in a synopsis of North American rocks and detritus. *American Journal of Science* 14: 145-159, 359-368.

Feldman, H. R. 1980. Level-bottom brachiopod communities in the Middle Devonian of New York. *Lethaia* 13: 27-46.

Feldman, H. R. 1985. Brachiopods of the Onondaga Limestone in central and southeastern New York. American Museum of Natural History Bulletin 179: 293-377.

Hall, J. 1841. Fifth annual report of the fourth geological district. New York Geological Survey Annual Report 5: 149-179.

Hall, J. 1843. Geology of New York. Part 4, comprising the survey of the fourth geological district. Albany.

Lindemann, R. H. 1980. Paleosynecology and paleoenvironments of the Onondaga Limestone in New York State. Unpublished PhD thesis, Rensselaer Polytechnic Institute, Troy, New York.

Lindemann, R. H., and Feldman, H. R. 1981. Paleocommunities of the Onondaga Limestone (Middle Devonian) in central New York State. *New York State Geological Association Guidebook, 53rd Annual Meeting, Binghamton, New York*, 79-96.

Oliver, W. A., Jr. 1954. Stratigraphy of the Onondaga Limestone (Devonian) in central New York. *Geological Society of America Bulletin* 65: 621-652.

Oliver, W. A., Jr. 1956. Stratigraphy of the Onondaga Limestone in eastern New York. Geological Society of America Bulletin 67: 1441-1474.

Oliver, W. A., Jr., and Klapper, G. (eds.). 1981. *Devonian Biostratigraphy of New York, Part 1, Text: International Union of Geological Sciences, Subcommission on Devonian Stratigraphy.*

Rickard, L. V. 1975. Correlation of the Silurian and Devonian rocks in New York State. *New York State Museum and Science Service Map and Chart Series* 24.

Shimer, H. W., and Shrock, R. R. 1944. *Index Fossils of North America.* Cambridge, MA: MIT Press.

Vanuxem, L. 1839. Third annual report of the geological survey of the third district. *New York Geological Survey Annual Report* 3: 241-285.

Vanuxem, L. 1842. *Geology of New York,* Part 3, comprising the survey of the third geological district. Albany.

Wells, J. W. 1963. Early investigations of the Devonian system in New York, 1656-1836. *Geological Society of America Special Paper* 74.

Ziegler, W., and Klapper, G. 1985. Stages of the Devonian system. *Episodes* 8: (2): 104-109.

Chapter Twelve

Paleogeography and Brachiopod Paleoecology of the Onondaga Limestone in Eastern New York

INTRODUCTION

During the decades since Oliver (1954) formally divided the Middle Devonian Onondaga Limestone into its Edgecliff, Nedrow, Moorehouse, and Seneca Members, the formation has been the focus of considerable attention. However, with the exception of subsurface investigations, research has concentrated on the formation as seen in outcrop (Figure 1) between Buffalo and the Helderberg/Catskill area. Few geologists have ever examined the Onondaga south of its classic exposure in Leeds Gorge. As an arbitrary measure of this, we note that between 1962 and 1986 more than ten NYGSA field trips concentrated on the Onondaga, while only one (Oliver, 1962) was sited in southeastern New York. As a result, stratigraphers and paleontologists alike tend to have a somewhat skewed impression of Onondaga paleogeography, a situation we hope this field trip will partially rectify.

LITHOSTRATIGRAPHY

In the central New York type area, the Onondaga unconformably overlies Lower Devonian carbonates of the Helderberg Group. In that area the formation's members are lithologically distinct from one another. The lowermost Edgecliff bed(s) is a calcareous and phosphatic quartz arenite that grades upward into the typical thick-bedded, light gray Edgecliff biosparites. The typical Nedrow lithology is a thin-bedded, dark gray, argillaceous, fine-grained limestone. This grades upward into the coarser grained, thin to medium-bedded, dark gray, cherty, and argillaceous limestone typical of the Moorehouse. The Moorehouse is overlain by the Tioga Bentonite, which is succeeded by dark gray, fine to medium-grained limestones of the Seneca Member. While the uppermost Seneca becomes increasingly argillaceous, it includes a regionally extensive "bone bed" and shows signs of erosional truncation prior to deposition of the overlying Union Springs Shale.

Type lithologies of the four members do not extend into eastern New York. Oliver (1954, 1956) found that the Lower Edgecliff beds are gradational with the Schoharie Formation in the east and that the contact is best recognized based on fauna. The argillaceous Nedrow beds extent eastward only to the approximate longitude of Cherry Valley, beyond which they grade into an Edgecliff lithology. Similarly, the Moorehouse is much like Edgecliff in the Helderberg area. The Seneca Member pinches out east of Cherry Valley, though Rickard (personal communication, 1980) found it to be present in the subsurface of southeastern New York.

Oliver (1956, 1962) found that changes in lateral facies toward eastern New York necessitated the use of faunal criteria in recognizing the individual members. He also found a pronounced thickening of the formation into eastern, and to an even greater degree into southeastern, New York (Table 1). Oliver recognized the eastern Edgecliff beds by the presence of large crinoid columnals, the Nedrow on the basis of numerous platyceratid gastropods, and the Moorehouse by its brachiopod-dominated fauna and abundant black chert. However, Lindemann (1979, 1980) found platyceratids to be more common in the cherty Moorehouse beds than in the Nedrow. Be that as it may, fossils remain the best criteria for correlation, particularly in southeastern New York, where outcrops are small and few and the entire formation is fine-grained. There, at least, the Edgecliff's crinoid columnals persist despite the non-typical lithology. These can be traced into the Buttermilk Falls Formation of Pennsylvania (Oliver, 1962). Pronounced facies changes in the Onondaga-equivalent strata of southeastern New York prompted Rickard (1975) to assign those strata to the Buttermilk Falls. While this is prudent lithostratigraphy, we will refer to them as Onondaga due to their significance in paleogeographic reconstruction.

TABLE 1. Representative Thicknesses (in ft) of the Onondaga Limestone and its Members in Eastern New York.

	Syracuse	Helderbergs	Leeds	Saugerties	Port Jervis
Seneca	25	–	–	–	–
Moorehouse	25	70	37+	100+	190+
Nedrow	15	15	43	34	?
Edgecliff	22	30	36	36	30?
Total	87	115	116	170	200+

LITHOFACIES

To facilitate the interpretation of depositional environments, Lindemann (1980) identified six carbonate lithofacies on the basis of relative abundances of calcisiltite, bioclasts, cement, argillaceous mud, and pyrite as pointcounted in thin section. While all of the lithofacies are described below, only three are well represented in eastern New York and figure directly in this discussion. As the lithofacies do not exactly correspond to formally named carbonate lithologies, they are referred to by Roman numerals and mean abundances of their constituents are shown in Table 2. Note that the Roman numerals used herein do not correspond to those of Lindemann (1979), but do correspond to those of Lindemann (1980) as well as Feldman and Lindemann (1986).

Lithofacies I: Lithofacies I consists of thick-bedded to massive, medium gray, sparse and packed biocalcisiltites. Terrigenous mud tends to be

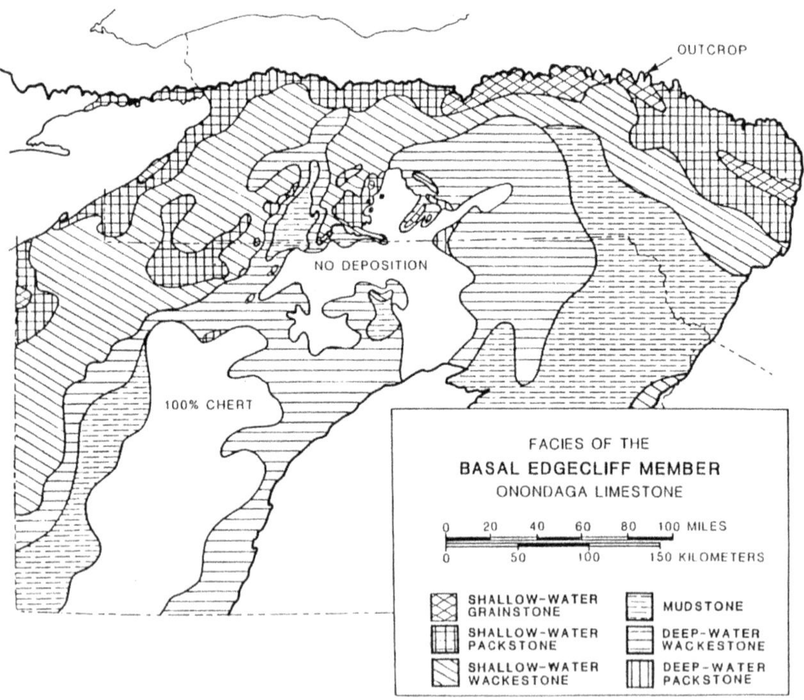

FIGURE 1. Map of the Onondaga outcrop belt and distribution of basal Edgecliff facies (Cassa and Kissling, 1982).

concentrated in styolites and is otherwise scarce. While crinoids and fenestrate bryozoans are dominant fossils in thin section, the ramose bryozoan *Fistulipora* dominates the fauna as seen in outcrop. Some specimens are encrusted by a calcareous alga similar in morphology to *Sphaerocodium*. *Chondrites*, small vertical burrows, and calcarenite-filled grooves on bedding surfaces dominate the ichnofauna. Intimately associated with Lithofacies II and VI, Lithofacies I occurs most commonly in the Leeds to Saugerties area. Except for occurrences in the Moorehouse of the Helderbergs and westernmost New York, it is not commonly encountered.

Lithofacies I is interpreted as having been deposited in fairly quiet, slightly turbid waters. Occasional reworking of the sediment by storm waves is indicated by thin shell layers and by the presence of spar cement in some samples. At Leeds this lithofacies alternates with Lithofacies VI, and both are associated with a *Fistulipora*-dominated community, which persisted from early Edgecliff to late Moorehouse time. As will be more thoroughly discussed later, the Onondaga in this area is interpreted as a sequence of bryozoan bafflestone biostromes, which grew at a rate comparable to that of basinal subsidence.

Lithofacies II: Lithofacies II consists of medium-bedded to massive, medium gray packed biocalcisiltites. Crinoids and bryozoans dominate the fossils in thin section and in the field. Intimately associated with Lithofacies I and VI, and occurring in the Edgecliff throughout the state and the Upper Onondaga to the east and west, Lithofacies II differs from I and VI in abundances of fossil debris and spar cement respectively. There are additional faunal differences, particularly in western New York, but these do not figure in this description.

Lithofacies II is interpreted as having been deposited under nonturbid carbonate shelf conditions quieter than, but similar to, those of Lithofacies VI. Stratigraphic distribution and association with other moderate energy lithofacies indicate that II was generally deposited just offshore from, or in slightly deeper water than, VI. With the possible exception of occurrences in the Clarence Member of western New York, lagoonal conditions are not indicated.

Lithofacies III: Lithofacies III consists of laminated to medium-bedded, dark gray, argillaceous calcisiltite, fossiliferous calcisiltite, and sparse biocalcisiltite. The pyrite content of this lithofacies does not exceed one percent in central New York. Comminuted crinoids and trilobites dominate the megafossils in thin section and the microfossil *Styliolina fissurella* reaches its maximum abundance. Fossils are occasionally concentrated in thin stringers associated with argillaceous laminae. However, most fossil fragments were scattered by intense bioturbation. This lithofacies predominates in the Nedrow of central New York and in the Moorehouse elsewhere in the state. It does not occur east of Cobleskill.

Lithofacies III is interpreted as having been deposited in quiet, moderately turbid water offshore from Lithofacies VI and II. Restricted circulation and low oxygen levels are not indicated. The sediment's fine-grained nature suggests a flocculent or soupy sediment-water interface, a condition not particularly conducive to colonization by the larvae of sessile organisms. This accounts for the relative abundance of calcisiltites and planktonic styliolines.

Lithofacies IV: Lithofacies IV consists of thin to medium-bedded, dark gray, moderately argillaceous, fossilferous to sparse biocalcisiltites. Fenestrate bryozoan and crinoid debris dominate the fossils seen in thin section while brachiopods and trilobites dominate the fauna seen in outcrop. The ichnofauna includes a diverse set of small horizontal burrows and *Chondrites*. Relatively coarse-grained lag deposits and cross laminae are present in some beds. Lithofacies IV is common only in the Clarence Member of western New York and the Moorehouse of the Cherry Valley-Schoharie area.

Lithofacies IV is interpreted as having been deposited in quiet, moderately turbid waters. Though calcisiltite abundances far exceed those of fossils, it appears that bottom conditions were not as quiet as those of Lithofacies III and that storm-generated waves often reworked the sediment. Occurrences of Lithofacies IV in the Clarence Member are interpreted as lagoonal deposits while occurrences in the Moorehouse at Cherry Valley are interpreted as shelf margin or transitional deposits between the shallow shelf to the east and the Appalachian Basin to the west.

Lithofacies V: Lithofacies V consists of laminated to medium-bedded, dark gray, highly argillaceous, fossiliferous calcisiltites and sparse biocalcisiltites. Trilobites dominate the fossils in thin section and share dominance with brachiopods in field observations. *Chondrites* and general signs of bioturbation are abundant. This facies is virtually restricted to the Moorehouse and Seneca members of central New York, where it is intimately associated with Lithofacies III. It differs from III in containing about twice as much argillaceous mud and slightly more pyrite (Table 2).

Lithofacies V is interpreted as having been deposited in quiet, relatively deep and turbid water in or near the subsiding axis of the Appalachian Basin. Because this facies occurs in the area representing the lowest rate of sedimentation for the formation (probably less than half that of some sites to the east, for example), the magnitude of real day-to-day turbidity required to attain its approximately 20 percent argillaceous content is uncertain. While fluctuations in argillaceous influx are evident as shale laminae, it appears that the depositional conditions of Lithofacies V differ from those of III primarily in geographic proximity to the relatively carbonate-starved and more restricted axis of the Appalachian Basin.

Lithofacies VI: Lithofacies VI consists of thick-bedded to massive, light gray, poorly washed to sorted biosparites. Varying abundances of quartz sand, glauconite, and phosphorite nodules are present in samples from the lower beds of the Edgecliff Member. Comminuted crinoids and bryozoans dominate the fossils seen in thin section and macrofossils vary, as this lithofacies occurs with both coral and bryozoan-dominated communities. Rare cryptalgal laminae and calcareous algae are present. While evidence of bioturbation is rarely observed, vertical burrows are common, as are shell lag concentrations and cross-laminae. This lithofacies is characteristic of the Edgecliff throughout the state and also occurs higher in the formation in eastern and western New York. It dominates the Nedrow and lower half of the Moorehouse in the Helderbergs and alternates with Lithofacies I at Leeds. It has not been observed at any horizon in the Onondaga south of Leeds.

Lithofacies VI is interpreted as having been deposited under shallow shelf conditions in wave-agitated waters of very low turbidity.

TABLE 2. Mean percent abundances of Unondaga lithofacies constituents.

Facies	Calcisiltite	Bioclasts	Cement	Detrital Mud	Pyrite
I	50	38	2	7	7
II	38	53	3	4	1
III	84	2	0	10	2
IV	74	16	<1	7	1
V	67	10	0	21	2+
VI	15	61	21	4	1

THE ONONDAGA/BAKOVEN CONTACT

Contacts between the Onondaga Limestone and overlying black shale units are few and far between. Only one (Locality 7) is known in the area of this field trip. This horizon is deserving of detailed consideration, as conclusions drawn from it have far-reaching significance. Oliver (1956) judged the Moorehouse/Bakoven contact to represent only a minor break in deposition. However, Chadwick (1944: 103) described the contact as a "calcarenyte of tiny crinoidal fragments, black in color like the shale and containing also comminuted fish remains with an occasional brachiopod shell seemingly reworked from the limestone beneath. The basal (Bakoven) contact here shows this bed bonded into solution pitting in the limestone, indicating a distinct break and disconformity."

The uppermost Moorehouse bed is a medium dark gray, sparse to packed bioclacisiltite containing trilobite, brachiopod, and crinoid fragments along with a few phosphatic particles. Authigenic quartz and silicified brachiopods are common, while unquestionably detrital quartz silt is uncommon. Terrigenous mud constitutes less than 6 percent of the rock's weight and organic matter makes up an additional 3 to 6 percent. Typical of the Upper Moorehouse in the mid-Hudson Valley, terrigenous mud is concentrated in microstyololites, giving weathered exposures a "shaly" appearance. This is the case at Locality 7, where this uppermost Onondaga bed is a Lithofacies I limestone typical of the area. It is significant that this does not resemble Lithofacies V of the Upper Onondaga in central New York.

Chadwick's crinoidal "calcarenyte" abruptly overlies the Moorehouse. This horizon is approximately 1 cm thick and bears little resemblance to the Onondaga below or the Bakoven above. Whereas the rock is packed with crinoid fragments, it is unlike Onondaga lithologies in that both spar cement and calcisiltite are absent. About 7 percent of the rock volume consists of fish remains; quartz silt and sand constitute an additional 5 percent. It is worth noting that quartz sand does not occur above or below this horizon. The original thickness and terrigenous percentage of this horizon remain uncertain due to an unusually intense intergranular pressure solution between crinoid particles. What does this "bone bed" represent?

The Upper Onondaga of the central Hudson Valley is a sequence of bryozoan bafflestones interbedded with normal sparse to packed biocalcisiltites, deposited marginal to a carbonate shelf. Open circulation in a fairly quiet environment near wave base are indicated by both fauna and lithology. The Bakoven, on the other hand, is a black, carbonaceous shale that emits a petroliferous odor from freshly broken specimens. The fauna is dominated by planktonic forms including *Styliolina fissurella* and "*Tentaculites*" cf. *gracilistriatus*. Signs of bioturbation are absent, to the extent that current-oriented styliolines were not disturbed. These characteristics are consistent with deposition in a stratified, dysaerobic, quiet-water environment. Pedersen et al. (1976) interpret the Bakoven as a distal basin deposit consisting of the first and stratigraphically lowermost muds of the Catskill Delta-complex. They also note that there is a problem with this interpretation. If there is validity to Walther's law of the correlation of facies (Middleton, 1973), as applied to vertical sedimentary sequences, the lithologically abrupt contact between a shallow-water to moderate-depth carbonate and a distal basin black shale facies must represent a disconformity of pronounced magnitude. It is certain that the contact is disconformable and clear that the "bone bed" was deposited on an already lithified Onondaga. However, many Bakoven styliolines contain pyritic steinkerns, indicating that they settled to the bottom with their cellular material intact. While we have no data on what might be a soft tissue compensation depth for styliolines, the extremes of distal basin habitat would almost certainly exceed it. We suggest that the Onondaga/Bakoven contact is a disconformity representing a relatively brief time span during which rapid crustal subsidence, to a shallow or proximal basin depth position beneath storm-wave base, resulted in stratification of the water column and a dysaerobic benthic condition conducive to the eradication of a benthic fauna and deposition of black shale. Savarese et. al. (1986) suggest a deepening event starting at 20-25 m and finishing at 100-150 m for roughly similar limestone/shale contacts in the Hamilton group of central and western New York. Further interpretation of the contact is presently premature.

PALEOGEOGRAPHIC SETTING

Buffalo to the Helderbergs

Understanding of Onondaga paleogeography, as studied in east-west outcrop, has not changed substantially since the work of Oliver (1954, 1956), Lindholm (1969), and Laporte (1971). A short-lived late Emsian regression of the sea to a position in eastern New York left the western and central areas of the state subaerially exposed. Early in the Eifelian, assuming that the entire Onondaga is Middle Devonian, a transgression submerged the region, initiating Edgecliff deposition in a shallow shelf environment. Shortly thereafter, subsidence in central New York, resulting from a northward extension of the Appalachian Basin, brought a deeper water, or offshore, environment to that area. The initial pulses of subsidence are recorded in the Nedrow Member, while continued subsidence is recorded in the Moorehouse and Seneca members of central New York. However, the eastern and western parts of the state remained in shallow shelf conditions throughout Onondaga deposition. Thus, post Edgecliff paleogeography, as seen in east-west outcrop, consists of a symmetric shelf-basin-shelf pattern. Subsurface studies (Kissling and Moshier, 1981; Cassa and Kissling, 1982) indicate that deposition took place on a carbonate ramp dipping predominantly southward into the Appalachian Basin and that east-west outcrop roughly parallels depositional strike for at least Edgecliff time (Figure 1).

The Helderbergs to Port Jervis

To understand Onondaga paleogeography in eastern and southeastern New York, it is helpful to establish a context in the mid-Silurian and proceed up the section. The Middle Silurian Shawangunk Conglomerate thins from approximately 1500 ft in northern New Jersey (Wolfe, 1977) to a pinchout just north of Rosendale, New York. A roughly similar pattern is evident throughout the Upper Silurian and Lower Devonian. As an arbitrary example of what this means in terms of depositional environments and an onshore-offshore orientation, Waines (1976) reports that the Silurian Binnewater Sandstone not only thickens to the south of Kingston, New York, but it also becomes increasingly dolomitic as the unit grades southward from supratidal to intertidal and shallow marine facies. During deposition of the Lower Devonian section, the onshore direction shifted to the northwest (Anderson, 1971). Lithostratigraphy (Rickard, 1975) and subsurface isopachs (Figure 2B) (Mesolella, 1978) indicate that this orientation persisted until the Middle Devonian.

It is generally acknowledged that during Onondaga deposition, the axis of the Appalachian Basin migrated into central New York. This shows up clearly in the distribution of Onondaga lithofacies and in subsurface isopachs (Figure 2A). However, to clarify eastern New York paleogeography it is

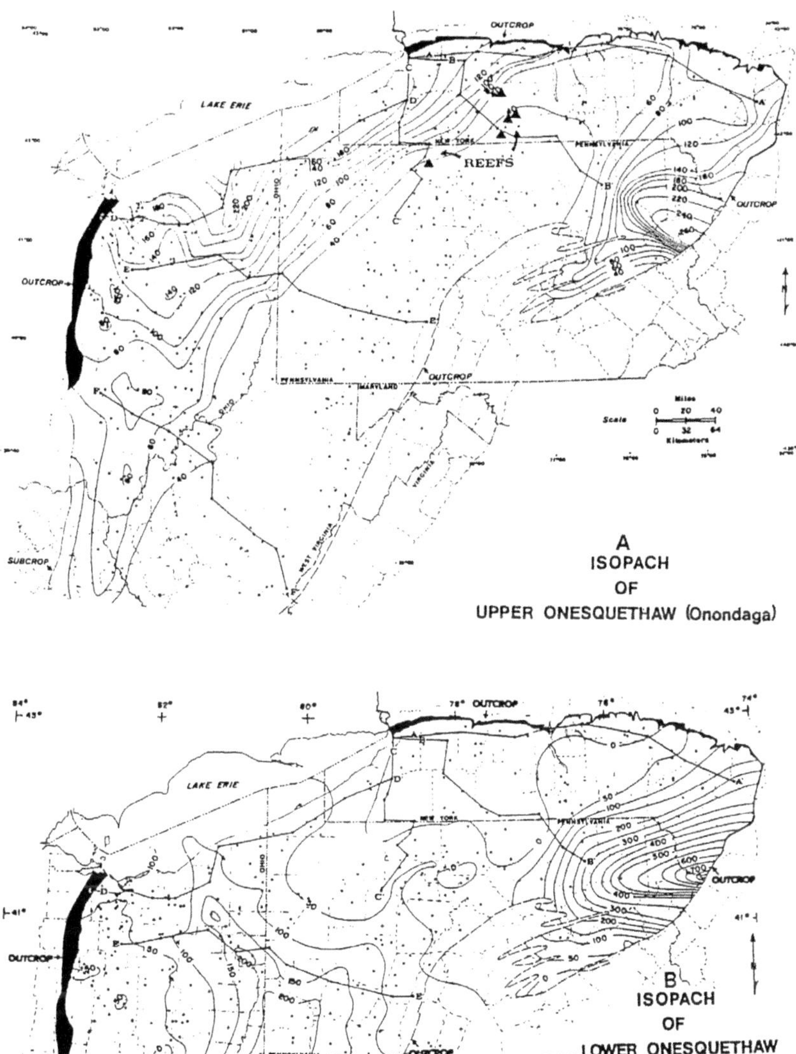

FIGURE 2. Isopach maps of Upper Onesquethaw (A) and Lower Onesquethaw (B) strata. C. I. = 50 ft. From Mesolella, 1978.

necessary to discriminate between "topographic" and "structural" basins. Mesolella (1978) defines a topographic basin as the area of deepest water and a structural basin as the area of greatest sediment accumulation, implying greatest crustal subsidence. As previously discussed, the preponderance of Lithofacies V in the Upper Onondaga in central New York is the direct result of deposition in proximity to the carbonate-starved axis of the Appalachian (topographic) Basin. However, Lithofacies V is absent from eastern New York at least as far south as the mid-Hudson Valley, where it is replaced by the coarser-grained and less argillaceous Lithofacies I and II. Even the most offshore Onondaga facies in eastern New York, the Buttermilk Falls Limestone, is interpreted as a "shallow marine basin" (Wolfe, 1977). However, the greatest sediment accumulation is centered in, or just south of, the Port Jervis (Tristates) area (Figure 2A). This was the focus of a subsiding structural basin not directly related to the topographic basin of central New York. This structural basin, which had existed since the Middle Silurian, exerted a pronounced influence on Onondaga deposition in eastern New York by establishing a paleogeographic pattern of a shallow carbonate shelf in the Helderberg-Coxsackie area, a thick accumulation of shelf-margin bryozoan bafflestones between Leeds and Saugerties, and an even thicker accumulation of sparse to packed biocalcisiltites deposited on a carbonate slope or ramp dipping into the Port Jervis area. Apparently water depths on this slope were never great, at least within the field trip area, prior to the subsidence event which set the stage for deposition of the Bakoven Shale.

MEGAFOSSILS OF THE ONONDAGA LIMESTONE IN SOUTHEASTERN NEW YORK

Collecting megafossils in the Onondaga Limestone in southeastern New York presents certain problems not inherent in the central part of the state. For example, there is no shaly Nedrow facies from which well-preserved specimens weather out, and there are few quarries which allow for extensive collecting on bedding surfaces. Most of the exposures in the mid-Hudson Valley are vertical and weather slowly. The limestone is quite dense with little shale; consequently megafossils, although observable in cross section, are difficult if not impossible to remove without damage to the specimen. However, one of the advantages of collecting in the south-eastern part of the state is the occurrence of silicified fossils in parts of the mid-Hudson Valley. During this trip we expect to sample some of these silicified outcrops and blocks may be taken for subsequent etching in hydrochloric (or muriatic) acid. Brachiopods (Table 3; Figures 3-4) and corals (Table 4) are the most dominant megafossils that we will collect from the Onondaga and therefore will be treated in more detail than other taxa (Table 5).

BRACHIOPODS

Acrospirifer duodenaria: Biconvex shells transversely subelliptical in outline; hinge line long and straight; medial open delthyrium with no preserved deltidial plates; pedicle valve bears narrow, triangular, moderately deep, noncostate sulcus; brachial valve bears corresponding fold; five to six rounded plications on each pedicle flank with U-shaped interspaces; anterior commissure uniplicate.

Ambocoelia sp.: Small, ventribiconvex shells; pedicle valve with weak sulcus; beak incurved; hinge line straight, delthyrium open; brachial valve slightly convex with no ornamentation; anterior commissure rectimarginate to uniplicate to slightly intraplicate.

Athyridacean indet.: Small, ovate shells with laterally directed spiralia; crura united with primary lamellae by pair of S-shaped loops; most closely resemble the Meristellidae.

Athyris sp. A: Shells transversely suboval in outline, subequally biconvex, with pedicle valve slightly deeper than brachial valve; ventral beak suberect, terminating in small round foramen; brachial beak smaller and less noticeable; pedicle valve bears shallow sulcus with corresponding low fold on brachial valve; anterior commissure weakly uniplicate; some forms nonsulcate and rectimarginate; fine, concentric growth lines on both valves.

Athyris sp. B: Differs from *Athyris* sp. A in its larger size, subparallel dental plates, and narrow muscle field.

Atribonium halli: Shells small, astrophic, impunctate, and subpentagonal in outline; beak short, curved, rounded, and suberect; commissure uniplicate with high brachial fold and deep pedicle sulcus; costae weak, rounded; small pedicle foramen and triangular delthyrium.

Atrypa "reticularis": Dorsibiconvex shells with well-rounded radial costellae, which increase in size and number anteriorly; costellae separated by U-shaped interspaces; concentric growth lamellae cross the costellae, becoming more distinct and frilly anteriorly; anterior commissure rectimarginate or slightly deflected toward brachial valve.

Atlanticocoelia acutiplicata: Subcircular in outline with length almost equal to width; brachial valve gently convex, pedicle valve slightly more so; weak pedicle sulcus sometimes noticeable on larger specimens; no corresponding dorsal fold; hinge line very short and becomes rounded anteriorly; no interareas present; anterior and lateral commissures crenulate; ten to twelve plications with U-shaped interspaces; concentric growth lines, two or three per shell, common on ephebic forms.

Coelospira camilla: Small, concavoconvex to planoconvex, subcircular to suboval in outline; small, distinct pedicle foramen on incurved pedicle beak; no interarea evident; maximum width about one-third valve length in adults; pedicle valve bears two medial plications usually at least as large as remaining radial plications on flanks; interspaces U-shaped; brachial valve bears medial plication, which generally bifurcates at one-third valve length; median interpace usually flat but sometimes bears small ridge; plications broader on flanks and thinner toward lateral commissure; several well defined, concentric growth lines evident near anterior commissure in adult forms.

Cupularostrum? sp. A: Shells small, equibiconvex, and subtrigonal to to transversely suboval in outline; pedicle beak erect to slightly incurved; delthyrium open, triangular wth small foramen located apically; pedicle valve with sulcus and brachial valve with corresponding fold considerably weaker than sulcus; about 15 simple plicae, U-shaped in cross section.

Cupularostrum? sp. B: Externally identical with *Cupularostrum* sp. A except for lack of sulcus and fold.

Cyrtina hamiltonensis: Shells small, hemipyramidal in outline with straight hinge line; ventral interarea high, smooth; convex psuedodeltidium covers triangular delthyrium in most specimens; pedicle valve bears triangular, smooth sulcus with two or three rounded plications along flanks; brachial valve bears fold with three to four lateral plications; ornamentation consists of concentric growth lamellae.

Cyrtina sp. A: May be differentiated from *Cyrtina hamiltonensis* by larger size and more robust appearance.

Dalejina aff. *alsa:* Shells ventribiconvex, transversely suboval to subcircular in outline; hinge line very short and straight in apical area but becomes rounded approaching lateral margins; maximum width at or just anterior to midlength; pedicle valve bears slight median depression; brachial valve often bears corresponding median ridge; anterior commissure most often rectimarginate to slightly sulcate; ventral interarea short, narrow; numerous radial costellae increase anteriorly both by intercalation and bifurcation; at anterior commissure there are 18 to 20 costellae per 5 mm, near midline; costellae medially

TABLE 3. Brachiopods of the Onondaga Limestone in southeastern New York (from AMNH Loc. 3132 [Thompson 's Lake] to AMNH Loc. 3151 [Wawarsing]; See Feldman, 1985, for index map of localities and locality descriptions.)

Taxon	Common	Rare	Very rare
Acrospirifer duodenaria	X		
Ambocoelia sp.			X
Athyridacean indet.			X
Athyris sp. A		X	
Athyris sp. B			X
Atribonium halli			X
Atrypa "reticularis"	X		
Atlanticocoelia acutiplicata			X
"Hallinetes" aff. *lineata*			X
Coelospira camilla	X		
Cupularostrum? sp. A		X	
Cupularostrum? sp. B			X
Cyrtina hamiltonensis		X	
Cyrtina sp. A			X
Dalejina aff. *alsa*		X	
Discomyorthis? sp.			X
Elytha fimbriata		X	
eospiriferid? indet.			X
Gypidula sp.			X
Leptaena aff. *"rhomboidalis"*	X		
Levenea aff. *subcarinata*		X	
Megakozlowskiella raricosta	X		
Megastrophia sp.		X	
Meristina cf. *nasuta*			X
"Mucrospirifer" cf. *macra*		X	
Nucleospira aff. *ventricosa*	X		
orthotetacid indet.			X
Pentagonia unisulcata		X	
Pentamerella arata		X	
Rhipidomella?			X
Rhynchospirina sp.			X
Schizophoria cf. *multistriata*		X	
"Schuchertella" sp.			X
Strophodonta cf. *demissa*		X	
Stropheodontid indet.			X

Note: Although this table denotes relative abundance of brachiopod taxa in the central part of the state in terms of common, rare, and very rare, it should be noted that some species are more abundant in specific horizons or beds and are relatively rare throughout the remainder of the formation. For example, *Levenea* occurs abundantly in Wawarsing, New York, but sporadically in the rest of the southeastern exposures of the Onondaga.

TABLE 4. Corals of the Onondaga Limestone in southeastern New York.

Taxa	Common	Rare
Tabulates		
Aulocystis (*Ceratopora*)	X	
Aulopora	X	
Favosites	X	
Striatopora	X	
Rugosans		
cf. *Amplexiphyllum*	X	
Acinophyllum	X	
Breviphrentis	X	
Cystimorph?		X
Heliophyllum	X	
"*Heterophrentis*"	X	
cf. *Syringaxon*	X	

TABLE 5. Other faunal constituents of the Onondaga Limestone in southeastern New York.

Taxa	Common	Rare	Very rare
Gastropods			
Platyceras domusum	X		
Platyceras (*Platystoma*)	X		
Platyceras sp.	X		
pseudophoracean indet.		X	
Ecculiomphalus			X
Loxonema			X
Trilobites			
Phacops cf. *cristata*	X		
dalmanitid fragments			X
indet. Fragments	X		
Crinoids			
non-pinnulate inadunate ossicles	X		
camerate columnals	X		
Bryozoans			
Dyoidophragma			X
Sponges			
Hindia	X		

grooved, flat, occasionally crossed by concentric growth lines near anterior margins.

Discomyorthis? sp.: Similar to *Dalejina* in general morphology but may be differentiated by more circular outline and larger ventral diductors; pedicle valve bears well developed pedicle callist and short, triangular hinge teeth; costellae medially grooved .

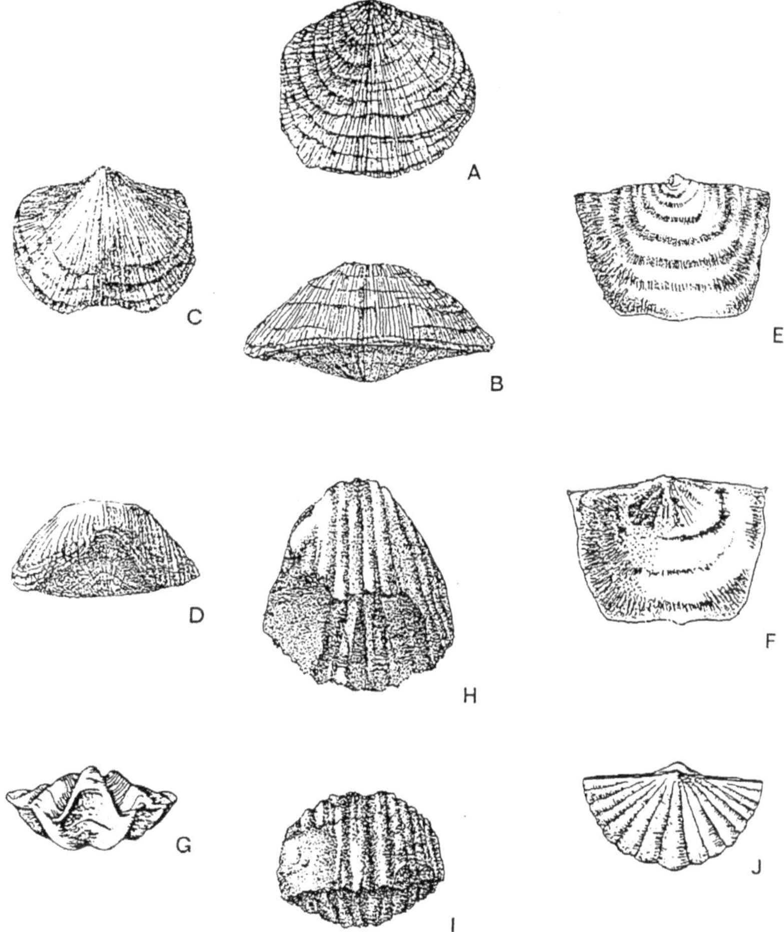

FIGURE 3. Brachiopods of the Onondaga Limestone in eastern New York. A, B. *Rhipidomella* sp., ventral and anterior views, x3. C, D. *Schizophoria* cf. *multistriata*, ventral and anterior views, x1.5. E, F. *Leptaena* aff. "*rhomboidalis*," ventral exterior and interior, x1.5. G. *Megakozlowskiella raricosta*, anterior view, x1.25. H, I. *Pentamerella arata*, ventral and anterior views, x1.75. J. *Acrospirifer duodenaria*, dorsal view, x3. Modified from Dunn and Rickard (1961).

Elita fimbriata: The shells are medium-sized, biconvex in lateral profile and transversely oval in outline; beak short and erect; pedicle valve bears shallow, triangular sulcus with corresponding low, rounded fold on brachial valve; faint plications cover lateral slopes; concentric growth lamallae cross plications and terminate in short, attenuated spines; anterior commissure uniplicate.

Eospiriferid indet.: Extremely rare in the formation and represented by only one pedicle valve, which is convex, moderately transverse, sulcate, plicate, and covered by fine radiating striae; delthyrium triangular with possible deltidial plates.

Gypidula sp.: Elongate oval to subcircular in outline; pedicle valve swollen; costate to multicostate; almost identical to *Pentamerella arata* (see description below) but can be differentiated by a pedicle fold and brachial sulcus, whereas *Pentamerella* has a pedicle sulcus and brachial fold.

"*Hallinetes*" aff. *lineatus*: Shells small, subsemicircular in outline and concavo-convex in lateral profile; interareas very narrow; no delthyrial structures preserved; greatest width at hinge line or anterior to midlength; valves covered with fine capillae which increase anteriorly by bifurcation.

Leptaena aff. "*rhomboidalis*": Transversely subqudrate in outline, concavo-convex to slightly biconvex with pedicle valve strongly geniculate at anterior and lateral commissures; brachial valve correspondingly geniculate within pedicle trail; hinge line straight, pedicle interarea flat; ornamentation consists of radial costellae, which extend past point of geniculation and continue on trail of valves; concentric rugae cross costellae becoming larger anteriorly.

Levenea aff. *subcarinata*: Shells small to medium sized, transversely suboval in outline, ventribiconvex in lateral profile; brachial valve bears shallow, rounded sulcus which broadens anteriorly; maximum width at or just anterior to midlength; ventral interarea short, slightly incurved; triangular dethyrium encloses angle of approximately 60 degrees; delthyrium often widens apically into small, circular foramen; ornamentation consists of rounded radial costellae, which increase in number anteriorly by bifurcation .

Megakozlowskiella raricosta: Shells subtransverse in outline, strophic, medium to large, ventribiconvex; hinge line straight; pedicle interarea moderately narrow with striae that parallel hinge line; brachial interarea extremely narrow; distinct slightly flattened fold on brachial valve and corresponding deep, U-shaped sulcus on pedicle valve; commonly three plications on flanks; delthyrium includes angle of approximately 60 degrees; no deltidial plates preserved; anterior commissure uniplicate; strong, concentric growth lamellae with anterior frills; radial ornamentation consists of very fine striae.

Megastrophia sp.: Medium-sized to large, subsemicircular to transversely suboval in outine; somewhat alate, concavoconvex in lateral profile; maximum width attained at hinge line; unequally parvicostellate to subuniformly costellate; psuedodeltidium flat, complete, with narrow median ridge; chilidium flat, complete, with median ridge; hinge entirely denticulate.

Meristina cf. *nasuta*: Convex, elongate and suboval in outline with no noticeable interarea; unequally biconvex with pedicle valve much deeper than brachial valve; maximum width commonly anterior to midlength; delthyrium broad, triangular, opens apically into semicircular foramen; faint pedicle sulcus modified by development of low, rounded medial plication that extends anterior commissure in tongue-like projection; concentric growth lamellae evident at anterior portion of valves but remainder of shell smooth.

"*Mucrospirifer*" cf. *macra*: Small to large alate shells transversely subtrigonal to subsemicircular in outline; biconvex in lateral profile with brachial valve slightly flatter than pedicle valve; ventral interarea moderately high, long, somewhat curved; ventral beak, posterior to interarea, short and stubby; open, triangular delthyrium present, which divides interarea medially; dorsal interarea long, thin, ribbon-like; brachial valve bears high, medial fold flattened at top; pedicle valve bears corresponding U-shaped sulcus; surface of shells covered by sharply defined plications ranging from U-shaped to subangular in cross section; numerous, concentric, frilly growth lines present; no fine radial ornamentation.

Nucleospira aff. *ventricosa*: Small, transversely suboval in outline, biconvex in lateral profile with pedicle valve slightly deeper than brachial valve; hinge line curved; brachial beak fits into anterior end of delthyrium, which is partially covered by concave pseudodeltidium in some specimens; both beaks erect, no interarea evident; shell surface lacks radial ornamentation; no fold or sulcus present; pedicle valve shows faint median depression in some specimens; concentric growth lamellae present, more concentrated toward rectimarginate anterior commissure.

Orthotetacid indet.: Small to medium sized shells, generally poorly preserved as internal impressions; hinge line straight; ornamentation finely costellate.

Pentagonia unisulcata: Medium-sized, nonstrophic, pentagonal in outline when viewed posteriorly; beak suberect, dorsibiconvex, with greatest width attained between midlength and anterior commissure; brachial valve cariniform due to presence of raised, rounded fold bearing narrow, median groove; in some forms groove widens slightly anteriorly, forming two parallel to subparallel ridges extending almost half

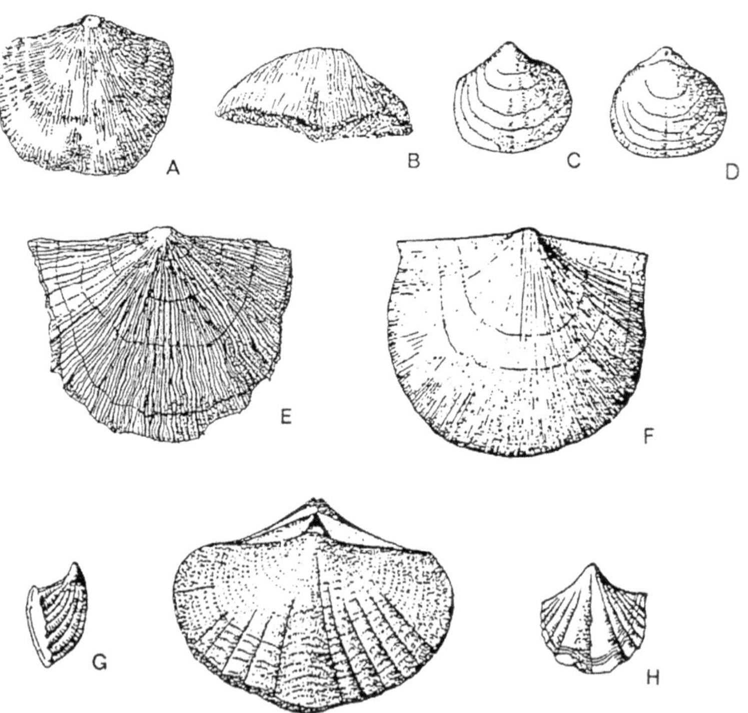

FIGURE 4. Brachiopods of the Onondaga Limestone in eastern New York. A, B. *Schuchertella* sp., ventral and anterior views, x1.75. C, D. *Athyris* sp. A, ventral and dorsal views, x2. E. *Megastrophia* sp., ventral view, x2. F. *Strophodonta demissa*, ventral view, x3. G, H. *Cyrtina hamiltonensis*, lateral and ventral views, x2. I. *Elita fimbriata*, dorsal view, x3. Modified from Dunn and Rickard (1961).

the valve length; flanks concave, dropping steeply away from sulcate fold; sulcus broad, shallow, with two distinct ridges that define sulcus laterally and extend from umbo across posterolateral margins of flanks to uniplicate anterolateral commissure; vague, concentric growth lines on anterior portion of shell.

Pentamerella arata: Subglobose and broadly pyriform in outline with strongly convex pedicle valve and weakly convex brachial valve; hinge line short, curved, narrow; no interarea evident; pedicle valve beak short, strong, incurved, not closely pressed against brachial beak; brachial beak small, less erect and less incurved; maximum width attained at or about midlength; weak sulcus present on anterior half of pedicle valve with corresponding fold on brachial valve (note: this morphological feature is the key to differentiating *Gypidula* from *Pentamerella*; see *Gypidula* above); both valves ornamented with numerous, rounded, bifurcating plications, which become narrower on lateral slopes than near midline; interspaces between plications U-shaped and wider than plications which tend to become slightly V-shaped in cross section on some specimens; about 5 plications in sulcus and 6 on fold; concentric growth lines more numerous anteriorly.

Rhipidomella?: Subcircular in outline, dorsibiconvex, delthyrium open; costellae cylindrical in cross section, not grooved; fold and sulcus weak, if present at all.

Rhynchospirina sp.: Shells small, pyriform in outline and biconvex in lateral profile; pedicle beak erect with small permesothyrid foramen; weak sulcus on pedicle valve but no corresponding fold on brachial valve; anterior commissure slightly uniplicate; normally about eight subangular plications with subangular interspaces.

Schizophoria cf. *multistriata*: Shells medium sized, suboval to subquadrate in outline, unequally biconvex; brachial valve deeper and more uniformly convex; in juveniles, both valves become almost equally biconvex; pedicle valve develops broad, shallow sulcus on adult forms; brachial valve bears indistinct fold; hinge line short, slightly rounded; maximum width attained at or just past midlength; ventral interarea triangular, fairly high in larger shells, relatively narrow in younger ones; dorsal interarea narrower; interareas of both valves equal to about one-half width; ornamentation consists of rounded to subangular radial costellae with broad, flat interspaces; about 11 costellae in a 5 mm space near anterior commissure at midline.

"*Schuchertella*" sp.: Medium-sized, planoconvex to biconvex, transversely subelliptical in outline but subpyramidal in umbonal region; exterior multicostellate with costellae added by intercalation; interarea flat and broadly triangular; delthyrium covered by convex pseudodeltidium.

Strophodonta cf. *demissa*: Subcircular to shield-shaped shells, concavo-convex in lateral profile; shells wider than long; point of maximum width at hinge line; lateral margins almost straight posteriorly; anterior margins evenly rounded; all margins crenulate; anterior commissure rectimarginate; costellae coarse, bifurcating with angular interspaces in cross section.

Stropheodontid indet.: Small, subcircular in outline, alate; Smooth exterior with irregularly spaced growth lines.

BRACHIOPOD LIFE STRATEGIES

The terminology used in this paper follows Bassett (1984), in which he reviewed the life strategies of Silurian brachiopods. Table 6 summarizes the main strategies under which the brachiopods of the Onondaga Limestone can be categorized, but it must be noted that the classification is flexible. It is possible that several taxa may fit into different categories as ontogeny progressed, since most brachiopods require an initial post-larval attachment to a hard bottom but may differ in post-larval development, especially in relation to the substrate. Ephebic or mature forms usually fall into a single category, while immature forms may pass through more than one category during development.

Brachiopod specimens used in this study were collected from a variety of depositional environments and modes of preservation vary from well to poorly to non-silicified forms. In addition, specimens were studied *in situ* in cases where removal from the field was impossible and successful extraction from the encasing matrix doubtful.

Quasi-infaunal Forms

Rudwick (1970) first used the term quasi-infaunal to describe strophomenid brachiopods that were partially buried or sank into sediment during ontogeny after initial hard-bottom attachment. These forms could become free-lying during burial or remain attached. A concavo-convex morphology is most typical of quasi-infaunal brachiopods found in the Onondaga Limestone. This occurred as a result of an alteration of growth rate later in ontogeny combined with an increased thickening of the convex valve (usually the ventral valve). The effect of this change in growth increased stabilization on the sea floor and prevented overturning by current action. In soft sediment, of course, the convex valve would have partially sunk in to a certain degree. During turbulence, sediment falling on the concave valve may have concealed the entire brachiopod except for the crescentic valve edges projecting above the surface of the sediment. If burial was too severe, a quick "snap" of the valves would have lifted it back and up, above the sediment-water interface. There are no known Recent examples of quasi-infaunal brachiopods. However, this mode of life would have been the closet to a truly infaunal habitat known for any of the articulates.

Some forms, such as *Cymostrophia* cf. *patersoni* and *Strophodonta demissa*, display a strong increase in curvature in adults, which raised the commissure as burial increased. The most extreme increase in curvature is shown by *Leptaena* cf. "*rhomboidalis*" and *Strophonella* sp., which are both geniculated. It is conceivable that these brachiopods lived almost buried within the sediment with the commissure extended for feeding. A "snapping" action may not have been periodically necessary.

Leptaena depressa from the Silurian of the Anglo-Baltic region displays both geniculation and folding, with the dorsally deflected anterior shell bearing a median fold. This fold is, in many instances, developed as a long trail able to extend well above the sediment-water interface. Indications of a trail are sometimes evident in *Leptaena* cf. "*rhomboidalis*" from the Onondaga. The brachiopod's efficiency in

TABLE 6. Life Strategies of Ephebic Brachiopods from the Onondaga Limestone.

Life Strategies	Nature of Substrate
Endofaunal habits	
Quasi-infaunal	Partial burial in soft bottom
Epifaunal habits	
Fixosessile	
Plenipedunculate	Usually hard bottom
Rhizopedunculate	Hard or soft bottom
Epiphytic	Plants or plant–like structures
Liberosessile	
Ambitopic	Hard or soft bottom
cosupportive	Mutual support in dense clusters

separating inhalant and exhalant water currents would have been increased by the presence of a trail.

Plenipedunculate Forms

The brachiopod pedicle was once thought to be a rather simple, relatively short, fleshy projection with a more or less constant diameter. However, recent workers have shed light on the tremendous variation in pedicle morphology (Bromley and Surlyk, 1973; Curry, 1981; Richardson, 1979, 1981). Variation in thickness and length is considerable, with expansion and contraction often occurring outside the shell so that foramen size is not necessarily a reliable guide to determine functional diameter or strength (Bassett, 1984). Nevertheless, the presence of an open pedicle foramen throughout life is indicative of a functional pedicle, presumably in all ontogenetic stages. Bassett (1984) used the term plenipedunculate for brachiopods in which the pedicle is a single, unbranched muscular structure apart from its distal tip (included in which are groups 1-4 of Bromley and Surlyk [1973: 350-351]).

Most Recent brachiopods attach to hard bottoms by mucal adhesion of the distal tip of the pedicle, which usually possesses hold-fast papillae or terminal rootlets that are able to etch and penetrate carbonate substrates for additional attachment strength (Bromley and Surlyk, 1973). Specimens of *Athyris* collected from the Onondaga Limestone near Saugerties, New York, have an open, rounded pedicle foramen throughout ontogeny. The structure of the pedicle opening is very similar to that of Recent terebratulids and rhynchonellids, and therefore suggests a similar pedicle function.

Rhizopedunculate Forms

Bassett (1984) uses the term rhizopedunculate for those brachiopods in which the pedicle is branched into fine filaments throughout much of its length rather than only at the distal tip (eqiuvalent to groups 6 and 7 of Bromley and Surlyk [1973: 351]). Bromley and Surlyk (1973) studied the pedicles of Recent brachiopods and found that they etch a very characteristic trace, composed of a number of pits, into hard calcareous substrates. The trace in rhizopedunculate forms consists of a series of widely scattered pits corresponding to the rootlets of the pedicles. The fact that the pedicle is so variable and that it is able to dissolve carbonates implies that many brachiopods are capable of attaching themselves to a wide variety of substrates. This, in effect, means that many brachiopod-substrate relationships in paleoecology must be re-evaluated.

A well-known Recent example of a rhizopedunculate brachiopod is *Chlidonophora*, a terebratulacean, in which the pedicle root lets have been found to penetrate *Globigerina* tests (Rudwick, 1970) and thereby become rooted into the foraminiferal ooze. The importance of this lies in the fact that the size of the pedicle foramen alone would not indicate the pedicle length (which is rather long in *Chlidonophora chuni,* for example) or the branching, rootlike character of the pedicle. It is clear that this type of attachment may have been more common among Devonian articulate brachiopods than earlier workers assumed. However, it is difficult to confirm that any Devonian forms were in fact rhizopedunculate, or even plenipedunculate, but it is possible to draw certain conclusions based on morphology and substrate.

The Moorehouse Member of the Onondaga Limestone in the mid-Hudson Valley is thought to represent a very soft-bottomed, lime mud substrate. It displays the greatest faunal diversity of all members of the Onondaga Limestone in the area. Many forms may have existed with rhizoid pedicles, which were able to attach to local hard bottoms, such as shell fragments. For example, *Atrypa "reticularis," Athyris*, and *Leptaena "rhombodalis"* showed traces of a small pedicle foramen early in ontogeny that may have accomodated pedicles that acted in the capacity of tethers and may have very well been rhizoid. This type of attachment would have permitted the brachiopods to utilize a wide range of substrates during Onondaga time.

All adult specimens of *Mucrospirifer* collected from the Onondaga Limestone had an open delthyrium with no evidence of a stegidium or modifying plates. It is therefore assumed that a functional pedicle was probably absent throughout all ontogenetic stages, although the dimensions of the pedicle are unknown. Cowen (1968) correlated the decrease in function of the pedicle with an increase in the development of alae, which acted to stabilize the shells on the substratum. *Mucrospirifer* may have possessed an inert pedicle, as described by Richardson (1981), similar to that of the Recent *Magadina cumingi* from southern Australia, in which the pedicle acted as a pivot around which the shell moved by contraction of the pedicle muscles. The inert pedicle may also have branched into fine filaments. Richardson's (1981) criteria for determining the relationships between pedicle and shell characters are useful for Recent forms, but are not yet proven valid for fossil specimens. She noted that a straight beak with a wide, high deltidium is characteristic of species with an inert motile pedicle. This description is compatible with *Mucrospirifer* in that the deltidium is relatively high but the beak ranges from straight to suberect.

Epiphytic Forms

There have been reports in the literature (Rudwick, 1961, 1970; Foster, 1974) of brachiopods attaching themselves to various structures for support. Rudwick (1961) found shells of *Terebratella sanguina* dredged from a muddy bottom off the coast of New Zealand, attached by their pedicles to the tangled, horny tubes of *Phyllochaetopterus socialis*, a chaetopterid worm. Hosts such as these would not normally be preserved as fossils, but it appears probable that they were used in the past as they are used now. Some workers noted that certain thin-shelled, lightweight brachiopods, such as *Aegira grayi*, could have floated or attached to drifting algae. Bergstrom (1968) believed that *Shagamella ludlovensis* was epiphytic on benthic algae. Silurian forms from England, such as *Dicoelosia biloba*, have been described (Wright, 1968) that were attached to a stick-like organic fragment (Bryozoan?).

Thin-shelled forms of *Atribomium halli*, *Coelospira camilla*, and *Ambocoelia* found in the Moorehouse Member may have been epiphytic. However, it seems clear that even if they were epiphytic, the majority of forms found in the Middle Devonian Onondaga Limestone were predominantly benthic.

Ambitopic Forms

Ambitopic brachiopods were attached at early growth stages and subsequently became detached and capable of resting on soft bottoms (Jaanussen, 1979). As adults all ambitopic forms were liberosessile, but some liberosessile forms are not adapted to living on soft bottoms. There are some Recent brachiopods that became detached and remained on hard bottoms, although these forms appear to have a reduced life expectancy, as noted by Doherty (1979).

Some brachiopods, such as *Costistrophonella punctulifera*, *Schuchertella*, and indeterminate strophodontids collected from the Nedrow and Moorehouse members of the Onondaga near Kingston, New York, display weak curvatures and are relatively thin-shelled, indicating that they most likely rested on the sediment surface. Thus, there was intergradation between those forms and the more strongly curved types, such as *Leptaena*, *Cymostrophia*, *Megastrophia*, and *Strophonella*, that sank into the sediment.

Atrypa "reticularis" from the Moorehouse Member of the Onondaga developed a frilly border thought to function as a snowshoe in preventing adults from sinking into the sediment. Specimens from the underlying Nedrow and Edgecliff members had no frills, possibly indicating a less muddy, higher energy environment.

An unnamed species of *Cyrtina* from the Moorehouse Member near Leeds, New York, appears to have an atrophied pedicle in addition to a broad ventral interarea upon which the animal rested. This, in effect, spread the weight in order to prevent sinking into the substrate. Alate species of *Mucrospirifer* from the same member show a more extreme variety of this morphotype.

The concavoconvex *Hallinetes lineatus*, found by the thousands in the "*Hallinetes*" Zone 10 ft above the Tioga Bentonite in central New York, but also recovered from Upper Moorehouse strata in the mid-Hudson Valley, apparently closed its pedicle opening early in ontogeny and rested convex side down with the spines acting as restraints in preventing sinking. Poor preservation precludes exact determination of spine morphology.

Pentagonia unisulcata from the Moorehouse Member of the Onondaga is a thick- shelled, biconvex brachiopod, which probably depended on weight to maintain stability on the seafloor. *Pentagonia* was posteriorly weighted and possessed a minute pedicle foramen in some ephebic specimens and none at all in others, indicating pedicle atrophy. Secondary shell material was developed posteriorly in mature individuals, further increasing stability. Curry (1981) noted that *Neothyris lenticularis*, a Recent brachiopod from New Zealand, had a posteriorly weighted shell, minute foramen, and atrophied pedicle and could be considered an ideal adaptation for the high energy subtidal habitats of the species, which is frequently disturbed by bottom currents. The morphology and adaptive features of *Pentagonia* may be indicative of a similar habitat in Moorehouse time

Cosupportive Forms

Bassett (1984) introduced the term cosupportive to describe those ambitopic brachiopods that maintain an umbo-down posture and are packed tightly together, often growing on one another. This type of growth afforded the brachiopods some degree of mutual support from the time of pedicle atrophy. Examples in the Onondaga Limestone are rare, with clusters of *Atrypa "reticularis"* possibly showing this type of life strategy. Occasional specimens retrieved from Moorehouse strata near Leeds, New York, display deformation indicative of crowding. *Pentamerella arata* from the Nedrow and Moorehouse members observed *in situ* near Saugerties, also show evidence of deformation, indicating a possible cosupportive life strategy.

FIELD TRIP OUTCROP LOCALITIES

Below is a list of out crop localities, mostly in the mid-Hudson Valley, which illustrate how Onondaga deposition in the southeastern part of New York varied from a shallow carbonate shelf in the Helderberg-Coxsackie area to shelf-margin bryozoan bafflestones between Leeds and Saugerties, and relatively thick, sparse to packed calcisiltites deposited on a shelf-to-basin ramp that deepens into the Tristates area. The trip begins in Wawarsing, New York, where we believe that the accumulations are indicative of the deeper part of a second basin, the first, well known to geologists, located across the state (east to west) with the basinal axis near Syracuse. As we progress northeast up the mid-Hudson Valley, the outcrops show a progressively shallower water facies, correlative with movement up a ramp toward the strand line.

Locality 1: The trip begins in Wawarsing, New York, approximately 0.5 mi northeast of Vernooy Kill, 100 ft north of Route 209, on the property of Steve and Sue Caruso (be sure to ask permission before entering outcrop area). The Onondaga-Schoharie contact can be observed in an abandoned quarry (AMNH Locality 3151B; see Feldman, 1985) where the Edgecliff Member with characteristic large crinoids, trilobites, and some *Amplexiphyllum* is accessible. The Edgecliff here is finer grained than in the mid-Hudson Valley and represents the deepest part of the basin to shelf lithology we will see today. Proceed north on Route 209 for 14 mi (note turnoff to Ulster County Highway 26) and continue for another 3 mi.

Locality 2: Pull off on the west side of Route 209 (wide shoulder). On the east side of the road note a transitional, deeper water facies, similar to that found at Locality 1, about 1 m thick, with 0.5 m of shallower water, "cleaner" Edgecliff Limestone. The fauna here consists of large crinoids, *Amplexiphyllum* (?), fenestrate (?) bryozoans (weathered), and *Syringopora*. Note storm layers and a sharp break between two facies. Continue on Route 209 north for 2.6 mi, where the Onondaga outcrops on the west side of the road.

Locality 3: Here we are higher in the Edgecliff (about 4 m thick). The water was shallower and the brachiopods larger. *Chondrites* is evident here but not further southeast, due to the fact that as the ramp deepened (to the southwest) the sediment became too "soupy" for tubelike or tunnel structures. Continue to Route 199 east and exit at Route 32 south (7.2 mi). (Note that along Route 199 we are passing through the entire Lower Devonian section, from the Thacher at the base, to the Schoharie, which underlies the Onondaga in this part of New York State.) Make a left turn at the stop sign and continue south on Route 32. On the west side of the road note complex thrust slices oblique to the section and, about one-eighth of a mile further south, an angular unconformity between the Wilbur Limestone Member of the Rosendale Formation (Late Silurian) on Normanskill (Austen Glen aspect) strata. After 2.7 mi, make a left turn onto Route 9W south (Frank Koenig Boulevard), continue for 1 mi, and stop at the Delaware Avenue sign.

Locality 4: Note the gradational nature of the Onondaga-Schoharie contact, which is placed at the uppermost buff-weathering band. Ranging through the Upper Schoharie into the Lower Edgecliff are massive cyclostome bryozoans, often encrusting (e.g. crinoids). Other faunal constituents here include brachiopods, particularly *Atrypa "reticularis,"* and *Fistulipora*. The stratigraphy at this locality is complicated by faulting and repetition due to thrusting (note slickensides and slip-fiber sheets), thus making the Edgecliff seem to be thicker than it actually is. According to Marshak (1986) there are two major thrusts that display a relatively large stratigraphic throw along Route 9W. The upper one, now covered by the exit ramp, emplaces Esopus Formation on Onondaga Limestone, while the lower one, exposed further to the north along the roadcut, emplaces Onondaga Limestone on Schoharie. The upper fault may be the continuation of the Fly Mountain Thrust. In general, there is an absence of favositids between here and Wawarsing, but north of here, especially in Leeds, they are abundant. Note the light-weathering chert. Exit at Delaware Ave., make a left turn at the light, and re-enter Route 9W north. Proceed to Route 32, turn right (north), and continue for 8.3 mi. Pull into McDonald's parking lot for a brief lunch stop.

LUNCH STOP: MCDONALD'S, ROUTE 32.

Upon exiting the lot, turn north and continue on Route 32 passing through the town of Saugerties. Cross the New York State Thruway and note

Howard Johnson's on the right, 3.4 mi from McDonald's. (If time permits we will pull into the parking lot and observe well weathered blocks of Onondaga with silicified fossils [bryozoan bafflestone].) Continue north on Route 32 for another 1.6 mi until the turnoff for Old King's Highway (Old King's Road). Pull off on the right just before the turnoff.

Locality 5: On the east side of Route 32, note a large outcrop of Edgecliff which shows, for the first time on this southwest-northeast transect, typical coarse-grained Edgecliff lithology. This is the southernmost exposure of the crinoidal biosparites which are so characteristic of the Edgecliff Member. Turn onto Old King's Highway (Greene County Route 47) and proceed 5.2 mi to High Falls Road. Make a left turn and pull over to the right just before the sharp bend (0.3 mi).

Locality 6: DO NOT COLLECT AT THIS STOP; IT IS ON PRIVATE PROPERTY AND THE OWNERS DO NOT WANT ANY SPECIMENS OR BLOCKS REMOVED! This outcrop, on the Kaaterskill (AMNH Locality 3137; see Feldman, 1985), known as Quatawichna-ach, takes its name from the Indian "place where all the water goes in a hole," referring to the chert seams and massive joints that take the water underground as it passes through the limestone (Chadwick, 1944). A stratigraphic placement of Upper Moorehouse is indicated by (1) shale chips in the adjacent woods (probably Bakoven, since it outcrops a short way downstream), (2) dark-weathering chert, and (3) faunal similarity with Upper Moorehose strata from the Leeds area, specifically, *Platyceras dumosum* and *Atrypa* "reticularis." The fauna here is moderately to well silicified, probably due to the diagenetic action of percolating groundwater through the numerous joints. At this locality, Feldman (1980) recognized a highly diverse *Atrypa-Coelospira-Nucleospira* Community containing the following morphotypes:

(1) Orthotetacids, *Schuchertella* (broad, flat)
(2) *Nucleospira, Athyris* (smooth spiriferids)
(3) *Schizophoria* (unequally biconvex)
(4) *Atrypa* (with frills)

There is a great similarity here with Lenz's (1976) Lower Lochkovian *Howellella-Protathyris* Community from the northern Canadian Cordillera, in which he found similar faunal elements in an offshore position. A similarity is also evident to Copper's (1966) biotope of primitive, abundant

Atrypidae, in which variably-sized atrypids occur with spiriferids and schizophorids in fine-grained sandstones, siltstones, and shales with thin limestone interfingerings. Proceed back up High Falls Road and turn left onto Old King's Highway. Continue for 2.1 mi to Route 23A. Turn left and continue for 0.3 mi, and just before crossing Kaaterskill Creek pull onto right-hand shoulder.

Locality 7: Descend steep embankment to Kaaterskill Creek, where the only known exposure of the Bakoven/Onondaga contact exists. Whereas Oliver (1956) considered the Onondaga/Bakoven contact to represent a minor break in deposition, we believe that the contact is a disconformity, albeit a minor one in this part of the state, representing a time period during which rapid crustal subsidence, to a shallow or proximal basin depth position below the storm wave base, resulted in stratification of the water column and dysaerobic bottom conditions. We found evidence of a more substantial disconformity in central New York, near Syracuse, where the Union Springs Shale (lateral equivalent of the Bakoven) truncates the westwardly dipping Seneca strata. At that contact there exists a substantial bone bed with reworked (?) crinoids and brachiopods, and current-sorted fish spines and/or teeth. Proceed up the hill (east) on Route 23 and pick up Route 9W north through Catskill, New York. Pass under the trestle and go 8.1 mi (from Locality 7), turn left (west) onto Route 23. Follow signs to New York State Thruway (exit and make right turn), and head into Leeds (Green County Route 23B). In Leeds, turn left on Gilfeather Park Road, drive to the end and park.

Locality 8: Overlooking Catskill Creek to the east, note the Schoharie/Onondaga contact, to the east of which the Onondaga lies in an (overturned?) thrust-faulted syncline. If time permits, we will examine the typical Edgecliff lithology at the waterfalls where trilobites, corals, and brachiopods are evident.

ACKNOWLEDGMENTS

We thank Russell Waines and Jack Epstein for discussion of mid-Hudson Valley stratigraphy. Sherrielyn Koye helped in the preparation of certain figures, Susan Feldman deserves thanks for typing portions of the manuscript, and Susan M. Klofak aided in proofreading.

REFERENCES

Anderson, E. J. 1971. Interpretation of calcarenite paleoenvironments. *Eastern Section Society of Economic Paleontologists Mineralogy Guidebook.*

Bassett, M. G. 1984. Life strategies of Silurian brachiopods. In M. G. Bassett and J. D. Lawson (eds.), *Autecology of Silurian organisms, Special Papers in Palaeontology* 32: 237-26.

Bergstrom, J. 1968. Some Ordovician and Silurian brachiopod assemblages. *Lethaia* 1: 230-237.

Bromley, R. G., and Surlyk, F. 1973. Borings produced by brachiopod pedicles, fossil and Recent. *Lethaia* 6: 349-365.

Cassa, M. R., and Kissling, D. L. 1982. Carbonate facies of the Onondaga and Bois Blanc Formations, Niagara Peninsula, Ontario. In E. J. Buehler and P. E. Calkin (eds.), *New York State Geological Association Guidebook, 54th Annual Meeting, Buffalo.*

Chadwick, G. H. 1944. Geology of the Catskill and Kaaterskill Quadrangles: Part II, Silurian and Devonian geology. *New York State Museum Bulletin* 336.

Copper, P. 1966. Ecological distribution of Devonian atrypid brachiopods. *Palaeogeography, Palaeoclimatology, Palaeoecology* 2: 245-266.

Cowen, R. 1968. A new type of delthyrial cover in the Devonian brachiopod *Mucrospirifer. Palaeontology* 11: 317-327.

Curry, G. B. 1981. Variable pedicle morphology in a population of the Recent brachiopod *Terebratulina septentrionalis. Lethaia* 14: 9-20.

Doherty, P. J. 1979. A demographic study of a subtidal population of the New Zealand articulate brachiopod *Terebratella inconspicua*. Marine Biology 52: 331-342.

Dunn, J. R., and Rickard, L. V. 1961. Silurian and Devonian rocks of the central Hudson Valley. In R. G. LaFleur (ed.), *New York State Geological Association Guidebook, 33rd Annual Meeting*, Troy, C1-C32.

Feldman, H. R. 1980. Level-bottom brachiopod communities in the Middle Devonian of New York. *Lethaia* 13: 27-46.

Feldman, H. R. 1986. Brachiopods of the Onondaga Limestone in central and southeastern New York. *American Museum of Natural History Bulletin* 179: 289-377.

Feldman, H. R., and Lindemann, R. H. 1986. Facies and fossils of the Onondaga Limestone in central New York. In *New York State Geological Association Guidebook, 58th Annual Meeting, Ithaca*, 145-166.

Foster, M. W. 1974. Recent Antarctic and Subantarctic brachiopods. *Antarctic Research Series, Washington* 21: 1-189.

Jaanusson, V. 1979. Ecology and faunal dynamics. In V. Jaanusson, S. Laufeld, and R. Skoglund (eds.), Lower Wenlock Faunal and Floral Dynamics: Vattenfallet section, Gotland. *Sveriges Geologiska Undersökning* 762: 253-294.

Kissling, D. L., and Moshier, S. U. 1981. The subsurface Onondaga Limestone: Stratigraphy, facies and paleogeography. In P. Enos (ed.), *New York State Geological Association Guidebook, 53rd Annual Meeting*, SUNY at Binghamton, 279-280.

Laporte, L. F. 1971. Paleozoic carbonate facies of the central Appalachian shelf. Journal of Sedimentary Petrology 41: 724-740.

Lenz, A. C. 1976. Lower Devonian brachiopod communities of the northern Canadian Cordillera. *Lethaia* 9: 19-28.

Lindemann, R. H. 1979. Stratigraphy and depositional history of the Onondaga Limestone in eastern New York. In G. F. Friedman (ed.), *New York State Geological Association Guidebook, 51st Annual Meeting*, Troy, 351-387.

Lindemann, R. H. 1980. Paleosynecology and paleoenvironments of the Onondaga Limestone in New York State. Unpublished PhD thesis, Rensselaer Polytechnic Institute, Troy, New York.

Lindholm, R. C. 1969. Carbonate petrology of the Onondaga Limestone (Middle Devonian), New York: A case for calcisiltite. *Journal of Sedimentary Petrology* 39: 268-275.

Marshak, S. 1986. Structure of the Hudson Valley fold-thrust belt between Catskill and Kingston, New York: A field guide. *Geological Society of America Northeastern Section Meeting*.

Mesolella, K. J. 1978. Paleogeography of some Silurian and Devonian reef trends, Central Appalachian Basin. *American Association of Petroleum Geologists Bulletin* 62: 1607-1644.

Middleton, G. V. 1973. Johannes Walther's law of the correlation or facies. *Geological Society of America Bulletin* 84: 979-988.

Oliver, W. A., Jr. 1954. Stratigraphy of the Onondaga Limestone (Devonian) in central New York. *Geological Society of America Bulletin* 65: 621-652.

Oliver, W. A., Jr. 1956. Stratigraphy of the Onondaga Limestone in eastern New York. *Geological Society of America Bulletin* 67: 1441-1474.

Oliver, W. A., Jr. 1962. The Onondaga Limestone in southeastern New York. In W. G. Valentine (ed.), *New York State Geological Association, 34th Annual Meeting*, Port Jervis, A1-A23.

Pederson, K., Sichko, M. J., Jr., and Wolff, M. 1976. Stratigraphy and structure of Silurian and Devonian rocks in the vicinity of Kingston, New York. *New York State Geological Association Guidebook*, B-4-1 to B-4-27.

Richardson, J. R. 1979. Pedicle structure of articulate brachiopods. *Journal of the Royal Society of New Zealand* 9: 415-436.

Richardson, J. R. 1981. Brachiopods and pedicles. *Paleobiology* 7: 87-95.

Rickard, L. V. 1975. Correlation of Silurian and Devonian rocks in New York State. *New York State Museum Science Service Map and Chart Series* 24: 1-16.

Rudwick, M. J .S. 1961. The anchorage of articulate brachiopods on soft substrata. *Palaeontology* 4: 475-476.

Rudwick, M. J. S. 1970. *Living and Fossil Brachiopods*. London: Hutchinson and Co.

Savarese, M., Gray, L. M., and Brett, C. E. 1986. Faunal and lithologic cyclicity in the Centerfield Member (Middle Devonian, Hamilton Group) of western New York: A reinterpretation of depositional history. In C. E. Brett (ed.), Dynamic Stratigraphy and Depositional Environments of the Hamilton Group (Middle Devonian) in New York State, Part 1. *New York State Museum Bulletin* 457: 5-56.

Waines, R. H. 1976. Stratigraphy and paleontology of the Binnewater Sandstone from Accord to Wilbur, New York. New York State Geological Association Guidebook, B-3-1 to B-3-15.

Wolfe, P. E. 1977. *The Geology and Landscapes of New Jersey*. Crane: Russak Co.

Chapter Thirteen

Understanding the East Central Onondaga Formation (Middle Devonian): An Examination of the Facies and Brachiopod Communities of the Cherry Valley Section, and Mt. Tom, a Small Pinnacle Reef

INTRODUCTION

The Onondaga Formation of New York and Ontario, Canada has been extensively studied (see Oliver, 1976 for references), and yet is still poorly understood. This unit is, in the western part of New York and in Ontario, clearly transgressive, and yet it lacks any of the classic peritidal facies associated with shallow water carbonates. It is a "reefy" unit, but the major reef building paleocommunity of the Middle Devonian (stromatoporoids and algae) are either extremely rare or absent. Finally, it has been described as an example of carbonate deposition along a gently subsiding ramp, which would seem to imply symmetry on either side of the basinal axis, but the pinnacle reefs—which are so highly sought after by explorationists—have been found on the western side of the basinal axis, but to date not on the eastern side. Kissling (1987) suggested that these unusual characteristics were due to deposition in deep water, possibly below the photic zone. As an alternative, Wolosz (1990, 1991) has argued that the Onondaga represents an example of a Devonian temperate water limestone.

On this trip we will attempt to come to our own conclusions by examining a nearly complete section of the Onondaga (Cherry Valley) and a small pinnacle reef (Mt. Tom). In the following text, Lindemann analyzes the stratigraphy and depositional environments, Feldman looks at the significance of the brachiopod communities, and Wolosz and Paquette address the depositional history the Mt. Tom reef.

STRATIGRAPHY AND DEPOSITIONAL HISTORY

The Onondaga Limestone is a 21 to over 50 m thick unit of lower Middle Devonian marine limestones, deposited during the final major phase of carbonate production prior to the influx of siliciclastic sediments shed from the Acadian mountain buildup. This component of the field trip is intended to provide an overview of Onondaga stratigraphy and depositional environments in Otsego County, New York. It centers on a nearly complete composite section of the formation exposed in road cuts on U. S. Route 20 at Cherry Valley (Sprout Brook, New York, 7.5 ft quadrangle). These exposures mark the easternmost extent of the "typical" Onondaga in central New York, as defined by Oliver (1954).

The concept of what is now the Onondaga Formation began to be developed prior to the First Geological Survey of New York (see Eaton, 1832). During the survey, Vanuxem (1842) recognized four "formations," the uppermost of which he named the "Seneca Limestone," a label that persists today. Hall (1843) recognized a three-fold division and applied the term "Onondaga Limestone" to those strata now known as the Edgecliff Member. Over time the "Onondaga Limestone" came to include the entire interval of limestone strata, which overlies formations of the Lower Devonian and is itself overlain by the black shales of the Marcellus Formation. Oliver (1954, 1956a) formally subdivided the Onondaga into four members. In the best tradition of the recently canonized Nicolas Steno, the members arranged from oldest to youngest are the Edgecliff, Nedrow, Moorehouse, and Seneca limestones. Descriptions of the members provided herein are specific to the exposures at Cherry Valley, NY. Lithologic terminology corresponds to that of Lindholm (1964).

BIOSTRATIGRAPHY AND CORRELATION

The biostratigraphic basis for correlation of the Onondaga Formation, particularly the Edgecliff Member, to the standard biozones and stages of Europe reads like a "who-done-it?" with the last few pages missing. Brachiopod and coral faunas have long served to place the formation at the base of the Middle Devonian Series (Rickard, 1975). Dutro (1981) reports that the Edgecliff Member coincides with the base of the *Frimbrispirifer divaricatus* Subzone of the *Amphigenia* Assemblage Zone, which marks the base of the Southwoodian (Upper Onesquethawan) Stage. Similarly, Oliver and Sorauf (1981) state that the Edgecliff base coincides with the base of the *Acinophyllum segreatum* Assemblage Zone and the bottom of the Southwoodian Stage. High in the formation, the Seneca Member is within the *Paraspirifer acuminatus* Assemblage Zone as well as an unnamed coral assemblage zone (Zone 8 of Oliver and Sorauf, 1981), both of which place the Seneca in the Cazenovian Stage. These fossils firmly establish the Onondaga Edgecliff Member as the lowermost Middle Devonian unit in New York State relative to the North American stages. However, Onondaga corals and brachiopods are geographically restricted to North America, a condition that precludes direct correlation with the Eifelian Stage of Europe. Cephalopods do little to facilitate this correlation. *Foordites* cf. *buttsi* from the Nedrow Member (Oliver, 1956b; House, 1962) suggests an Eifelian age. However, House (1981) notes that *Foordites* is a long-ranging genus, and that European zonal taxa are not known from the Onondaga. Furthermore, *Foordites* has not been reported from the Edgecliff and cannot lend its support to interpretation of an Eifelian age for the lowermost Onondaga member.

The International Union of Geological Sciences recently ratified the decision of the Subcommission on Devonian Stratigraphy to drive the golden spike marking the base of the Middle Devonian Series and the Eifelian Stage at the first occurrence of the conodont *Polygnathus costatus partitus* (Ziegler and Klapper, 1985). The subspecies *partitus* is the second in a lineage of three. The bottom of the *P. c. patulus* Zone is high in the Emsian stage and the bottom of the *P. c. costatus* Zone is well within the Eifelian. Klapper (1981) reports that the Upper Nedrow beds at Cherry Valley yield both *P. c. costatus* and *P. c. patulus*, placing the member's top well within the *partitus* Zone. Noting this along with the fact the *P. c. partitus* is unknown from the Onondaga, Ziegler and Klapper (1985) suggest, with question marks, that the Edgecliff Member is within the *patulus* Zone and correlative to the Emsian stage of the Lower Devonian series. At the very least, this obfuscates the Lower-Middle Devonian boundary in New York state and speaks for a "handle with care" approach in transatlantic correlation of the North American stages which abut that boundary.

Recent studies of the Onondaga's styliolinid and tentaculitid faunas have done little to improve upon this situation. A previously unknown nowakid fauna has been discovered in the Nedrow and Moorehouse Members at Cherry Valley, but the taxonomic status of the species is currently undetermined. Lindemann and Yochelson (1984) reported that the first occurrence of *Styliolina fissurella* (Hall) in the Devonian of New York is coincident with the base of the Edgecliff Member. Indeed, at Cherry Valley this enigmatic microfossil is present in the lowermost bed of the Edgecliff and absent from the subjacent Carlisle Center. *S. fissurella* (Hall) was a zooplankter reputed to have had a nearly worldwide distribution

(Boucek, 1964). The potential for correlation is obvious. However, Lindemann and Yochelson (1994) have found that many, possibly all, reports of the species from the Devonian of Europe are incorrect. Thus, without specific confirmation, reports of *S. fissurella* (Hall) from the Lower Devonian must be regarded with suspicion. To date, the chronostratigraphic placement of the lowermost Onondaga member remains uncertain.

DESCRIPTIONS OF THE MEMBERS

Edgecliff Member: The Edgecliff is 7 m thick and is divisible into two components, which correspond to the C1 and C2 zones of Oliver (1956a). The lowermost beds contain quartz sand and silt, glauconite sand, and phosphatic nodules. The limestones associated with these particles, which overlie the beds containing them, are thin to medium-bedded, dark gray, argillaceous, packed biocalcisiltite. The Middle and Upper Edgecliff consists of thick to very thick-bedded, medium gray, poorly washed to unsorted biosparites. While corals dominate the macrofauna, pelmatozoan ossicles and fenestrate bryozoans volumetrically dominate the sediment. The uppermost Edgecliff bed is a poorly washed biosparite which contains an abundance of pyrite. This bed is abruptly overlain by the basal Nedrow.

Nedrow Member: The Nedrow is a 4 m thick package of what might be described as coarsening upward cycles. More accurately, they are cycles of progressive carbonate enrichment without pronounced textural cyclicity. A cycle begins abruptly with a thickly laminated, argillaceous and pyritic, fossiliferous biocalcisiltite and grades vertically into medium-bedded, dark gray, sparse biocalcisiltite. The sediments are extensively bioturbated. While pelmatozoans and trilobites are the most abundant biogenic particles, ramose bryozoans and styliolines reach maximal abundances. Crushed styliolines in the more argillaceous beds indicate an overall thickness loss of approximately 75 percent due to soft sediment compaction. Thus, the original Nedrow sediment may have been 15-20 m thick. Laminae within the Nedrow beds are the result of compaction of the sediment. They are not primary sedimentary structures.

Moorehouse Member: The Nedrow/Moorehouse contact coincides with the first occurrence of black chert (Oliver, 1956a). Nodules, anastomosing masses, and thin beds of dark gray to black chert are characteristic of the Lower and Middle Moorehouse. Limestones associated with the chert are a sequence of medium-bedded, dark gray, fossiliferous to sparse biocalcisiltites. Terrigenous mud occurs as thin laminae and pyrite is rare. Bioturbation is abundant to pervasive, though individual burrows are indistinctly defined. Pelmatozoans, trilobites, and brachiopods variously dominate the sediment. Fenestrate bryozoans increase in abundance to become a major component high in the Middle Moorehouse. Corals such as *Aulopora* and *Thamnopora* also increase in abundance, as do goniatite cephalopods.

The uppermost Moorehouse is distinct from the lower and middle sections. Chert is rare to absent. Terrigenous mud is minimal. The limestone itself consists of thickly bedded, medium gray, packed biocalcisiltites and poorly washed biosparites. Cross stratification is present, though not common. While the macrofauna is dominated by the encrusting cyclostome *Fistulipora* and other bryozoans of ramose form, the sediment matrix is dominated by pelmatozoans and fenestrate bryozoans.

Seneca Member: The lowermost bed of the Seneca Member is the Tioga Bentonite (Oliver, 1954). At Cherry Valley the Tioga is 10 cm thick. It is extremely weathered, producing a deep re-entrant between the more resistant Moorehouse and Seneca limestones. The Seneca proper consists of approximately 2 m of thick-bedded, medium gray, packed biocalcisiltites and poorly washed biosparites. High angle cross laminae are present and the majority of disarticulated brachiopod valves are in a convex-up orientation. Pyrite is virtually absent and terrigenous mud attains a formational minimum for this locality. While the macrofauna is dominated by atrypid brachiopods, the sediment matrix is predominantly pelmatozoan debris and fenestrate bryozoans. The top of the Seneca is approximated though not attained.

DEPOSITIONAL HISTORY

The Onondaga has long been interpreted as a sequence of limestones deposited in progressively deepening waters and terminated by the progradation of the Marcellus black shales derived from the rising Acadian orogen. Within this model, the Nedrow Member represents an influx of terrigenous mud; a hint of greater things to come. The Tioga Bentonite is a single event horizon. The Moorehouse and Seneca members become increasingly argillaceous as the sea gradually deepened and the Marcellus muds slowly advanced from east to west across the state. Sir Charles Lyell would have found comfort in this model. Its ponderous unfolding would have appealed to his aesthetic tastes. However, to badly paraphrase Mark Twain—recent study has cast much darkness upon the subject.

To begin with, the fidelity of the Tioga Bentonite has been called into question. For some years now it has been known that there are three separate bentonites high in the Onondaga of western New York. It has been supposed that they converge to one in the vicinity of Syracuse due to a relatively low rate of sedimentation in that area. However, on an NYSGA field trip in 1986 a second bentonite was discovered at Jamesville, New York. How many more are there? The recent report of multiple bentonites in the Lower Devonian Kalkberg Limestone of eastern New York (Shaw et.al., 1991) suggests that there may be several.

During the above-mentioned field trip (Feldman and Lindemann, 1986) a classic Devonian bone bed was found high in the Seneca Member. It was also observed that there is no lithologic gradation between the Seneca and the Marcellus and that the contact between the two units is an erosional truncation surface involving up to three beds of the uppermost Seneca. Lindemann and Feldman (1987) described a comparable disconformity at the top of the Onondaga in the central Hudson Valley of eastern New York. At Cherry Valley the precise top of the Seneca is not exposed, but the beds which can be seen give no indication of gradually giving way to shale. As in the case in the Hudson Valley, a relatively brief time of rapid crustal subsidence and an interruption of sedimentation would seem to be indicated.

Abrupt fluctuations in water depth are also indicated at the lower end of the column. Glauconite sand and phosphatic gravel in the lowermost beds of the Edgecliff Member at Cherry Valley indicate an interruption in sedimentation. It was during this unrecorded interval that deposition of the Carlisle Center ended as the depositional environment shifted to one favoring carbonate production. Since there is no definitive interpretation for the depositional history of the Carlisle Center, it is difficult to ascertain what might have transpired during the unrecorded interval. Quartz siltand well-rounded grains of quartz sand at the base of the Edgecliff could suggest relatively high levels of water energy, but the abundances of terrigenous mud and calcisilt with which they occur suggest otherwise. Furthermore, the phosphatic gravels at the base of the Edgecliff appear to have been involved in multiple generations of exhumation and reburial. There would seem to be more involved here than was previously supposed.

Considering the absence of a biostratigraphic basis for the correlation of the Edgecliff to either the Upper Emsian or the Lower Eifelian, this phosphatic diastem is intriguing. Hopefully an ongoing study of this interval will soon yield results. The remainder of the Edgecliff at Cherry Valley is equally intriguing, though less cryptic, from a paleoenvironmental point of view. The Edgecliff consists of dark gray packed biocalcisiltites (C1 zone of Oliver, 1956a) overlain by medium gray, coraliferous, biosparites (C2 zone of Oliver, 1956a). Obviously deposition did not begin in a high energy environment. Wolosz (1985) reported that Edgecliff reefs of the Hudson Valley exhibited evidence of a brief lowering of relative sea level. This was followed by a sea level rise and a resumption of reef growth. Wolosz and Lindemann (1986) correlated the shallowing event to the abrupt onset of biosparite deposition in the Edgecliff throughout eastern New York. This interpretation remains appropriate for the Edgecliff at Cherry Valley.

The top of the Edgecliff is anomalously pyritic. It is immediately overlain by the argillaceous biocalcisiltites of the Nedrow Member. The contact between the two is interpreted to be a diastem resulting from a pulse of crustal subsidence. Pyrite in the Nedrow and the Lower Moorehouse indicate relatively low concentrations of oxygen, possibly due to stratification of the water column. The Nedrow sediments do not suggest an influx of terrigenous mud, but rather a shift to an offshore position coupled with a drastic reduction in carbonate production. The sediments' fine-grained nature suggests a flocculent or soupy sediment-water interface, a condition not particularly conducive to colonization by the larvae of sessile organisms. Thus, the Edgecliff reefs were drowned in deep water rather than suffocated in mud.

Moorehouse deposition marks a return to enhanced carbonate production by benthic organisms living at depths well in excess of the wave base. This is quite different from the top of the Moorehouse, where a carbonate bank environment near the wave base is indicated. Unfortunately the Moorehouse is not fully exposed and the transitional beds are not available for study. However, detailed study of polished slabs and thin sections through the Lower and Middle Moorehouse reveals a symmetry in the sequence of lithologic changes that centers around beds about 10 m from the base of the member. The beds below indicate a progressive increase in water depth and soupiness of the substrate. The beds above show the exact opposite trend. Unlike the remainder of the Onondaga at Cherry Valley, it appears likely that the shallowing upward trend was gradual and not a punctuational event. Lyell would have preferred it this way.

BRACHIOPOD COMMUNITIES

WHAT IS A COMMUNITY?

Communities are often defined as recurrent associations of taxa that were presumably controlled by a set of env ironmental factors such as: substrate, salinity, temperature, pressure, current action, wave action, light penetration, nutrients, dissolved oxygen, and water chemistry. Ecologists are not necessarily in agreement as to what the definition of a community is, nor how to recognize one. Boucot (1981) notes major subdivisions of current conceptualizations of community definition, including those who define "community" as a superorganism that has a virtual life of its own—a living and breathing community. At the other extreme are those who hold that communities are no more than chance aggregations of organisms conducting their affairs quite independently of one another—ships that pass in the night; apartment dwellers who have never been introduced to their neighbors.

Paleoecologists are at a distinct disadvantage in attempting to reconstruct ancient communities, since it is extremely difficult to determine the various relationships of taxa in terms of parasitism, commensalism, mutualism, and other dependent and interdependent variables that are not readily apparent in the fossil record. As Boucot (1981) notes, the paleoecologist is reduced to examining statistical data on relative abundance and presence or absence of taxa in an attempt to infer ecological interaction. There is much biological information important in community reconstruction which cannot be retrieved from the rock record, and this must be kept in mind when coming to conclusions about community make-up. Ecologists who study Recent communities have a distinct advantage in this regard over paleoecologists, and are able avoid dependence solely on hard part data.

BRACHIOPOD COMMUNITIES OF THE ONONDAGA LIMESTONE

When studying the brachiopod communities of the Onondaga Limestone in New York State, other faunal constituents and their fragments, such as trilobites, corals, and gastropods were tabulated (Feldman, 1980; Feldman and Lindemann, 1986; Lindemann and Feldman, 1987). Numbers of brachiopods were determined by counting the most abundant valve. Relative abundance was variable, depending on geographic area and member sampled. For example, collecting in the shaly Nedrow Member in central New York was much more productive than in the dense Moorehouse Member. However, in eastern New York, the silicified Moorehouse yielded many more well-preserved taxa than did the nonsilicified Nedrow. Therefore, relative abundance seems to be a function of: (1) lithology, (2) rate of weathering, and (3) silicification. The Onondaga Limestone is most productive, in terms of brachiopods, when well silicified. Unfortunately, this occurs rarely, notable localities being in the mid-Hudson and Genesee valleys. In the mid-Hudson Valley, heavy jointing is associated with silicification. There are many outcrops that show evidence of weak silicification and collecting from these areas can range from excellent to poor, depending on the degree of silicification (whether surficial or deep). Beekite rings on shells observed in outcrop are usually indicative of weak silicification. The Onondaga brachiopod communities recognized in New York State are briefly described below.

Atrypa-Coelospira-Nucleospira Community. The ACN (*Atrypa-Coelospira-Nucleospira*) Community ranges from Leeds to just south of Kingston, New York and occurs predominantly in the Moorehouse Member. Diversity here is great (29 brachiopod genera), but only 13 genera comprise the bulk of the community (Feldman, 1980). Of those, three genera (*Atrypa, Coelospira, Nucleospira*) represent a trophic nucleus of low-level epifaunal suspension feeders. A similar fauna is found in Lenz's (1976) Lower Lochkovian *Howellella-Protathyris* Community in an offshore position. Taxa in common include: *Atrypa, Schizophoria, Ambocoelia, Coelospira, Nucleospira*, and "*Schuchertella*." Lenz's fauna is characterized by. similar morphotypes (Table 1).

Atrypa-Megakozlowskiella Community. The AM (*Atrypa-Megakozlowskiella*) Community, recognized from Clarksville to Cherry Valley, New York, is lower in diversity than the ACN Community (22 compared to 29 genera). This may be indicative of a position closer to shore and consequently nearer to wave base. A major faunal element that appears here is the robust spiriferid *Megakozlowskiella raricosta*, which had a large, triangular delthyrium in the ephebic stage, with lateral bordering ridges indicative of a deltidial plate. If the pedicle had no way of protruding, the

TABLE 1. A Comparison of Lenz's (1976) *Howellella-Protathyris* Community with the ACN Community of the Onondaga Limestone.

Morphotype	*Howellella-Protathyris* Community	ACN Community
Broad, flat	*"Schuchertella"*	*Schuchertella*
Relatively smooth	*Protathyris, Cryptatrypa*	*Nucleospira, Athyris*
Broad, unequally biconvex	*Schizophoria*	*Schizophoria*
Frilly	*Atrypa*	*Atrypa*

brachiopod would have lived free on the sea floor. The pedicle valve had deeper ribs and was more convex than the flatter brachial valve, which would have provided a more hydrodynamically stable position for the animal if the pedicle valve was in an "up" position. Some gerontic shells had secondary shell material deposited in the umbonal region as a counterweight, serving to keep the anterior commissure above the sediment-water interface.

Atrypa Community: The Atrypa Community occurs from the mid-Hudson Valley to Cherry Valley and is dominated by Hudson Valley "*reticularis*" (52.4%), with a relatively high diversity of 18 brachiopod genera (compare with the diversity of the AM Community of 22 genera). This community is very similar to Copper's (1966) European Eifel magnafacies, which is composed of calcareous shales, muddy limestones, and rare dolomites. Although there are no dolomites within the Onondaga, the Nedrow and Moorehouse members certainly contain a fair amount of mud. Other similarities, in addition to lithology, include the presence of varied brachiopod genera in both environments (such as spiriferids, rhynchonellids, athyrids, meristellids, and gypidulids) and the occurrence of rugose and tabulate corals, stromatoporoids, and crinoids.

Leptaena-Megakozlowskiella Community: The LM (*Leptaena-Megakozlowskiella*) Community is recognized in the Syracuse area of central New York and is dominated by *Megakozlowskiella raricosta* and the ubiquitous *Leptaena "rhomboidalis."* Within the Onondaga, *Leptaena* occurs more frequently in "muddier" limestone units and is relatively rare in the Edgecliff Member. A distinct association between the two genera is very evident on bedding plane surfaces in the shaly Nedrow Member, where they comprise a trophic nucleus of low-level suspension feeders. The diversity is fairly high, with 17 brachiopod genera represented. Crinoidal fragments and platyceratid gastropods, which are common in the ACN Community, are absent here; in their place are other gastropod genera, such as *Straparollus*, *Liospira*, and *Ecculiomphalus*.

"*Pacificocoelia*" Community: This community has been found at only one outcrop in the Nedrow Member near Syracuse, New York, and is similar to the LM Community in two respects: (1) There is a close association between *Leptaena* (9.7%) and *Megakozlowskiella* (8.6%), and (2) both communities are typically found in the shaly rather than the "cleaner" lime units. They differ in that in the "*Pacificocoelia*" Community the brachiopod diversity is low (10 genera) and no corals were recovered, whereas in the LM Community 17 brachiopod genera were found, as well as 3 rugose and 3 tabulate coral genera.

Levenea Community I: This community occurs in the Edgecliff Member from Cherry Valley southeast to Kingston, New York and is dominated by *Levenea* sp. A (67.4%), with minor occurrences of *Atrypa*, *Levenea* sp. B, *Leptaena*, *Pentamerella*, and *Elita*. In general, the brachiopods are poorly represented in the Edgecliff Member. This may somehow be related to the large amount of chert present in the east, which seems to correlate with a reduced coral fauna. In central New York there is a large coral fauna, relatively little chert, and more brachiopods.

Levenea Community II: Found only in the Moorehouse Member of southeastern New York, at an abandoned quarry in Wawarsing, the *Levenea* Community II consists exclusively of *Levenea* sp. A (100%). It differs from *Levenea* Community I in two respects: (1) the diversity is extremely low, and (2) the lithology is very different, consisting of "muddy" rocks interpreted to represent deposition in a more offshore position. This is consistent with the interpretation of a deepening structural basin in Onondaga times southwest toward Port Jervis (Lindemann and Feldman, 1987).

Amphigenia? Community: The *Amphigenia*? Community is found in the basal Edgecliff near Syracuse, New York, and is based on the recovery of only 12 specimens. There is a possibility that these fragmental shells were reworked and transported, since the occurrence of *Amphigenia* in the sandy facies of the Edgecliff is not compatible with Boucot's (1975) placement of the genus in a Benthic Assemblage 3-5 position.

Hallinetes Community: The *Hallinetes* Community (Racheboeuf and Feldman, 1990), formerly recognized as a *Chonetes* Community (Feldman, 1980), occurs only in the Seneca Member of the Onondaga Limestone. Three taxa comprise the chonetacean brachiopods in this community: *Hallinetes lineatus* (92%), *Longispina mucronata* (5.4%) and "*Eodevonaria*" *hemispherica* (2%). Other brachiopod taxa are present but represent minor faunal constituents (see Feldman, 1980, p. 40). Based on new observations, it is apparent that the *Hallinetes* Community is most accurately represented by the shells within the dark mudstone matrix rather than by those distributed on bedding plane surfaces. The community is a low-diversity, "highly dominated" (although not monospecific) community within a quiet water environment (Racheboeuf and Feldman, 1990).

COMMUNITIES OF WESTERN NEW YORK

Based on preliminary analysis of material collected from the Moorehouse Member of the Onondaga Limestone in the Genesee Valley of western New York, an almost identical ACN Community to the one found in the mid-Hudson Valley is recognized. Similarities include dominance by the low-level epifaunal suspension feeders *Atrypa*, *Coelospira*, and *Nucleospira* as well as very high diversity (39 brachiopod genera, including two new athyrids). The two communities differ in that the ACN Community of the Genesee Valley has a significantly larger proportion of strophomenids, including some genera absent in southeastern New York: "*Brachiprion*" aff. *mirabilis*, *Protoleptostrophia perplana*, *Plicostropheodonta*? sp. and *Costistrophonella ampla*. Also, there are other taxa in the west not recovered from southeastern New York: *Camarospira*? sp., *Alatiformia*? sp., *Mediospirifer* sp.A and B, *Paraspirifer* sp., *Cranaena* sp., and *Cryptonella* sp.

COMMUNITY PALEOGEOGRAPHY

Work is currently in progress which will clarify the relationships of these various communities to one another across New York state. However, a general pattern can be observed (Table 2). Most data have been collected from the Nedrow and Moorehouse members, therefore, by omitting those communities found only in the Edgecliff and Seneca members (*Amphigenia*, *Levenea* Community I, *Hallinetes*), it appears that during Nedrow-Moorehouse time, (1) there was a trend towards increasing diversity away from the basinal axis and, (2) diversity decreased towards a subsiding structural basin.

Lindemann and Feldman (1987) note that in central New York, a transgression submerged the region initiating Edgecliff deposition in a shallow shelf environment. Soon thereafter subsidence in central New York, resulting from a northward extension of the Appalachian Basin, brought deeper water and an offshore environment to the area. The initial pulses of subsidence are recorded in the Nedrow Member, while conti nued subsidence is evidenced in the Moorehouse and Seneca members of the central region. The eastern and western areas, that is, those areas away from the basinal axis, remained in shallow shelf conditions resulting in a symmetric shelf-basin-shelf pattern as seen in east-west outcrop.

A subsiding structural basin in the Tristates area, (not directly related to the topographic basin of central New York), which had existed since the Middle Silurian was noted by Lindemann and Feldman, 1987). This basin greatly influenced Onondaga deposition in southeastern New York by creating a carbonate slope, or ramp, dipping into the Port Jervis area. It is from this ramp that the *Levenea* Community II was recovered, indicating a trend toward lower diversity in the direction of the deep waters of the structural basin. Further collecting and analysis of brachiopod communities along the ramp will help support or reject this proposed paleogeographic distribution and correlation with basin depth.

TABLE 2. Brachiopod Communities in the Onondaga Limestone of New York.

Brachiopod community	Number of Genera*	Member	Geographic Location
ACN	39	Moorehouse	Genesee Valley
ACN	29	Moorehouse	Hudson Valley
AM	22	Moorehouse	Cherry Valley
Atrypa	18	Nedrow-Moorehouse	Hudson Valley, Cherry Valley
LM	17	Nedrow-Moorehouse	Syracuse
Hallinetes	10	Seneca	Syracuse
"*Pacificocoelia*"	10	Nedrow	Syracuse
Levenea Community I	6	Edgecliff	Hudson Valley
Levenea Community II	1	Moorehouse	Wawarsing
Amphigenia	1	Edgecliff	Syracuse

*Refers to brachiopod genera; ACN = *Atrypa-Coelospira-Nucleospira* community; AM = *Atrypa-Megakozlowskiella* Community; LM = *Leptaena-Megakozlowskiella* Community.

MT. TOM: A SMALL EDGECLIFF PINNACLE REEF

Oliver (1956c) described the location and size of Mt. Tom, labeled it Mt. Tom #1, and included it among the seven reef exposures comprising the Mt. Tom Reef Group, which are scattered over an approximately 9 square mile area at the boundary of the East Springfield, Richfield Springs, Jordanville, and Van Hornesville 7.5 minute quadrangles. Mt. Tom is the largest reef exposure in the group, forming a prominent hill in the northwest corner of the East Springfield 7.5 minute Quadrangle (it is, in fact, the thickest known surface exposure of an Edgecliff reef [Oliver, 1956c: 21]).

Whereas Oliver considered all seven Mt. Tom Group exposures to represent separate reefs, Paquette and Wolosz (1987) noted that the two

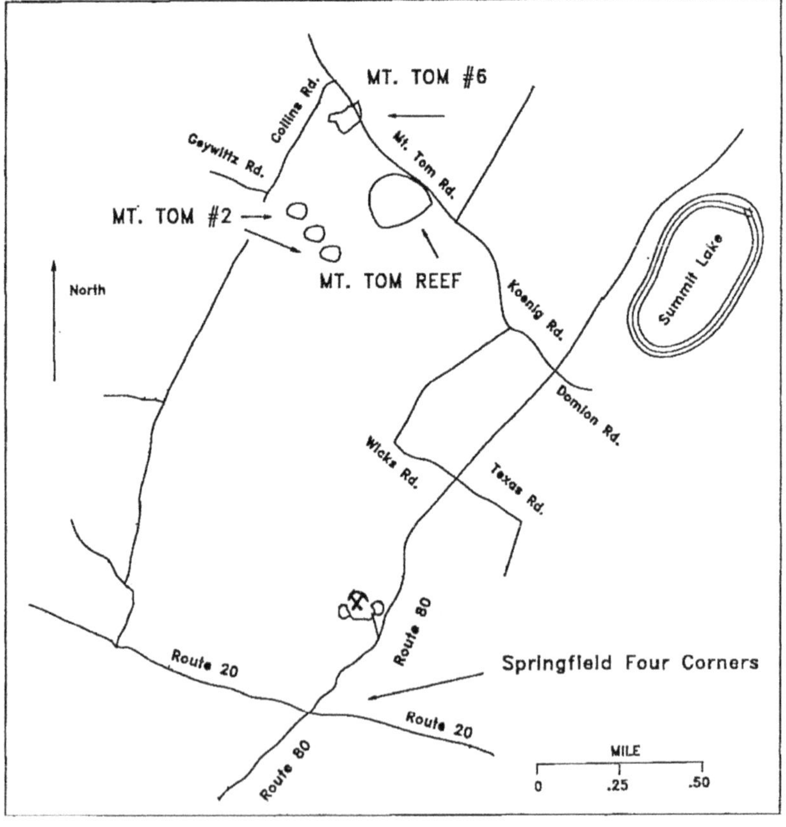

FIGURE 1. Location map of Mt. Tom reef #1, 2, and 6. Note relative positions of reefs.

exposures closest to Mt. Tom #1 (see Figure 1)—Mt. Tom #2 reef (approximately due west of Mt. Tom) and Mt. Tom #6 (northwest of Mt. Tom)—were comprised mainly of crinoidal grainstone/packstone, which dipped away from Mt. Tom #1. They argued that these three exposures represent the erosionally dissected remains of a small pinnacle reef, approximated at 150 acres.

REEF COMMUNITIES

Typical of Edgecliff reefs, Mt. Tom is made up of two distinct paleocommunities—the phaceloid colonial rugosan paleocommunity and the favositid/crinoidal sand paleocommunity.

The phaceloid colonial rugosan paleocommunity consists almost exclusively of colonial rugosans. Common genera include *Acinophyllum*, *Cylindrophyllum*, and *Cyathocylindrium*, with *Eridophyllum*, *Synaptophyllum*, and rare phaceloid colonies of *Heliophyllum* as accessories. The dense growth of these rugosan colonies appears to have restricted most other organisms to only minor roles, with favositids (both domal and branching) being small and rare, brachiopods uncommon, and bryozoans mainly fragmentary encrusters.

The favositid/crinoidal sand paleocommunity displays a much higher diversity than the rugosan paleocommunity. This paleocommunity is more biostromal than biohermal. Large sheetlike to domal favositids are abundant, but never form a constructional mass. Solitary rugose corals are also extremely abundant, as are fenestrate bryozoan colonies. Single colonies of the mound building phaceloid rugosans are occasionally found. Brachiopods and other reef dwellers are also common, although never extremely abundant. Stromatoporoids and massive colonial rugosans, while extremely rare in the Edgecliff reefs, when found are part of this paleocommunity. The crinoids were the greatest contributor to this paleocommunity—ossicles making up the bulk of the rock and indicating abundant growth of these organisms—but complete calyces are never found.

MT. TOM REEFS #1, 2, AND 6

Wolosz (1990) and Wolosz and Paquette (1995) presented a classification of Edgecliff reef types based on the relative importance of the two paleocommunities to the development of the reef structure. Mt. Tom reef is an example of a mound/bank composite structure. Mounds are distinct high relief buildups of the phaceloid colonial rugosan paleocommunity, which occur as either small (generally not more than 1-3 m thick) monogeneric to mixed faunal buildups, or Successional Mounds up to roughly 15 m thick, which display an internal succession of mound building colonial rugosan genera. The term "bank" follows the definition of Nelson et al. (1962: 242): "a skeletal limestone deposit formed by organisms which do not have the ecologic potential to erect a rigid, wave resistant structure." Hence, mound/bank reefs are large structures resulting from the repetitive intergrowth of rugosan mounds and the favositid/crinoidal sand facies. Pinnacle reefs found in the subsurface in New York and Pennsylvania also represent this type of structure and reach thicknesses of up to 60 m.

The mound\bank nature of Mt. Tom #1 is displayed in the cliff face along the southeast side of the hill (Figure 2). The reef is underlain by the basal Edgecliff calcisiltite (C1 unit of Oliver, 1956a), with the base of the reef marked by thickets of *Aginophyllum*. Small phaceloid colonial rugosan mounds (again, mainly *Acinophyllum*) can be observed along the cliff near the base of the reef. These small mounds and thickets coalesced to begin the formation of the larger structure. Dominance of the initial large mound shifted between *Acinophyllum* and *Cylindrophyllum* prior to onlapping by the crinoidal sands of the favositid/crinoidal sand paleocommunity. A second mound stage made up of *Cylindrophyllum* thickets overlies these grainstones and packstones. In turn, the second mound stage is itself onlapped and eventually swamped by the favositid/crinoidal sand paleocommunity (exposed further back on the top of the hill, not shown in Figure 2). Overall, Mt. Tom #1 is roughly 18 m thick as preserved.

Wolosz and Paquette (1988) have interpreted this mound/bank/mound/bank pattern as catch-up/fall back cycles controlled by fluctuations in water depth above the top of the reef. It is important to note that the second mound building stage at Mt. Tom #1 (Figure 2) does not drape the entire preexisting structure, but is instead restricted to the top of that structure. In effect, during bank stage, the reef was a high relief platform on the sea floor, with its top within the ecologic mound building zone of the colonial rugosans. Upward growth of the reef is mainly due to the repetitive establishment of new mounds on the top of the platform. As sea level was approached, the mound building colonial rugosans were overwhelmed by increased turbulence conditions and the mounds onlapped by encroaching crinoidal sands, producing a bank stage; but with sea level rise the mounds became re-established. This shifting between rugosan mound/thicket construction and the favositid/crinoidal sand paleocommunity has been attributed to a water turbulence controlled community succession (Wolosz, 1989a, 1989b, 1995).

Following the initial mound building stage, lateral growth of Mt. Tom appears to have been due mainly to deposition of crinoidal debris flanks, with occasional small mound structures (satellite mounds) growing in those flanks (see discussion of Mt. Tom #2). A similar but less well developed mound/hank/mound sequence has been described at Roberts Hill Reef south of Albany (Wolosz, 1985).

To the northwest, Mt. Tom #6 is a small ridge that consists mainly of crinoidal grainstone/packstone, but with more abundant fossils. Small overturned favositids are common, as are both solitary and phaceloid rugosans,

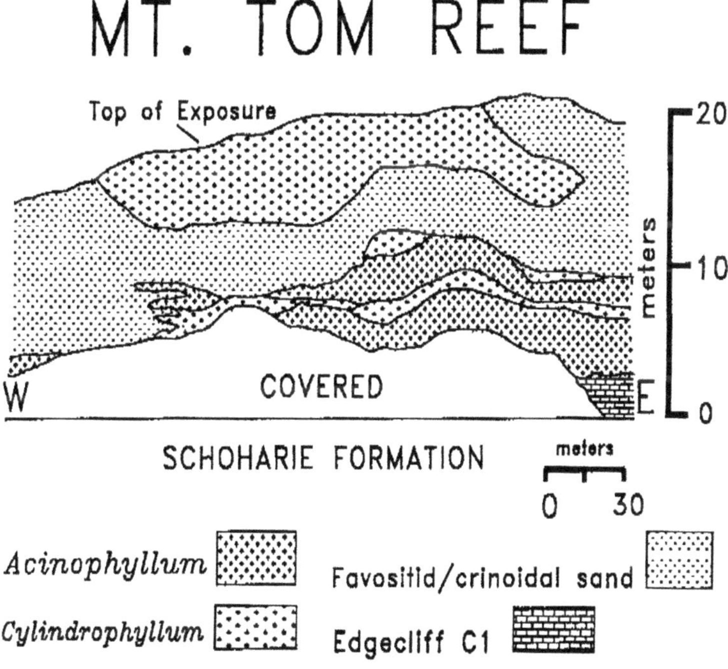

FIGURE 2. Cross-section of cliff face at Mt. Tom #1 reef illustrating mound/bank structure. Two rugosan mound stages are separated by favositid/crinoidal sand facies (bank stage). Note that second rugosan mound stage does not drape entire reef structure (after Wolosz, 1992).

but no evidence of mound formation is present. However, when one observes Mt. Tom #6 from Collins Road (see map, Figure 1), the questa-like nature of this small ridge is evident, with the dip slope pointing to the north-northwest, directly away from the main mass of Mt. Tom.

Topographically, Mt. Tom #6 is at the same elevation as the present top of Mt. Tom. since the regional southwest dip of about 18 m per km (Rickard and Zenger, 1964: 5) would not greatly alter this topographic relationship, the elevations of the Mt. Tom #6 exposure and the top of Mt. Tom were probably also equivalent at the time of deposition. Paquette and Wolosz (1987) cited this as evidence that the two exposures are parts of one reef, with Mt. Tom #6 consisting of distal flank beds. Mt. Tom reef would then be at least 0.8 km long on a northwest axis from Mt. Tom #1 to Mt. Tom #6.

In contrast, Mt. Tom #2 lies to the west of Mt. Tom #1 and is topographically roughly 18 m below #6. Stratigraphically older beds can be examined here, with the Edgecliff/Carlisle center contact marked by the appearance of a spring just east of the intersection of Collins and Geywittz Roads. A small quarry, visible from the road, exposes bedded Edgecliff with overturned colonial coral. To the southeast of this quarry is an exposure of a small colonial rugosan mound roughly 17 m across and of indeterminate thickness. East from the quarry, along the south side of the creek, there are numerous outcrops of bedded crinoidal grainstone/packstone with abundant favositids. Small patches or lenses of colonial rugosans within the bedded packestones are common, and represent small satellite thickets or mounds that appear to range stratigraphically from near the C1/C2 contact (roughly the point at which growth of Mt. Tom #1 began), upward to about 6 m above that contact. The packstones surrounding these upper mounds dip away from Mt. Tom #1 at roughly 15 degrees.

TYING THE EXPOSURES TOGETHER: DEVELOPMENT OF THE MT. TOM PINNACLE REEF

Figure 3 illustrates an interpreted developmental history for the Mt. Tom (small) pinnacle reef. As sea level dropped from possible deep water conditions of Carlisle Center deposition through the early Edgecliff (C1), abundant small rugosan thickets and mounds began to form in the late Cl calcisilts. By the beginning of C2 deposition, these thickets and small mounds had begun to coalesce to form the initial large mound at Mt. Tom #1 (Mound Stage I), while an abundance of other small mounds dotted the crinoidal sand sea floor as satellites to the growing reef. Crinoidal debris of the favositid/crinoidal sand paleocommunity lapped up onto the large mound, eventually forming flank beds that spread outward from the main mass of the reef. Small satellite mounds continued to develop along distal flank beds (Mt. Tom #2), contributing to the overall volume of the reef structure, but never coalescing into a large central structure similar to Mt. Tom #1. Continued sea level drop resulted in the cessation of rugosan mound growth and the eventual swamping of the mound by the crinoidal sand beds, resulting in Bank Stage I. A second cycle of sea level rise resulted in the establishment of new rugosan thickets and mounds on the top of the bank

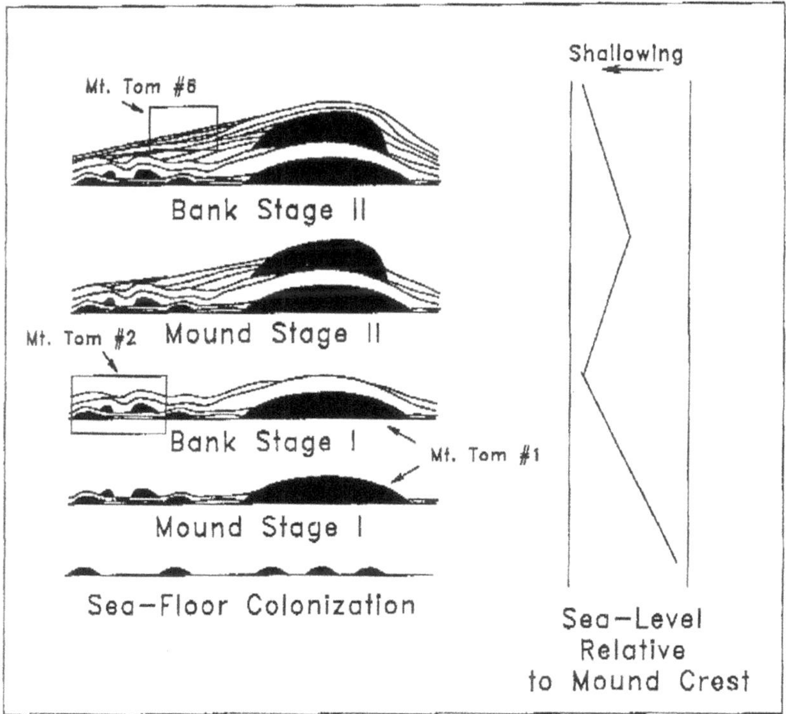

FIGURE 3. Sequential model for growth of the Mt. Tom pinnacle reef. Interpreted sea level changes shown at right. Boxes indicate interpreted position of Mt. Tom # 2 and 6 exposures. Main mound is Mt. Tom #1 exposure. See text for further details.

(Mound Stage II), but later shallowing over the crest of the reef again caused the demise of the colonial rugosans and the re-establishment of the favositid/crinoidal sand paleocommunity in Bank Stage II.

In Figure 3, Mt. Tom #6 is illustrated as distal flank beds. While this is correct, the illustration is not to scale and somewhat misleading. When the distance from Mt. Tom #6 to #1 is considered (roughly 0.8 km), along with the 15 to 25 degree dip of the beds at #6, and the already noted paleo-topographic equivalence of the present top of #1 and #6, the conclusion that, if totally preserved, Mt. Tom reef would be much thicker than the present erosional remnant is easily drawn. Unfortunately, there appears to be no way to achieve a valid estimate of that thickness.

Any attempt to directly correlate the reef growth cycles preserved at Mt. Tom with the non-reefal Onondaga (for instance at Cherry Valley) would require a detailed micro-stratigraphy, which is, unfortunately, not available. However, the following statements can serve as a basis for discussion and further research.

The first mound/bank cycle at Mt. Tom follows the shallowing trend from C1 to C2 deposition in the Edgecliff. The initial pattern here is similar to that described at Roberts Hill (Wolosz, 1985; Wolosz and Lindemann, 1986). However, as sea level begins to rise, leading to the second mound/bank cycle, the environment at the crest of the reef (or the top of the bank) becomes decoupled from that of the surrounding sea floor. In order to produce the large pinnacle structure, the top of the bank must be maintained within a fairly narrow environmental range suitable for the two reef building paleocommunities (see Wolosz, 1995 for discussion). If we assume that by the second bank stage (current top of Mt. Tom #1) the reef was roughly 18 m thick with the bank top at least 10 m above the surrounding ocean floor (given that the C2 at East Springfield is roughly 4 m thick [Oliver, 1956a], and allowing for a 50 percent compaction of the carbonate sediments), and also that the Edgecliff-Nedrow contact marks a starvation boundary (see discussion of stratigraphy), then at this point much lateral growth of the bank would occur, since large amounts of crinoidal debris from the favositid/crinoidal sand community would be washed off the bank onto the flanks while upward growth would be limited by sea level. Such a scenario would leave a well-developed bank with the potential for continued upward growth once renewed subsidence led to the onset of Nedrow deposition. In effect, environmental conditions characteristic of the Edgecliff would continue on the bank top while Nedrow sediments were being deposited on the surrounding sea floor.

THE EDGECLIFF REEFS: COOL WATER STRUCTURES?

As mentioned in the introduction, Kissling and his students have pointed to the lack of stromatoporoids and calcareous algae, in conjunction with the absence of clear peritidal deposits to suggest that the Edgecliff reefs may have been deposited in deep water. An alternative hypothesis to the deep water model is that the Edgecliff was deposited under cool water conditions. Wolosz and Paquette (1988) suggested a cool water environment for the Edgecliff, as have Koch and Boucot (1982) based on the Edgecliff brachiopod fauna, Blodgett et al. (1988) based on gastropod faunas, and Wolosz (1990b, 1991) based on stromatoporoid abundance trends.

The cool water model for Edgecliff deposition supplies answers to many of the questions listed in the introduction. The C2 facies is a shallow water facies, but one more akin to modern FORAMOL deposition (Lees, 1975) than to tropical carbonate deposition. The reefs are then analogous to modern ahermatypic coral banks, built by relatively slow-growing colonial rugosans poorly adapted to high energy conditions—hence their replacement by the favositid/crinoidal sand community under high energy conditions. The cool waters would also explain the rarity of stromatoporoids and the absence of algae—both groups are restricted to warm waters.

In conclusion, the paleo-biological evidence appears to support a model of the Edgecliff as a temperate water carbonate.

ROAD LOG

Cumulative Miles	Mileage from Last Point	Route Description
0.0	0.0	Intersection of Routes 20 and 1a, Sharon Springs. Proceed west along Route 20.
7.0	7.0	Stop 1. Cherry Valley Section. Park along road at top of west end of road cut. An almost complete section of the Onondaga Limestone is exposed along this cut. (See discussion of stratigraphy and Brachiopod Communities). Return to cars, proceed west along Route 20.
13.2	6.2	Right turn on Route 80 (see Figure 1).
14.7	1.5	Left turn onto Koenig Road.

15.3	0.6	Bear left onto Mt. Tom Road.
15.4	0.1	Stop 2. Mt. Tom reef makes up the large hill to the south of the road (See discussion of Mt. Tom). Return to oars and continue northwest on Mt. Tom Road.
15.9	0.5	Stop 3. Mt. Tom #6 forms the low, wooded ridge to the southwest of the road (See discussion in text). Return to cars and continue northwest on Mt. Tom Road.
16.05	0.15	Left turn onto Collins Road.
16.55	0.5	Stop 4. Intersection of Collins and Geywittz Roads. Leave cars and proceed east from the intersection. Mt. Tom #2 forms the low hill to the south of the small creek, and numerous small outcrops may be examined along the south side of the creek valley or on the hill itself. A small quarry on the northwest edge of the hillside exposes bedded Edgecliff facies, while a small rugosan mound is located just to the southeast of the quarry among the trees. (See text for discussion). Return to cars follow Collins Road back to Route 20.

ACKNOWLEDGMENTS

Study of the Mt. Tom reefs was supported by The U. S. Department of Energy Special Research Grants Program Grant #DE-FG02-87ER13747. A000 to T. H. Wolosz.

REFERENCES

Blodgett, R. B., Rohr, D. M., and Boucot, A. J. 1988. Lower Devonian gastropod biogeography of the western hemisphere. In N. J. McMillan, A. F. Embry, and D. J. Glass (eds.), *Devonian of the World, proceedings of the Second International Symposium on the Devonian system*, CSPG Memoir 14, vol. III, 281-294.

Boucek, B. 1964. *The Tentaculites of Bohemia*. Prague: Publishing House of the Czechoslovak Academy of Sciences.

Boucot, A. J. 1975. *Evolution and Extinction Rate Controls*. Amsterdam: Elsevier.

Boucot, A. J. 1981. *Principles of Benthic Marine Paleoecology*. New York: Academic Press.

Copper, P. 1966. Ecological distribution of Devonian atrypid brachiopods. *Palaeogeography, Palaeoclimatology, Palaeoecology* 2: 245-266.

Dutro, J. T., Jr. 1981. Devonian brachiopod stratigraphy of New York State. In W. A. Oliver, Jr. and G. Klapper (eds.), *Devonian Biostratigraphy of*

New York, Part 1. International union of Geological Sciences, Subcommission on Devonian Stratigraphy, 67-82.

Eaton, A. 1832. *Geological textbook.* Albany: printed by Webster and Skinners.

Feldman, H. R. 1980. Level-bottom brachiopod communities in the Middle Devonian of New York. *Lethaia* 13: 27-46.

Feldman, H. R. 1985. Brachiopods of the central and southeastern New York. *American Museum of Natural History Bulletin* 179: 289-377.

Feldman, H. R., and Lindemann, R. H. 1986. Facies and fossils of the Onondaga Limestone in central New York. *New York State Geological Association Field Trip Guidebook, 58th Annual Meeting,* Cornell University, Ithaca, New York, 145-166.

Hall, J. 1843. Natural History of New York. Geology, pt. 4. Albany: Carroll and Cook Printers.

House, M. R. 1962. Observations on the ammonoid succession of the North American Devonian. *Journal of Paleontology* 36: 247-284.

House, M. R. 1981. Lower and Middle Devonian goniatite biostratigraphy. In W. A. Oliver, Jr. and G. Klapper (eds.), *Devonian Biostratigraphy of New York, Part 1,* 33-37. International Union of Geological Sciences, Subcommission on Devonian Stratigraphy.

Kissling, D. L. 1987. Middle Devonian Onondaga pinnacle reefs and bioherms, Northern Appalachian Basin. *Second International Symposium on the Devonian System, Calgary, Alberta, Canada, Program and Abstracts,* 131. (abstract)

Klapper, G. 1981. Review of New York Devonian conodont biostratigraphy. In W. A. Oliver, Jr. and G. Klapper (eds.), *Devonian Biostratigraphy of New York, Part 1,* 57-66. International Union of Geological Sciences, Subcommission on Devonian stratigraphy.

Koch, W. F., II, and Boucot, A. J. 1982. Temperature fluctuations in the Devonian Eastern Americas Realm. *Journal of Paleontology* 56: 240-243.

Lees, A. 1975. Possible influences of salinity and temperature on modern shelf carbonate sedimentation. *Marine Geology* 19: 159-198.

Lenz, A. C. 1976. Lower Devonian brachiopod communities of the northern Canadian Cordillera. *Lethaia* 9: 19-28.

Lindemann, R. H., and Feldman, H. R. 1987. Paleogeography and brachiopod paleoecology of the Onondaga Limestone in eastern New York. *New York State Geological Association Field Trip Guidebook, 59th Annual Meeting,* State University College of New York at New Paltz, D1-D30.

Lindemann, R. H. and Yochelson, E. L. 1984. Styliolines from the Onondaga Limestone (Middle Devonian) of New York. *Journal of Paleontology* 58: 1251-1259.

Lindemann, R. H., and Yochelson, E. L. 1994. Redescription of Styliolina [INCERTAE SEDIS]: Styliolina fissurella (Hall) and the type species S. nucleata (Karpinsky). In E. Landing (ed.), *Studies in Stratigraphy and Paleontology in Honor of Donald W. Fisher.* New York State Museum Bulletin, 481: 149-160.

Lindholm, R. C. 1967. Petrology of the Onondaga Limestone (Middle Devonian), New York. PhD dissertation, Johns Hopkins University, Baltimore, MD.

Nelson, H. F., Brown, C. W., and Brineman, J. H. 1962. Skeletal limestone classification. In W. E. Ham (ed.), *Classification of Carbonate Rocks, A Symposium: AAPG Memoir* 1: 224-252.

Oliver, W. A., Jr. 1954. Stratigraphy of the Onondaga Limestone (Devonian) in central New York. *Geological Society of America Bulletin* 65: 621-652.

Oliver, W. A., Jr. 1956a. Stratigraphy of the Onondaga Limestone in eastern New York. *Geological Society of America Bulletin* 67: 1441-1474.

Oliver, W. A., Jr. 1956b. Tornoceras from the Devonian Onondaga Limestone of New York. *Journal of Paleontology* 30: 402-405.

Oliver, W. A., Jr. 1956c. Biostromes and bioherms of the Onondaga Limestone in eastern New York. *New York State Museum Circular* 45.

Oliver, W. A., Jr. 1976. Noncystimorph colonial rugose corals of the Onesquethaw and Lower Cazenovia stages (Lower and Middle Devonian) in New York and adjacent areas. *U. S. Geological Survey Professional Paper* 869.

Oliver, W. A., Jr., and Sorauf, J. E. 1981. Rugose coral biostratigraphy of the Devonian of New York and adjacent areas. In W. A. Oliver, Jr. and G. Klapper (eds.), *Devonian Biostratigraphy of New York, Part 1*, 97-105. International Union of Geological Sciences, Subcommission on Devonian Stratigraphy.

Paquette, D. E., and Wolosz, T. H. 1987. Mt. Tom Reefs #'s 1, 2 & 6: A possible erosional remnant of an Edgecliff pinnacle reef (mid. Devonian, Onondaga formation of New York). *Geological Society of America, Abstract with Programs* 19 (1): 50.

Racheboeuf, P. R., and Feldman, H. R. 1990. Chonetacean brachiopods of the "Pink *Chonetes*" Zone, Onondaga Limestone (Devonian, Eifelian), central New York. *American Museum Novitates* 2982: 1-16.

Rickard, L. V. 1975. Correlation of the Silurian and Devonian rocks in New York State. *New York State Museum and Science Service Map and Chart Series* 24.

Rickard, L. V., and Zenger, D. H. 1964. Stratigraphy and paleontology of the Richfield Springs and Cooperstown Quadrangles, New York. *New York State Museum and Science Service, Bulletin* 396.

Shaw, G. H., Chen, Y.-a., and Scott, J. 1991. Multiple K-bentonite layers in the Lower Devonian Kalkberg Formation—Cobleskill, NY. *Geological Society of America Abstract with Programs* 23 (1): 126.

Vanuxem, L. 1842. *Natural History of New York: Geology*, pt. 3. Albany: W. and A. White and J. Visscher.

Wolosz, T. H. 1991. Edgecliff reefs: Devonian temperate water carbonate deposition. *American Association of Petroleum Geologists Bulletin* 75 (3): 696. (abstract)

Wolosz, T. H. 1990. Shallow water reefs of the Middle Devonian Edgecliff member of the Onondaga Formation, Port Colborne, Ontario, Canada. *New York State Geological Association, 62nd Annual Meeting, Field Trip Guidebook*, Sun.E1-Sun.E17.

Wolosz, T. H. 1989a. Water turbulence: The controlling factor in colonial rugosan successions within Edgecliff reefs. *Geological Society of America, Abstract with Programs* 21 (2): 77. (abstract)

Wolosz, T. H. 1989b. Thicketing events: A key to understanding the ecology of the Edgecliff reefs (Middle Devonian Onondaga Formation of New York). *Geological Society of America, Abstract with Programs* 21 (2): 77. (abstract)

Wolosz, T. H. 1985. Roberts Hill and Albrights Reefs: Faunal and sedimentary evidence for an eastern Onondaga sea-level fluctuation. *New York State Geological Association, 57th Annual Meeting, Field Trip Guidebook,* 169-185.

Wolosz, T. H. 1992. Patterns of reef growth in the Middle Devonian Edgecliff Member of the Onondaga Formation of New York and Ontario, Canada and their ecological significance. *Journal of Paleontology* 66: 8-15.

Wolosz, T. H. 1995. Turbulence controlled succession in Middle Devonian reefs of eastern New York State. *Lethaia* 25: 283-290.

Wolosz, T. H., and Lindemann, R. H. 1986. Correlation of a sea-level drop recorded on patch reefs of the Edgecliff Member, Onondaga Formation in eastern New York. *Geological Society of America, Abstract and Programs* 18: 77.

Wolosz, T. H., and Paquette, D. E. 1988. Middle Devonian reefs of the Edgecliff Member of the Onondaga Formation of New York. In N. J. McMillan, A. F. Embry, and D. J. Glass (eds.), *Devonian of the World, Proceedings of the Second International Symposium on the Devonian System, Canadian Society of Petroleum Geologists Memoir* 14, vol. II: 531-539.

Wolosz, T. H., and D. E. Paquette. 1995. Middle Devonian temperate water bioherms of eastern New York State (Edgecliff member, Onondaga Formation), *New York State Geological Association Field Trip Guidebook, 67th Annual Meeting*, 227-250.

Ziegler, W., and Klapper, G. 1985. Stages of the Devonian system. *Episodes* 8: 104-109.

APPENDIX

Original Publications and Contributing Authors

1. Notes on the Geology of the Shawangunk Ridge on the Mohonk Preserve and Environs Howard R. Feldman, John A. Smoliga, and Brian A. Feldman Northeast Natural History Conference 2011: Selected Papers 2012. *Northeastern Naturalist* 19, Special Issue, 6: 3-12.

Reproduced with permission of the Eagle Hill Institute where this paper was original published (Northeast Natural History Conference 2011: Selected Papers 2012 Northeastern Naturalist 19, Special Issue, 6: 3–12). Any further reproduction or distribution other than for personal research purposes is forbidden without the written consent of the Eagle Hill Institute.

Note that the running title has Epstein listed as coauthor, but that was written in error; he was not a coauthor.

2. Level-bottom Brachiopod Communities in the Middle Devonian of New York Howard R. Feldman *Lethaia* 12 (1980): 27-46.

3. Brachiopods of the Onondaga Limestone in Central and Southeastern New York Howard R. Feldman *American Museum of Natural History Bulletin* 179 (1985): 289-377.

4. Chonetacean Brachiopods of the "Pink *Chonetes*" Zone, Onondaga Limestone (Devonian, Eifelian), Central New York Patrick R. Racheboeuf and Howard R. Feldman *American Museum Novitates* 2982 (1990): 1-16.

5. Brachiopods of the Onondaga Formation, Moorehouse Member (Devonian, Eifelian), in the Genesee Valley, Western New York Howard R. Feldman *Bulletins of American Paleontology* 107 (1994): 1-56.

Reproduced with permission from Paleontological Research Institution, Ithaca, New York.

6. Paleoecology and Morphologic Variation of a Paleocene Terebratulid Brachiopod (*Oleneothyris harlani*) from the Hornerstown Formation of New Jersey Howard R. Feldman *Journal of Paleontology* 51 (1977): 86-107.

7. Notes on and Description of *Oleneothyris fragilis* (Morton) 1828 (Brachiopoda, Terebratulidae) Howard R. Feldman *American Museum Novitates* 2621 (1977): 1-16.

8. Paleontological Note: *Oleneothyris subfragilis* (d'Orbigny, 1850), a Replacement Name for the Brachiopod *Oleneothyris fragilis* (Morton, 1828) Howard R. Feldman *Journal of Paleontology* 59 (1985): 1485.

9. The Shawangunk and Martinsburg Formations Revisited: Sedimentology, Stratigraphy, Mineralogy, Geochemistry, Structure, and Paleontology Howard R. Feldman, J. B. Epstein, and John A. Smoliga *New York State Geological Association Guidebook for Field Trips*, SUNY, New Paltz, NY 81 (2009): 12.1-12.12.

10. Paleocommunities of the Onondaga Limestone (Middle Devonian) in Central New York State Richard H. Lindemann and Howard R. Feldman *New York State Geological Association Guidebook for Field Trips in South-central New York, 53rd Annual Meeting* 53 (1981): 79-96.

11. Fossils and Facies of the Onondaga Limestone in Central New York Howard R. Feldman and Richard H. Lindemann *New York State Geological Association Guidebook for Field Trips, 58th Annual Meeting* 58 (1986): 145-166.

12. Paleogeography and Brachiopod Paleoecology of the Onondaga Limestone in Eastern New York Richard H. Lindemann and Howard R. Feldman *New York State Geological Association Guidebook for Field Trips, 59th Annual Meeting* 59 (1987): 1-30.

13. Understanding the East Central Onondaga Formation (Middle Devonian): An Examination of the Facies and Brachiopod Communities of the Cherry Valley Section, and Mt. Tom, a Small Pinnacle Reef Thomas H. Wolosz, Howard R. Feldman, Richard H. Lindemann, and Donald E. Paquette *New York State Geological Association Guidebook for Field Trips, 63rd Annual Meeting* 63 (1991): 373-412.

INDEX

D

P

T

www.ingramcontent.com/pod-product-compliance
Ingram Content Group UK Ltd.
Pitfield, Milton Keynes, MK11 3LW, UK
UKHW052138220526
471278UK00018BA/62